HYDROSTATICS

HARI KISHAN

ATLANTIC
PUBLISHERS & DISTRIBUTORS (P) LTD

ATLANTIC
PUBLISHERS & DISTRIBUTORS (P) LTD
7/22, Ansari Road, Darya Ganj, New Delhi
Tel.: +91-11-4077 5252, 2327 3880
E-mail: orders@atlanticbooks.com
Web: www.atlanticbooks.com

Published by Atlantic Publishers & Distributors (P) Ltd 2024

Printed & bound in India by Atlantic Print Services

Preface

Hydrostatics, a fascinating branch of mathematics, explores the behaviour of fluids at rest under the influence of gravity. It is a subfield of fluid mechanics which plays a crucial role in understanding various natural and man-made phenomena. From the buoyancy that keeps ships afloat to the calculations behind the stability of dams and the functioning of hydraulic systems, and from the hydraulic systems that power machines to the breathtaking beauty of a floating balloon, hydrostatics is at the core of countless natural phenomena and technological innovations. It is a subject that captivates our curiosity and has been a cornerstone of scientific inquiry for centuries. It plays a crucial role in various fields, from engineering and physics to geology and environmental science.

The present endeavour is a comprehensive exploration of the subject, designed to provide readers with a solid foundation in hydrostatics, providing with a solid foundation in the mathematical principles and concepts that underpin this field. Fluids, whether they be water, oil, or air, exhibit intriguing behaviours when at rest, as they respond to the forces acting upon them. The chapters within this book guide the readers through the intricate world of hydrostatics, from the basic principles to more advanced concepts. The first chapter is introductory which sets the stage by introducing the fundamental concepts and principles of hydrostatics. It gives insight into why fluids behave the way they do when they are at rest and how mathematical modelling plays a pivotal role in understanding these behaviours. The second chapter delves into the fundamental properties of fluids, including their density and specific gravity. Here readers will learn how to calculate these important parameters and understand their implications in hydrostatics. The next chapter introduces the core principles of hydrostatics, exploring the fundamental theorem of hydrostatics and Pascal's law. These principles lay the groundwork for understanding fluid pressure in various situations. The fourth chapter takes a deep dive into the conditions that must be met for a fluid to be in equilibrium, helping to grasp the conditions under which fluids remain stable without any motion. It explores the concept of pressure variation with depth and its implications on fluid behaviour. The fifth chapter provides a comprehensive analysis of how fluid pressure acts on flat surfaces, both submerged and exposed. It discusses the method to calculate the force exerted by fluids on such surfaces. The next chapter investigates the intriguing concept of the center of pressure and how it affects the equilibrium of submerged objects. The seventh chapter delves into the complexities of calculating the thrust on curved surfaces submerged in fluids, expanding the knowledge of fluid dynamics. The last chapter of the book explores the equilibrium conditions of floating bodies, which have practical applications in ship design and more. Here, one will learn about the principles of buoyancy and how to analyse floating objects.

Throughout the book, a number of solved examples have been provided to illustrate the application of mathematical principles to real-world hydrostatic problems. These examples have been designed to enhance the understanding of the subject and provide with the tools to tackle similar problems independently. Additionally, each chapter concludes with a set of exercises to test and reinforce readers' comprehension of the material. These practical applications will help to bridge the gap between theory and real-world scenarios.

This book will serve as a valuable resource for both students and professionals in the field of mathematics, physics, engineering, and beyond. Whether one is preparing for an examination, seeking to expand his/her knowledge, or simply intrigued by the wonders of hydrostatics, this book will be his/her trusted guide.

Hari Kishan

Contents

Detailed Contents

Chapter 1

Introduction

1. Matter

Anything in the universe which occupies space and has mass is called matter or substance. For example: wood, iron, stone, coal, ice, water, oil, honey, air, steam, etc.

In nature, matter exists in two forms namely solid and fluid. In the above examples, wood, iron, stone, coal, ice belong to the first form whereas water, oil, honey, air, steam belong to the second form.

2. Solid

A solid is that form of matter which has definite shape and size and retains its shape and size when left to itself until and unless some external force is applied to it. For example: wood, iron, etc. are solids.

3. Fluid

A fluid is that form of matter which requires some support to retain its shape and size. In other words, fluid is a substance which flows or is capable of flowing. For example: water, oil, air, steam, etc. are fluids.

Further, fluids are also of two types namely liquid and gas.

4. Liquid

A liquid is that type of fluid which is incompressible and requires external support to maintain its shape. [Note that an incompressible fluid is defined as a fluid which cannot occupy a small or large space under pressure.] In other words, we mean that the volume of a given quantity of liquid does not alter though it may change in shape and moulds itself to the shape of the vessel in which it is poured. Thus, a liquid has a definite size (volume) but has no definite shape. For example: water, oil are liquids.

5. Gas

A gas is that type of fluid which is compressible and a given portion of a gas can be made to expand indefinitely by increasing sufficiently the space to which it has access. [Note that a compressible fluid is defined as a fluid which can be made to occupy a small or large space by increasing or decreasing the pressure.] In other words, we mean that a gas has neither a fixed volume nor a fixed shape, i.e., the volume and shape of a gas are the same as those of the vessel in which it is contained. For example: air, steam are gases.

Note: Gases are also called elastic fluids as they can be compressed or expanded. On the other hand, liquids are called inelastic fluids as they cannot be compressed or expanded.

6. Distinction between Solid, Liquid and Gas

(i) A solid has a definite shape and a definite size.

(ii) A liquid has a definite size but not a definite shape.

(iii) A gas has neither a definite shape nor a definite size.

7. Hydrostatics

It is that branch of mathematics which deals with the conditions of equilibrium of masses of fluids or of solids in contact with the fluids at rest.

8. Shearing Stress

Let V be the region bounded by a matter and P be any point inside it. Imagine a small plane surface α of unit area placed at P. The plane α separates the region at P into two parts and each part exerts a force on the other across α.

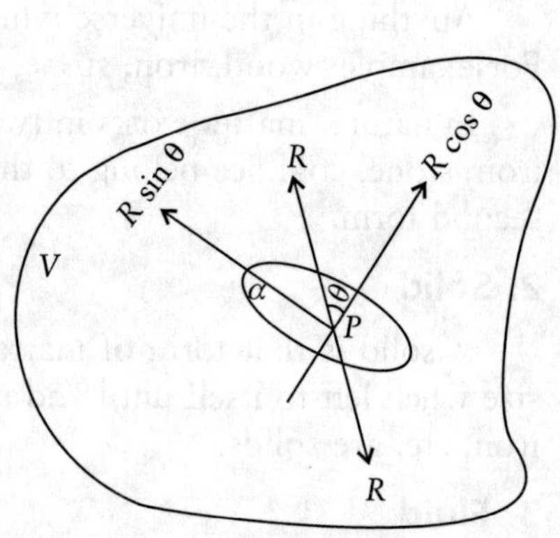

By Newton's Third Law of Motion, we know that the action and reaction are equal and opposite. Let the two parts of the region on opposite sides of α exert on one another equal and opposite forces of magnitude R. Thus, there is a mutual action and reaction R between them. This mutual action and reaction is called mutual stress or simply as *stress*.

The stress R is resolved into two components $R\cos\theta$ and $R\sin\theta$ acting along the normal and tangent to the plane α. The normal component $R\cos\theta$ is called the *thrust*.

Also, the tangential component $R\sin\theta$ is called the *shearing stress*.

9. Types of Fluid in Terms of Shearing Stress

In terms of shearing stress, fluids can be classified in the following two types:

(a) Perfect fluid

(b) Viscous fluid

10. Perfect Fluid

A perfect fluid is an ideal substance which, whether at rest or in motion, is incapable of exerting (or offering) resistance to shearing stress, i.e., the shearing stress is zero; the only direction along which it can exert an action will be along the normal to the surface with which it is in contact. A fluid at rest does not exert shearing stress but in motion, it may do so. But a perfect fluid is not expected to exert shearing stress even when it is in motion.

Thus, a perfect fluid is a substance such that its shape can be altered by any tangential force, however small, if applied sufficiently for a long time, of which portions can be separated from the rest of the mass and between different portions of which these is no tangential force of the nature of friction. When water is at rest, it is practically equivalent to a perfect fluid.

11. Viscous Fluid

When the particles of a fluid exert resistance to one another or it offers resistance to the motion of a body in it, then the fluid is called viscous. Honey, tar, treacle are examples of such a fluid.

The magnitude of such forces depends upon the stickiness of the fluid. The mathematical measurement of stickiness of fluid is called the viscosity of the fluid. The viscosity of a fluid does not affect its equilibrium, but it does affect its motion.

12. Homogeneous and Heterogeneous Matter

A matter is called homogeneous, if equal volumes of it, however small and from wherever they may be taken from the substance have equal masses. Matter is assumed to be heterogeneous when the contrary is stated explicitly.

13. Fluid Pressure

Suppose a hole P is made in the side of a vessel containing fluid, and that this hole P is covered by a plate which exactly fits into the hole. We may observe that the plate will only remain at rest when some external force is applied to it. This shows that the fluid exerts some force on the plate which pushes it outwards. Moreover, by definition, the fluid can only exert a force perpendicular to each element of area of this plate. If F be the external force required keep the plate in position by counterbalancing the action of the fluid, then F is the measure of the fluid pressure or fluid thrust on the whole area of the plate.

14. Uniform Pressure

If the fluid pressure over any portion of a plane area be proportional to the area of that portion, i.e., the force exerted by the fluid on each element of same area be the same, then the fluid pressure on the area is said to be uniform.

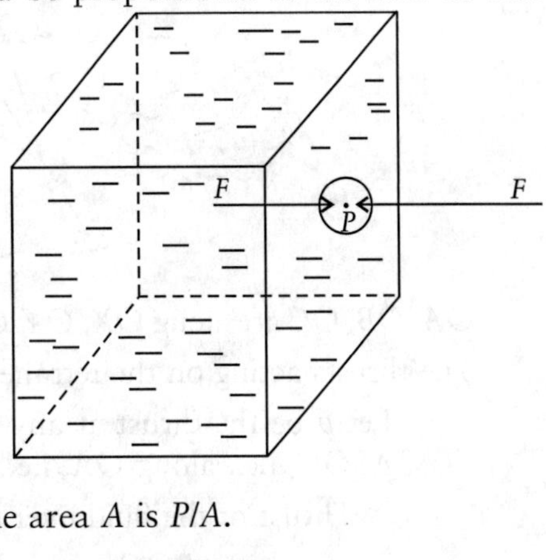

15. Varying Pressure

If the pressure of the fluid on elements of equal areas is different, then the pressure in said to be varying.

16. Mean Pressure

The mean pressure on a plane area is defined as the uniform pressure on it which will give the same resultant thrust as the actual thrust on it. Thus, if A is the area and P is the fluid thrust on it, then the mean pressure over the area A is P/A.

17. Pressure at a Point

If the fluid pressure is uniform over an area, then the measure of the pressure at a point inside the fluid is defined as the force exerted on a unit area.

If the fluid pressure is not uniform, then the pressure at a point P inside the fluid is defined as the limit of the mean pressure on a small plane surface around that point, as the area of the plane surface becomes infinitesimally small in the limit, i.e., approaches zero.

$$\therefore \text{ Pressure at } P = \lim_{A\to 0} T/A$$

where A is the area of the small plane surface inside the fluid and containing the point P and T is the total thrust on the surface.

18. Properties of Fluid Pressure

(a) The pressure of a fluid is always along the normal to the surface with which it is in contact.

(b) The pressure at a point of a fluid at rest is the same in all directions.

19. Theorem

Prove that the pressure at any point of a fluid at rest in the same in all directions.

Proof: Let O be any point in the fluid. Imagine a small tetrahedron O, ABC with three mutually perpendicular faces OBC, OCA and OAB. Note that

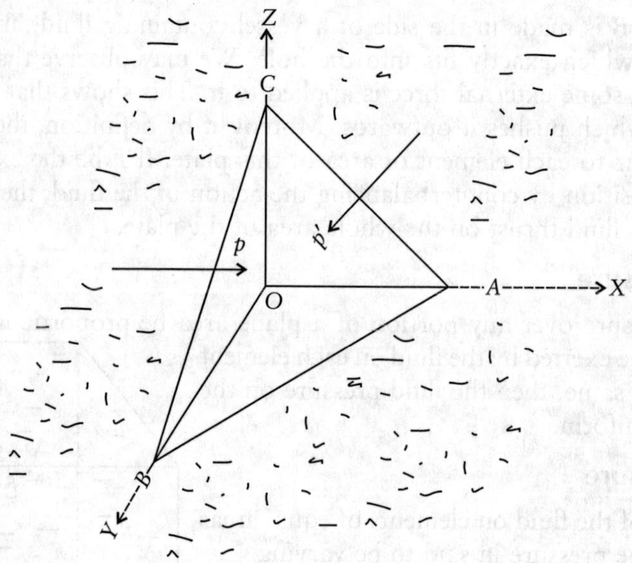

OA, OB, OC are along OX, OY, OZ respectively which are three mutually perpendicular lines.

The thrusts acting on the tetrahedron along OX are as follows:

(i) Let p be the thrust at any point of ΔBOC. This thrust will act along the normal of ΔBOC, i.e., along OA, i.e., along OX.

$\therefore$ Thrust of the fluid over ΔBOC

$= p \times \Delta BOC$ ...(1)

(ii) The thrust of fluid over the area ΔAOB is along OZ which is perpendicular to OX.

$\therefore$ The thrust of fluid over the area ΔAOB along $OX = 0$ ...(2)

(iii) The thrust of fluid over the area ΔAOC is along OY which is perpendicular to OX.

$\therefore$ The thrust of fluid over the area ΔAOC along $OX = 0$ ...(3)

(iv) Let p' be the thrust at any point on ΔABC. Let the plane ABC make an angle θ with the plane of ΔBOC.

$\therefore$ Fluid thrust on ΔABC

$= p' \times \Delta ABC$

The direction of the thrust is perpendicular to the plane ABC. Resolved part of this along OX

$= p' \times \Delta ABC \times \cos\theta$

$= p' \times (\Delta ABC \times \cos\theta)$

$= p' \times \Delta BOC$ | $\because$ θ is the angle between the plane ABC and plane BOC = angle between their normals $\therefore \Delta ABC \times \cos\theta = \Delta BOC$ = Projection of ΔABC on plane BOC

(v) External field of force of gravity acts on the fluid. Let X be the force acting on the fluid along OX.

$\therefore$ Component of the external force on the fluid in tetrahedron (O, ABC) in the direction along OX

$= X$. mass of the fluid in the tetrahedron

$= X$. volume of tetrahedron $\times$ density

$= X \,.\, 1/3\, (\Delta\, BOC) \,.\, OA \times \rho$

where ρ = density of the fluid.

Since the fluid in the tetrahedron is in equilibrium, therefore, the algebraic sum of resolved parts along OX is zero.

$$\therefore p\, \Delta BOC - p'\, \Delta BOC + X \,.\, 1/3\, \Delta BOC \,.\, OA \times \rho = 0$$

$$\Rightarrow p - p' + X \,.\, 1/3\, OA \,.\, \rho = 0$$ | Cancelling the common factor ΔBOC throughout

When every dimension of the tetrahedron is diminished, then OA $\to$ 0.

$$\therefore p - p' = 0$$

$$\Rightarrow p = p'$$

This shows that the pressure at any point of a fluid at rest is the same in all directions.

20. Dimensions of Pressure

We know that the pressure is defined as the force per unit area. Therefore, its dimensions in terms of the fundamental units of mass, length and time are

$$M\, L\, T^{-2}/L^2$$

$$= M\, L^{-1}\, T^{-2}$$

21. Principle of Transmissibility of Liquid Pressure or Pascal's Law

Pascal gave an important principle, know as Pascal's Law. It may be stated as follows:

Theorem: An increase of pressure at any point of a liquid at rest under given external forces is transmitted without change to every other point of the liquid.

Proof: Let A and B be any two point in the liquid.

Case I: *When the straight line AB lies wholly in the liquid.*

Construct a cylinder of small cross-section α taking the straight line AB as axis whose plane ends one perpendicular to AB. Let p and p' be the fluid thrusts at any point on the plane ends A and B respectively.

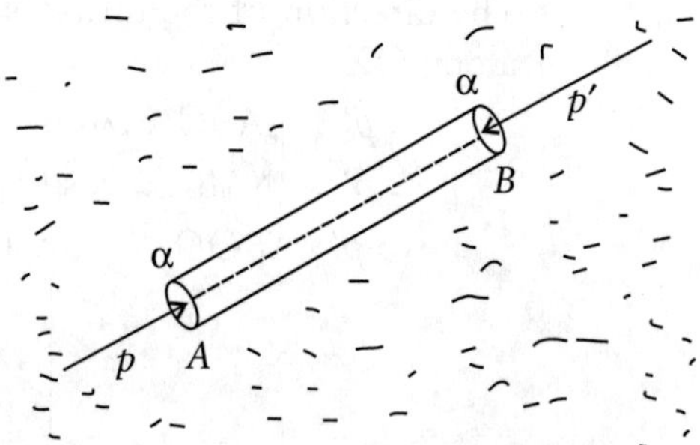

The cylinder of liquid is in equilibrium under the action of the following forces:

(i) The fluid pressures $p\alpha$ and $p'\alpha$ on the plane ends A and B respectively acting along AB.

(ii) The pressure at each point on the curved surface of the cylinder acting perpendicular to AB.

(iii) The component X of the resultant external force on the liquid within the cylinder along AB.

Since the liquid in the cylinder is in equilibrium, therefore, the algebraic sum of the forces acting along AB is zero.

$$\therefore\ p\alpha - p'\alpha + X = 0 \qquad \text{...(1)}$$

Suppose an extra (additional) pressure of intensity P is applied at the end A and let the corresponding increase in pressure at the end B be of intensity P'.

Again, the system is in equilibrium. Therefore, the algebraic sum of the forces acting along AB is zero.

$$\therefore\ (p + P)\,\alpha - (p' + P')\,\alpha + X = 0 \qquad \text{...(2)}$$

Subtracting (1) from (2), we have,

$$P\alpha - P'\alpha = 0$$

$$\Rightarrow\ (P - P')\,\alpha = 0$$

$$\Rightarrow\ P - P' = 0 \qquad |\ \because \alpha \neq 0$$

$$\Rightarrow\ P = P'$$

$\Rightarrow$ Increase of pressure at A = Increase of pressure at B.

Thus we see that a pressure of intensity P applied at the end A is transmitted with the same intensity P to the other end B.

Case II: *When the straight line AB does not lie wholly in the liquid.*

Join AB by means of straight lines AC, CD, DE and EB each of which lies wholly in the liquid.

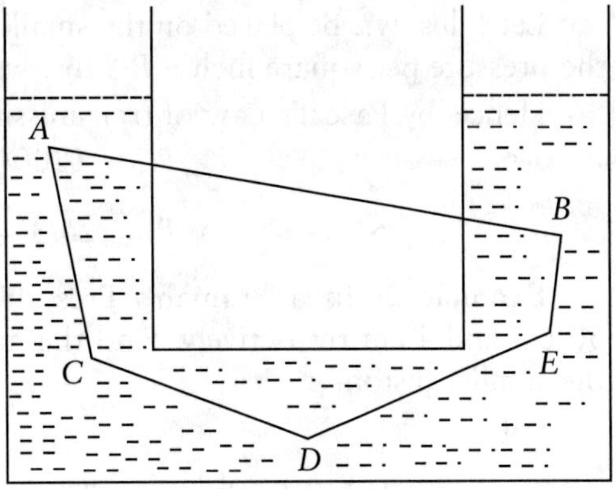

Then, from case I above, we get, increase of pressure at A.

= increase of pressure at C.
= increase of pressure at D.
= increase of pressure at E.
= increase of pressure at B.

Thus, increase of pressure at A = increase of pressure at B.

22. Hydraulic or Bramah's Press

Hydraulic press is constructed on the principle of transmissibility of liquid pressure. This machine consists of two vertical cylinders communicating with one another by a tube near their bases and each fitted with a water-tight piston and the space below these pistons being filled with water. The cross-section of one cylinder is much greater than that of the other. Let them be A and a respectively.

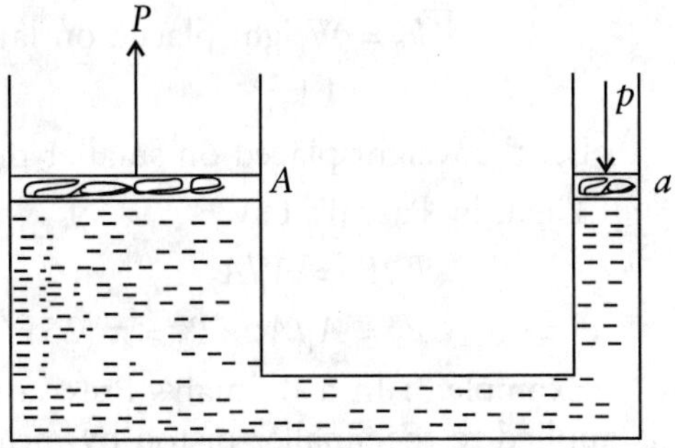

Let a downward force p be applied to the piston of area a. This force will create an additional pressure, whose value per unit area will be p/a which will be transmitted throughout the liquid by Pascal's Law. Consequently, the resultant upward thrust upon the piston of area A will be represented by

$$P = A\, p/a$$

$$\Rightarrow P : p = A : a$$

This shows that the thrusts on the pistons are proportional to their areas of cross-sections.

Hence, by suitably adjusting the ratio of the areas of cross-sections, any force, however small, may by its transmission through a liquid be made to support any weight however large. This is known as "Hydrostatics Paradox".

Note: In practice, we find that the strength of the sides of the vessel cannot be so strong as to support indefinitely great amount of fluid pressure which may be transmitted to them. So, a limit is, however, supposed to be imposed keeping the above fact in mind.

ILLUSTRATIVE EXAMPLES

Example 1: In a Bramah's Press the piston can safely bear a pressure of 1200 lbs. wt. per square foot. What will be the greatest weight that can be placed on the smaller piston, its cross-section being S square inches.

Sol. Pressure P_1 to be borne by larger piston

= 1200 lbs. wt. per square foot
= 1200/12 × 12 lbs. wt. per square inch.

Let P lbs. wt. be placed on the smaller piston whose cross-section is S square inches. Then, the pressure per square inch = P/S lbs. wt.

Hence, by Pascal's Law of transmissibility of pressure, we get

$$P/S = 1200/12 \times 12$$

$$\Rightarrow \quad P = 125/3 = 41\frac{2}{3} \text{ lbs. wt.}$$

Example 2: In a Bramah's Press, the diameters of the larger and smaller pistons are 50 cm and 4 cm respectively. Find the mass which can be supported by a kilogram placed on the smaller piston.

Sol.

A_1 = Area of larger piston
$= \pi (25)^2 \text{ cm}^2$
A_2 = Area of smaller piston
$= \pi (2)^2 \text{ cm}^2$
P_2 = Weight placed on larger piston
$= 1$ kg

Let the weight placed on smaller piston be P_1 kg

Then, by Pascal's Law,

$$P_1/P_2 = A_1/A_2$$

$$\Rightarrow \quad P_1 = A_1/A_2 \times P_2 = \pi (25)^2/ \pi (2)^2 \times 1 = 625/4 = 156.25 \text{ kg}$$

Example 3: In a Bramah's Press a total pressure of 2240 kg is produced when a pressure is applied to the smaller piston by means of a force 5 kg placed at the end of the lever. If the diameters of the pistons are as 8 : 1, find the ratio of arms of the lever employed to work the piston.

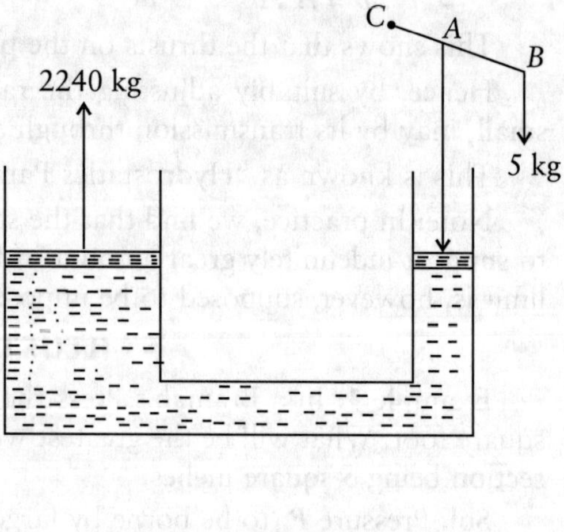

Sol.

Ratio of the diameters of the piston.
= Ratio of the radii of the piston

Let $8r$ and r be the radii of the pistons.

$\therefore$ Areas A_1 and A_2 are

$$A_1 = \pi (8r)^2 = 64 \pi r^2$$

$$A_2 = \pi r^2$$

Pressure P_1 on larger piston
= 2240 kg

Let P_2 be the pressure on the smaller piston

By Pascal's Law,

$$P_1/P_2 = A_1/A_2$$

$$\Rightarrow \quad P_2 = A_2/A_1\, P_1$$

$$= \pi r^2/64\, \pi\, r^2 \times 2240 = 2240/64 = 35 \text{ kg}$$

The weight 5 kg is placed at point B of the lever whereas a force 35 kg acts at A.

Taking moments about C,

$$CA \times 35 = CB \times 5$$

$$\Rightarrow CA/CB = 5/35 = 1/7$$

Example 4: In a Bramah's Press the area of the larger piston is 100 square inches and that of the smaller one is 1/4 square inch. Find the force that must be applied to the latter so that the former may lift 1 ton.

Sol. By Pascal's Law,

$$\frac{\text{Pressure on smaller piston}}{\text{Pressure on larger piston}} = \frac{\text{Area of smaller piston}}{\text{Area of larger piston}}$$

$$\Rightarrow \frac{\text{Pressure on smaller piston}}{1 \times 2240} = \frac{\frac{1}{4}}{100} \qquad | \because 1 \text{ ton} = 2240 \text{ lbs. wt.}$$

$$= \text{Pressure on smaller piston}$$

$$= \frac{2240 \times 1}{4 \times 100} = \frac{28}{5} = 5.6 \text{ lbs. wt.}$$

Example 5: A vessel full of water is fitted with a tight cork. How is it that a slight blow on the cork may be sufficient to break the vessel?

Sol. Let A be the area of cross-section of the cork and S the area of surface of the vessel. Let p be the intensity of the slight blow on the cork. Since water is incompressible, therefore, by Pascal's Law, the pressure p/A per unit area will be transmitted to each point on the surface of the vessel. Thus, the total pressure transmitted will be $(p/A)\ S$ which, S being large, is very large. On account of this pressure on the surface of the vessel being very large, this pressure may be sufficient to break the vessel.

Hence a slight blow on the cork may be sufficient to break the vessel. Obviously, the chance of breaking increases if the surface of the vessel S is large and the area of the cork A is small.

EXERCISE

1. In a Bramah's Press the piston can safely bear a pressure of 1200 gms. wt. per square decimetre. What will be the greatest weight that can be placed on the smaller piston of cross-section 5 sq. cm?

   Ans. 60 gms. wt.

2. In a Bramah's Press the piston can safely bear a pressure of 1000 gms. wt./sq. decimetre. What will be the greatest weight that can be placed on the smaller piston, its cross-section being 5 sq. cm.

   Ans. 50 gms. wt.

3. In a Bramah's Press the diameter of the larger piston is 20 cm and that of the smaller one is 2.5 cm. What force must be applied to raise a weight of 1000 tons at the other piston?

Ans. 15.625 tonnes weight

4. In a Bramah's Press a total pressure of 1000 kg is produced when a pressure is applied to the smaller piston by means of a force of 5 kgs. placed at the end of the lever. If the diameters of the pistons are 10 : 1, find the ratio of the arms of the lever employed to work the piston.

   Ans. 1 : 1

5. In a Bramah's Press a total pressure of 1 ton is produced when a pressure is applied to the smaller piston by means of force of 5 lbs. placed at the end of the lever. If the diameters of the pistons are as 8 : 1, find the ratio of the arms of the lever employed to work the piston.

   Ans. 1 : 7

6. In a Bramah's Press the diameters of the larger and smaller pistons are 50 cm and 4 cm respectively. Find the mass which can be supported by a kilogram placed on the smaller piston.

   Ans. 156.25 kg.

7. In a Bramah's Press the diameter of the larger piston is 80 inches and that of the smaller one is 20 inches. What force must be applied to the smaller piston to raise a weight of 1000 tons at the other piston?

   Ans. 15.625 tons wt.

8. If the sections of the cylinders of a Bramah's Press be 18 sq. inches and 1 sq. ft. respectively, what pressure must be applied to the smaller cylinder to produce a pressure of two tons upon the larger?

   Ans. 560 lbs. wt.

9. A cylindrical pipe which is filled with water opens into another pipe, the diameter of which is three times its own diameter. If a force of 20 lbs. wt. be applied to the water in the smaller pipe, find the force on the open end of the larger pipe, which is necessary to keep the water at rest.

   Ans. 180 lbs. wt.

Chapter

2

Density and Specific Gravity

1. Density

The density of a homogeneous substance is defined as its mass per unit volume. The mass of 1 cubic centimetre of water at 4°C is 1 gram. Hence, the density of water at 4°C in C.G.S. system is 1 gram per cubic centimetre.

The mass of 1 cubic foot of water at 4°C is 62.5 lbs. Hence, the density of water at 4°C in F.P.S. system is 62.5 lbs. per cubic foot.

Symbolically, if ρ, M and V denote density, mass and volume respectively, then

$$\rho = M/V$$

The dimension of density is ML^{-3}.

2. Homogeneous and Heterogeneous Substances

A substance is said to be homogeneous or of uniform density if equal volumes of it, however small, have equal masses.

A substance is said to be heterogeneous or of variable density if equal volumes of it, however small, have different masses.

3. Weight in Terms of Density

If W is the weight of a given substance in dynes, ρ is its density in gms. per cubic cm., V is its volume in cubic cm and g is the acceleration due to gravity in cm per second^2, then

$$W = Vg\,\rho$$

Let M be the mass of the substance. Then,

$$W = Mg \qquad \text{| By Dynamics} \qquad ...(1)$$

Again,

$$M = \text{Mass of } V \text{ cubic cm of the substance}$$

$$= V \times \text{mass of 1 cubic cm of the substance}$$

$$= V\rho \qquad \text{| By definition of density} \qquad ....(2)$$

From (1) and (2), we get

$$W = Vg\rho$$

Note: If W is expressed in poundals, ρ in pounds per cubic foot, V in cubic foot and g in foot per second^2, then also

$$W = Vg\rho$$

4. Specific Gravity or Relative Density

The specific gravity of a substance is defined as the ratio of the weight of any volume of that substance to the weight of an equal volume of a standard substance.

The standard substance is usually taken to be pure water at a temperature of 4°C. The specific gravity in short is written as sp. gr. It is simply a number. So it has no dimensions.

5. Weight in terms of Specific Gravity

It W is the weight of a volume V of a substance whose specific gravity is s and w is the weight of a unit volume of the standard substance, then

$$W = V\,s\,w$$

We know that

$$s = \frac{\text{Weight of a unit volume of the substance}}{\text{Weight of a unit volume of the standard substance}}$$

| By definition of specific gravity

$$\Rightarrow \quad s = \frac{\text{Weight of a unit volume of the substance}}{w}$$

where w is the weight of a unit volume of the standard substance

$\Rightarrow$ weight of a unit volume of the substance $= sw$

$\therefore$ Weight of V units of volume of the substance $= V\,sw$

Aliter: We know that

$$W = V\,g\,\rho \qquad \text{| Refer § 3} \qquad \text{...(1)}$$

Let ρ_0 be the density of the standard substance. Then,

$$s = \frac{\rho}{\rho_0} \qquad \text{| By definition of specific gravity} \qquad \text{...(2)}$$

Also, $W = g\,\rho_0$ ...(3)

From (1) and (2), we get

$$W = V\,g\,s\,\rho_0$$

$$\Rightarrow \quad W = V\,s\,(g\,\rho_0)$$

$$\Rightarrow \quad W = V\,sw \qquad \text{| From (3)} \qquad \text{...(4)}$$

6. Specific Gravity of Mixtures (Volumes)

To find the specific gravity of mixture of different substances whose volumes and specific gravities are given:

Let $v_1, v_2, v_3, \ldots$ be the volumes and $s_1, s_2, s_3, \ldots.$ be the specific gravities of different substances.

Let w be the weight of a unit volume of the standard substance. Then, the weights of the different substances are:

$v_1s_1w, v_2s_2w, v_3s_3w, \ldots\ldots$ | Refer § 5

Let s be the specific gravity of the mixture. Then,

total weight of the mixture

$$= (v_1 + v_2 + v_3 + \ldots\ldots)\, sw \quad \ldots(1)$$

Also, sum of the weights of the different substances

$$= v_1s_1w + v_2s_2w + v_3s_3w + \ldots. \quad \ldots(2)$$

These two weights must be equal.

Therefore, equating (1) and (2), we get,

$$(v_1 + v_2 + v_3 + \ldots\ldots)\, s\, w = v_1s_1w + v_2s_2w + v_3s_3w + \ldots.$$

$$\Rightarrow \quad s = \frac{v_1s_1 + v_2s_2 + v_3s_3 + \ldots.}{v_1 + v_2 + v_3 + \ldots.}$$

7. Specific Gravity of Mixtures (Weights)

To find the specific gravity of a mixture of different substances whose weights and specific gravities are given:

Let $w_1, w_2, w_3, \ldots$ be the weights and $s_1, s_2, s_3, \ldots.$ be the specific gravities respectively of different substances which constitute the mixture. Let w be the weight of a unit volume of the standard substance. Then, the respective volumes of the different substances will be:

$\frac{w_1}{s_1w}, \frac{w_2}{s_2w}, \frac{w_3}{s_3w}, \ldots.$ | Refer § 5

Let s be the specific gravity of the mixture. Then, the weight of the mixture

$$= \left(\frac{w_1}{s_1w} + \frac{w_2}{s_2w} + \frac{w_3}{s_3w} + \ldots.\right) sw \qquad | \text{ Refer § 5}$$

$$= \left(\frac{w_1}{s_1} + \frac{w_2}{s_2} + \frac{w_3}{s_3} + \ldots.\right) s \quad \ldots(1)$$

Also, total weight of the mixture

= sum of the weights of the substances which constitute the mixture

$$= w_1 + w_2 + w_3 + \ldots. \quad \ldots(2)$$

These two weights must be equal.

Therefore, equating (1) and (2), we get,

$$\left(\frac{w_1}{s_1} + \frac{w_2}{s_2} + \frac{w_3}{s_3} + \ldots.\right) s = w_1 + w_2 + w_3 + \ldots. \quad \Rightarrow \quad s = \frac{w_1 + w_2 + w_3 + \ldots}{\frac{w_1}{s_1} + \frac{w_2}{s_2} + \frac{w_3}{s_3} + \ldots}$$

Aliter

Sum of the volumes of the constituents $= \frac{w_1}{s_1 w} + \frac{w_2}{s_2 w} + \frac{w_3}{s_3 w} + \ldots$ ...(1)

Also, total volume of the mixture $= \frac{(w_1 + w_2 + w_3 + \ldots)}{sw}$ ...(2)

Since there is no change in the total volume, therefore, the total volume of the mixture must be equal to the sum of the volumes of the constituents. Hence, equating (1) and (2), we get,

$$\frac{w_1 + w_2 + w_3 + \ldots}{sw} = \frac{w_1}{s_1 w} + \frac{w_2}{s_2 w} + \frac{w_3}{s_3 w} + \ldots$$

$$\Rightarrow \frac{w_1 + w_2 + w_3 + \ldots}{s} = \frac{w_1}{s_1} + \frac{w_2}{s_2} + \frac{w_3}{s_3} + \ldots \qquad | \text{ Cancelling } 1/w \text{ throughout}$$

$$\Rightarrow s = \frac{w_1 + w_2 + w_3 + \ldots}{\frac{w_1}{s_1} + \frac{w_2}{s_2} + \frac{w_3}{s_3} + \ldots}$$

8. Density of Mixture

Analogous to § 7, if ρ_1, ρ_2, ρ_3, ... are the respective densities of different substances which constitute the mixture and w_1, w_2, w_3, ... be the respective weights of the constituents, then the density ρ of the mixture is given by:

$$\rho = \frac{w_1 + w_2 + w_3 + \ldots}{\frac{w_1}{\rho_1} + \frac{w_2}{\rho_2} + \frac{w_3}{\rho_3} + \ldots}$$

ILLUSTRATIVE EXAMPLES

Example 1: Five litres of a liquid whose specific gravity is 1.3 are mixed with 7 litres of a liquid whose specific gravity is 0.78. Find the specific gravity of the mixture.

Sol. $v_1 = 5$ litres, $v_2 = 7$ litres

$s_1 = 1.3$, $s_2 = 0.78$

Let $\bar{s}$ be the specific gravity of the mixture. Then,

$$\bar{s} = \frac{v_1 s_1 + v_2 s_2}{v_1 + v_2}$$

$$= \frac{5 \times 1.3 + 7 \times 0.78}{5 + 7} = \frac{6.5 + 5.46}{12}$$

$$= \frac{11.96}{12} = 0.9967$$

Example 2: The specific gravities of pure gold and pure copper are 19.3 and 8.62 respectively. Find the specific gravity of standard gold which is an alloy of gold and copper in the proportion of 11 : 1.

Sol. Let the volumes of pure gold and copper be $11v$ and v respectively.

$\therefore$ Specific gravity of the alloy

$$= \frac{v_1s_1 + v_2s_2}{v_1 + v_2} \qquad = \frac{(11v)\times 19.3 + v\times 8.62}{11v + v}$$

$$= \frac{212.3 + 8.62}{12} \qquad = \frac{220.92}{12}$$

$$= 18.41$$

Example 3: What quantity of water must be mixed with a litre of milk to reduce its specific gravity from 1.03 to 1.02?

Sol. Let the volume of water to be mixed with the milk be v_2 litres.

$v_1 = 1$ litre $\qquad s_1 = 1.03$

$s_2 = 1$ $\qquad$ | specific gravity of water = 1

$\bar{s} = 1.02$

Now, $$\bar{s} = \frac{v_1s_1 + v_2s_2}{v_1 + v_2}$$

$$\Rightarrow \quad 1.02 = \frac{1\times 1.03 + v_2\times 1}{1 + v_2}$$

$$\Rightarrow \quad 1.02\,(1 + v_2) = 1.03 + v_2$$

$$\Rightarrow \quad 0.02\,v_2 = 1.03 - 1.02 = 0.01$$

$$\Rightarrow \quad v_2 = \frac{0.01}{0.02} = \frac{1}{2} = 0.5 \text{ litre}$$

Example 4: How much water must be added to 20 c.c. of a salt solution whose sp. gr. is 1.08 so that sp. gr. of the mixture may be 1.05?

Sol. Let x c.c. of water must be added to the salt solution. Then,

$v_1 = 20$ c.c. $\qquad s_1 = 1.08$

$v_2 = x$ c.c. $\qquad s_2 = 1$ $\qquad$ | $\because$ specific gravity of water = 1

$\bar{s}$ = specific gravity of mixture = 1.05

Now, $$\bar{s} = \frac{v_1s_1 + v_2s_2}{v_1 + v_2}$$

$$\Rightarrow \quad 1.05 = \frac{20\times 1.08 + x\times 1}{20 + x}$$

$$\Rightarrow \quad (20 + x)\times 1.05 = 20\times 1.08 + x$$

$$\Rightarrow \quad 21 + 1.05x = 21.6 + x$$

$$\Rightarrow \quad 0.5x = 0.6$$

$$\Rightarrow \quad x = 12$$

Example 5: Two metals of which the specific gravities are 11.22 and 7.25 when mixed in certain proportion without condensation form an alloy whose specific gravity is 8.72. Find the proportion by volume of the metals in the alloy.

Sol. Let the required volumes of the two metals be v_1 and v_2. Then,

$$\bar{s} = \frac{v_1 s_1 + v_2 s_2}{v_1 + v_2}$$

$$\Rightarrow \quad 8.72 = \frac{v_1 \times 11.22 + v_2 \times 7.25}{v_1 + v_2}$$

$$\Rightarrow \quad 8.72\,(v_1 + v_2) = 11.22\,v_1 + 7.25\,v_2$$

$$\Rightarrow \quad (8.72 - 7.25)\,v_2 = (11.22 - 8.72)\,v_1$$

$$\Rightarrow \quad 1.47\,v_2 = 2.50\,v_1$$

$$\Rightarrow \quad v_1/v_2 = 147/250 \quad \Rightarrow \quad v_1 : v_2 = 147 : 250$$

Example 6: When equal volumes of two substances are mixed, the specific gravity of the mixture is 4. When equal weights of the same substances are mixed, the specific gravity of the mixture is 3. Find the specific gravities of the substances.

Sol. Let the required specific gravities of the two substances be s_1 and s_2.

Let v, v be the equal volumes of the two substances which are mixed. Then, specific gravity of the mixture is given by

$$\bar{s} = \frac{v_1 s_1 + v_2 s_2}{v_1 + v_2}$$

$$\Rightarrow \quad 4 = \frac{v s_1 + v s_2}{v + v} \quad \Rightarrow \quad 4 = \frac{s_1 + s_2}{2}$$

$$\Rightarrow \quad s_1 + s_2 = 8 \qquad \text{...(1)}$$

Let w, w be the equal weights of the two substances which are mixed. Then specific gravity of the mixture is given by

$$\bar{s} = \frac{w_1 + w_2}{\frac{w_1}{s_1} + \frac{w_2}{s_2}} \quad \Rightarrow \quad 3 = \frac{w + w}{\frac{w}{s_1} + \frac{w}{s_2}}$$

$$\Rightarrow \quad \frac{1}{s_1} + \frac{1}{s_2} = \frac{2}{3} \quad \Rightarrow \quad \frac{s_1 + s_2}{s_1 s_2} = \frac{2}{3}$$

$$\Rightarrow \quad \frac{8}{s_1 s_2} = \frac{2}{3} \qquad | \text{ From (1)}$$

$$\Rightarrow \quad s_1 s_2 = 12 \qquad \text{...(2)}$$

Now, $(s_1 - s_2)^2 = (s_1 + s_2)^2 - 4 s_1 s_2$

$$= 64 - 48 \qquad | \text{ From (1) and (2)}$$

$$= 16$$

$$\Rightarrow \quad s_1 - s_2 = 4 \qquad \text{...(3)}$$

Solving equations (1) and (2), we get

$$s_1 = 6,\ s_2 = 2$$

Example 7: What is the volume of a mass of wood of specific gravity 0.5 so that when attached to 500 ozs. of iron of specific gravity 7, the mean specific gravity of the whole may be unity?

Sol. Let W ozs. wt. be the weight of the wood which is attached to the given iron.

We know that

$$\bar{s} = \frac{w_1 + w_2}{\frac{w_1}{s_1} + \frac{w_2}{s_2}} \quad \Rightarrow \quad 1 = \frac{500 + W}{\frac{500}{7} + \frac{W}{0.5}}$$

$$\Rightarrow \quad \frac{500}{7} + \frac{W}{0.5} = 500 + W \quad \Rightarrow \quad W\left(1 - \frac{1}{0.5}\right) = \frac{500}{7} - 500$$

$$\Rightarrow \quad W = \frac{500 \times 6}{7} \quad \Rightarrow \quad W = \frac{3000}{7} \text{ ozs.wt.}$$

Let the required volume be v cu. ft.

Then,

$$W = v\, s\, w$$

$$\Rightarrow \quad 3000/7 \times 1/16 = v \times (0.5)\,(62.5)$$

| weight of 1 cubic ft. of water = 62.5 lbs.

$$\Rightarrow \quad v = \frac{3000}{7 \times 16 \times 0.5 \times 62.5}$$

$$\Rightarrow \quad v = 6/7 \text{ cu. ft.}$$

Example 8: An alloy of zinc of specific gravity 7.2 and copper of specific gravity 8.95 has a mass of 467 grammes. Its volume is 60 cubic centimetres. Find the volume of each component.

Sol. Specific gravity of the mixture

= Mass/Volume

= 467/60

Let the volumes of the zinc and copper be v_1 and v_2 respectively.

Also,

$$s_1 = 7.2 \qquad s_2 = 8.95$$

We know that specific gravity of the mixture $= \dfrac{v_1 s_1 + v_2 s_2}{v_1 + v_2}$

$$\Rightarrow \quad \frac{467}{60} = \frac{v_1(7.2) + v_2(8.95)}{v_1 + v_2}$$

$$\Rightarrow \quad 467\,(v_1 + v_2) = 432\, v_1 + 537\, v_2$$

$\Rightarrow \quad (467 - 432)\, v_1 = (537 - 467)\, v_2$

$\Rightarrow \quad 35\, v_1 = 70\, v_2$

$\Rightarrow \quad v_1 = 2\, v_2$ ...(1)

Again, $v_1 + v_2 = 60$

$\Rightarrow \quad 2v_2 + v_2 = 60$ | From (1)

$\Rightarrow \quad 3v_2 = 60$

$\Rightarrow \quad v_2 = 20$ c.c.

$\therefore \quad v_1 = 40$ c.c.

Hence, the volumes of zinc and copper are 40 c.c. and 20 c.c. respectively.

Example 9: Three litres of a liquid whose specific gravity is 0.8 are mixed with 5 litres of another liquid whose specific gravity is 1.04. Find the specific gravity of the mixture if there is a contraction of 5% on the joint volume.

Sol. Let w be the weight per litre of the standard substance. Then, the sum of the weights of the given liquids

$= 3 \times 0.8\, w + 5 \times 1.04\, w$ ...(1)

Also, the volume of the mixture after 5% reduction

$= 95/100\ (3 + 5)$

$= 38/3$ litres

Let $\bar{s}$ be the specific gravity of the mixture. Then, the weight of the mixture

$= 38/5\ \bar{s}\, w$ ...(2)

Since these two weights are equal, therefore, from (1) and (2), we get

$38/5\ \bar{s}\, w = 3 \times 0.8\, w + 5 \times 1.04\, w$

$= 7.6\, w$

$\Rightarrow \quad \bar{s} = 7.6 \times 5/38 = 1$

Hence, the specific gravity of the mixture is 1.

Example 10: A mixture is made by 7 cu. cms. of sulphuric acid (specific gravity = 1.843) and 3 cu. cms. of distilled water and its specific gravity when cold is found to be 1.615. What contraction has taken place?

Sol. Let the volume of the mixture when cold be $1/n\ (7 + 3)$, i.e., $10/n$ cu. cms. ...(1)

We know that if $\bar{s}$ be the specific gravity of the mixture of volumes v_1 and v_2 of two liquids of specific gravity s_1 and s_2 and if the volume after contraction is $v_1 + v_2/n$, then

$$\bar{s} = \frac{v_1 s_2 + v_2 s_2}{\frac{1}{n}(v_1 + v_2)}$$

$$\therefore \quad 1.615 = \frac{7 \times 1.843 + 3 \times 1}{\frac{1}{n}(7 + 3)}$$

$\Rightarrow \quad 10\ (1.615) = n\ [12.901 + 3]$

$\Rightarrow \quad 16.15 = 15.901\ n$

$\Rightarrow \quad n = 16150/15901$ ...(2)

From (1) and (2), the contracted

volume = 15901/16150 (10) cu. cms.

= 9.85 cu. cms.

$\therefore$ The percentage of contraction

= 10 – 9.85/10 × 100

= 0.15 × 100/10 = 1.5%

Example 11: Having given the specific gravity ρ of the mixture formed of equal volumes of two fluids and also the specific gravity σ of a mixture formed by taking a volume of one fluid double that of the other, prove that the specific gravities of the fluids are $3\sigma - 2\rho$ and $4\rho - 3\sigma$.

Sol. Let the specific gravities of the fluids be s_1 and s_2.

Case I $\quad \rho = \dfrac{(v)s_1 + (v)s_2}{v + v}$

$$\Rightarrow \quad \rho = \frac{s_1 + s_2}{2}$$

$$\Rightarrow \quad s_1 + s_2 = 2\rho \qquad \text{...(1)}$$

Case II $\quad \sigma = \dfrac{(2v)s_1 + (v)s_2}{2v + v}$

$$\Rightarrow \quad \sigma = \frac{2s_1 + s_2}{3}$$

$$\Rightarrow \quad 2s_1 + s_2 = 3\sigma \qquad \text{...(2)}$$

Solving (1) and (2), we get

$$s_1 = 3\sigma - 2\rho$$

and $\quad s_2 = 4\rho - 3\sigma$

Example 12: Two liquids of specific gravities s and s' and of volumes v and v', having been mixed, the specific gravity of the mixture is found to be σ. Find the volume of the mixture.

Sol. Let w be the weight per unit volume of the standard substance and v be the volume of the mixture. Then,

weight of the mixture = $v\ \sigma\ w$ ...(1)

Sum of the weights of two liquids

= $v\ sw + v's'\ w$ ...(2)

Since the two weights are equal, therefore, equating (1) and (2), we get

$$v\ \sigma\ w = v\ s\ w + v's'\ w$$

$$v = vs + v's' / \sigma$$

Example 13: A mixture is formed of two fluids. The density ρ of the mixture, the ratio $m : 1$ of the volumes, and the ratio $n : 1$ of densities are given. Prove that the densities of fluids are:

$$\frac{mn+n}{mn+1}\rho \text{ and } \frac{m+1}{mn+1}\rho$$

Sol. Let the volumes of two fluids be mv and v and their densities be nk and k. Then, the density ρ of the mixture is given by

$$\rho = \frac{v_1\rho_1 + v_2\rho_2}{v_1 + v_2} = \frac{(mv)(nk) + vk}{mv + v}$$

$$= \frac{(mn+1)k}{m+1} \quad \Rightarrow \quad k = \frac{(m+1)\rho}{mn+1}$$

$\therefore$ Densities of the fluids are

$$\frac{n(m+1)}{mn+1}\rho \text{ and } \frac{m+1}{mn+1}\rho$$

Example 14: The density of milk in a vessel is ρ and the density of pure milk is σ. Prove that the ratio of volumes of water and milk in the vessel is $\sigma - \rho / \rho - 1$.

Sol. Let v_1 and v_2 be volumes of water and pure milk in the vessel. Then, the density of the mixture $= v_1 + v_2\, \sigma / v_1 + v_2$

According to the question,

$$\frac{v_1 + v_2\sigma}{v_1 + v_2} = \rho$$

$$\Rightarrow \quad v_1 + v_2\,\sigma = v_1\,\rho + v_2\,\rho \quad \Rightarrow v_1\,(\rho - 1) = v_2\,(\sigma - \rho) \quad \Rightarrow v_1 / v_2 = \sigma - \rho / \rho - 1$$

Example 15: If equal weights of two different substances be mixed, prove that the specific gravity of the mixture is the harmonic mean of the specific gravities of the component substances.

Sol. Let w, w be the equal weights of two different substances. Let s_1 and s_2 be the specific gravities of the two substances. Let $\bar{s}$ be the specific gravity of the mixture. Then,

$$\bar{s} = \frac{w+w}{\frac{w}{s_1} + \frac{w}{s_2}}$$

$$\Rightarrow \quad \bar{s} = \frac{2s_1s_2}{s_1 + s_2}$$

$\Rightarrow$ $\bar{s}$ is the harmonic mean of s_1 and s_2.

Example 16: Three equal vessels A, B and C are half full of liquids of density ρ_1, ρ_2 and ρ_3 respectively. If now B is filled from A and then C from B, prove that the density of the liquid now contained in C is $1/4\ (\rho_1 + \rho_2 + 2\rho_3)$, the liquids being supposed to mix completely.

Sol. Let the volume of each of the three vessels A, B and C be $2V$. Then,

Volume of liquid of density ρ_1 in vessel A = V

Volume of liquid of density ρ_2 in vessel B = V

Volume of liquid of density ρ_3 in vessel C = V

When B is filled from A, then density ρ of the mixture is given by

$$\rho = \frac{V\rho_1 + V\rho_2}{V+V} = \frac{\rho_1 + \rho_2}{2}$$

Now,

Volume of mixture in vessel B = $2V$.

Density of mixture in vessel B = $\rho_1 + \rho_2 /2$.

When C is filled from B, then the density σ of the mixture is given by

$$\sigma = \frac{V\left(\frac{\rho_1+\rho_2}{2}\right) + V\rho_3}{V+V}$$

∵ We can pour a volume V of the mixture in B into volume V of liquid in C.

$$\Rightarrow \quad \sigma = \frac{\rho_1 + \rho_2 + 2\rho_3}{4}$$

Example 17:

(a) A mixture is formed of equal volumes of n liquids, the densities of which are in the ratio of the numbers 1, 2, 3, ..., n. Find the density of the mixture.

(b) Also, find the density of the mixture when the volumes are in the ratio of the numbers

(i) 1, 2, 3, ..., n

(ii) n, $(n-1)$, $(n-2)$, ..., 1

Sol.

(a) Let v be the volume of each of the n liquids. Let their densities be ρ, 2ρ, 3ρ, ..., nρ. Then, density of the mixture (σ)

$$= \frac{v(\rho) + v(2\rho) + v(3\rho) + \ldots + v(n\rho)}{v + v + v + \ldots\ldots + v} = \frac{1+2+3+\ldots.+n}{n}\rho$$

$$= \frac{n(n+1)}{2n}\rho = \frac{n+1}{2}\rho$$

(b) (i) Let the volumes of n liquids be v, $2v$, $3v$, ..., nv

Then, density of the mixture

$$= \frac{v(\rho) + 2v(2\rho) + 3v(3\rho) + \ldots + nv(n\rho)}{v + 2v + 3v + \ldots + nv} = \frac{1^2 + 2^2 + 3^2 + \ldots + n^2}{1+2+3+\ldots+n}\rho$$

$$= \frac{\frac{n(n+1)(2n+1)}{6}}{\frac{n(n+1)}{2}}\rho = \frac{2n+1}{3}\rho$$

(ii) Let the volumes of n liquids be $nv, (n-1)v, (n-2)v, \ldots, v$. Then, density of the mixture

$$= \frac{nv(\rho)+(n-1)v(2\rho)+(n-2)v(3\rho)+\ldots+v(n\rho)}{nv+(n-1)v+(n-2)v+\ldots+v}$$

$$= \frac{n.1+(n-1).2+(n-2).3+\ldots+1.n}{n+(n-1)+(n-2)+\ldots+1}\rho \qquad = \frac{n.1+(n-1).2+(n-2).3+\ldots+1.n}{\frac{n(n+1)}{2}}\rho$$

General term (rth term) of the series in the numerator

$$= T_r = (n-r+1)\,r$$

$$\Rightarrow \qquad T_r = (n+1)\,r - r^2$$

Putting $r = 1, 2, 3, \ldots, n$, we get,

$$T_1 = (n+1).\,1 - 1^2;\ T_2 = (n+1).2 - 2^2;\ T_3 = (n+1).3 - 3^2$$

$$\vdots$$

$$T_n = (n+1)\,n - n^2$$

$\therefore$ Sum of the series

$$= T_1 + T_2 + T_3 + \ldots + T_n = (n+1)(1+2+3+\ldots+n) - (1^2+2^2+3^2+\ldots n^2)$$

$$= (n+1)\,n(n+1)/2 - n\,(n+1)\,(2n+1)/6$$

$$= n(n+1)/6\,[3(n+1)-(2n+1)] \qquad = n(n+1)(n+2)\,/\,6$$

$\therefore$ Density of the mixture

$$= \frac{\frac{n(n+1)(n+2)}{6}}{\frac{n(n+1)}{2}}\rho \qquad = \frac{n+2}{3}\rho$$

Example 18: The specific gravity of a mixture of equal volumes of two liquids is a and that of a mixture of equal weights of the same liquids is b. Show that the specific gravities of the liquids are given by $x^2 - 2ax + ab = 0$

Sol. Let the volume of each of the two liquids be v. Let w be the weight of each of the two liquids. Let s_1 and s_2 be the specific gravities of the two liquids. Then, according to the question,

$$a = \frac{vs_1 + vs_2}{v+v} \quad \Rightarrow \quad a = \frac{s_1+s_2}{2}$$

$$\Rightarrow \qquad s_1 + s_2 = 2a \qquad \ldots(1)$$

and, $$b = \frac{w+w}{\frac{w}{s_1}+\frac{w}{s_2}} \quad \Rightarrow \quad b = \frac{2s_1s_2}{s_1+s_2}$$

$$= \frac{2s_1s_2}{2a} \quad |\text{From (1)} \qquad = \frac{s_1s_2}{a}$$

$$\Rightarrow \qquad s_1\,s_2 = ab \qquad \ldots(2)$$

From (1) and (2), it follows that s_1, s_2 are the roots of the quadratic equation $x^2 - 2ax + ab = 0$.

Example 19: Equal volumes of three liquids are mixed and the mixture separated into three parts; to each part is added its own volume of the original liquids, and the densities so formed are in the ratio 3 : 4 : 5. Prove that the densities of the liquids are as 1 : 2 : 3.

Sol. Let v be the volume of each of the three liquids. Let the densities of the three liquids be ρ_1, ρ_2 and ρ_3. Let ρ be the density of the mixture. Then,

$$\rho = \frac{v\rho_1 + v\rho_2 + v\rho_3}{v + v + v}$$

$$\Rightarrow \qquad \rho = \frac{\rho_1 + \rho_2 + \rho_3}{3} \qquad ...(1)$$

Let the volumes of the three parts of the mixture be v_1, v_2 and v_3. To first part of the mixture, a volume v_1 of first liquid of density ρ_1 is added. Let the density of the mixture be 3σ. Then,

$$3\sigma = \frac{v_1\rho_1 + v_1.1/3(\rho_1 + \rho_2 + \rho_3)}{v_1 + v_1}$$

$$\Rightarrow \qquad 3\sigma = \frac{4\rho_1 + \rho_2 + \rho_3}{6} \qquad ...(2)$$

Similarly to second part of the mixture, a volume v_2 of second liquid of density ρ_2 is added. Let the density of the mixture be 4σ. Then,

$$4\sigma = \frac{v_2\rho_2 + v_2.1/3(\rho_1 + \rho_2 + \rho_3)}{v_2 + v_2}$$

$$\Rightarrow \qquad 4\sigma = \frac{\rho_1 + 4\rho_2 + \rho_3}{6} \qquad ...(3)$$

and, to third part of the mixture, a volume v_3 of third liquid of density ρ_3 is added. Let the density of the mixture be 5σ. Then,

$$5\sigma = \frac{v_3\rho_3 + v_3.1/3(\rho_1 + \rho_2 + \rho_3)}{v_3 + v_3}$$

$$\Rightarrow \qquad 5\sigma = \frac{\rho_1 + \rho_2 + 4\rho_3}{6} \qquad ...(4)$$

Solving equations (2), (3) and (4), we get,

$$\rho_1 = 2\sigma, \ \rho_2 = 4\sigma, \ \rho_3 = 6\sigma$$

Hence, $\rho_1 : \rho_2 : \rho_3 = 2\sigma : 4\sigma : 6\sigma = 1 : 2 : 3$

Example 20: The mixture of a quantity of a liquid A with n lbs. of B has a specific gravity s, with $2n$ lbs. of B a specific gravity s', with $3n$ lbs. of B a specific gravity s''. Show that $2(s - s_1)(s_2 - s') = (s_2 - s)(s' - s_1)$, and $3(s - s_1)(s_2 - s'') = (s_2 - s)(s'' - s_1)$, where s_1 and s_2 are the specific gravities of the liquids.

Sol. We know that

$$s = \frac{w_1 + w_2}{\dfrac{w_1}{s_1} + \dfrac{w_2}{s_2}}$$

Let w be the weight of liquid A. Then,

$$s = \frac{w+n}{\frac{w}{s_1}+\frac{n}{s_2}} \qquad \Rightarrow \qquad s\left(\frac{w}{s_1}+\frac{n}{s_2}\right) = w+n$$

$$\Rightarrow \qquad w\left(\frac{s}{s_1}-1\right) = n\left(1-\frac{s}{s_2}\right)$$

$$\Rightarrow \qquad \frac{w(s-s_1)}{s_1} = \frac{n(s_2-s)}{s_2} \qquad \text{...(1)}$$

Similarly,

$$\frac{w(s'-s_1)}{s_1} = \frac{2n(s_2-s')}{s_2} \qquad \text{...(2)}$$

and, $$\frac{w(s''-s_1)}{s_1} = \frac{3n(s_2-s'')}{s_2} \qquad \text{...(3)}$$

Dividing (1) by (2), we get,

$$\frac{s-s_1}{s'-s_1} = \frac{s_2-s}{2(s_2-s')}$$

$$\Rightarrow \qquad 2(s-s_1)(s_2-s') = (s_2-s)(s'-s_1) \qquad \text{...(4)}$$

Dividing (1) by (3), we get,

$$\frac{s-s_1}{s''-s_1} = \frac{s_2-s}{3(s_2-s'')}$$

$$\Rightarrow \qquad 3(s-s_1)(s_2-s'') = (s_2-s)(s''-s_1) \qquad \text{...(5)}$$

With the help of equations (4) and (5), we can determine s_1 and s_2.

Example 21: Show that the specific gravity of a mixture of n liquids is greater when equal volumes are taken than when equal weights are taken, assuming no change in volume as the result of mixing.

Sol. Let s be the specific gravity of the mixture when mixture of equal volumes is taken. Let s' be the specific gravity of the mixture when mixture of equal weights is taken. Let $s_1, s_2, ..., s_n$ be the specific gravities of the different substances.

Then, we have

$$s = \frac{s_1+s_2+...+s_n}{n} \qquad \text{...(1)}$$

and, $$\frac{1}{s'} = \frac{1}{n}\left(\frac{1}{s_1}+\frac{1}{s_2}+...+\frac{1}{s_n}\right) \qquad \text{...(2)}$$

Also, we know that the Arithmetic Mean (A.M.) of n unequal positive numbers is greater than their Harmonic Mean (H.M.).

Now, arithmetic mean of $s_1, s_2, ..., s_n = \dfrac{s_1 + s_2 + ... + s_n}{n} = s$ | From (1)

and, harmonic mean of $s_1, s_2, ..., s_n = \dfrac{n}{\dfrac{1}{s_1} + \dfrac{1}{s_2} + ... + \dfrac{1}{s_n}} = s'$ | From (2)

Hence, $s > s'$

Aliter: We know that the Arithmetic Mean (A.M.) of n unequal positive numbers is greater than their Geometric Mean (G.M.)

So, we have,

$$\frac{s_1 + s_2 + ... + s_n}{n} > (s_1 . s_2 ... s_n)^{\frac{1}{n}} \quad ...(3)$$

and,
$$\frac{1}{n}\left(\frac{1}{s_1} + \frac{1}{s_2} + ... + \frac{1}{s_n}\right) > \left(\frac{1}{s_1} . \frac{1}{s_2} ... \frac{1}{s_n}\right)^{\frac{1}{n}} \quad ...(4)$$

$\therefore$
$$s > (s_1 . s_2 ... s_n)^{1/n} \quad ...(5)$$

and,
$$\frac{1}{s'} > \frac{1}{(s_1 . s_2 ... s_n)^{\frac{1}{n}}} \quad ...(6)$$

Multiplying the results (5) and (6), we get,

$$s\frac{1}{s'} > (s_1 . s_2 ... s_n)^{\frac{1}{n}} . \frac{1}{(s_1 . s_2 ... s_n)^{\frac{1}{n}}} = 1$$

Hence, $s > s'$

This proves the assertion.

EXERCISE

1. 12 kgms. weight of liquid of specific gravity 1.1 is mixed with 20 kgms. weight of a liquid of specific gravity 0.9. What is the specific gravity of the mixture?
   Ans. 0.9658.
2. The specific gravity of a cork being 0.24, find what volume of water weighs as much as a cubic decimetre of cork?
   Ans. 240 cubic cm.
3. The specific gravity of a cork being 0.24, find what volume of water weighs as much as a cubic yard of cork?
   Ans. 6.48 cu. ft.
4. 10 lbs. weight of a liquid of specific gravity 1.25 is mixed with 6 lbs. weight of a liquid of specific gravity 1.15. What is the specific gravity of the mixture?
   Ans. 1.2105.

5. Find the specific gravity of an alloy of gold and copper made in the ratio 5 : 2 by volume, their specific gravity being 19.4 and 8.84 respectively.
   Ans. 16.383 nearly.
6. When equal volumes of two substances are mixed, the specific gravity of the mixture is 8. When equal weights of the same substances are mixed, the specific gravity of the mixture is 6. Find the specific gravities of the substances.
   Ans. 12, 4.
7. When equal volumes of two substances are mixed, the specific gravity of the mixture is 6. When equal weights of the same substances are mixed, the specific gravity of the mixture is 4. Find the specific gravities of the substances.
   Ans. $6 \pm 2\sqrt{3}$
8. The specific gravities of gold and copper are 19.3 and 8.62 respectively. These metals are mixed in certain proportion to form an alloy of specific gravity 10.2. Find the proportion by volumes of metals in the alloy.
   Ans. 79 : 455.
9. If the volume of 10 c.c. of a liquid of density 0.8 gramme per c.c. be mixed with 15 c.c. of a liquid of density 0.6 gramme per c.c., find the density of the mixture.
   Ans. 0.68 gramme/c.c.
10. What weight of water must be added to 30 gm. of a solution whose specific gravity in 1.06 so that the specific gravity of the mixture may be 1.03?
    Ans. $28\frac{16}{53}$ gm.
11. The weight of an empty vessel is 3 kg. When filled with water it is 9 kg and when filled with oil, it is 8.4 kg. Find the specific gravity of the oil.
    Ans. 0.9.
12. Five litres of a liquid whose specific gravity is 1.3 are mixed with 7 litres of a liquid whose specific gravity is 0.78. If the bulk of the liquid shrinks 1 per cent on mixing. Find the specific gravity of the mixture.
    Ans. $1\frac{2}{297}$.
13. The density of mixture of equal volumes of three liquids is A and that of a mixture of equal weights of the same three liquid is B. Show that A > B, if there is no change in volume on mixing.

Chapter 3

Theorems on Fluid Pressure Under Gravity

1. Light Fluid and Heavy Fluid

A mass of fluid may be maintained at rest, i.e., in equilibrium, under the action of a number of external forces, uniform or variable. Here, we shall be limited to the case when the only external force acting on the mass of fluid is uniform and it is the weight of the fluid. In other words, we shall regard the external force as force due to gravity only.

In the above light, we may divide a fluid into two categories. These are light or heavy.

A fluid is said to be light if the force due to gravity is not taken into account. On the contrary, a fluid is said to be heavy if the force due to gravity is taken into consideration.

2. Theorem 1

The pressure of a heavy homogeneous fluid at rest under gravity is the same at all points in the same horizontal plane.

Proof: Let A and B be any two points within the fluid in the same horizontal plane. Let the straight line AB lie entirely within the fluid. With AB as axis, construct a fluid cylinder whose cross-section S is very small and the plane ends are perpendicular to AB.

Let p and p' be the pressures per unit area at A and B respectively. The fluid inside this cylinder is in equilibrium under action of the following forces.

(1) The weight W of the fluid contained acting vertically downwards.

(2) The pressures ps and $p's$ at the plane ends A and B acting perpendicular to AB along $\overrightarrow{AB}$ and $\overrightarrow{BA}$ respectively.

(3) The pressure of the surrounding fluid at each point of the curved surface of the cylinder acting in a direction perpendicular to AB. Pressures on the vertical curved sides of the cylinder balance one another.

Hence, for the sake of equilibrium, resolving all the forces in the sense of the horizontal line AB, we get

$$p\,S - p'\,S = 0$$

$$\Rightarrow \quad p = p'$$

Note: In the case when the line AB does not lie entirely within the fluid, the points A and B can be joined by a broken horizontal line $ACDB$, the parts AC, CD and DB lying entirely within the fluid and then we have by above theorem,

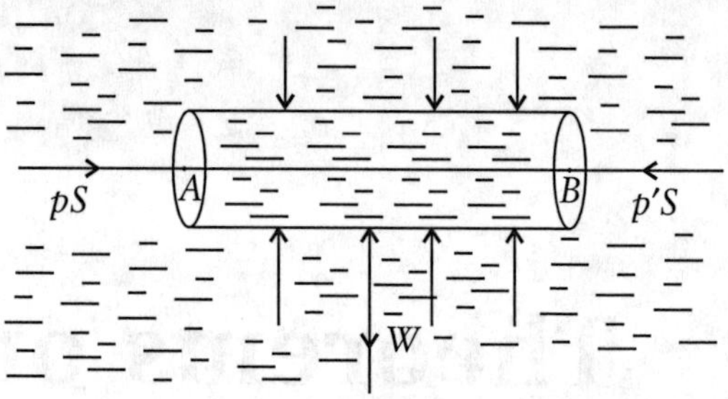

Pressure at A = Pressure at C

= Pressure at D

= Pressure at B

Hence, the pressures at A and B are equal.

3. Theorem 2

In a heavy homogeneous fluid at rest under gravity, the difference between the pressures at two points varies is the difference of their vertical depths below the free surface.

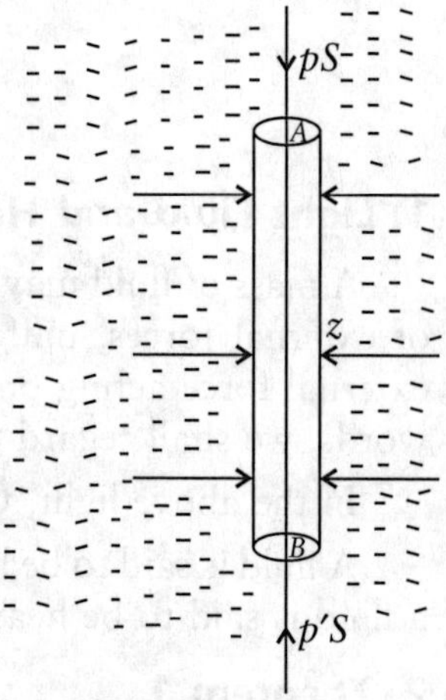

Let A and B be two points in the fluid.

Case I: When the straight line AB is vertical and lies entirely within the fluid.

Let z be the vertical distance between A and B. Then, $AB = z$. With AB as axis, construct a vertical fluid cylinder whose cross section S is very small and the plane ends are perpendicular to AB. Let p and p' be the pressures per unit area at A and B respectively.

The fluid inside this small fluid cylinder is in equilibrium under the action of the following forces.

(1) The weight $S\,z\,\rho\,g$ of the fluid contained acting vertically downwards, ρ being the density of the fluid.

(2) The pressure $p\,S$ acting vertically downwards on the upper end at A.

(3) The pressure $p'S$ acting vertically upwards on the lower end at B.

(4) The pressure of the surrounding fluid at each point of the curved surface of the cylinder acting horizontally. Pressures on the vertical curved sides of the cylinder balance one another.

Hence, for the sake of equilibrium, resolving the forces in the sense of the vertical line AB, we have

$$pS + S\,z\,\rho\,g - p'S = 0$$

$$\Rightarrow \quad p' - p = z\,\rho\,g$$

$$\Rightarrow \quad p' - p \propto z$$

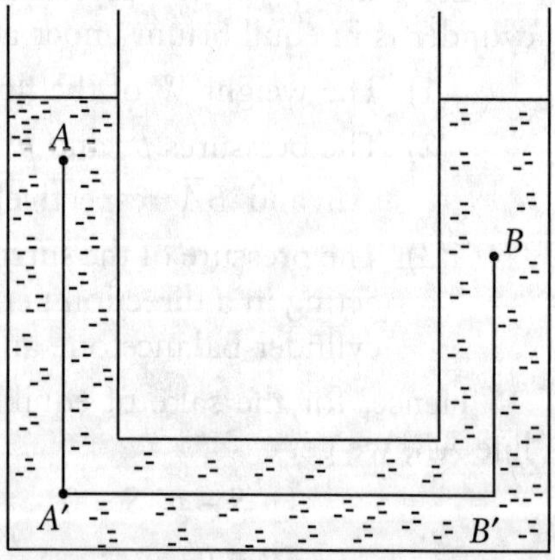

Case II: When the straight line AB is not vertical and does not lie entirely within the fluid.

Let A and B be two points, not in the same vertical line, in the fluid contained in a vessel as shown aside.

Let A' and B' be the points in the same horizontal line vertically below A and B respectively.

Then, by Theorem 2 above,

Pressure at A' – Pressure at A = $\rho g\ AA'$

and, Pressure at B' – Pressure at B = $\rho g\ BB'$

But the pressure at A' and B' are equal since the points A' and B' lie in the same horizontal line.

[Refer Theorem 1]. Therefore,

Pressure at B – Pressure at A = $\rho g\ (AA' - BB')$

= ρgz where z is the difference of the vertical depths of A and B

4. Pressure at a Depth z below the Surface of Liquid

(a) If the liquid is homogeneous, then the pressure at a depth z below the surface of liquid is given by $p = \Pi + wz$, where Π is the atmospheric pressure and w is the weight of unit volume of the liquid.

Let us take a point P within the liquid at a depth z below the surface of the liquid. Draw PQ perpendicular to the surface of the liquid. Taking PQ as axis, construct a fluid cylinder of small cross-section S. This liquid cylinder is in equilibrium under the action of the following forces:

(1) Pressure $\Pi\ S$ on the plane face at Q, acting vertically downwards.

(2) Pressure $p\ S$ on the plane face at P, acting vertically upwards.

(3) Pressure of the surrounding liquid at each point on the curved surface of the cylinder acting horizontally. These pressures on the vertical curved sides of the cylinder balance one another.

(4) Weight Szw of the liquid inside the cylinder acting vertically downwards.

Hence, for the sake of equilibrium, resolving the forces in the sense of vertical line QP, we have

$$\Pi\ S + Szw - Sp = 0$$

$$\Rightarrow \quad p = \Pi + wz$$

If these is no atmospheric pressure or if the atmospheric pressure is neglected, then, we have

$$p = wz$$

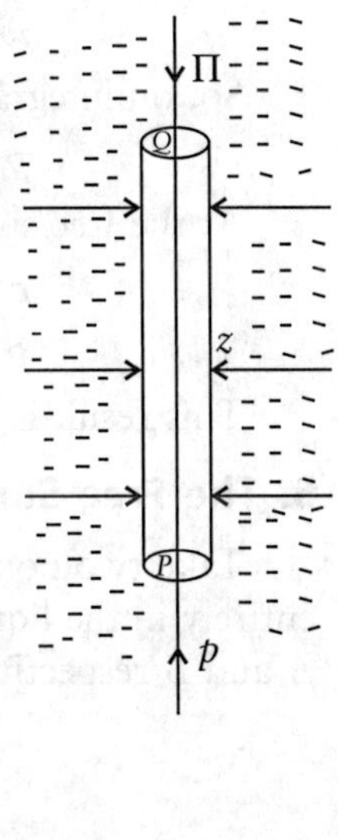

This shows that the pressure at any point varies as the depth of the point below the surface of the liquid.

Note: Let p_1 and p_2 be the pressures at depths z_1 and z_2 respectively where $z_1 > z_2$.

Then,

$$p_1 = \Pi + wz_1$$

$$p_2 = \Pi + wz_2$$

$$\therefore \quad p_1 - p_2 = w\ (z_1 - z_2)$$

If the liquid is uniform, then w is constant.

$$\therefore \quad p_1 - p_2\ \alpha\ z_1 - z_2$$

If a homogeneous liquid is at rest under gravity, then, the difference between the pressures at any two points of the liquid varies as the vertical distance between them.

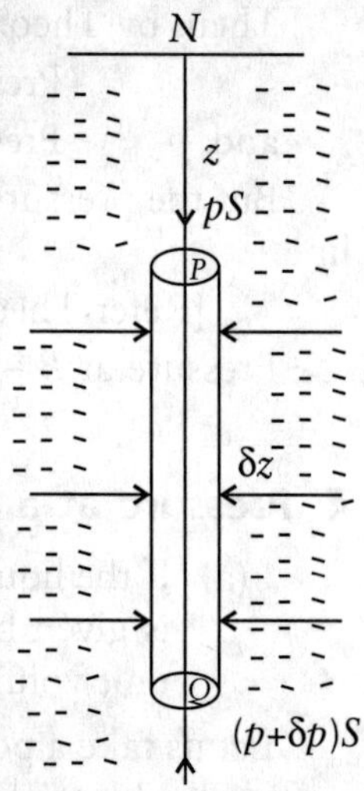

(b) *When the Liquid is Heterogeneous*: Let P be a point in a heterogeneous liquid at a depth z below the surface of liquid. Draw a perpendicular PN to the surface. Let NP be produced to Q such that $PQ = \delta z$. Taking PQ as axis, construct a fluid cylinder of a very small cross-section S.

Since the liquid cylinder is very small, therefore, the density of the liquid inside this liquid cylinder may be assumed to be uniform. Let the weight of unit volume of it be w. [It is to be noted here that w is not the same at every point of the liquid but varies from point to point].

Let the intensity of pressure at P be p. Then, the intensity of pressure at $Q = p + \delta p$.

Now, this elementary liquid cylinder is in equilibrium under the action of the following forces:

(1) Pressure pS on the plane face at P acting vertically downwards.
(2) Pressure $(p + \delta p)\ S$ on the plane face at Q acting vertically upwards.
(3) Weight $S\delta zw$ of the liquid inside the cylinder acting vertically downwards.
(4) Pressure of the surrounding liquid at each point of the curved surface of the cylinder acting horizontally. Pressures on the vertical curved sides of the cylinder balance one another.

Hence, for equilibrium, resolving the forces in the sense of vertical line PQ we have

$$pS + S\delta zw - (p + \delta p)\ S = 0$$

$$\Rightarrow \quad \delta p = w\delta z$$

In limit,

$$dp/dz = w$$

where $w = g\rho$, ρ being the density of the liquid at a depth z

Particular Case

If the liquid is homogeneous, then,

$$w = \text{constant.}$$

So, on integration, we get

$$p = wz + C$$

At the free surface, $z = 0$, $p = \Pi$ (atmospheric pressure)

$$\therefore \quad c = \Pi$$

$$\therefore \quad p = \Pi + wz$$

This result is the same as proved in part (a).

5. The Free Surface of a Heavy Liquid at Rest is a Horizontal Plane

Take two points P and Q in a liquid lying in the same horizontal plane. Let the line PQ lie entirely in the liquid. From P and Q, draw vertical lines PA and QB to meet the free surface in A and B respectively.

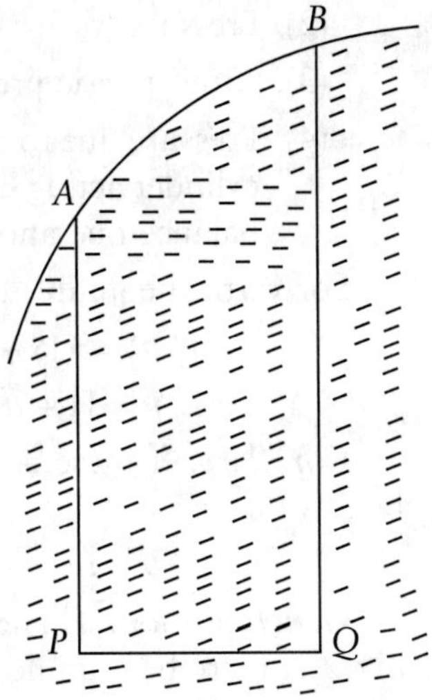

Since the points P and Q are in the same horizontal plane, therefore

Pressure at A = Pressure at B | Refer Theorem 1

$\Rightarrow \Pi + w\,AP = \Pi + w\,BQ$ | Refer § 4

where Π is the atmospheric pressure

$\Rightarrow AP = BQ$ ...(1)

Since PQ is horizontal, therefore AB must be horizontal. | By (1)

Since A and B are any two points in the surface of the liquid, therefore, it follows that the free surface of a liquid at rest is a horizontal plane.

Note: The result five confirms the popular saying that "*water finds its own level.*"

6. The Densities at Two Points in a Fluid at Rest under Gravity and in the Same Horizontal plane are Equal

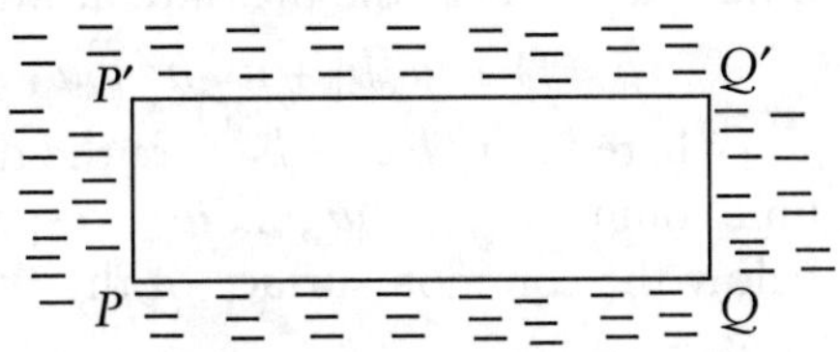

Let P and Q be two points in the fluid, which is at rest under gravity, in the same horizontal line. Let P' and Q' be two points at a short distance vertically above P and Q respectively such that $PP' = QQ'$.

Let ρ be the mean density of the fluid between P and P' and let ρ' be the mean density of the fluid between Q and Q'. Then,

Pressure at P – Pressure at $P' = g\,\rho\,PP'$

Pressure at Q – Pressure at $Q' = g\,\rho'\,QQ'$

Also, Pressure at A = Pressure at Q

Pressure at P' = Pressure at Q'

$g\,\rho\,PP' = g\,\rho'\,QQ'$

$\Rightarrow \rho = \rho'$ | $\because PP' = QQ'$

7. To Find the Pressure at any Point in the Lower of the Two given Heavy Homogeneous Liquids which do not Mix

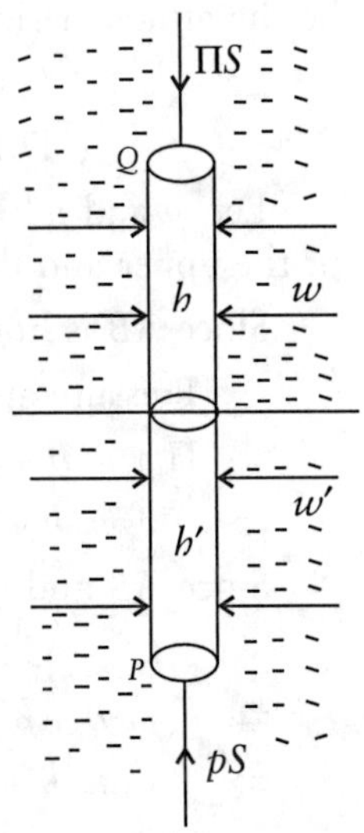

Let w and w' be the weights per unit volume of the upper and lower liquids respectively. Let the depth of the upper liquid be h. Let P be any point in the lower liquid at a depth h' below the common surface, i.e., surface of separation of the two liquids. Through P, draw perpendicular PQ to the free surface. Construct a small liquid cylinder with PQ as axis, its lower end at P and upper end in the free surface of the upper liquid.

Let the small area of cross-section of this cylinder be S. Let p be the intensity of pressure at P. Let Π be the atmospheric pressure.

This liquid cylinder is in equilibrium under the action of the following forces:

(1) Weight $(whS + w'h'S)$ of the liquid within the liquid cylinder acting vertically downwards.

(2) Pressure (thrust) pS on the plane face at P acting vertically upwards.

(3) Atmospheric pressure ΠS on the plane face at Q acting vertically downwards.

(4) Pressure due to the surrounding liquid at each point of the curved surface of the liquid cylinder acting horizontally. These pressures on the vertical curved sides of the cylinder balance one another.

For vertical equilibrium, resolving the forces in the sense of vertical line PQ, we have

$$pS = \Pi S + w\,h\,S + w'\,h'\,S$$

$$\Rightarrow \quad p = \Pi + w\,h + w'\,h'$$

Corollary: If these is no atmospheric pressure or if the atmospheric pressure is neglected, then, we have

$$p = w\,h + w'\,h'$$

Generalisation: This result can be easily extended to more than two liquids which do not mix. Let there be n liquids whose weights per unit volume are $w_1, w_2, w_3, ..., w_n$ $(w_1 < w_2 < w_3 < .... < w_n)$. Then, the pressure at any point P in the lowermost liquid is given by

$$p = \Pi + w_1\,h_1 + w_2\,h_2 + w_3\,h_3 + ... + w_n\,h_n$$

where $h_1, h_2, h_3, ..., h_{n-1}$ are the thicknesses of the layers (strata) of the liquids of weights per unit volume $w_1, w_2, w_3, ..., w_{n-1}$ respectively and h_n is the depth of the point under consideration below the common surface of the $(n-1)^{th}$ and n^{th} liquids.

8. The Surface of Separation of Two Heavy Homogeneous Liquids that do not Mix and are at Rest under Gravity is a Horizontal Plane

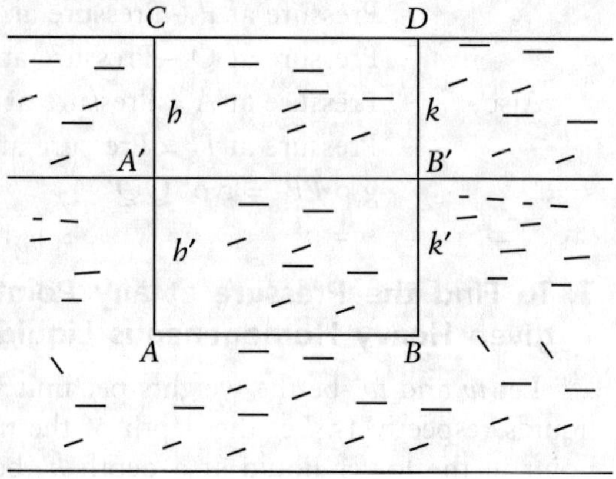

Let A and B be any two points same horizontal plane in the lower liquid. Let the vertical lines drawn through A and B meet the common surface of the two liquids at A' and B' and the free surface of the upper liquid at C and D respectively. Let Π be the atmospheric pressure.

Let $\quad CA' = h',\ A'A = h'$

$\quad DB' = k,\ B'B = k'$

Let w and w' be the weights per unit volume of the upper and the lower liquids respectively.

Since AB is horizontal, therefore,

Pressure at A – Pressure at B | Refer theorem 1

$\Rightarrow \Pi + w\,h + w'\,h' = \Pi + wk + w'\,k'$ | Refer § 7

$$\Rightarrow \quad w\,h + w'h' = wk + w'k' \qquad ...(1)$$

Since AB and CD both are horizontal, therefore,

$$CA = DB$$

$$\Rightarrow \quad h + h' = h + k'$$

$$\Rightarrow \quad w'h + w'h' + w'k + w'k \qquad ...(2)$$

Subtracting (1) from (2), we get

$$(w' - w)h = (w' - w)k$$

$$\Rightarrow \quad h = k \quad \mid \because w' \neq w$$

Hence, the common surface $A'B'$ is horizontal.

9. Effecting Surface

We know that the pressure at a point P at a depth z below the free surface of a heavy homogeneous liquid which is at rest under gravity is given by

$$p = \Pi + wz \quad \mid \text{Refer § 4}$$

where Π is the atmospheric pressure and w is the weight per unit volume of the liquid.

Let us now suppose that there is no atmosphere (or the atmosphere is neglected) and we superimpose a stratum of the same liquid of thickness Π/w (= z' say) on the free surface above the original liquid.

Then, the depth of this point P below the free surface of the superimposed liquid = $z + z'$

$\therefore$ Pressure at the point P

$$= w\,(z + z') \quad = w\,(z + \Pi/w) \quad = wz + \Pi$$

which is the same as before when the atmosphere was not removed.

The upper surface of the assumed superimposed liquid of thickness z' (= Π/w) is termed as the *effective surface* or *surface of zero pressure.*

Hence, we may conclude that the pressure at any point of a heavy homogeneous liquid is proportional to its depth below the effective surface.

10. Head of Liquid

When we say that the pressure at a point in a liquid is due to head of h metres of the liquid, then it means that the point under consideration is at a depth of h metres below the effective surface of that liquid.

11. Characteristics of Liquids

The characteristics of liquids are as follows:

(1) The surface of a heavy liquid at rest is horizontal.

(2) The common surface of two heavy liquids that do not mix is horizontal.

(3) If there are two liquids that do not mix, then the density of the lower liquid is greater than the density of the upper liquid.

(4) A liquid always finds its own level.

ILLUSTRATIVE EXAMPLES

Example 1: Pressure at the bottom of a well, filled completely with water, is three times the pressure at a depth of 3 feet. If the height of the water barometer be 3 feet, find the depth of the well.

Sol. Let the depth of the well be h feet. Let the density of water in the well be ρ.

Atmospheric pressure $\Pi = 30\ \rho g$ units

According to the question,

Pressure at the bottom = 3 pressure at a depth of 3 feet

$\Rightarrow\ \Pi + \rho g h = 3\ (\Pi + 3\ \rho g) \Rightarrow \quad 30\ \rho g + \rho g h = 3\ (30\ \rho g + 3\ \rho g)$

$\Rightarrow\ 30 + h = 99 \quad \Rightarrow \quad h = 69$ feet

Hence, the depth of the well is 69 feet.

Example 2: The pressure in a water-pipe at the base of a building is 39 lbs. wt. per sq. inch and on the roof it is 19 lbs. wt. per sq. inch. Find the height of the roof. (1 cu. ft. of water weighs 62.5 lbs.)

Sol. Let the height of the roof be h inches.

p = pressure at the base = 39 lbs. wt. per sq. inch

p' = pressure on the roof = 19 lbs. wt. per sq. inch

Weight of 1 cu. ft. of water = 62.5 lbs.

$\Rightarrow$ Weight of 1 cu. inch of water

$$= \frac{62.5}{12 \times 12 \times 12} \text{lb.}$$

We know that

$$p - p' = wh$$

$$\Rightarrow \quad 39 - 19 = \frac{62.5}{12 \times 12 \times 12} h$$

$$\Rightarrow \quad h = \frac{12 \times 12 \times 12 \times 20}{62.5} \text{ inches}$$

$$= \frac{12 \times 12 \times 20}{62.5} \text{ ft.} \qquad = 46.08 \text{ ft.}$$

Hence, the height of the roof is 46.08 ft.

Example 3: Find the depth in water at which the pressure per sq. inch is 140 lbs. assuming the atmospheric pressure to be 15 lbs. per sq. inch and the weight of 1 cubic ft. of water is 1000 ozs.

Sol. Let the depth of water be z inches.

Weight of water per cubic ft.

= 1000 oz. wt.

$\therefore$ Weight of water per cubic inch

$$= \frac{1000}{12 \times 12 \times 12} \text{ oz. wt.}$$

$$= \frac{1000}{12 \times 12 \times 12 \times 16} \text{ lbs. wt.} \qquad | \because 16 \text{ ozs.} = 1 \text{ lb.}$$

According to the question,

$$140 = 15 + \frac{1000}{12\times12\times12\times16} z$$

$$\Rightarrow \qquad z = \frac{125\times12\times12\times12\times16}{1000}$$

$$= 3456 \text{ inches}$$

$$= 288 \text{ ft.} \qquad | \because 12 \text{ inches} = 1 \text{ ft.}$$

Example 4: A layer of mercury (sp. gr. 13.16) 25 cm. deep is covered by one of water of the same depth. Above this there is a layer 50 cm. in depth of oil (sp. gr. 0.9). Find the pressure at the bottom in gms. wt. per sq. cm., assuming the atmospheric pressure to be due to a column of 76 cm. of mercury.

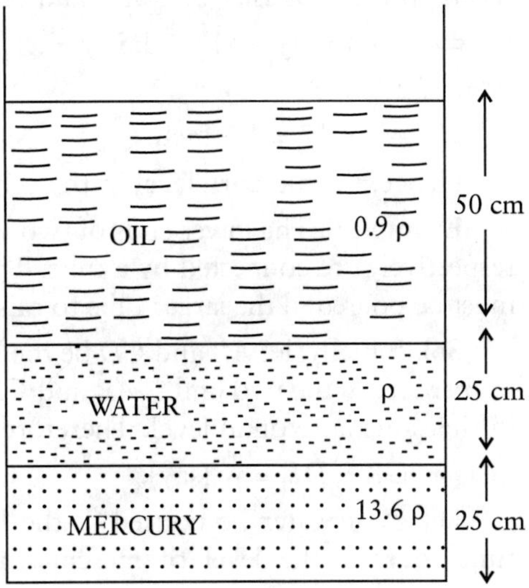

Sol. Let ρ be the density of water.

Then, weight of 1 cubic centimetre of water = $g\rho$; Also $g\rho$ = 1 gm. wt.

∴ Weight of 1 cubic centimetre of oil = $0.9\, g\rho$

and, weight of 1 cubic centimetre of mercury = $13.6\, g\rho$

Atmospheric pressure $\Pi = 76 \times 13.6\, g\rho$

∴ Required pressure

$= 76 \times 13.6\, g\rho + 50 \times 0.9\, g\rho + 25\, g\rho + 25 \times 13.6\, g\rho$

$= 1443.6\, g\rho = 1443.6$ gms. wt.

Example 5: In a U-tube of uniform cross-section these is some mercury. Water is then poured into one limb and it occupies a length of 9 inches. If the sp.gr. of mercury is taken to be 13.5, find the distance through which the mercury level in the other limb is raised.

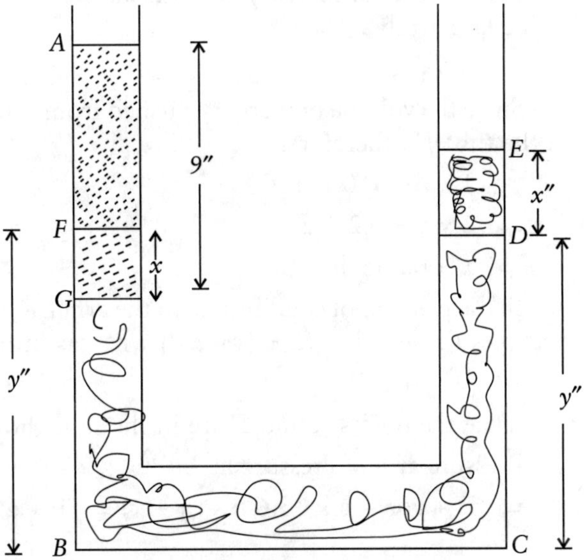

Sol. Let the initial height of the mercury in each tube be y''. Then,

$$FB = DC = 9''$$

Water is poured into the limb AB of the U-tube. Since it occupies a length of 9 inches, therefore,

$$AG = 9''$$

As a result, the level of mercury sinks through x'' (say) in the arm AB and the level of mercury rises through x'' in the arm AC because the cross-section of the U-tube is uniform.

Thus, $FG = x' = ED$.

Therefore, in the position of equilibrium, in the arm AB, the water occupies a length $AG = 9''$ and mercury occupies a length $GB = (y - x)''$ whereas in the arm EC, mercury occupies the length $(y + x)''$.

Since the point B and C lie in the same horizontal level, therefore,

Pressure at B = Pressure at C

$\Rightarrow \Pi + 9\,g\rho + (y - x) \times 13.5\,g\rho = \Pi + (y + x) \times 13.5\,g\rho$

where, ρ is the density of water and Π is the atmospheric pressure.

$\Rightarrow 9 + 13.5(y - x) = 13.5\,(y + x)$

$\Rightarrow 27x = 9$

$\Rightarrow x = 9/27 = 1/3$ inch

Hence, the mercury level in the other limb will rise through 1/3 inch.

Example 6: The lower ends of two vertical tubes whose cross-sections are 2 and 0.2 sq. inches respectively are connected by a tube. The tube contains mercury (sp. gr. 13.6). How much water must be poured in the larger tube to raise the level of the mercury in the smaller tube by 2 inches?

Sol. Initially, let AB and CD be the levels of mercury in the two limbs. AB and CD are in the same horizontal level. Therefore,

$GK = DL = y''$ (say).

Let z'' of water be poured in the larger tube to raise the level of mercury in the smaller tube by 2 inches.

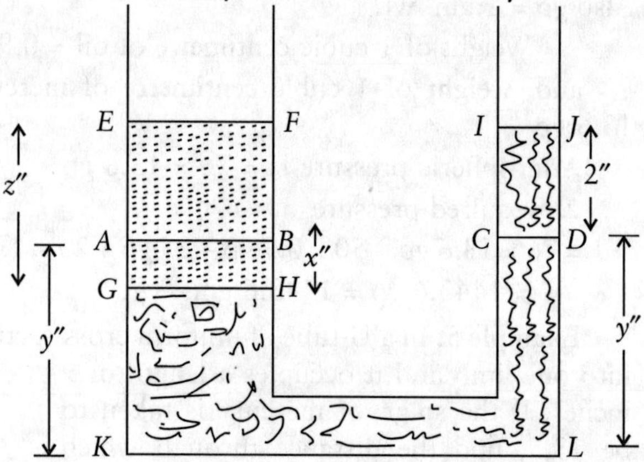

Then, $EG = z''$ and $DJ = 2''$

Let the level of mercury sink through x'' in the larger tube. Then,

$AG = x''$

Since the volume of mercury pushed from tube EK is equal to the volume of mercury entered in the tube JL, therefore,

$2 \times AG = 0.2 \times IC$

$\Rightarrow 2 \times x = 0.2 \times 2$

$\Rightarrow x = 0.2$ inch ...(1)

In the position of equilibrium, in the arm EK, water occupies the length EG $(= z'')$ and mercury occupies the length GK $[= (y - x)'']$ whereas in the arm JL, mercury occupies the length LJ $[= (y + 2)'']$.

Since the points K and L are in the same horizontal level, therefore,

Pressure at K = Pressure at L

$\Rightarrow \Pi + g\rho z + g \times 13.6\,\rho + (y - x) = \Pi + g \times 13.6\,\rho\,(y + 2)$

where, ρ is the density of water and Π is the atmospheric pressure.

$\Rightarrow\ z + 13.6(y - 0.2) = 13.6\,(y + 2)$ | From (1)

$\Rightarrow\ z = 13.6(2 + 0.2) = 13.6 \times 2.2 = 29.92$ inches

Hence, 29.92 inches of water must be poured in the larger tube to raise the level of the mercury in the smaller tube by 2 inches.

Example 7: A U-tube of uniform cross-section, open at both ends, contains 27 inches of water and such a quantity of mercury that the water is entirely in one of the straight portions of the tube. The straight portions are connected by a fine horizontal tube which passes between them half an inch above the surface of separation of the water and mercury, the passage through which is originally closed. If the passage is now opened, find in what direction flow takes place through it, and the volume of the liquid which passes before equilibrium is again established.

(specific gravity of mercury being 13.5)

Sol. Since the connecting tube is fine, therefore, its capacity in negligible.

Case I: *When the passage is closed*

Let the height of mercury in the tube AB be x''.

Since the points B and D are in the same horizontal level, therefore,

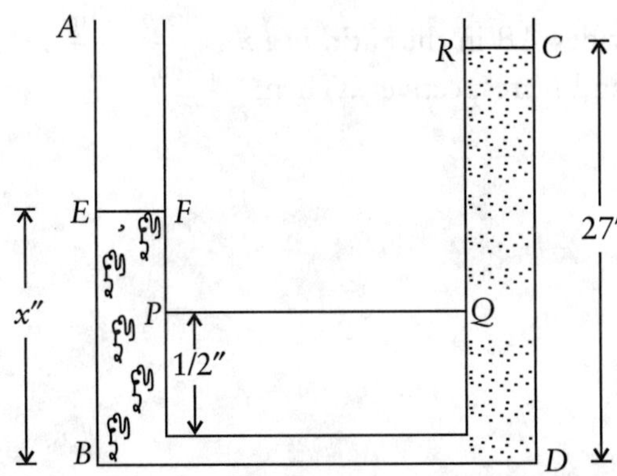

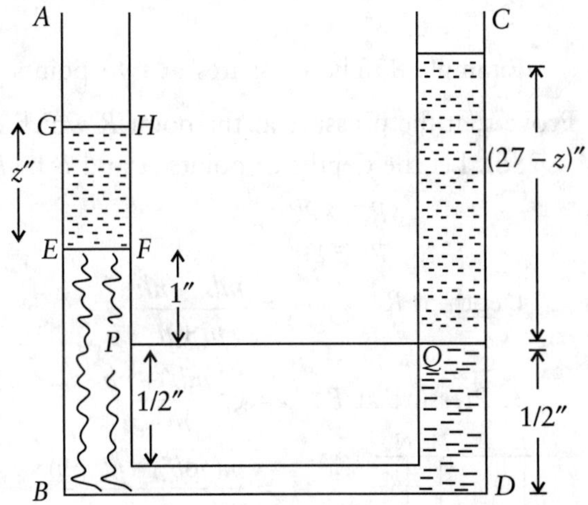

Pressure at B = Pressure at D

$\Rightarrow\ \Pi + g \times 13.5\ \rho \times x = \Pi + g \times\ \rho \times 27$

where ρ is the density of water and Π is the atmospheric pressure.

$\Rightarrow\qquad x = 2$ inches ...(1)

Pressure at $Q = \Pi + g\rho\ (27 - 1/2) = \Pi + g\rho\ (53/2)$

$= \Pi + g\rho\ (26.5)$ ...(2)

Pressure at $P = \Pi + g \times 13.5\ \rho\ (x - 1/2)$

$= \Pi + g \times 13.5\ \rho\ (1.5)$

$= \Pi + g\rho\ (20.25)$ ...(3)

From (2) and (3), it is evident that

Pressure at $Q >$ Pressure at P ...(4)

Case II: *When the passage is opened*

As soon as the passage is opened, the water will flow into the other arm of the U-tube through the fine tube *PQ* towards *P* from *Q*. The pressure of the liquid will increase in the arm *AB*. As a result, the mercury will flow through the tube *BD* into the other arm *CD*. The level of mercury in the water tube *CD* will be at least 1/2″ because the densities at *P* and *Q* will be the same, *P* and *Q* being in the same horizontal line.

Let the water enter in the other tube *AB* by *z*″.

Now the tube *AB* contains *z*″ of water and (2 – 1/2), i.e., $1\frac{1''}{2}$ of mercury whereas the tube *CD* contains (27 – *z*)″ of water and 1/2″ of mercury.

Since the points *P* and *Q* are in the same horizontal level, therefore,

Pressure at *P* = Pressure at *Q*

$$\Rightarrow \Pi + g\rho z + g \times 13.5\,\rho \times (1\tfrac{1}{2} - 1/2)$$
$$= \Pi + g\rho\,(27 - z)$$
$$\Rightarrow z + 13.5 = 27 - z$$
$$\Rightarrow 2z = 13.5$$
$$\Rightarrow z = 6.75''$$

Example 8: The pressures at two points *A* and *B* in a homogeneous liquid are p and p'. Prove that the pressure at the point *R* which divides *AB* in the ratio $m : n$ is $\frac{mp' + np}{m+n}$.

Sol. Let the depths of points *A* and *B* be h and h' respectively. Then,

$$p = g\rho h \quad \text{...(1)}$$
$$p' = g\rho h' \quad \text{...(2)}$$

Depth of *R* $= \frac{mh' + nh}{m+n}$

$\therefore$ Pressure at *R* $= g\rho\frac{mh' + nh}{m+n}$

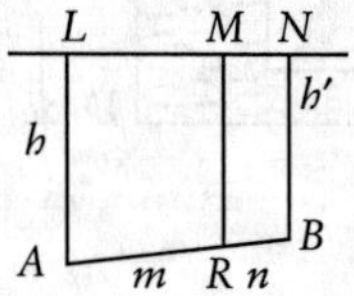

$$= \frac{m(g\rho h') + n(g\rho h)}{m+n}$$
$$= \frac{mp' + np}{m+n} \qquad \Big|\ \text{From (1) and (2)}$$

Example 9: The pressures at depths h and h_2 in a liquid whose surface is exposed to atmospheric pressure are p_1 and p_2 respectively. Find the pressure at depth h.

Sol. $p_1 = \Pi + wh_1$ ...(1)

$p_2 = \Pi + wh_2$ ...(2)

Subtracting (2) from (1), we get

$$p_1 - p_2 = w\,(h_1 - h_2)$$
$$\Rightarrow \quad w = \frac{p_1 - p_2}{h_1 - h_2} \quad \text{...(3)}$$

From (1) and (3)

$$p_1 = \Pi + \frac{p_1 - p_2}{h_1 - h_2} h_1$$

$$\Rightarrow \qquad \Pi = p_1 - \frac{p_1 - p_2}{h_1 - h_2} h_1$$

$$= \frac{p_2 h_1 - p_1 h_2}{h_1 - h_2} \qquad \text{...(4)}$$

Now, pressure at depth h

$$= p\ \Pi + wh$$

$$= \frac{p_2 h_1 - p_1 h_2}{h_1 - h_2} + \frac{p_1 - p_2}{h_1 - h_2} h \qquad \text{| From (3) and (4)}$$

Example 10: Two liquids A and B do not mix and have different densities. When A is poured in a vessel to a vertical height h, the pressure on the bottom is the same as when A stands to a height h_1 and B to height h_2 above it. Show that the ratio of the density of A to that of B is $h_2/h - h_1$.

Sol. Let ρ_1 ρ_2 be the densities of the liquids A and B respectively.

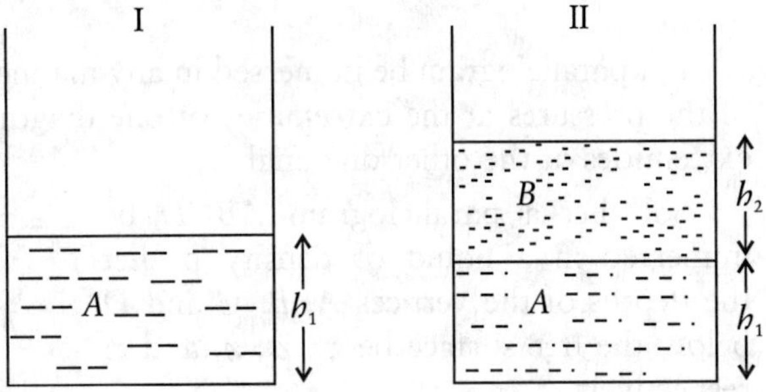

Let π be the atmospheric pressure. Let S be the area of cross-section of the vessel. Pressure on the bottom when liquid A stands to a height h

$$= \pi + g\ \rho_1\ h\ S \qquad \text{...(1)}$$

Pressure on the bottom when liquid A stands to a height h_1 and B to a height h_2 above A.

$$= \pi + g\ \rho_2\ h_2\ S + g\ \rho_1\ h_1\ S \qquad \text{...(2)}$$

Since the two pressures are given to be equal, therefore,

$$\pi + g\ \rho_1\ h\ S = \pi + g\ \rho_2\ h_2\ S + g\ \rho_1\ h_1\ S$$

$$\Rightarrow \rho_1 h = \rho_2 h_2 + \rho_1 h_1$$

$$\Rightarrow \rho_1 (h - h_1) = \rho_2 h_2$$

$$\Rightarrow \frac{\rho_1}{\rho_2} = \frac{h_2}{h - h_1}$$

Example 11: Prove that the pressure at the centre of gravity of a triangular lamina, wholly immersed in a homogeneous liquid, in any manner, is one-third the sum of the pressures at the angular points.

If a triangle be immersed in a homogeneous liquid, prove that the sum of the pressures at the vertices is equal to three times the pressure at the centroid of the triangle.

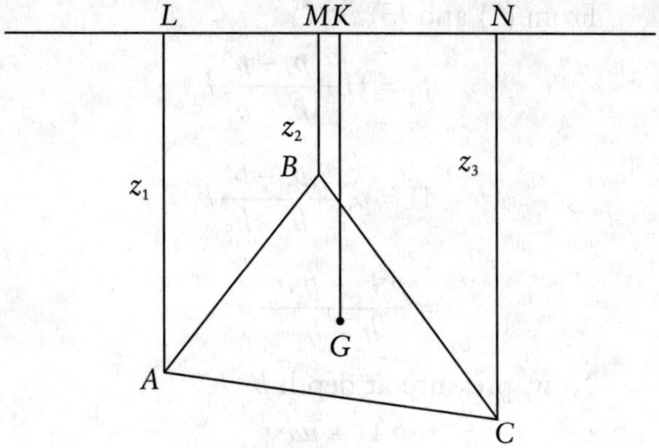

Sol. Let ABC be a triangular lamina immersed in a liquid of density ρ. Let the depths of the vertices A, B and C of this lamina from the free surface be z_1, z_2 and z_3 respectively.

Then, depth of the C.G. (G) of the lamina from the free surface = $1/3\ (z_1 + z_2 + z_3)$.

$\therefore$ Pressure at the C.G. (G)

$= g\ \rho\ 1/3\ (z_1 + z_2 + z_3)$

$= 1/3\ (g\rho z_1 + g\rho z_2 + g\rho z_3)$

= 1/3 (Pressure at A + Pressure at B + Pressure at C)

= 1/3 (sum of the pressures at the angular points)

$\Rightarrow$ Sum of the pressures at the angular points (vertices)

= 3 Pressure at the centroid of the triangle

Example 12: Prove that if a parallelogram be immersed in any manner in a homogeneous liquid, the sum of the pressures at the extremities of each diagonal is the same.

or

If a parallelogram be immersed in any manner in a homogeneous liquid, prove that the sum of the pressures at the extremities of one diagonal is equal to the sum of the pressures at the extremities of the other diagonal.

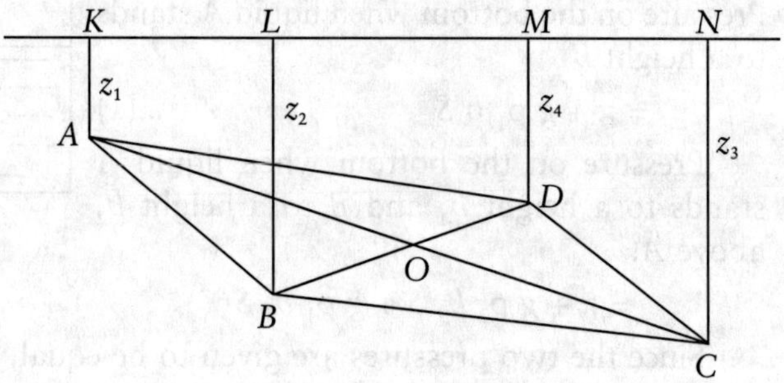

Sol. Let a parallelogram $ABCD$ be immersed in a liquid of density ρ. Let the depths of the vertices A, B, C and D below the free surface be z_1, z_2, z_3 and z_4 respectively.

Then, depth of the middle point O of the diagonal $AC = 1/2\ (z_1 + z_3)$ and, depth of the middle point O of the diagonal $BD = 1/2\ (z_2 + z_4)$

$\because$ Diagonals of a parallelogram bisect each other

$\therefore$ Middle point of the diagonal AC

= Middle point of the diagonal BD

$\therefore$ Depth of the middle point O of the diagonal AC

= Depth of the middle point O of the diagonal BD

$\Rightarrow\ 1/2\ (z_1 + z_3) = 1/2\ (z_2 + z_4)$

$\Rightarrow\ z_1 + z_3 = z_2 + z_4$ ...(1)

Now, sum of the pressures at the extremities A and C of the diagonal AC

$= g\rho z_1 + g\rho z_3$

$= g\rho\ (z_1 + z_3)$ ...(2)

and, sum of the pressures at the extremities B and D of the diagonal BD

$$= g\rho z_2 + g\rho z_4$$

$$= g\rho\,(z_2 + z_4) \qquad \ldots(3)$$

From (1), we find that these two sums of pressures are equal.

Example 13: If there be n lines arranged in strata of equal thickness h, and the density of the uppermost being ρ, that of the next 2ρ, and so on, that of the lowest being $n\rho$, prove that the pressure at the lowest point of the lowest stratum is

$$1/2\; n(n + 1)\; g\rho h$$

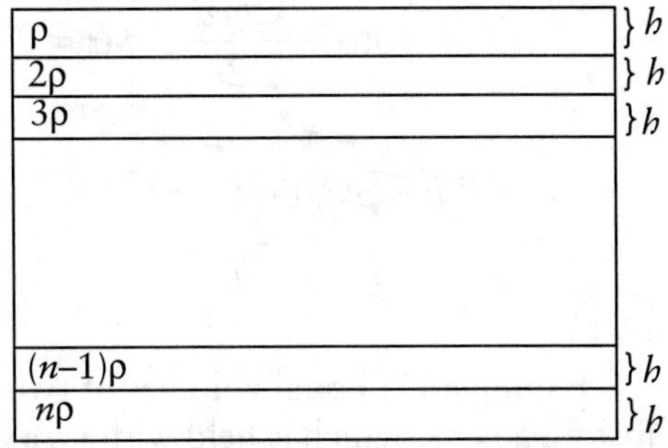

Sol. Pressure at the lowest point of the lowest stratum.

$$= g\rho h + g\,2\rho h + g\,3\rho h + \ldots. + g\,(n - 1)\,\rho h + gn\rho h$$

$$= g\rho h\,[1 + 2 + 3 + \ldots. + (n - 1) + n]$$

$$= g\rho h\; 1/2\; n(n + 1)$$

$$= 1/2\; n(n + 1)\; g\rho h$$

Example 14: If a liquid be heterogeneous and of density kz at depth z, show now the pressure is

$$\pi + 1/2\; gk\, z^2$$

where π is the atmospheric pressure.

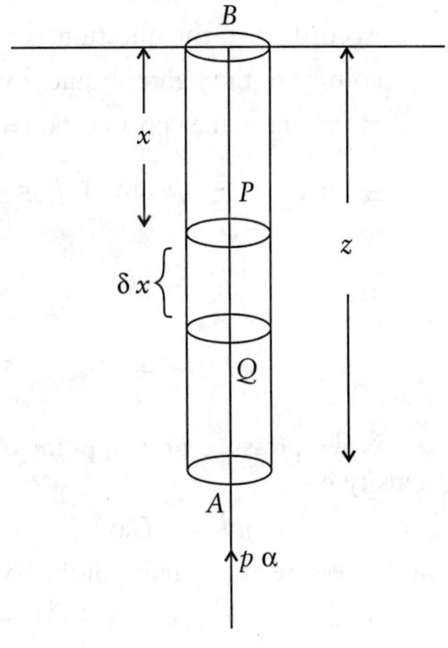

Sol. Let A be a point in the liquid at a depth z below the free surface. From A, draw a perpendicular AB to the free surface. On OA, take two points P and Q at depths x and $(x + \delta x)$ respectively below B.

Taking AB as axis, construct a liquid cylinder whose area of cross section is α (say). Then, weight of the elementary small liquid cylinder with PQ as axis

$$= \alpha\,\delta x\,(kx)\,g \qquad \Big|\ \text{Density at } P \text{ at depth } x = kx$$

$$= k\,\alpha\,g\,x\,\delta x$$

$\therefore$ Weight of the liquid within the cylinder with AB as axis

$$= \int_0^z k\alpha\, gx\, dx \;= \frac{k\alpha g z^2}{2}$$

This liquid cylinder is in equilibrium under the action of the following forces:

(1) The atmosphere pressure $\pi\alpha$ on the upper face at B acting vertically downwards.

(2) The pressure $\rho\alpha$ on the lower face at A acting vertically upwards, where ρ is the intensity of pressure at A.

(3) The pressure of the surrounding liquid at each point of the curved surface of the cylinder acting horizontally. These pressures on the vertical curved sides of the cylinder balance one another.

For equilibrium, resolving the forces vertically in the sense of the vertical line BA, we get

$$\pi\alpha + \frac{k\alpha g z^2}{2} - p\alpha = 0$$

$$\Rightarrow \quad p = \pi + 1/2\, gkz^2$$

Note: If $k = \rho/a$, then

$$p = \pi + \frac{g\rho z^2}{2a}$$

Example 15: Equal volumes of two liquids of densities ρ and 3ρ, which do not mix together, just fill a cone, which is held with its axis vertical and vertex uppermost. Prove that the pressure at any point of the base is $3-\sqrt[3]{4}$ times the pressure at the same point when the cone is filled with the lighter liquid.

Sol. Let h be the height and α be the semi-vertical angle of the cone. Let the depth of the heavier liquid filled in the cone be $(h - h^1)$. Then, the lighter liquid occupies the upper portion of height h' of the cone.

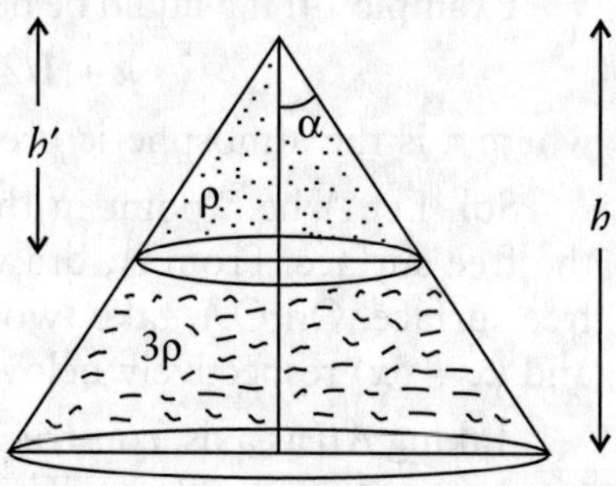

According to the question,

volume of the lighter liquid = volume of the heavier liquid.

⇒ volume of the upper portion = 1/2 volume of the whole cone

$$\Rightarrow \quad \frac{1}{3}\pi(h'\tan\alpha)^2\, h' = \frac{1}{2}\frac{1}{3}\pi(h\tan\alpha)^2 h$$

$$\Rightarrow \quad h'^3 = \frac{1}{2}h^3$$

$$\Rightarrow \quad h'^3 = \frac{1}{2^{\frac{1}{3}}}h \qquad \text{...(1)}$$

Now, pressure at any point of the base when the cone is filled with the lighter liquid of density ρ

$$= g\rho h = p \text{ (say)} \qquad \text{...(2)}$$

and, pressure at any point of the base when the cone is filled with two liquids of densities ρ and 3ρ

$$= g\rho h' + g\,3\rho\,(h - h') \qquad = g\rho\,(3h - 2\,h')$$

$$= g\rho\left(3h - \frac{2h}{2^{\frac{1}{3}}}\right) \qquad = g\rho h\left(3 - \frac{2}{2^{\frac{1}{3}}}\right)$$

$$= g\rho h\left(3 - 2^{\frac{2}{3}}\right) \qquad = g\rho h\left[3 - \left(4^{\frac{1}{2}}\right)^{2/3}\right]$$

$= g\rho h\left(3 - 4^{\frac{1}{3}}\right) \qquad\qquad = g\rho h(3 - \sqrt[3]{4})$

$= p(3 - \sqrt[3]{4})$ | From (2) $= (3 - \sqrt[3]{4})p$

EXERCISE 1

1. The pressure at the bottom of a well is four times that at a depth of 2 m. Find the depth of the well, it the atmospheric pressure be equivalent to that of 30 m of water.

   Ans. 98 m.

2. Pressure at the bottom of a well, filled completely with water, is three times the pressure at a depth of 6 feet. If the height of the water barometer be 30 feet, find the depth of the well.

   Ans. 78 feet.

3. The pressure of the atmosphere is equivalent to that of 10 m of water. Assuming this to be true, find the depth of a well the pressure at the bottom of which is 3 times that at a depth of 1 m. The well is full of water.

   Ans. 23 m.

4. The pressure at the bottom of a well is four times that at a depth of 2 feet. What is the depth of the well if the pressure of the atmosphere be equivalent to that of 30 feet of water?

   Ans. 98 feet.

5. The pressure at the bottom of a well is six times that at a depth of 3 m. Find the depth of the well, if the atmospheric pressure be equivalent to that of 15 m of water.

   Ans. 93 m.

6. If the height of water barometer be 34 feet, find the depth of a point below the surface of water such that the pressure at it may be twice the pressure at a point 10 feet below the surface.

   Ans. 54 feet.

7. The pressure in a water-pipe at the base of a building is 34 kg. weight per square cm. and on the roof it is 18 kg. weight per square cm. Find the height of the roof.
   1 cu. cm. of water weighs 1 kg.

   Ans. 16 m.

8. The pressure in the water-pipe at the basement of a building is 35 lbs. wt. per square inch, whereas at the third floor, it is only 19.375 lbs. wt. per square inch. Find the height of the third floor.
   1 cubic foot of water weighs 62.5 lbs.

   Ans. 36 feet.

9. The pressure in the water-pipe at the basement of a building is 35 lbs. wt. per square inch, and at the upper floor is 18.75 lbs. wt. per square inch. Find the height of the upper floor if one cubic ft. of water weighs 62.5 lbs.

   Ans. 37.44 feet.

10. The pressure in a water-pipe at the basement of a building is 34 lbs. wt. per square inch and at the third floor it is 18 lbs. wt. per square inch. Find the height of this floor above the basement. (1 cu. ft. of water weighs 62.5 lbs.)

    Ans. 36.864 feet.

11. The atmospheric pressure at the surface of a lake is 15 lbs. wt. per square inch. Find at what depth the pressure will be 45 lbs. wt. per square inch, the weight of a cubic foot of water being taken to be 1000 ozs.

    Ans. 69.12 feet.

12. The specific gravity of mercury is 13.6. At what depth in mercury will the pressure be equal to that at 500 metres in water?

    Ans. $36\frac{13}{17}$ metres

13. If a cubic metre of water weighs 1000 kgs., what is the pressure per square cm at the depth of a kilometre below the surface of water?

    Ans. 100 kg.wt.

14. A water tank is 60 m above the level of the ground-floor of a house. Find the pressure of the water in a pipe at a height of 10 m above the ground-floor.

    Ans. 5 kg.wt. per square inch.

15. A vessel whose bottom is horizontal contains mercury whose depth is 75 cm and water floats on the mercury to a depth of 60 cm. Find the pressure at a point on the bottom of the vessel, the specific gravity of mercury being 13.6.

    Ans. 1020 gr.wt.

16. Two liquids *A* and *B* do not mix and have densities in the ratio 3 : 2. When *A* is poured in a vessel to a vertical height 1 m, the pressure on the bottom is the same as when *A* stands to a height 70 cm and *B* to a height *h* above it. Find the value of *h*.

    Ans. 45 cm.

17. A glass tube of inform base is bent into the form of a U-tube, whose arms are vertical. It contains mercury of specific gravity 13.6. Water is poured into one arm to occupy a height of 408 cms of it. Find the rise of the level of mercury in the other arm.

    Ans. 15 cms.

18. The lower ends of two vertical tubes, whose cross-sections are 6.25 square cm and 5/8 square cm respectively, are connected by a tube. The tube contains mercury of specific gravity 13.596. How much water must be poured into the larger tube to raise the level in the smaller tube 2.5 cm?

    Ans. 233.68125 cu.cm.

19. A layer of mercury (specific gravity 13.6) 25 cm deep is covered by one of water of the same depth. Above this there is a layer 50 cm in depth of oil (specific gravity 0.9). Find the pressure at the bottom in centimetres of mercury, assuming the atmosphere to be due to a column of 76 cm of mercury.

    Ans. 106.15 gr.wt.

20. Prove that in a fluid at rest under gravity, horizontal planes are surfaces of equal pressure and equal density.
21. Two liquids that do not mix together meet in a bent tube. Prove that the heights of their surfaces above their common surface will be inversely proportional to their densities.

MORE ILLUSTRATIVE EXAMPLES

Example 1: If ρ, ρ′ be the densities of fluids (ρ < ρ′) and the lengths of the arms of a U-tube in which they meet be m and n inches respectively, prove that in order that the tube may be completely filled, the height of the column of the lighter fluid above the horizontal plane in which they meet must be

$$\frac{\rho'(m-n)}{\rho'-\rho} \text{ inches.}$$

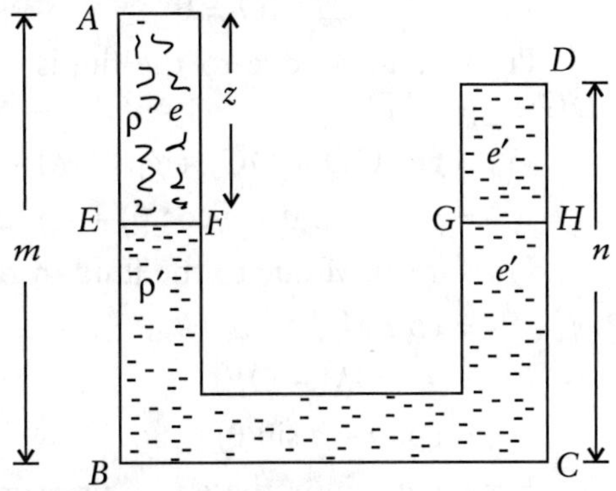

Sol.

$AB = m,\ DC = n$

EF is the surface of separation of the two fluids. AE is the column of the lighter liquid whose density is ρ.

Let $AE = z$

Let H be a point in the column DC which is in level with point E.

Since the points E and H are in the same horizontal level, therefore,

Pressure at E = Pressure at H

$$\Rightarrow \quad g\rho\, AE = g\rho'\, DH \qquad \Rightarrow \quad g\rho z = g\rho' (n - EB)$$

$$\Rightarrow \quad g\rho z = g\rho' \{n - (m - z)\} \qquad \Rightarrow \quad g\rho z = g\rho' \{(n - m) + g\rho' z\}$$

$$\Rightarrow \quad (\rho' - \rho)\, z = \rho' (m - n) \qquad \Rightarrow \quad z = \frac{\rho'(m-n)}{\rho'-\rho} \text{ inches.}$$

Example 2: A small uniform tube is bent into the form of a circle whose plane is vertical. Equal volumes of two fluids whose densities are ρ, σ fill half the tube. Show that the radius passing through the common surface makes with the vertical an angle θ given by

$$\tan\theta = \frac{\rho-\sigma}{\rho+\sigma}$$

Sol. Let the portion AB of the tube be occupied by the fluid of density σ and let the portion BMC of the tube be occupied by the fluid of density ρ, where M is the lowest point of the tube. Let the radius passing through the through the common surface make an angle θ with the vertical. Then,

$\angle BOM = \theta$

Let a be the radius of the tube.

Draw perpendiculars HD, CE and FG from H, C and F respectively to the vertical diameter MN.

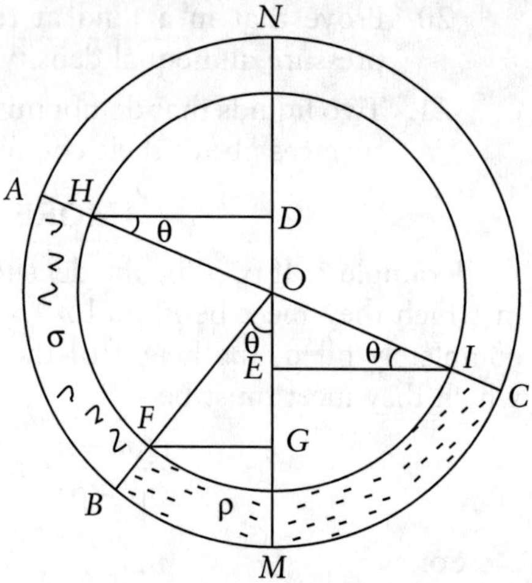

Volume of fluid in AB = Volume of fluid in BC | Given

$\Rightarrow$ arc AB = arc BC

But total arc = 1/2 × circumference

$\therefore \quad \angle AOB = \angle COB = \pi/2$

$\angle MOC = \pi/2 - \theta$

$\therefore \quad \angle OIE = \theta$

$\angle DHO = \theta$

Pressure at M due to the fluids in tube $ABM = g\sigma\ DG + g\rho\ GM$

$= g\sigma\ (OD + OG) + g\rho\ (OM - OG)$

$= g\sigma\ (a \sin\theta + a\cos\theta) + g\rho\ (a - a\cos\theta)$

Pressure at M due to the fluid in tube CM

$= g\rho\ EM$

$= g\rho\ (OM - OE)$

$= g\rho\ (a - a\sin\theta)$

For equilibrium, these two pressures must be equal

$\therefore \quad g\sigma\ (a\sin\theta + a\cos\theta) + g\rho\ (a - a\cos\theta) = g\rho\ (a + a\sin\theta)$

$\Rightarrow \sigma\ (\sin\theta + \cos\theta) + \rho\ (1 - \cos\theta) = \rho\ (1 + \sin\theta)$

$\Rightarrow \sigma\ (\sin\theta + \cos\theta) = \rho\ (\cos\theta + \sin\theta)$

$\Rightarrow \sin\theta\ (\rho + \sigma) = (\rho - \sigma)\cos\theta$

$\Rightarrow \tan\theta = (\rho - \sigma)/\ (\rho + \sigma)$

Example 3: Three liquids, whose densities are in A.P., fill a semi-circular lamina whose bounding diameter is horizontal. Prove that the depth of the one of the common surfaces is double that of the other.

Sol. Let the densities of the three fluids in A.P. be $\rho - \sigma$, ρ and $\rho + \sigma$. O is the centre of the semi-circular lamina whose bounding diameter AD is horizontal. BE and CF are the two common surfaces. Let portion AB be occupied by fluid of density ρ, portion BC be occupied by fluid of density $\rho + \sigma$ and portion CD be occupied by fluid of density $\rho - \sigma$. N is the lowest point of the lamina. Draw perpendiculars EM and FL from E and F respectively to ON.

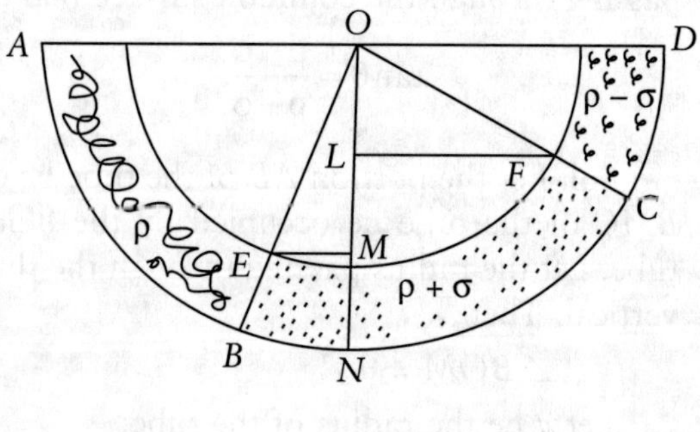

Let a be the radius the semi-circular lamina.

Pressure at N due to the fluids in tube ABN

$= g\rho\ OM + g\ (\rho + \sigma)\ MN$

$= g\rho\ OM + g\ (\rho + \sigma)\ (ON - OM)$

Pressure at N due to the fluids in tube DCN

$= g\ (\rho - \sigma)\ OL + g\ (\rho + \sigma)\ (LN)$

$= g\ (\rho - \sigma)\ OL + g\ (\rho + \sigma)\ (ON - OL)$

For equilibrium, these two pressures must be equal.

$\therefore\ g\rho\ OM + g\ (\rho + \sigma)\ (ON - OM) = g\ (\rho - \sigma)\ OL + g\ (\rho + \sigma)\ (ON - OL)$

$\Rightarrow\ \rho\ OM + (\rho + \sigma)\ (ON - OM) = (\rho - \sigma)\ OL + (\rho + \sigma)\ (ON - OL)$

$\Rightarrow\ \rho\ OM - (\rho + \sigma)\ OM = (\rho - \sigma)\ OL - (\rho + \sigma)\ OL$

$\Rightarrow\ -\sigma\ OM = -2\ \sigma\ OL$

$\Rightarrow\ OM = 2\ OL$

Hence, the depth of E is double that of F.

Example 4: In the lower half of a uniform circular tube, one quadrant is occupied by a liquid of density 2ρ, and the other by two liquids of density ρ and 3ρ. Prove that the volume of the lower of the two latter liquids is twice that of the other.

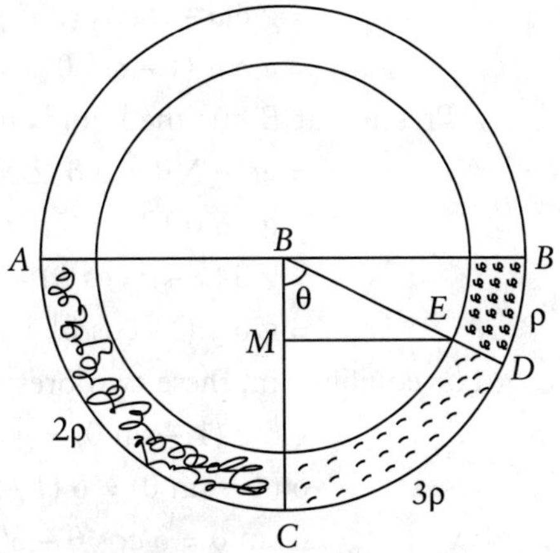

Sol. O is the centre of the circular tube. C is the lowest point. The portions AC, CD and CB of the tube are occupied by liquids of density 2ρ, 3ρ and ρ respectively.

AB is a diameter. Draw perpendicular EM from E to OC. Let a be the radius of the tube. Let $\angle COD = \theta$

Pressure at C due to the liquid in tube

$AC = g\ .\ 2\rho\ .\ OC = 2\ g\rho a$

Pressure at C due to the liquids in tube

$BDC = g\ .\ 3\rho\ .\ MC + g\ .\rho\ .\ OM$

$= 3g\rho\ (OC - OM) + g\ .\rho\ .\ OM$

$= 3g\rho\ (a - a\cos\theta) + g\ .\rho a\cos\theta$

For equilibrium, these two pressures must be equal

$\therefore\quad 2g\rho a = 3g\rho\ (a - a\cos\theta) + g\rho a\cos\theta$

$\Rightarrow\quad 2 = 3\ (1 - \cos\theta) + \cos\theta \quad\Rightarrow\quad \cos\theta = 1/2 = \cos 60°$

$\Rightarrow\quad \theta = 60° \quad\Rightarrow\quad \angle COD = 60°$

$\therefore\quad \angle BOD = 30° \quad\Rightarrow\quad \angle COD = 2\ \angle BOD$

$\Rightarrow$ arc CD = 2 arc DB

$\Rightarrow$ Volume of the lower of the two latter liquids = 2 Volume of the other

Example 5: A fine circular tube in the vertical plane contains a column of liquid, of density δ, which subtends a right angle at the centre, and a column of density δ' subtending angle α. If θ is the angle that the radius through the common surface makes with the vertical, prove that

$$\tan\theta = \frac{\delta - \delta' + \delta'\cos\alpha}{\delta + \delta'\sin\alpha}$$

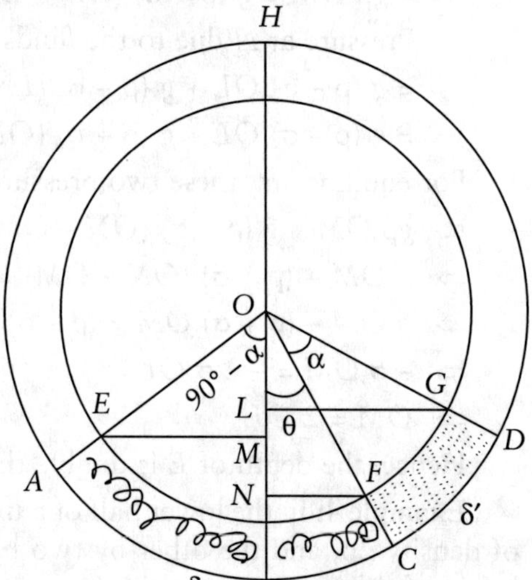

Sol. O is the centre and BH is the vertical diameter of the circular tube. Let the radius of the circular tube be a. The portion AC of the tube is occupied by the liquid of density δ. The portion CD of the tube is occupied by the liquid of density δ'. Given,

$$\angle AOC = 90°$$
$$\angle COD = \alpha$$
$$\angle BOD = \theta$$
$$\therefore \quad \angle AOD = 90° - \theta$$

Draw perpendiculars EM, FN and GL from E, F and G respectively to OB.

Pressure at B due to the liquid in the tube AB = $g\,\delta\, MB$

$$= g\,\delta(OB - OM)$$
$$= g\,\delta\{a - a\cos(90° - \theta)\}$$
$$= g\,\delta\, a\,(1 - \sin\theta)$$

Pressure at B due the liquids in the tube BCD

$$= g\,.\,\delta\, NB + g\,\delta'\, LN$$
$$= g\,.\,\delta\,(OB - ON) + g\,\delta'\,(ON - OL)$$
$$= g\,\delta\,(a - a\cos\theta) + g\,\delta'\,\{a\cos\theta - a\cos(\theta + \alpha)\}$$
$$= g\,\delta\, a\,(1 - \cos\theta) + g\,\delta'\, a\,\{\cos\theta - \cos(\theta + \alpha)\}$$

For equilibrium, these two pressures must be equal.

$$\therefore \quad g\,\delta\, a\,(1 - \sin\theta) = g\,\delta\, a\,(1 - \cos\theta) + g\,\delta'\, a\,\{\cos\theta - \cos(\theta + \alpha)\}$$
$$\Rightarrow \quad \delta\,(1 - \sin\theta) = \delta\,(1 - \cos\theta) + \delta'\,\{\cos\theta - \cos(\theta + \alpha)\}$$
$$\Rightarrow \quad \delta\sin\theta = \delta\cos\theta - \delta'\,\{\cos\theta - \cos\theta\cos\alpha + \sin\theta\sin\alpha\}$$
$$\Rightarrow \quad \sin\theta\,(\delta + \delta'\sin\alpha) = \cos\theta\,(\delta - \delta' + \delta'\cos\alpha)$$
$$\Rightarrow \quad \tan\theta = \frac{\delta - \delta' + \delta'\cos\alpha}{\delta + \delta'\sin\alpha}$$

Example 6: Equal volumes of three fluids of different densities, which do not mix together, completely fill a circular tube which is kept in a vertical plane. Prove that if the densities of the fluids are in A.P., the common surface of the lightest and heaviest fluids is at an extremity of a horizontal diameter of the circle.

Sol. Let the densities of three fluids in A.P. be $\rho - \delta$, ρ and $\rho + \delta$. Let the common surface of the lightest fluid, i.e., fluid of density $\rho - \delta$ and the heaviest fluid, i.e., fluid of density $\rho + \delta$ be at B, which is an extremity of the horizontal diameter AB.

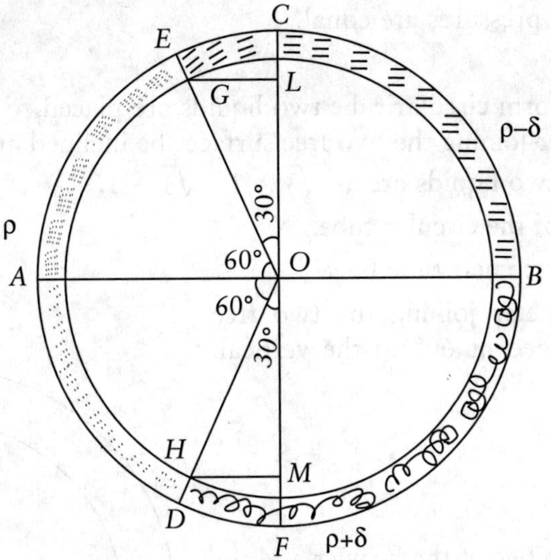

Since the volumes of the three liquids arc equal, therefore,

arc BD = arc BE = arc ED

$\therefore\ \angle DOB = \angle BOE = \angle DOE = 120°$

Obviously,

$\angle FOD = 30°$

$\angle COE = 30°$

$\angle AOE = 60°$

$\angle AOD = 60°$

Let the radius of the circular tube be a.

Draw perpendiculars HM and GL from H and G respectively to the vertical diameter CF.

Pressure at F due to liquids to the left of the vertical diameter CF

$$= g\,(\rho + \delta)\,MF + g\,\rho\,LM + g\,(\rho - \delta)\,CL$$

$$= g\,[(\rho + \delta)\,(OF - OM) + \rho\,(OL + OM) + (\rho - \delta)\,(OC - OL)]$$

$$= g\,[(\rho + \delta)\,(a - a\cos 30°) + \rho\,(a\cos 30° + a\cos 30°) + (\rho - \delta)\,(a - a\cos 30°)]$$

$$= ga\left[(\rho + \delta)\left(1 - \frac{\sqrt{3}}{2}\right) + \sqrt{3}\,\rho + (\rho - \delta)\left(1 - \frac{\sqrt{3}}{2}\right)\right]$$

$$= ga\left[\rho\left(1 - \frac{\sqrt{3}}{2} + \sqrt{3} + 1 - \frac{\sqrt{3}}{2}\right) + \delta\left(1 - \frac{\sqrt{3}}{2} - 1 + \frac{\sqrt{3}}{2}\right)\right]$$

$$= ga\,[2\rho]$$

$$= 2ga\rho \qquad ...(1)$$

Pressure at F due to liquids to the right of the vertical diameter CF

$$= g\,(\rho + \delta)\,OF + g\,(\rho - \delta)\,OC$$

$$= g\,(\rho + \delta)\,a + g\,(\rho - \delta)\,a$$

$$= 2ga\rho \qquad ...(2)$$

We find that the two pressures are equal.

Hence the result.

Example 7: In a uniform circular tube two liquids are placed so as to subtend 90° each at the centre. If the diameter joining the two free surfaces be inclined at 60° to the vertical, prove that the densities of the two liquids are as $\sqrt{3} + 1 : \sqrt{3} - 1$.

Sol. *O* is the centre of the circular tube.

Let the radius of the circular tube be *a*.

Let *AB* be the diameter joining the two free surface. Let *AB* be inclined at 60° to the vertical. Then,

$$\angle AOD = 60°$$
$$\angle DOE = 30°$$
$$\angle EOB = 90°$$

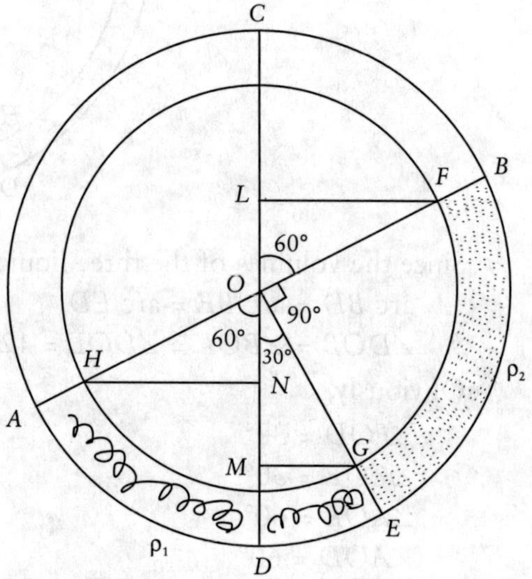

Let ρ_1, ρ_2 be the densities of the heavier and the lighter liquids respectively. Draw perpendiculars *FL*, *HN* and *GM* from *F*, *H* and *G* respectively to the vertical diameter.

Pressure at *D* due to liquid in tube $AD = g\rho_1 ND$

$= g\rho_1 (OD - ON) = g\rho_1 (a - \cos 60°) = g\rho_1 (a - a/2) = g\rho_1 a/2$

Pressure at *D* due to liquids in tube *DEB*

$= g\rho_1 MD + g\rho_2 LM$

$= g\rho_1 (OD - ON) + g\rho_2 (OL + OM)$

$= g\rho_1 (a - a \cos 30°) + g\rho_2 (a \cos 60° + a \cos 30°)$

$= ga [\rho_1 (1 - \frac{\sqrt{3}}{2}) + \rho_2 (1/2 + \frac{\sqrt{3}}{2})]$

For equilibrium, these two pressures must be equal.

$\therefore \quad g\rho_1 a/2 = ga [\rho_1 (1 - \frac{\sqrt{3}}{2}) + \rho_2 (1/2 + \frac{\sqrt{3}}{2})]$

$\Rightarrow \rho_1 = \rho_1 (2 - \sqrt{3}) + \rho_2 (1 + \sqrt{3})$

$\Rightarrow \rho_1 (1 - 2 + \sqrt{3}) = \rho_2 (1 + \sqrt{3})$

$\Rightarrow \rho_1 (\sqrt{3} - 1) = \rho_2 (1 + \sqrt{3})$

$\Rightarrow \rho_1 / \rho_2 = \sqrt{3} + 1 / \sqrt{3} - 1 = \rho_2 = \sqrt{3} + 1 : \sqrt{3} - 1$

Example 8: A circular tube of uniform thin bore is half filled with equal volumes of three liquids of specific gravities 3, 4 and 6, and is kept with its plane vertical. Prove that the diameter joining the two free surfaces makes an angle θ to the vertical such that

$$\tan\theta = \frac{19}{\sqrt{3}}$$

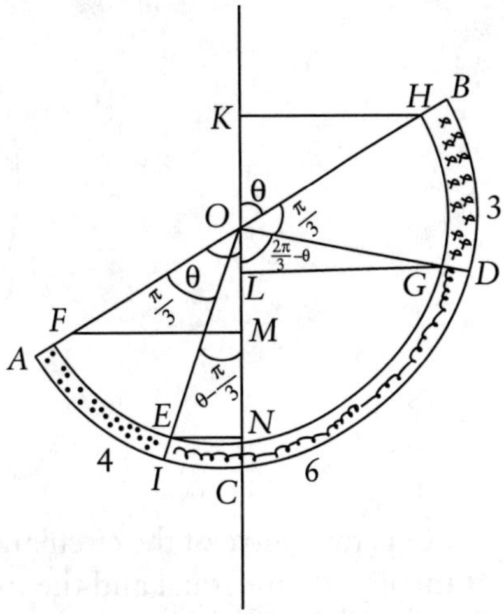

Sol. O is the centre of the circular tube. Let a be the radius of the circular tube.

$$\angle AOC = \theta$$

Draw perpendiculars EN, FM, GL and HK from E, F, G and H respectively to the vertical line KC.

$\because$ Volumes of the three liquids are equal

$\therefore \angle AOI = \angle IOD = \angle BOD = 180°/3 = 60° = \pi/3$

Pressure at C due to liquids in the tube AIC

$= g.4\rho.\ MN + g.6\rho.\ NC$

where ρ is the density of water.

$= g.4\rho.\ (ON - OM) + g.6\rho.\ (OC - ON)$

$= g.4\rho.\ \{a \cos (\theta - \pi/3) - a \cos \theta\} + g.6\rho.\ \{a - a \cos (\theta - \pi/3)\}$

$= g\rho a\ [-\ 2 \cos (\theta - \pi/3) - 4 \cos\theta + 6]$

Pressure at C due to liquids in tube BDC

$= g.6\rho.\ (OC - OL) + g.\ 3\rho\ (OL + OK)$

$= g.6\rho.\ \{a - a \cos (2\pi/3 - \theta)\} + g\ 3\rho\ \{a \cos (2\pi/3 - \theta 1) + a \cos\theta\}$

$= g.\rho a.\ [6 - 3 \cos (2\pi/3 - \theta) + 3 \cos\theta]$

For equilibrium, these two pressures must be equal.

$\therefore g\rho a\ [-\ 2 \cos (\theta - \pi/3) + 4 \cos\theta + 6]$

$\therefore g\rho a\ [6 - 3 \cos (2\pi/3 - \theta) + 3 \cos\theta]$

$\Rightarrow -2 \cos (\theta - \pi/3) - 4 \cos\theta = 3 \cos (2\pi/3 - \theta) + 3 \cos\theta$

$$\Rightarrow -2\ (1/2 \cos\theta - \frac{\sqrt{3}}{2} \sin\theta) - 4 \cos\theta = -3\ (-\ 1/2 \cos\theta + \frac{\sqrt{3}}{2} \sin\theta) + 3 \cos\theta$$

$$\Rightarrow -\cos\theta - \sqrt{3}\ \sin\theta - 4 \cos\theta = 3/2 \cos\theta - 3\frac{\sqrt{3}}{2} \sin\theta + 3 \cos\theta$$

$$\Rightarrow \left(8+\frac{3}{2}\right)\cos\theta = \sqrt{3}\left(\frac{3}{2}-1\right)\sin\theta$$

$$\Rightarrow 19/2 \cos\theta = \frac{\sqrt{3}}{2} \sin\theta$$

$$\Rightarrow \tan\theta = 19/\sqrt{3}$$

Example 9: A circular tube of fine uniform bore is half filled with equal volumes of four liquids which do not mix and whose densities are as 1:4:8:7 and is held with its plane vertical. Show that the diameter joining the free surfaces makes an angle θ with the vertical such that $\tan\theta = 2$.

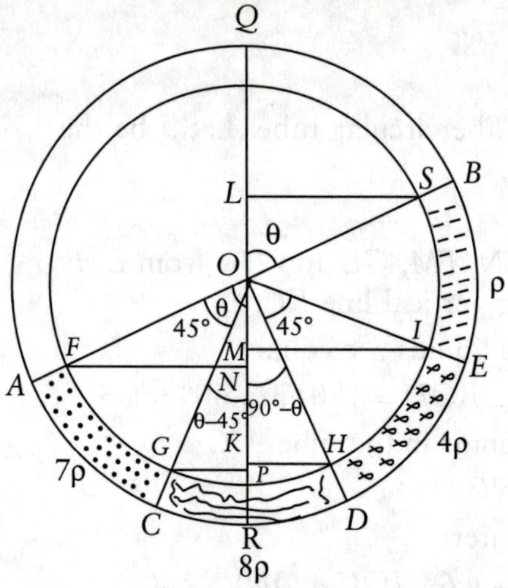

O is the centre of the circular tube. Let a be the radius of the circular tube. Since the volumes of the liquids are equal and the tube is half filled, therefore, they will subtend equal angles, each equal to 180°/4 = 45° at the centre so that

$$\angle AOC = \angle COD = \angle DOE = \angle EOB = 45°$$

Let the densities of the four liquids be ρ, 4ρ, 8ρ and 7ρ respectively. These liquids are arranged according to their densities as shown in the figure. Let the diameter joining the free surfaces A and B of the liquids make an angle θ with the vertical.

From A, G, H, I and S, draw perpendiculars AN, GP, HK, IM and SL respectively to the vertical diameter QR, whose R is the lowest point of the tube.

$$\angle QOB = \theta$$

$$\therefore \quad \angle AOR = \theta$$

$$\angle COR = \angle AOR - \angle AOC = \theta - 45°$$

$$\angle ROD = \angle AOD - \angle AOR = 90° - \theta$$

Pressure at R due to the liquids in the tube ACR

$= g.\ 7\rho\ .\ NP + g.\ 8\rho.\ PR$

$= g\rho\ [\{7\ (OP - ON)\} + \{8\ (OR - OP)\}]$

$= g\rho\ [7\ \{a \cos(\theta - 45°) - a \cos\theta\} + 8\ \{a - a \cos(\theta - 45°)\}]$

$= g\rho a\ [-7 \cos\theta - \cos(\theta - 45°) + 8]$ ...(1)

Pressure at R due to the liquids in the tube $RDEB$

$= g.\ 8\rho\ .\ KR + g.\ 4\rho\ .\ MK + g\rho\ (OM + OL)$

$= g\rho\ [8\ (OR - OK) + 4\ (OK - OM) + (OM + OL)]$

$= g\rho\ [8\ (OR - OK) + 4\ (OK - OM) + (OM + OL)]$

$= g\rho\ [8\ \{a - a \cos(90° - \theta)\} + 4\ \{a \cos(90° - \theta) - a \cos(135° - \theta)\} + \{a \cos(135° - \theta) + a \cos\theta\}]$

$= g\rho a\ [8 - 8 \sin\theta + 4 \sin\theta + 4 \sin(45° - \theta) - \sin(45° - \theta) + \cos\theta]$

$= g\rho a\ [8 - 4 \sin\theta + \cos\theta + 3\ \{1/\sqrt{2}\ \cos\theta - 1/\sqrt{2}\ \sin\theta\}]$

$= g\rho a\,[8 - 4\sin\theta + \cos\theta + 3/\sqrt{2}\,\cos\theta - 3/\sqrt{2}\,\sin\theta]$ ...(2)

For equilibrium, these two pressures must be equal, therefore,

$g\rho a\,[-7\cos\theta - \cos(\theta - 45°) + 8]$

$= g\rho a\,[8 - 4\sin\theta + \cos\theta + 3/\sqrt{2}\,\cos\theta - 3/\sqrt{2}\,\sin\theta]$

$\Rightarrow\; -7\cos\theta - \cos\theta/\sqrt{2} - \sin\theta/\sqrt{2} + 8] = 8 - 4\sin\theta + \cos\theta + 3\cos\theta/\sqrt{2} - 3\sin\theta/\sqrt{2}$

$\Rightarrow\; -8\cos\theta - 2\sqrt{2}\,\cos\theta = -4\sin\theta - \sqrt{2}\,\sin\theta$

$\Rightarrow\; 2\sqrt{2}\,\cos\theta = (1 + 2\sqrt{2})\,\sqrt{2}\,\sin\theta\,(1 + 2\sqrt{2})$

$\Rightarrow\; \tan\theta = 2$

Example 10: A tube of the form of a complete circle fixed in a vertical plane, contains equal lengths of four liquids which do not mix and whose specific gravities are as 1:2:4:3 filling the tube in this order. Prove that the inclinations to the vertical of the two diameters joining the points of division are $\tan^{-1} 1/2$ and $\tan^{-1} 2$.

Sol.

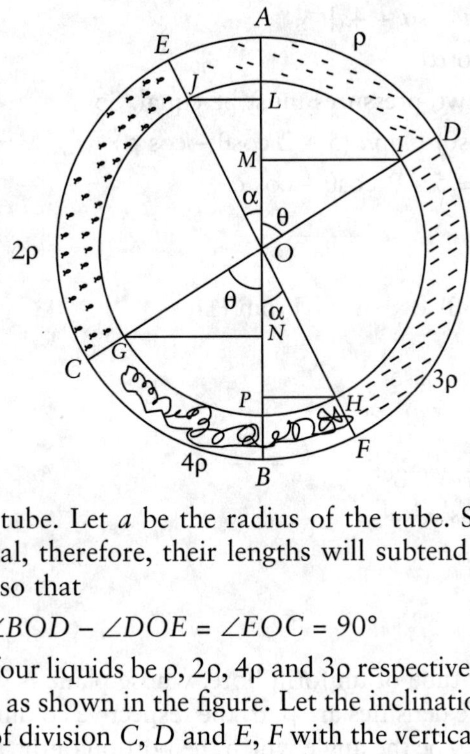

O is the centre of the tube. Let a be the radius of the tube. Since the lengths of the four liquids in the tube are equal, therefore, their lengths will subtend equal angles, each equal to $360°/4 = 90°$ at the centre so that

$$\angle COB = \angle BOD - \angle DOE = \angle EOC = 90°$$

Let the densities of the four liquids be ρ, 2ρ, 4ρ and 3ρ respectively. These liquids are arranged according to their densities as shown in the figure. Let the inclinations of the two diameters CD and EF joining the points of division C, D and E, F with the vertical be θ and α respectively. So,

$\angle COB = \theta$

$\angle FOB = \alpha$

Clearly, $\theta + \alpha = 90°$ ...(1)

From G, J, I and H, draw perpendiculars GN, JL, IM and HP respectively to the vertical diameter AB, where B is the lowest point of the tube.

Pressure at B due to the liquids in the tube $AECB$

$= g\rho\, AL + g.\, 2\rho.\, LN + g.\, 4\rho.\, NB$

$= g\rho\, [AL + 2\, LN + 4\, NB]$

$= g\rho\, [OA - OL + 2\, (OL + ON) + 4\, (OB - ON)]$

$= g\rho\, [a - a \cos \alpha + 2\, (a \cos \alpha + a \cos \theta) + 4\, (a - a \cos \theta)]$

$= g\rho a\, [1 - \cos \alpha + 2 \cos \alpha + 2 \cos \theta + 4 - 4 \cos \theta]$

$= g\rho a\, [5 + \cos \alpha - 2 \cos \theta]$ ...(2)

Pressure at B due to the liquids in the tube $ADFB$

$= g\rho\, AM + g\, 3\rho\, MP + g\, 4\rho\, PB$

$= g\rho\, AM + g\, 3\rho\, (OM + OP) + g\, 4\rho\, .\, PB$

$= g\rho\, [AM + 3\, OM + 3\, OP + 4\, PB]$

$= g\rho\, [OA - OM + 3\, OM + 3\, OP + 4\, (OB - OP)]$

$= g\rho\, [OA + 2\, OM - OP + 4\, OB]$

$= g\rho\, [a + 2\, a \cos\theta - a \cos\alpha + 4a]$

$= g\rho a\, [5 + 2 \cos\theta - \cos\alpha]$ ...(3)

For equilibrium, these two pressures must be equal. So,

$g\rho a\, (5 + \cos \alpha - 2 \cos\theta) = g\rho a\, (5 + 2 \cos\theta - \cos \alpha)$

$\Rightarrow\ 5 + \cos \alpha - 2 \cos\theta = 5 + 2 \cos\theta - \cos \alpha$

$\Rightarrow\ 4 \cos\theta = 2 \cos \alpha$

$\Rightarrow\ 2 \cos\theta = \cos \alpha$ ...(4)

$\Rightarrow\ 2 \cos\theta = \cos\, (90° - \theta)$ | From (1)

$\Rightarrow\ 2 \cos\theta = \sin\theta$

$\Rightarrow\ \tan\theta = 2$

$\Rightarrow\ \theta = \tan^{-1} 1/2$ ...(5)

Again, from (4) and (1), we get

$2 \cos\, (90° - \alpha) = \cos \alpha$

$\Rightarrow\ 2 \sin \alpha = \cos \alpha$

$\Rightarrow\ \tan \alpha = 1/2$

$\Rightarrow\ \alpha = \tan^{-1} 1/2$

Example 11: A circular tube of uniform bore, whose plane is vertical, contains columns of two liquids whose respective densities are ρ, σ, the respective columns subtending angles 2α, 2β at the centre of the circle. If θ be the angle which the portion of the tube intercepted between the lowest point and the common surface of the liquids subtends at the centre of the tube, prove that

$$\rho \sin \alpha \sin\, (\alpha \pm \theta) = \sigma \sin \beta \sin\, (\beta \mp \theta)$$

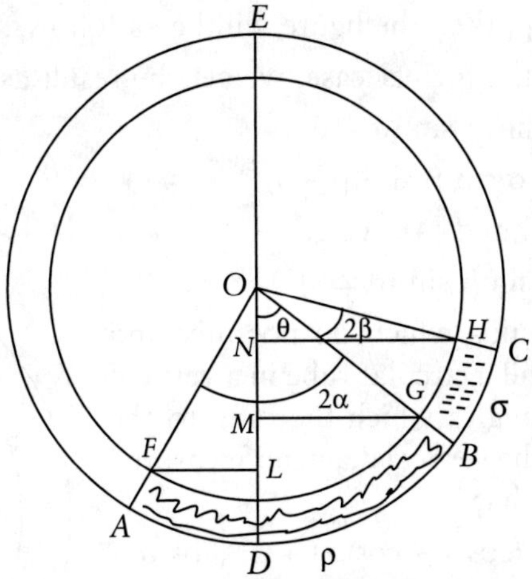

O is the centre of the circular tube. Let a be the radius of the circular tube. Let AB be the column of the liquid of density ρ subtending angle 2α at the centre of the circular tube. Let BC be the column of the liquid of density σ subtending angle 2β at the centre of the circular tube. The portion of the tube intercepted between the lowest point D and the common surface BG of the two liquids subtends angle θ at the centre of the tube.

$$\therefore \quad \angle BOD = \theta$$
$$\angle AOB = 2\alpha$$
$$\angle BOC = 2\beta$$

From F, G and H draw perpendiculars FL, GM and HN to the vertical diameter ED, where D is the lowest point of the tube.

Pressure at D due to liquid column AD

$= g\rho\ LD = g\rho\ (OD - OL)$

$= g\rho\ \{a - a\cos(2\alpha - \theta)\}$

$= g\rho a\ \{1 - \cos(2\alpha - \theta)\}$ ...(1)

Pressure at D due to liquids column DC

$= g\rho\ MD + g\sigma\ NM$

$= g\rho\ (OD - OM) + g\sigma\ (OM - ON)$

$= g\rho\ (a - a\cos\theta) + g\sigma\ \{a\cos\theta - a\cos(\theta + 2\beta)\}$

$= g\rho a\ (1 - \cos\theta) + g\sigma a\ \{\cos\theta - \cos(\theta + 2\beta)\}$ ...(2)

For equilibrium, these two pressures must be equal. So,

$g\rho a\ \{1 - \cos(2\alpha - \theta)\} = g\rho a\ (1 - \cos\theta) + g\sigma a\ \{\cos\theta - \cos(\theta + 2\beta)\}$

$\Rightarrow \quad \rho - \rho\cos(2\alpha - \theta) = \rho - \rho\cos\theta + \sigma\ \{2\sin(\theta + \beta)\sin\beta\}$

$\Rightarrow \quad \rho\ \{\cos\theta - \cos(2\alpha - \theta\} = \sigma\ .\ 2\sin(\theta + \beta)\sin\beta$

$\Rightarrow \quad \rho\ 2\sin\alpha\sin(\alpha - \theta\} = \sigma\ .\ 2\sin(\theta + \beta)\sin\beta$

$\Rightarrow \quad \rho\sin\alpha\sin(\alpha - \theta\} = \sigma\sin\beta\sin(\beta + \theta)$ ...(3)

If we interchange α and β, then the figure will be as follows.

Proceeding exactly as in the former case, we get the result as follows:

$\sigma \sin \beta \sin (\beta - \theta\} = \rho \sin \alpha \sin (\alpha + \theta)$

$\Rightarrow \rho \sin \alpha \sin (\alpha + \theta\} = \sigma \sin \beta \sin (\beta - \theta)$ ...(4)

Combining the results (3) and (4), we get

$\rho \sin \alpha \sin (\alpha \pm \theta\} = \sigma \sin \beta \sin (\beta \mp \theta)$

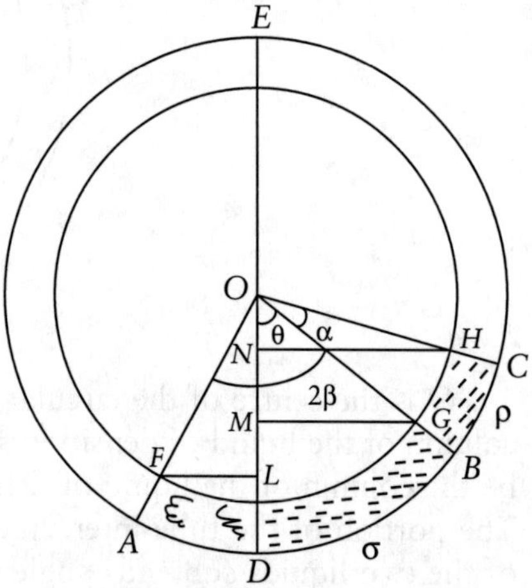

Example 12: If three liquids which do not mix and whose densities are ρ_1, ρ_2, ρ_3 fill a circular tube in a vertical plane and if α, β, γ are the angles which the radii to the common surfaces make with the vertical diameter measured in the same direction, prove that

$\rho_1 (\cos \beta - \cos \gamma) + \rho_2 (\cos \gamma - \cos \alpha) + \rho_3 (\cos \alpha - \cos \beta) = 0$

O is the centre of the circular tube. Let *a* be the radius of the circular tube. The portions *BD*, *DE* and *EB* of the tube are filled by liquids of densities ρ_1, ρ_2 and ρ_3 respectively. These liquids are arranged as shown in the figure. Let $\rho_1 > \rho_2 > \rho_3$.

$\angle EOC = \alpha$ | clockwise

$\therefore \angle EOA = \pi - \alpha$

$\angle BOC = \beta$ | clockwise

$\therefore \angle BOC = 2\pi - \beta$ | anti-clockwise

$\angle DOC = \gamma$ | clockwise

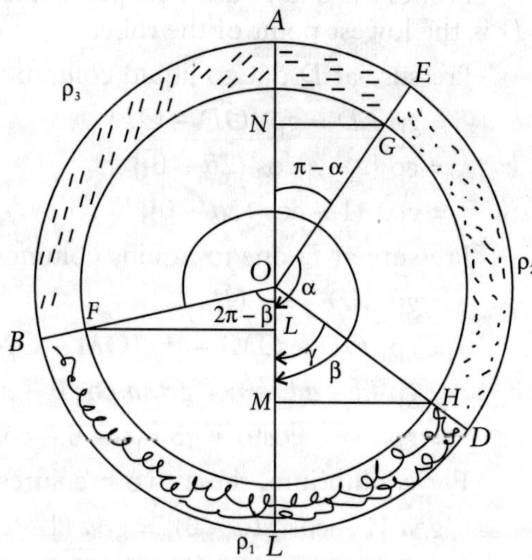

From *F*, *G* and *H*, draw perpendiculars *FL*, *GN* and *HM* respectively to the vertical diameter *AC*, where *C* is the lowest point of the tube.

Pressure at C and to liquids in the portion *ABC* of the tube

$= g\rho_3 AL + g \rho_1 LC$

$= g\rho_3 (OA + OL) + g\rho_1 (OC - OL)$

$= g\rho_3 \{a + a \cos (2\pi - \beta)\} + g\rho_1 \{a - a \cos (2\pi - \beta)\}$

$= g\rho_3 (a + a \cos \beta) + g\rho_1 (a - a \cos \beta)$ ...(1)

Pressure at *C* due to liquids in the portion *AEC* of the tube

$= g\rho_3 NA + g\rho_2 NM + g\rho_1 MC$

$= g\rho_3 (OA - ON) + g\rho_2 (ON + OM) + g\rho_1 (OC - OM)$

$= g\rho_3 \{a - a \cos (\pi - \alpha)\} + g\rho_2 \{a \cos (\pi - \alpha) + (a \cos \gamma)\} + g\rho_1 (a - a \cos \gamma)$

$= g\rho_3 (a + a \cos \alpha) + g\rho_2 (- a \cos \alpha + a \cos \gamma) + g\rho_1 (a - a \cos \gamma)$ ...(2)

For equilibrium, these two pressures must be equal. So,

$$g\rho_3 (a + a \cos \beta) + g\rho_1 (a - a \cos \beta)$$
$$= g\rho_3 (a + a \cos \alpha) + g\rho_2 (-a \cos \alpha + a \cos \gamma) + g\rho_1 (a - a \cos \gamma)$$
$$\Rightarrow \rho_3 (1 + \cos \beta) + \rho_1 (1 - \cos \beta)$$
$$= \rho_3 (1 + \cos \alpha) + \rho_2 (-\cos \alpha + \cos \gamma) + \rho_1 (1 - \cos \gamma)$$
$$\Rightarrow \rho_3 \cos \beta - \rho_1 \cos \beta = \rho_3 \cos \alpha - \rho_2 \cos \alpha + \rho_2 \cos \gamma - \rho_1 \cos \gamma$$
$$\Rightarrow \rho_1 (\cos \beta - \cos \gamma) + \rho_2 (\cos \gamma - \cos \alpha) + \rho_3 (\cos \alpha - \cos \beta) = 0$$

Example 13: The two arms of a U-tube are close together. In one arm there is water and in the other mercury, so that their common surface is at the lowest point. One quarter of water is taken out and is poured into the other arm over the mercury. Prove that in the new equilibrium position, the difference of heights of the upper surfaces is one-half of what it was formerly.

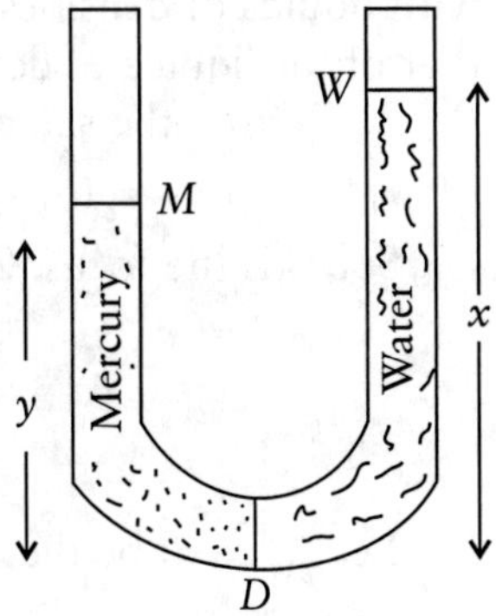

Sol. Initially let x and y be the heights of columns of water and mercury respectively.

Then, difference between the heights of water column and mercury column

$$= x - y$$

Now let us take out $x/4$ water from the arm containing water and pour over mercury in the second arm. Then, the new figure becomes looks like.

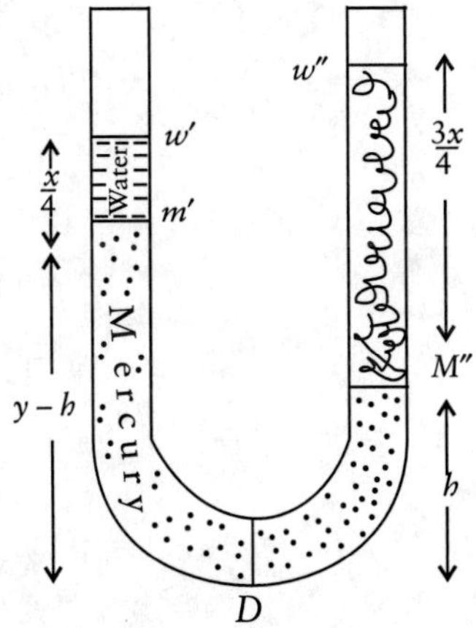

Let the level of mercury sink by h in the arm containing mercury and consequently rise by h in the arm containing water.

Equating the pressures at the lowest point D due to liquids from left and right, we get

$$g\rho\, x/4 + g\, 13.6\, \rho (y - h) = g\rho \,.\, 3x/4 + g \,.\, 13.6\, \rho h$$
$$\Rightarrow 13.6 (y - 2h) = 3x/4 - x/4$$
$$\Rightarrow 13.6\, y - (13.6)\, 2h = x/2$$

But $13.6\, y = x$ | Initially from former figure equating the pressures at D due to liquids from left and right

$$\therefore 13.6\, y - (13.6)(2h) = (13.6)y/2$$
$$\Rightarrow (13.6)\, y/2 = (13.6)(2h)$$
$$\Rightarrow y/2 = 2h$$
$$\Rightarrow y = 4h$$
$$\Rightarrow h = y/4$$

Hence, the new heights of the arms are as follows:

Right arm $= 3x/4 + h = 3x/4 + y/4$

Left arm $= x/4 + y - h - \gamma = x/4 + y - y/4 = x/4 + 3y/4$

∴ Difference of the heights of the upper surfaces of the two arms

$$= \left(\frac{3x}{4}+\frac{y}{4}\right)-\left(\frac{x}{4}+\frac{3y}{4}\right)$$

$$= \frac{x-y}{2}$$

= 1/2 (Initial difference)

Example 14: A closed tube in the form of an ellipse with its major axis vertical, is filled with liquids of densities ρ_1, ρ_2, ρ_3. P_1 is the point of separation of liquids of densities ρ_2, ρ_3; P_2 that of the liquids of densities ρ_3, ρ_1; and P_3 that of liquids of densities ρ_1, ρ_2. If the distances of P_1, P_2, P_3 from the same focus be r_1, r_2, r_3 respectively, prove that

$r_1(\rho_1 - \rho_2) + r_2(\rho_2 - \rho_3) + r_n(\rho_n - \rho_1) = 0$

Sol. S is the focus. ZZ' is the directrix of the ellipse. Then,

$$S P_1 = r_1$$
$$S P_2 = r_2$$
$$S P_3 = r_3$$

Let z_1, z_2, z_3 be the depths of the points P_1, P_2, P_3 respectively below the directrix ZZ'.

Then,
$$P_1 K = z_1$$
$$P_2 N = z_2$$
$$P_3 L = r_3$$

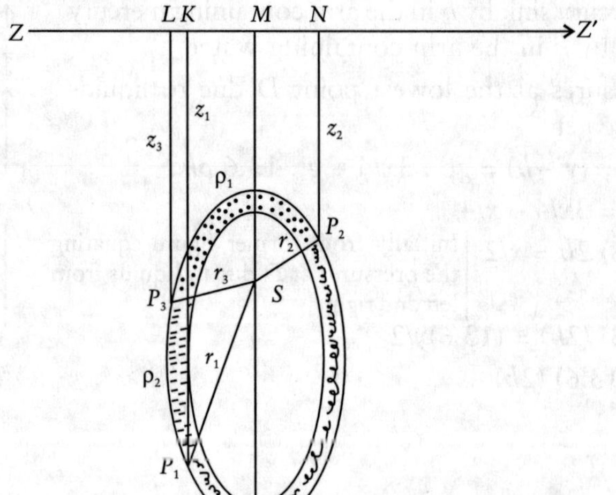

Let p_1, p_2, p_3 be the pressures at the points of separation P_1, P_2, P_3 respectively. Then,

$$p_1 - p_2 = (z_1 - z_2)\,\rho_3\, g$$
$$p_3 - p_2 = (z_3 - z_2)\,\rho_1\, g$$
$$p_1 - p_3 = (z_1 - z_3)\,\rho_2\, g$$

∵ In a homogeneous liquid at rest under gravity, the difference between the pressures at two points varies as the vertical distance between them.

$$\Rightarrow p_1 - p_2 = (z_1 - z_2)\,\rho_3\, g$$
$$p_2 - p_3 = (z_2 - z_3)\,\rho_1\, g$$
$$p_3 - p_1 = (z_3 - z_3)\,\rho_2\, g$$

Adding these equations, we get

$$\rho_1 (z_2 - z_3) + \rho_2 (z_3 - z_1) + \rho_3 (z_1 - z_2) = 0 \qquad ...(1)$$

We know that the ratio of the distances of a point on the ellipse from the focus and the directrix is constant.

Therefore,

$$r_1/z_1 = \text{constant} = \lambda \text{ (say)}$$
$$r_2/z_2 = \text{constant} = \lambda \text{ (say)}$$
$$r_3/z_3 = \text{constant} = \lambda \text{ (say)}$$
$$\Rightarrow z_1 = r_1/\lambda$$
$$z_2 = r_2/\lambda$$
$$z_3 = r_3/\lambda$$

Substituting these values of z_1, z_2 and z_3 in equation (1) and simplifying, we get

$$\rho_1 (r_2 - r_3) + \rho_2 (r_3 - r_1) + \rho_3 (r_1 - r_2) = 0$$
$$\Rightarrow r_1 (\rho_2 - \rho_3) + r_2 (\rho_3 - \rho_1) + r_3 (\rho_1 - \rho_2) = 0$$

Example 15: A fine tube bent in the form of an ellipse is held with the major axis vertical and is held with n liquids whose densities are $\rho_1, \rho_2, \rho_3 ...., \rho_n$ taken in order round the elliptic tube. If $r_1, r_2, r_3, ..., r_n$ be the distances of the points of separation from either focus, prove that

$$r_1 (\rho_1 - \rho_2) + r_2 (\rho_2 - \rho_3) + ... + r_n (\rho_n - \rho_{n-1}) = 0$$

Sol. Let S be the focus and ZZ' be the directrix of the ellipse. Let P_1, P_2, P_3, ..., P_n, be the points of separation of the liquids.

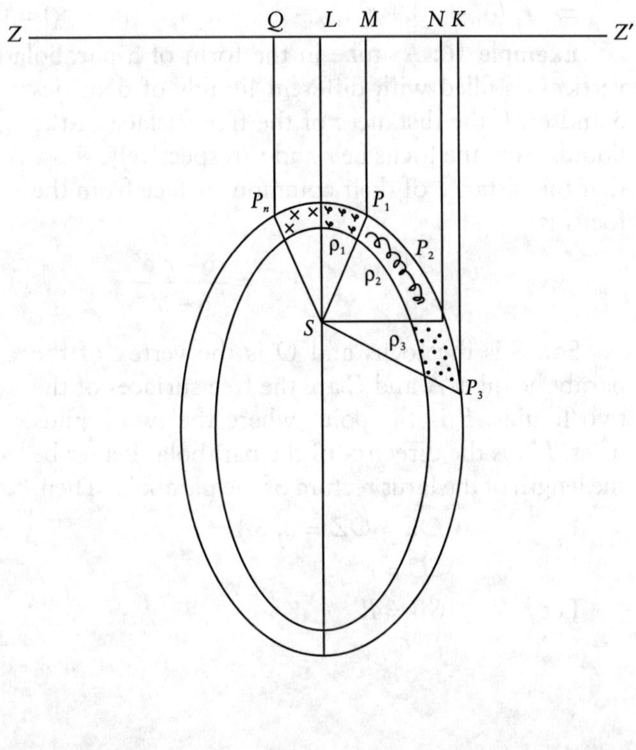

Then,

$$SP_1 = r_1$$
$$SP_2 = r_2$$
$$SP_3 = r_3$$
$$SP_n = r_n$$

Let $z_1, z_2, z_3, ..., z_n$ be the depths of the points $P_1, P_2, P_3 ..., P_n$ respectively below the directrix ZZ'.

Then,

$$P_1M = z_1$$
$$P_2N = z_2$$
$$P_3K = z_3$$
$$\vdots \qquad \vdots$$
$$P_nQ = z_n$$

Let $p_1, p_2, p_3 ..., p_n$ be the pressures at the points of separation. Then,

$$p_2 - p_1 = g\rho_2 (z_2 - z_1)$$
$$p_3 - p_2 = g\rho_3 (z_3 - z_2)$$
$$\vdots$$
$$p_n - p_{n-1} = g\rho_n (z_n - z_{n-1})$$
$$p_1 - p_n = g\rho_1 (z_1 - z_n)$$

| ∵ In a homogeneous liquid at rest under gravity, the difference between the pressures at two points varies as the vertical distance between them.

Adding, we get

$$g [\rho_2 (z_2 - z_1) + \rho_3 (z_3 - z_2) + ...\rho_1 (z_1 - z_n) = 0$$

$$\Rightarrow \rho_2 (z_2 - z_1) + \rho_3 (z_3 - z_2) + ...\rho_1 (z_1 - z_n) = 0 \quad ...(1)$$

$$\Rightarrow \rho_2 (ez_2 - ez_1) + \rho_3 (ez_3 - ez_2) + ... + \rho_1 (ez_1 - ez_n) = 0$$

| Multiply (1) by e, where e is the eccentricity of the ellipse.

$$\Rightarrow \rho_2 (r_2 - r_1) + \rho_3 (r_3 - r_2) + ...+ \rho_1 (r_1 - r_n) = 0$$

We know that the ratio of the distances of a point on the ellipse from the focus and the directrix is equal to the eccentricity e (constant) of the ellipse.

So, $r_1/e = z_1 \qquad r_2/e = z_2$

$$\vdots$$

$$r_n/e = z_n$$

$$\Rightarrow r_1 = ez_1 \qquad r_2 = ez_2$$

$$\vdots$$

$$r_n = ez_n$$

$$\Rightarrow r_1 (\rho_1 - \rho_2) + r_2 (\rho_2 - \rho_3) + ... + r_n (\rho_n - \rho_1) = 0$$

Example 16: A tube in the form of a parabola held with its vertex downwards and axis vertical, is filled with different liquids of densities δ and δ^1. If the distances of the free surface of the liquids from the focus be r and r' respectively, show that the distance of their common surface from the focus is

$$= \frac{r\delta - r'\delta'}{\delta - \delta'}$$

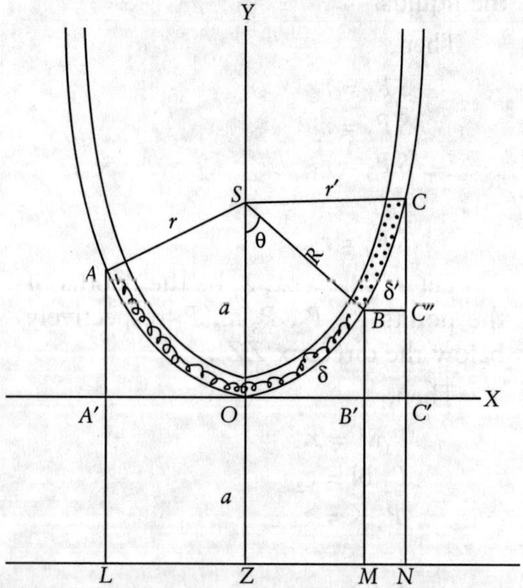

Sol. S is the focus and O is the vertex of the parabolic tube. A and C are the free surfaces of the two liquids. B is the point where the two liquids meet. LN is the directrix of the parabola. Let $4a$ be the length of the latus rectum of the parabola. Then,

$$OS = OZ = a, \; SA = r$$

$$SC = r'$$

Let $\qquad SB = R$

The portion AOB of the tube is filled with liquid of density δ whereas the portion BC of the tube is filled with liquid of density δ'.

We know that in a parabola, the distance of a point on the parabola from the focus is equal to the distance of that point from the directrix. From A, B and C, draw perpendiculars AL, BN and CN to the directrix.

Then,

$$AS = AL = r$$

$$BS = BM = R$$

$$CS = CN = r'$$

$$\because \quad OS = OZ = a$$

$$\therefore \quad A'L = B'M = C'N = a$$

Now, pressure at O due to the liquid in the tube OA

$$= g\delta \quad AA'$$

$$= g\delta \quad (AL - A'L)$$

$$= g\delta \quad (r - a)$$

and, pressure at O due to liquids in the tube OC

$$= g\delta\, BB' + g\delta'\, CC''$$

$$= g\delta\, (BM - B'M) + g\delta'\, (CN - C''N)$$

$$= g\delta\, (BM - B'M) + g\,\delta'\, (CN - BM)$$

$$= g\delta\, (R - a) + g\delta'\, (r' - R)$$

For equilibrium, these two pressures must be equal, therefore,

$$g\delta\, (r - a) = g\delta\, (R - a) + g\,\delta'\, (r' - R)$$

$$\Rightarrow \delta r = \delta R + \delta'(r' - R)$$

$$\Rightarrow \delta r - r'\delta' = R(\delta - \delta')$$

$$\Rightarrow R = \frac{r\delta - r'\delta'}{\delta - \delta'}$$

Aliter: Let SA and SC make angles α and β respectively with the axis OY of the parabola. Let SB make an angle δ with the axis of the parabola

Then,

$$A \to (r, \alpha), \qquad C \to (r', \beta), \qquad B \to (R, \theta)$$

Let the polar equation of the parabola be

$$\ell/R = 1 + \cos\theta \qquad \text{...(1)}$$

Since A, C and B lie on the parabola, therefore,

$$\ell/r = 1 + \cos\alpha \qquad \text{...(2)}$$

$$\ell/r' = 1 + \cos\beta \qquad \text{...(3)}$$

$$\ell/R = 1 + \cos\theta \qquad \text{...(4)}$$

Pressure at O from left

$$= \text{Pressure at O from right}$$

$\Rightarrow g\delta\, AA' = g\delta\, BB' + g\delta'\, CC''$

$\Rightarrow g\delta\,(a - r\cos\alpha) = g\delta\,(a - R\cos\theta) + g\delta'\,(R\cos\theta - r'\cos\beta)$

$\Rightarrow -\delta r\cos\alpha = -\delta R\cos\theta + \delta' R\cos\theta - \delta' r'\cos\beta$

$\Rightarrow -\delta r\,(\ell/r - 1) = -\delta R\,(\ell/R - 1) + \delta' R\,(\ell/R - 1) - \delta' r'\,(\ell/r' - 1)$

$\Rightarrow -\delta\ell + \delta r = -\delta\ell + \delta R + \delta'\ell - \delta' R - \delta'\ell + \delta' r'$

$\Rightarrow \delta r = \delta R - \delta' R + \delta' r'$

$\Rightarrow r\delta - r'\delta' = R\,(\delta - \delta')$

$\Rightarrow R = \dfrac{r\delta - r'\delta'}{\delta - \delta'}$

Example 17: A thin uniform cycloidal tube contains equal weights of two fluids. If it be placed with its axis vertical, prove that the heights of the free surfaces of the fluids above the vertex of the tube are as

$$(3a + b)^2 : (3b + a)^2$$

where a and b are the lengths of the tube which the fluids occupy.

Sol. O is the lowest point (vertex) of the cycloidal tub. Let the two liquids of densities ρ and σ ($\rho > \sigma$) occupy the lengths AOC and CB of the tube. Then,

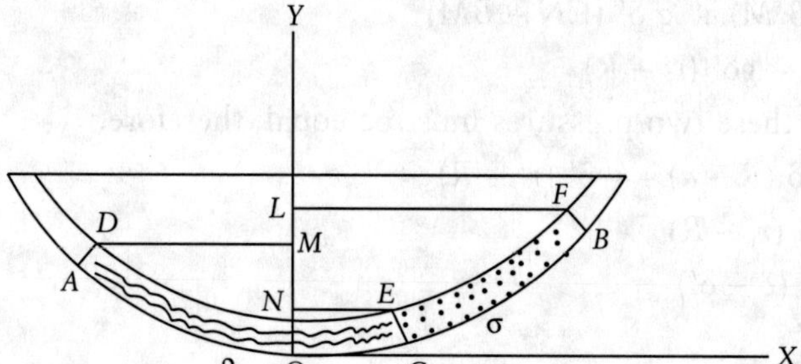

arc $AC = a$

arc $CB = b$

Let arc $OC = s$

Let the vertical heights of A, C and B above O be y_1, y_2 and y_3 respectively.

Since the weights of the two liquids are equal, therefore,

$$a\rho = b\sigma \qquad \text{...(1)}$$

From D, E and F, draw perpendiculars DM, EN and FL respectively to the Y-axis.

Pressure at O due to liquid in the portion OA of the tube $= g\rho y_1$

Pressure at O due to liquids in the portion OB of the tube

$$= g\rho y_2 + g\sigma\,(y_3 - y_2)$$

For equilibrium, these two pressures must be equal. Therefore,

$$g\rho y_1 = g\rho y_2 + g\sigma (y_3 - y_2)$$

$$\Rightarrow \quad \frac{\sigma}{\rho} = \frac{y_1 - y_2}{y_3 - y_2} \qquad ...(2)$$

Let the equation of the cycloid be

$$s^2 = 8\ R\ y \qquad ...(3)$$

where, R is the radius of the generating circle of the cycloid and y is the vertical height of a point above the vertex O whose arcual distance from O is s.

For A, $\quad (a - s)^2 = 8R\ y_1 \qquad ...(4)$

For C, $\quad s^2 = 8R\ y_2 \qquad ...(5)$

For B, $\quad (b - s)^2 = 8R\ y_3 \qquad ...(6)$

Subtracting (5) from (4), we get

$$a^2 - 2as = 8R\ (y_1 - y_2) \qquad ...(7)$$

Subtracting (5) from (6), we get

$$b^2 + 2bs = 8R\ (y_3 - y_2) \qquad ...(8)$$

Dividing (7) by (8), we get

$$\frac{a^2 - 2as}{b^2 + 2bs} = \frac{y_1 - y_2}{y_3 - y_2}$$

$$\Rightarrow \quad \frac{a^2 - 2as}{b^2 + 2bs} = \frac{\sigma}{\rho} \qquad | \text{ From (2)}$$

$$\Rightarrow \quad \frac{a^2 - 2as}{b^2 + 2bs} = \frac{a}{b} \qquad | \text{ From (1)}$$

$$\Rightarrow \quad b\ (a^2 - 2as) = a\ (b^2 + 2bs)$$

$$\Rightarrow \quad ab\ (a - 2s) = ab\ (b + 2s)$$

$$\Rightarrow \quad a - 2s = b + 2s$$

$$\Rightarrow \quad 4s = a - b$$

$$\Rightarrow \quad s = a - b/4 \qquad ...(9)$$

From (4) and (6), we get

$$\frac{y_1}{y_3} = \frac{(a-s)^2}{(b+s)^2}$$

$$= \frac{\left(a - \frac{a-b}{4}\right)^2}{\left(b + \frac{a-b}{4}\right)^2} \qquad | \text{ From (9)}$$

$$= \frac{(3a+b)^2}{(3b+a)^2}$$

$$\Rightarrow \quad y_1 : y_3 = (3a + b)^2 : (3b + a)^2$$

Example 18: Three liquids, whose densities ρ_1, ρ_2, ρ_3 $(\rho_1 > \rho_2 > \rho_3)$ are in A.P., completely fill a circular tube in vertical plane and occupy lengths of arcs subtending at the centre angles $2\alpha, 2\beta, 2\gamma$ respectively. If the radius to the point midway between the ends of the lightest liquid makes with the vertical angle θ, prove that

$$\cot \theta = \cot \alpha + 2 \cot \beta.$$

Sol. O is the centre of the circular tube. Let a be the radius of the circular tube. F is the lowest point of the circular tube. Let the arc AFB, arc BC and arc $CEDA$ of the tube be occupied by the liquids of density ρ_1, ρ_2 and ρ_3 as shown in the figure. These liquids have been arranged as $\rho_1 > \rho_2 > \rho_3$. E is the highest point of the tube. D is the midpoint of the arc EDA

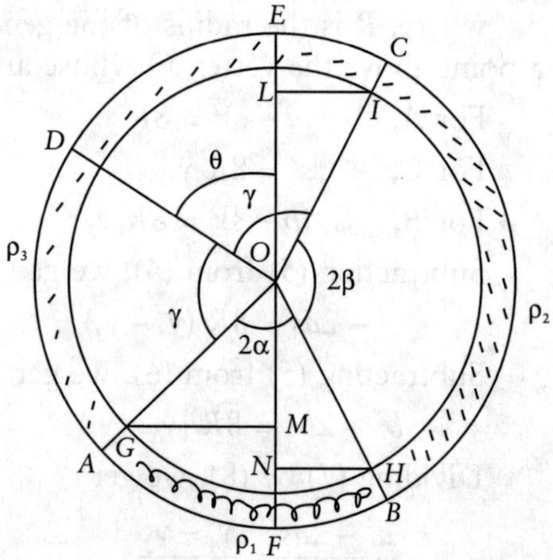

$$\angle DOE = \theta$$

$$\angle AOD = COD = \gamma$$

$$\angle AOB = 2\alpha$$

$$\angle BOC = 2\beta$$

From G, H and I, draw perpendiculars GM, HN and IC respectively to the vertical diameter.

Pressure at A from the left side

$$= g\rho_3\,(EM) + g\rho_1\,MF$$

$$= g\rho_3\,(OE + OM) + g\rho_1\,(OF - OM)$$

$$= g\rho_3\,\{a + a\cos(\pi - \theta - \gamma)\}\; g\rho_1\,\{a - a\cos(\pi - \theta - \gamma)\}$$

$$= ga\,[\rho_3\,\{1 - \cos(\theta + \gamma)\} + \rho_1\,\{1 + \cos(\theta + \gamma)\} \qquad ...(1)$$

Pressure at A from the right side

$$= g\rho_3\,LE + g\rho_2\,LN + g\rho_1\,NF$$

$$= g\rho_3\,(OE - OL) + g\rho_2\,(OL + ON) + g\rho_1\,(OF - ON)$$

$$= g\rho_3\,\{a - a\cos(\gamma - \theta)\} + g\rho_2\,\{a\cos(\gamma - \theta) + a\cos(\theta + \gamma + 2\alpha - \pi)\}$$

$$+ g\rho_1\,\{a - a\cos(\theta + \gamma + 2\alpha - \pi)\}$$

$$= ga\,[\rho_3\,\{1 - \cos(\gamma - \theta)\} + \rho_2\,\{\cos(\gamma - \theta) - \cos(\theta + \gamma + 2\alpha\} + \rho_1\,\{1 + \cos(\theta + \gamma + 2\alpha)\} \qquad ...(2)$$

For equilibrium, these two pressures must be equal. Therefore,

$$ga\,[\rho_3\{1 - \cos(\theta + \gamma)\} + \rho_1\,\{1 + \cos(\theta + \gamma)\}]$$

$$= ga\,[\rho_3\{1 - \cos(\gamma - \theta)\} + \rho_2\,\{\cos(\gamma - \theta) - \cos(\theta + \gamma + 2\alpha) + \rho_1\,\{1 + \cos(\theta + \gamma + 2\alpha)\}]$$

$$\Rightarrow \rho_3\{1 - \cos(\theta + \gamma)\} + \rho_1\,\{1 + \cos(\theta + \gamma)\}$$

$$= \rho_3\{1 - \cos(\gamma - \theta)\} + \rho_1\,\{\cos(\gamma - \theta) - \cos(\theta + \gamma + 2\alpha)\} + \rho_1\,\{1 + \cos(\theta + \gamma + 2\alpha)\}$$

$$\Rightarrow \quad (\rho_1 - \rho_3)\cos(\theta + \gamma) = (\rho_2 - \rho_3)\cos(\gamma - \theta) + (\rho_1 - \rho_2)\cos(\theta + \gamma + 2\alpha) \qquad ...(3)$$

$\because \quad \rho_1, \rho_2, \rho_3$ arc in A.P.

$$\therefore \quad \rho_2 = \frac{\rho_1 + \rho_3}{2}$$

$\therefore$ (3) gives

$$(\rho_1 - \rho_3)\cos(\theta+\gamma) = \left(\frac{\rho_1 - \rho_3}{2}\right)\cos(\gamma-\theta) + \left(\frac{\rho_1 - \rho_3}{2}\right)\cos(\theta+\gamma+2\alpha)$$

$$\Rightarrow \quad 2\cos(\theta+\gamma) = \cos(\gamma-\theta) + \cos(\theta+\gamma+2\alpha)$$

$$\Rightarrow \quad 2\cos(\theta+\gamma) = 2\cos(\gamma+\alpha)\cos(\alpha+\theta)$$

$$\Rightarrow \quad \cos(\theta+\gamma) = \cos(\gamma+\alpha)\cos(\alpha+\theta) \qquad ...(4)$$

But $\quad 2\alpha + 2\beta + 2\gamma = 2\pi \qquad$ | From figure

$$\Rightarrow \quad \alpha + \beta + \gamma = \pi$$

$$\Rightarrow \quad \gamma = \pi - (\alpha + \beta)$$

$\therefore$ (4) gives

$$\cos(\theta + \pi - \alpha - \beta) = \cos(\pi - \beta)\cos(\alpha + \theta)$$

$$\Rightarrow \cos\{\pi + (\theta - \alpha - \beta)\} = -\cos\beta\cos(\alpha+\theta)$$

$$\Rightarrow -\cos(\theta - \alpha - \beta) = -\cos\beta\cos(\alpha+\theta)$$

$$\Rightarrow \cos\{\theta - (\alpha+\beta)\} = \cos\beta\cos(\alpha+\theta)$$

$$\Rightarrow \cos\theta\cos(\alpha+\beta) + \sin\theta\sin(\alpha+\beta) = \cos\beta\cos\alpha\cos\theta - \cos\beta\sin\alpha\sin\theta$$

$$\Rightarrow \cos\theta\{\cos(\alpha+\beta) - \cos\theta\cos\beta\} = -\sin\theta\{\sin(\alpha+\beta) + \sin\alpha\cos\beta\}$$

$$\Rightarrow \cos\theta\sin\alpha\sin\beta = \sin\theta\{2\sin\alpha\cos\beta + \cos\alpha\sin\beta\}$$

$\Rightarrow \cos\theta = 2\cos\beta + \cot\alpha \qquad$ | dividing both sides by $\sin\alpha\sin\beta$

Example 19: A circular tube with centre O is filled with three fluids of densities ρ_1, ρ_2, ρ_3 (in descending order of magnitude) and placed in a vertical plane. If $2\alpha, 2\beta, 2\gamma$ be the angles subtended at the centre by the fluids and P be the point on the circumference midway between the ends of the lightest fluid, then angle θ which OP make with the vertical is given by

$$\frac{\rho_2 - \rho_3}{\rho_1 - \rho_3} = \frac{\sin\alpha}{\sin(\alpha+\theta)} \cdot \frac{\sin(\beta-\theta)}{\sin\beta}$$

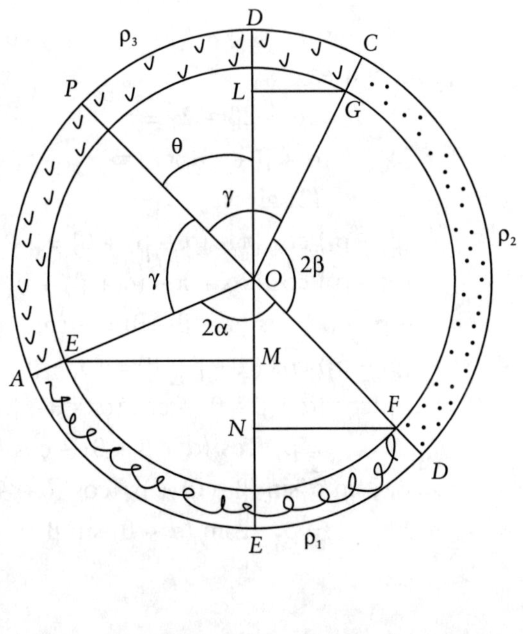

Sol. O is the centre of the circular tube. Let a be the radius of the circular tube. D and E are respectively the highest and the lowest points of the tube.

The portion AEB of the tube is filled by liquid of density ρ_1, the portion BC by liquid of density ρ_2 and the portion CA by the liquid of density ρ_3 as shown in the figure. These liquids have been arranged accordingly as $\rho_1 > \rho_2 > \rho_3$.

$$\angle AOB = 2\alpha$$
$$\angle BOC = 2\beta$$
$$\angle COA = 2\gamma$$

P is the point on the circumference between the ends of the lightest fluid (i.e., fluid of density ρ_3)

$\therefore \qquad \angle POA = \angle POC = \gamma$

$$\angle POD = \theta$$

From E, F and G, draw perpendiculars EM, FN and GL to the vertical diameter DE.

Pressure at E from the left side

$= g\rho_3\, DM + g\rho_1\, ME$

$= g\rho_3\,(OD + OM) + g\rho_1\,(OE - OM)$

$= g\rho_3\{a + a\cos\overline{\pi - (\gamma+\theta)}\} + g\rho_1\{a - a\cos\overline{\pi - (\gamma+\theta)}\}$

$= ga[\rho_3\{1 - \cos(\gamma+\theta)\} + \rho_1\{1 + \cos(\gamma+\theta)\}]$ ...(1)

Pressure at E from the right side

$= g\rho_3\, LD + g\rho_2\, LN + g\rho_1\, NE$

$= g\rho_3\,(OD - OL) + g\rho_2\,(OL + ON) + g\rho_1\,(OE - ON)$

$= g\rho_3\,\{a - a\cos(\gamma - \theta)\} + g\rho_2\,\{a\cos(\gamma - \theta) + a\cos(2\alpha + \gamma + \theta - \pi)\}$

$+ g\rho_1\,\{a - a\cos(2\alpha + \gamma + \theta - \pi)\}$

$= ga\,[\rho_3\,\{1 - \cos(\gamma - \theta) + \rho_2\,\{\cos(\gamma - \theta) - \cos(2\alpha + \gamma + \theta)\}$

$\qquad + \rho_1\,\{1 + \cos(2\alpha + \gamma + \theta)\}]$

For equilibrium, those two pressures must be equal. Therefore,

$\rho_3\,\{1 - \cos(\gamma + \theta)\} + \rho_1\,\{1 + \cos(\gamma + \theta)\}$

$= \rho_3\,\{1 - \cos(\gamma - \theta)\} + \rho_2\,\{\cos(\gamma - \theta) - \cos(2\alpha + \gamma + \theta)\} + \rho_1\,\{1 + \cos(2\alpha + \gamma + \theta)\}$

$\Rightarrow \quad (\rho_1 - \rho_3)\cos(\gamma + \theta) = (\rho_1 - \rho_2)\cos(2\alpha + \gamma + \theta) + (\rho_2 - \rho_3)\cos(\gamma - \theta)$ ...(2)

From figure,

$$2\alpha + 2\beta + 2\gamma = \pi$$

$\Rightarrow \qquad \alpha + \beta + \gamma = \pi \quad \Rightarrow \qquad \gamma = \pi - (\alpha + \beta)$

$\therefore \qquad$ (2) gives

$(\rho_1 - \rho_3)\cos\{\pi - (\alpha + \beta) + \theta\} =$

$(\rho_1 \;\; \rho_2)\cos[2\alpha + \pi \;\; (\alpha + \beta) + \theta] + (\rho_2 \;\; \rho_3)\cos[\pi \;\; (\alpha + \beta) \;\; \theta]$

$\Rightarrow -(\rho_1 - \rho_3)\cos(\alpha + \beta - \theta) = -(\rho_1 - \rho_2)\cos(\alpha + \theta - \beta) - (\rho_2 - \rho_3)\cos(\alpha + \beta + \theta)$

$\Rightarrow (\rho_1 - \rho_3)\cos(\alpha + \beta - \theta) = (\rho_1 - \rho_2)\cos(\alpha + \theta - \beta) + (\rho_2 - \rho_3)\cos(\alpha + \beta + \theta)$

$\Rightarrow \rho_1\,\{\cos(\alpha + \beta - \theta) - \cos(\alpha + \theta - \beta)\} + \rho_3\,\{\cos(\alpha + \beta + \theta) - \cos(\alpha + \beta - \theta)\}$

$\qquad = \rho_2\,\{\cos(\alpha + \beta + \theta) - \cos(\alpha + \theta - \beta)\}$

$\Rightarrow \rho_1\, 2\sin\alpha\,\sin(\beta - \theta) + \rho_3\,[\cos(\alpha+\beta+\theta) - \cos(\alpha+\theta-\beta) - \cos(\alpha+\beta-\theta) + \cos(\alpha+\theta-\beta)]$

$\qquad = \rho_2 \cdot 2\sin(\alpha + \theta)\sin\beta$

$\Rightarrow$ $\rho_1 \cdot 2\sin\alpha \sin(\beta-\theta) + \rho_3[2\sin(\alpha+\theta)\sin\beta - 2\sin\alpha\sin(\beta-\theta)] = \rho_2\, 2\sin(\alpha+\theta)\sin\beta$

$\Rightarrow$ $\rho_1 \sin\alpha \sin(\beta-\theta) + \rho_3[\sin(\alpha+\theta)\sin\beta - \sin\alpha\sin(\beta-\theta)] = \rho_2 \sin(\alpha+\theta)\sin\beta$

$\Rightarrow$ $(\rho_1-\rho_3)\sin\alpha\sin(\beta-\theta) = (\rho_2-\rho_3)\sin(\alpha+\theta)\sin\beta$

$$\Rightarrow \frac{\rho_2-\rho_3}{\rho_1-\rho_3} = \frac{\sin\alpha}{\sin(\alpha+\theta)}\cdot\frac{\sin(\beta-\alpha)}{\sin\beta}$$

Example 20: A fine glass tube in the form of an equilateral triangle is filled with equal volumes of three liquids which do not mix and whose densities are in A.P. The tube is held in a vertical plane, and the side that contains portions of the heaviest and lightest liquids makes an angle θ with the vertical. Show that the surface of separation divides the sides in the ratio.

$$\cos\left(\frac{\pi}{6}-\theta\right) : \cos\left(\frac{\pi}{6}+\theta\right)$$

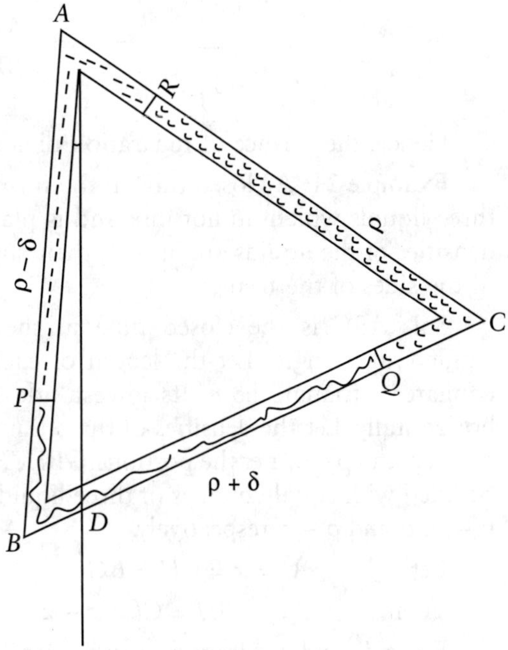

Sol. Let ABC be a fine glass tube in the form of an equilateral triangle. Let the side of this triangle be a. Let the portion PBQ of the tube be filled with liquid of density $\rho + \delta$, the portion QCR by liquid of density ρ and the portion RAP by the liquid of density $\rho - \delta$. Let $AP = x$.

Pressure at B due to liquids in the side AB

$= g\cdot(\rho-\delta)\,x\cos\theta + g(\rho+\delta)(a-x)\cos\theta$ ... (1)

Pressure at B due to liquids in the sides BC and CA

$= g\cdot(\rho+\delta)\,x\cos(\theta+\pi/3) + g\rho(a-x)\cos(\theta+\pi/3)$

$+ g\rho\, x\cos(\pi/3-\theta) + (\rho-\delta)(a-x)\cos(\pi/3-\theta)$ ... (2)

For equilibrium, these two pressures must be equal. Therefore,

$g(\rho-\delta)\,x\cos\theta + g(\rho+\delta)(a-x)\cos\theta$

$= g(\rho+\delta)\,x\cos(\theta+\pi/3) + g\rho(a-x)\cos(\theta+\pi/3)$

$+ g\rho\, x\cos(\pi/3-\theta) + (\rho-\delta)(a-x)\cos(\pi/3-\theta)$

$\Rightarrow$ $[(\rho-\delta)x + (\theta+\delta)(a-x)]\cos\theta$

$= [(\rho+\delta)x + \rho(a-x)]\cos(\theta+\pi/3) + [\rho x + (\rho-\delta)(a-x)]\cos(\pi/3-\theta)$

$\Rightarrow$ $[(a-x)\delta + \rho a - x\delta]\cos\theta = [\rho a + x\delta]\cos(\theta+\pi/3) + [\rho a - \delta(a-x)]\cos(\pi/3-\theta)$

$\Rightarrow$ $(a-x)\delta\cos\theta - x\delta\cos\theta = \delta x\cos(\theta+\pi/3) - \delta(a-x)\cos(\pi/3-\theta)$

$\Rightarrow$ $(a-x)\cos\theta - x\cos\theta = x\cos(\theta+\pi/3) - (a-x)\cos(\pi/3-\theta)$

$\Rightarrow$ $a[\cos\theta + \cos(\pi/3-\theta)] = x[2\cos\theta + \cos(\theta+\pi/3) + \cos(\pi/3-\theta)]$

$$\Rightarrow \frac{a}{x} = \frac{2\cos\theta + \cos\left(\theta+\frac{\pi}{3}\right) + \cos\left(\frac{\pi}{3}-\theta\right)}{\cos\theta + \cos\left(\frac{\pi}{3}-\theta\right)} \quad \Rightarrow \quad \frac{a-x}{x} = \frac{\cos\theta + \cos\left(\theta+\frac{\pi}{3}\right)}{\cos\theta + \cos\left(\frac{\pi}{3}-\theta\right)}$$

$$\Rightarrow \quad = \frac{\frac{3}{2}\cos\theta - \frac{\sqrt{3}}{2}\sin\theta}{\frac{3}{2}\cos\theta + \frac{\sqrt{3}}{2}\sin\theta} \qquad = \frac{\frac{\sqrt{3}}{2}\cos\theta - \frac{1}{2}\sin\theta}{\frac{\sqrt{3}}{2}\cos\theta + \frac{1}{2}\sin\theta}$$

$$= \frac{\cos\frac{\pi}{6}\cos\theta - \sin\frac{\pi}{6}\sin\theta}{\cos\frac{\pi}{6}\cos\theta + \sin\frac{\pi}{6}\sin\theta} \qquad = \frac{\cos\left(\frac{\pi}{6}+\theta\right)}{\cos\left(\frac{\pi}{6}-\theta\right)}$$

$$\Rightarrow \quad \frac{BP}{AP} = \frac{\cos\left(\frac{\pi}{6}+\theta\right)}{\cos\left(\frac{\pi}{6}-\theta\right)} \qquad \Rightarrow \quad \frac{AP}{BP} = \frac{\cos\left(\frac{\pi}{6}-\theta\right)}{\cos\left(\frac{\pi}{6}+\theta\right)}$$

Hence, the surface of reparation divides the side in the ratio cos $(\pi/6 - \theta)$: cos $(\pi/6 + \theta)$

Example 21: A closed tube in the form of an equilateral triangle contains equal volumes of three liquids which do not mix and is placed with its lowest side horizontal. Prove that if the densities of the liquids are in A.P., their surfaces of separation will be at the points of trisection of the sides of the triangle.

Sol. *ABC* is the closed tube in the form of an equilateral triangle. Let the length of each side of the equilateral triangle be *a*. Its lowest side *BC* is placed horizontally. Let the densities of three liquids in A.P. be $\rho + \sigma$, ρ and $\rho - \sigma$. Let the portions *GBH*, *HCI* and *IAG* be filled with equal volumes of three liquids of densities $\rho + \sigma$, ρ and $\rho - \sigma$ respectively.

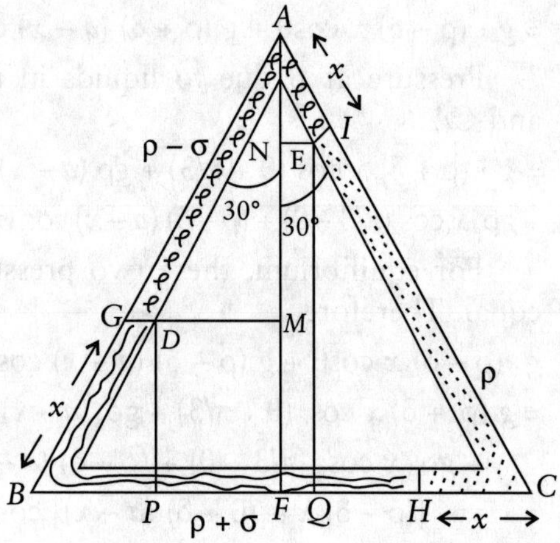

Let $\quad AI = x = CH = BG$

Then, $\quad AG = BH = CI = a - x$

From *D* and *E*, draw perpendiculars *DM* and *EN* respectively to the vertical through *A*.

Pressure at *B* due to the liquids in the tube *AGB*.

$= g\,(\rho - \sigma)\,AM + g\,(\rho + \sigma)\,MF$

$= g\,(\rho - \sigma)\,AG \cos 30° + g\,(\rho + \sigma)\,BG \sin 60°$

$= g\,(\rho - \sigma)\,(a - x)\,\sqrt{3}/2 + g\,(\rho + \sigma)\,x\,\sqrt{3}/2$ ...(1)

Pressure at *B* due to the liquids in the tube *AIC*

$= g\,(\rho - \sigma)\,AI \cos 30° + g\rho\,NF$

$= g\,(\rho - \sigma)\,x\,\sqrt{3}/2 + g\rho\,IC \sin 60°$

$= g\,(\rho - \sigma)\,x\,\sqrt{3}/2 + g\rho\,(a - x)\,\sqrt{3}/2$ ...(2)

For equilibrium, these two pressures must be equal. Therefore,

$$g(\rho-\sigma)(a-x)\sqrt{3}/2 + g(\rho+\sigma)\sqrt{3}/2 = g(\rho-\sigma)x\sqrt{3}/2 + g\rho(a-x)\sqrt{3}/2$$

$$\Rightarrow (\rho-\sigma)(a-x) + (\rho+\sigma)x = (\rho-\sigma)x + \rho(a-x)$$

$$\Rightarrow \rho a - \rho x - \sigma a + \sigma x + \rho x + \sigma x = \rho x - \sigma x + \rho a - \rho x$$

$$\Rightarrow 3\sigma x = \sigma a$$

$$\Rightarrow 3x = a \qquad \Rightarrow \qquad x = a/3$$

Hence, their surfaces of reparation will be at the points of trisection of the sides of the triangle.

Example 22: A fine tube of uniform bore is bent into the form of a square and filled with equal volumes of three heavy liquids of densities ρ_1, ρ_2, ρ_3 ($\rho_1 < \rho_2 < \rho_3$). If the tube is placed with one side vertical, show that a side of the square will be filled with liquid of the first, third or second kind only according as ρ_2 is $> 1/3\,(2\rho_3 + \rho_1)$ or $< 1/3\,(\rho_3 + 2\rho_1)$ or lies between these values.

Sol. Let the side of the square be a. Then, the length of the tube occupied by each liquid = $4a/3$.

Case I: Let one complete side of the tube be filled by the liquid of density ρ_1. Since the liquid with density ρ_1 is the lightest, therefore, this liquid will occupy the upper horizontal side of the tube. Now,

Pressure at A = Pressure at B

Pressure at D = Pressure at C

$\therefore$ Pressure at D – Pressure at A

= Pressure at C – Pressure at B.

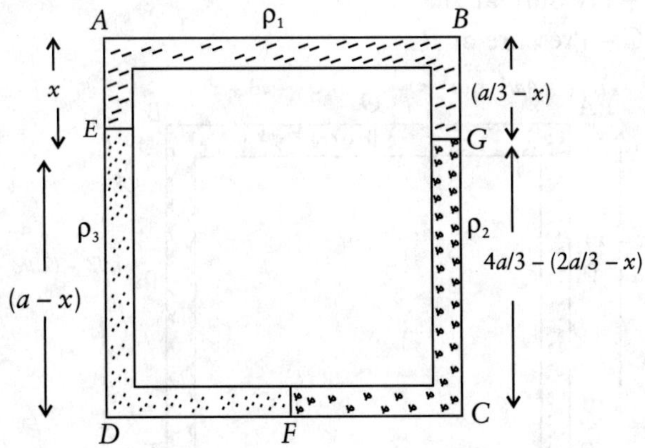

$$\therefore\ g\rho_3\,DE + g\rho_1\,AE = g\rho_2\,CG + g\rho_1\,GB$$

$$\Rightarrow g\rho_3(a-x) + g\rho_1(x) = g\rho_2(2a/3 + x) + g\rho_1(a/3 - x)$$

$$\Rightarrow \rho_3(a-x) + \rho_1 x = \rho_2(2a/3 + x) + \rho_1(a/3 - x)$$

$$\Rightarrow x(-\rho_3 + \rho_1 - \rho_2 + \rho_1) = a/3\,(2\rho_2 + \rho_1 - 3\rho_3)$$

$$\Rightarrow \frac{3x}{a} = \frac{2\rho_2 + \rho_1 - 3\rho_3}{-\rho_3 + 2\rho_1 - \rho_2} \qquad \Rightarrow \frac{3x}{a} = \frac{3\rho_3 - \rho_1 - 2\rho_2}{\rho_3 + \rho_2 - 2\rho_1}$$

The lightest liquid will fill the upper horizontal side of the tube if and only if

$0 < x < 4a/3 - a$

$\Rightarrow\ 0 < x < a/3 \quad \Rightarrow 0 < 3x/a < 1$

$\Rightarrow\ 0 < \dfrac{3\rho_3 - \rho_1 - 2\rho_2}{\rho_3 + \rho_2 - 2\rho_1} < 1$

$\Rightarrow\ 0 < 3\rho_3 - \rho_1 - 2\rho_2 < \rho_3 + \rho_2 - 2\rho_1$

$\because\ \rho_1 < \rho_2 < \rho_3$

$\therefore\ \rho_3 + \rho_2 - 2\rho_1 > 0$

$\Rightarrow\ 3\rho_3 - \rho_1 - 2\rho_2 < \rho_3 + \rho_2 - 2\rho_1$

$\because\ \rho_1 < \rho_2 < \rho_3$

$\therefore\ 3\rho_3 - \rho_1 - 2\rho_2 > 0$

$\Rightarrow\ 2\rho_3 + \rho_1 - 3\rho_2 < 0 \quad \Rightarrow 2\rho_3 + \rho_1 < 3\rho_2 \quad \Rightarrow \rho_2 > 1/3\ (2\rho_3 + \rho_1)$

Case II: Let one complete side of the tube be filled by the liquid of density ρ_3.

Since the liquid with density ρ_3 is the heaviest, therefore, this liquid will occupy the lower horizontal side of the tube.

Now,

Pressure at A = Pressure at B

Pressure at D = Pressure at C

$\therefore$ Pressure at D – Pressure at A

= Pressure at C – Pressure at B

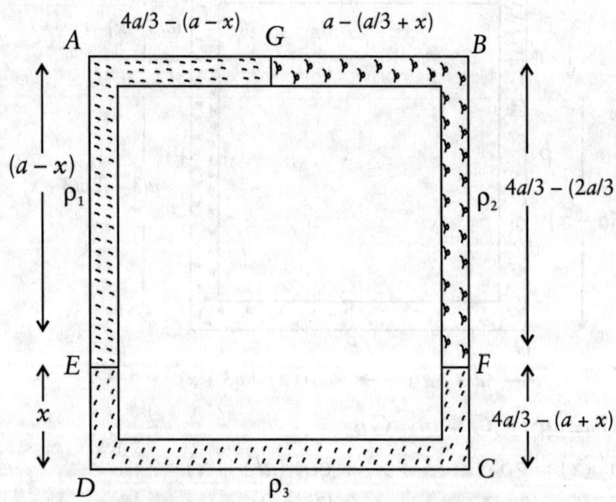

$\therefore\ g\rho_3 x + g\rho_1 (a - x) = g\rho_3 \{4a/3 - (a + x)\} + g\rho_2 (2a/3 + x)$

$\Rightarrow\ \rho_3 x + \rho_1 (a - x) = \rho_3 (a/3 - x) + \rho_2 (2a/3 + x)$

$\Rightarrow\ x(\rho_3 - \rho_1 + \rho_3 - \rho_2)$

$= a/3(\rho_3 + 2\rho_2 - 3\rho_1)$

$$\Rightarrow \frac{3x}{a} = \frac{2\rho_2 + \rho_3 - 3\rho_1}{2\rho_3 - \rho_1 - \rho_2}$$

The heaviest liquid will fill the lower horizontal side if and only if

$0 < x < 4a/3 - a$

$\Rightarrow 0 < x < a/3 \quad \Rightarrow 0 < 3x/a < 1$

$$\Rightarrow 0 < \frac{2\rho_2 + \rho_3 - 3\rho_1}{2\rho_3 - \rho_1 - \rho_2} < 1$$

$\Rightarrow 0 < 2\rho_2 + \rho_3 - 3\rho_1 < 2\rho_3 - \rho_1 - \rho_2$

$\Rightarrow 2\rho_2 + \rho_3 - 3\rho_1 < 2\rho_3 - \rho_1 - \rho_2$

$\Rightarrow 3\rho_2 - \rho_3 - 2\rho_1 < 0$

$\Rightarrow 3\rho_2 < \rho_3 + 2\rho_1$

$\Rightarrow \rho_2 < 1/3\ (\rho_3 + 2\rho_1)$

Case III: Let the right vertical side of the tube be filled by the liquid of density ρ_2.

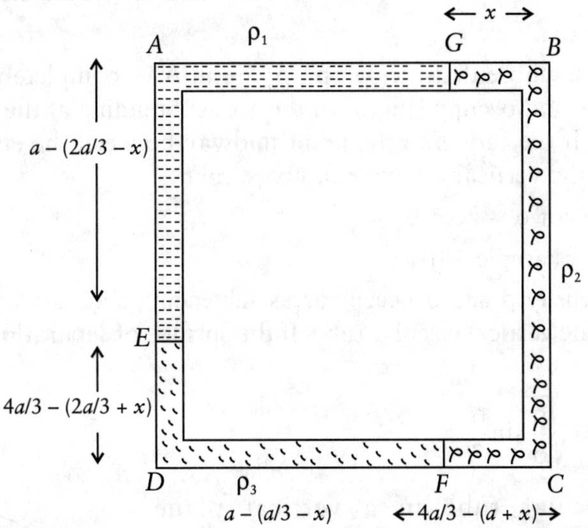

We have,

Pressure at D = Pressure at C

Pressure at A = Pressure at B

$\therefore$ Pressure at D – Pressure at A = Pressure at C – Pressure at B

$\Rightarrow g\rho_3\ (2a/3 - x) + g\rho_1\ (a/3 + x) = g\rho_2 \cdot a$

$\Rightarrow \rho_3\ (2a/3 - x) + \rho_1\ (a/3 + x) = \rho_2\ a$

$\Rightarrow x\ (-\rho_3 + \rho_1) = a/3\ (3\rho_2 - 2\rho_3 - \rho_1)$

$$\Rightarrow \frac{3x}{a} = \frac{3\rho_2 - 2\rho_3 - \rho_1}{\rho_1 - \rho_3} \qquad \Rightarrow \frac{3x}{a} = \frac{2\rho_3 + \rho_1 - 3\rho_2}{\rho_3 - \rho_1}$$

The liquid with density ρ_2 will fill the right vertical side if and only if

$0 < x < a/3$

$\Rightarrow \; 0 < 3x/a < 1$

$$\Rightarrow \; 0 < \frac{2\rho_3 + \rho_1 - 3\rho_2}{\rho_3 - \rho_1} < 1$$

$\Rightarrow \; 0 < 2\rho_3 + \rho_1 - 3\rho_2 < \rho_3 - \rho_1$

$\Rightarrow \; -2\rho_3 - \rho_1 < -3\rho_2 < \rho_3 - \rho_1 - 2\rho_3 - \rho_1$

$\Rightarrow \; 2\rho_3 + \rho_1 > 3\rho_2 > \rho_3 + 2\rho_1$

$\Rightarrow \; 1/3\,(\rho_3 + 2\rho_1) < \rho_2 < 1/3\,(2\rho_3 + \rho_1)$

EXERCISE 2

1. A fine circular tube of uniform thin bore is completely filled with equal volumes of three fluids which do not mix and whose densities are ρ, 2ρ and 3ρ. Prove that when the tube is held in a vertical plane, the surface of separation of the liquids of densities ρ and 3ρ is an extremity of the horizontal diameter of the circle.
   [**Hint:** Refer Example 6]
2. Three liquids, whose densities ρ_1, ρ_2, ρ_3 are in A.P., completely fill a circular tube in a vertical plane and occupy lengths of the arcs subtending at the centre angles 2α, 2β, 2γ respectively. If the radius to the point midway between the ends of the lightest liquid makes with the vertical an angle θ, prove that
   $$\cot\theta = \cot\alpha + 2\cot\beta$$
   [**Hint:** Refer Example 18]
3. Liquids of density ρ and σ occupy areas subtending angles α and β respectively at the centre of a fine vertical circular tube. If the surface of separation be at the lowest point, prove that
   $$\sin\frac{\beta}{2} = \sqrt{\frac{\rho}{\sigma}}\sin\frac{\alpha}{2}$$
4. A fine circular tube in a vertical plane contains two liquids of densities ρ and σ ($\rho > \sigma$) which subtend angles α, β respectively at the centre of the tube. Find the angle which the radius through the common surface of the liquids makes with the vertical.
   [**Hint**: Equating the pressures at B, we get
   $g\rho\,[a - a\cos(\alpha - \theta)] = g\rho\,[a - a\cos\theta] + g\sigma\,[a\cos\theta - a\cos(\theta + \beta)]$
   This gives θ]

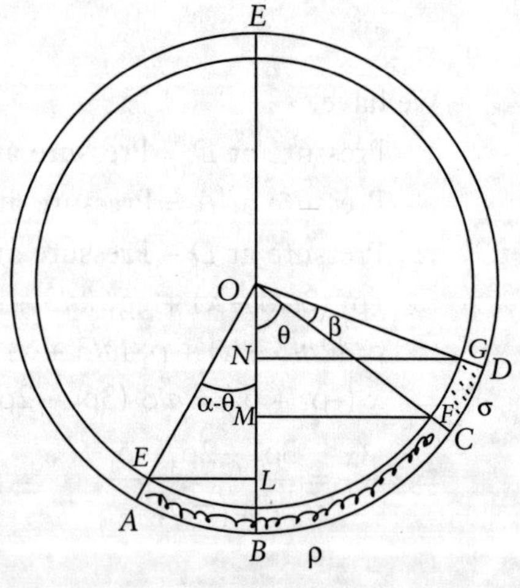

5. A fine tube of uniform bore, bent into the form of an arc of a parabola bounded by a double ordinate equal to twice the latus rectum is held,

with its axis vertical and vertex downwards. It is filled with three liquids (which do not mix) whose densities taken in order are as 10 : 13 : 2. If the line joining the points of separation of the liquids passes through the focus of the parabola, show that the indication to the vertical is 60°.

6. A tube of fine uniform bore of cross-section unity and length $(a + b)$, whose plane ends are perpendicular to its length, is bent into the form of an angle. The longer arm whose length a is kept vertical and the other arm is closed by a loose fitting cap of weight w, at the upper end. The tube contains liquid which fills the oblique arm and part of the vertical one. It is observed that the cap is just being lifted when the lengths of the liquid in the two arms are equal. Show that $w = bg\rho\ (\sec \alpha - 1)$, whose α is the angle between the arms and ρ is the density of the liquid.

[Hint:

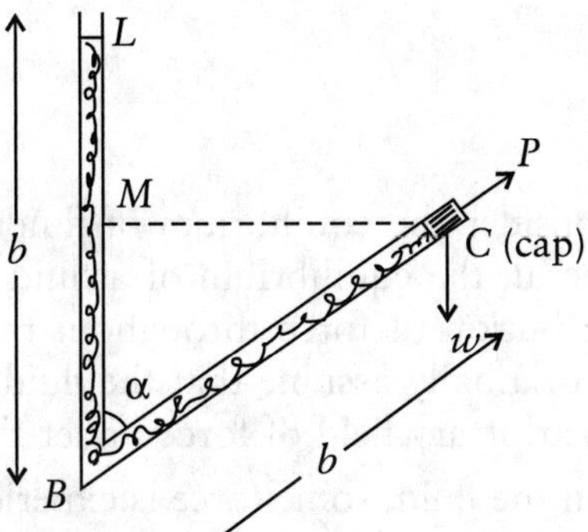

$$P = g\rho\ LM$$
$$= g\rho\ (b - b \cos \alpha)$$
$$= g\rho\ b\ (1 - \cos \alpha)$$

Now $\quad w \cos \alpha = g\rho\ b\ (1 - \cos \alpha)$

$\Rightarrow \quad w = bg\rho\ (\sec \alpha - 1)$]

7. A fine tube of uniform bore is bent into the form of a square and filled with equal volumes of three heavy liquids of densities, 1, ρ, 1.2. If the tube is placed with one side vertical, show that a side of the square will be filled with liquid of first, third or second kind only according as $\rho > \dfrac{17}{15}$ or $< \dfrac{16}{15}$ or lies between these values.

Chapter

4

Conditions of Equilibrium of Fluids

1. Introduction

In this chapter, we shall consider the equilibrium of fluids which are under the action of any given field of force. In general, the equilibrium of a fluid is not possible when it is under the action of an arbitrary distribution of force throughout the fluid. We shall proceed in the reverse direction, i.e., we shall originally assume that the fluid is in equilibrium and derive the condition to be satisfied by the arbitrary field of force under this assumption.

If at any point of a region in the fluid, some force is experienced, then that region is said to under a field of force. The magnitude of force on a unit particle placed at a point in a field of force is called the intensity of the field of force at the point under consideration. We shall also consider those surfaces in the fluid, on which the pressure has the same value. These surfaces are called surfaces of equi-pressure. Moreover, we will draw our attention to the lines of force also. A line of force in a given field of force is defined as a curve, the tangent at any point of which gives the direction of the field of force at that point.

2. Theorem

The pressure at any point of a fluid is the same in every direction.

Proof.

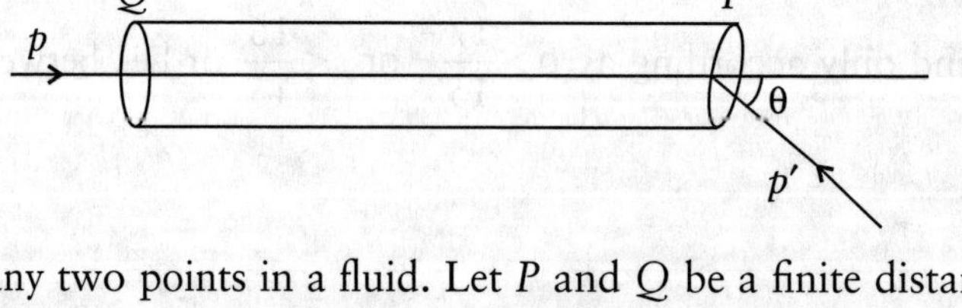

Let P and Q be any two points in a fluid. Let P and Q be a finite distance apart from each other. Taking PQ as axis, construct a fluid cylinder of a very small area of cross-section. Draw a plane through Q perpendicular to QP. Draw any arbitrary plane through P. Let the angle between QP and the normal line to the arbitrary plane at P be θ.

Let us consider the equilibrium of this fluid cylinder. This cylinder is in equilibrium under the action of impressed forces acting on it and the pressures on its curved surface.

Let p and p' be the intensities of pressure at Q and P respectively. Let α be the area of cross-section at Q and α' be the area of cross-section at P. Then, pressure on α is $p\alpha$ along QP and pressure on α' is $p'\alpha'$ perpendicular to α'. Let f be acceleration due to the impressed forces along QP. Then, the impressed force on an element ds is $f\ \rho\alpha\ ds$, where ρ is the density of the fluid. Hence, the total impressed force along QP is $\int f\ \rho\alpha\ ds$.

For equilibrium, resolving the forces along QP, we have

$$p'\alpha - p'\alpha\cos\theta + \int f\rho\alpha\, ds = 0$$

But $\alpha' \cos\theta = \alpha$

$$\therefore \quad p' = p + \int f\rho\, ds$$

We see that the above expression for p' is independent of θ. This shows that p' will be the same in whatsoever the direction, the face at P is taken into consideration.

3. Rate of Variation of Pressure

Theorem: *The rate of variation of pressure in any direction is equal to the density multiplied by the resolved part of the acceleration of the external forces in that direction.*

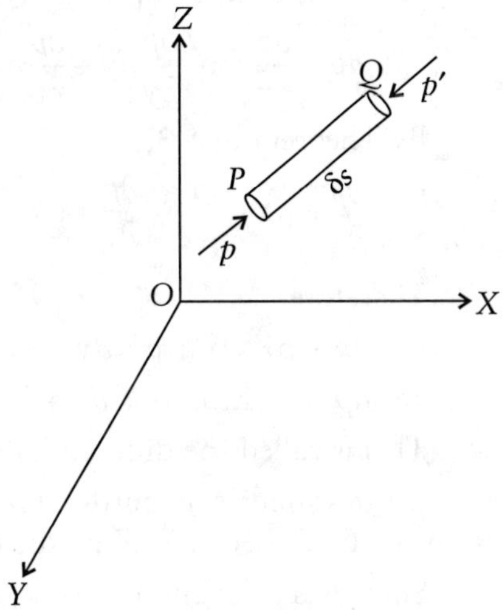

Let a mass of fluid be in equilibrium under the action of a given system of external forces. Let P be any arbitrary but fixed point in the fluid. Let Q be a point very near to it such that $PQ = \delta s$ is a short length in an assigned direction. About PQ as axis, describe a very small circular fluid cylinder having its plane ends perpendicular to PQ. Let α be the area of cross-section of the cylinder. Clearly α is very small.

Let p be the pressure at P and p' be the pressure at Q. Therefore, thrust on the plane face at P is $p\alpha$ along PQ and thrust on the plane face at Q is $p'\alpha$ along QP.

Let us consider the equilibrium of the fluid in the cylinder. The forces acting on it are as follows:

(1) Thrust $p\alpha$ on the face at P along PQ
(2) Thrust $p'\alpha$ on the face at Q along QP
(3) Thrust on the curved surface on each side of the axis PQ, which is everywhere perpendicular to PQ. As such they balance one another.
(4) *External forces*: Let f be acceleration of the external forces along PQ. Let ρ be the mean density of the fluid inside the cylinder. Then, the component of the external forces along $PQ = f\ \rho\alpha\ \delta s$

Hence, for equilibrium of the fluid in the cylinder, resolving the forces along PQ, we get

$$p\alpha - p'\alpha + f\ \rho\alpha\ \delta s = 0$$

$$\Rightarrow \quad p' - p = f\rho\delta s \qquad \text{...(1)}$$

$$\Rightarrow \quad \frac{p' - p}{\delta s} = \rho f$$

Proceeding to limit as $\delta s \to 0$, we get

$$\Rightarrow \quad \frac{\partial p}{\partial s} = \rho f \qquad \text{...(2)}$$

The result (1) shows that difference of pressures at two points very close to each other = Density × virtual work done by forces on unit mass for the displacement PQ.

The result (2) shows that the rate of variation of pressure in any direction = Density × resolved part of the acceleration in that direction.

Note: Equation (2) is known as the equation of pressure.

4. Differential Equation of Pressure

(1) In Cartesian coordinates (x, y, z): Let (x, y, z) be the Cartesian coordinates of a point P in the fluid referred to the rectangular axes OX, OY and OZ. Let X, Y, Z be the components of acceleration (or, external force per unit mass) parallel to the axes at the point P.

Since p is a function of (x, y, z), therefore, we have

$$dp = \frac{\partial p}{\partial x}dx + \frac{\partial p}{\partial y}dy + \frac{\partial p}{\partial z}dz$$

By Theorem of § 3,

$$\frac{\partial p}{\partial x} = \rho X \qquad \frac{\partial p}{\partial y} = \rho Y \qquad \frac{\partial p}{\partial z} = \rho Z$$

Therefore,

$$dp = \rho X\, dx + \rho Y\, dy + \rho Z\, dz$$

$$\Rightarrow \quad dp = \rho\,(X\, dx + Y\, dy + Z\, dz)$$

This is called the differential equation of pressure in the Cartesian coordinates (x, y, z).

(2) In cylindrical coordinates (r, ϕ, z): Let (r, ϕ, z) be the cylindrical coordinates of a point P in the fluid. Let P, T, Z be the components of acceleration in the directions of r, ϕ, z.

Since p is a function of (r, ϕ, z) therefore, we have

$$dp = \frac{\partial p}{\partial r}dr + \frac{\partial p}{\partial \phi}d\phi + \frac{\partial p}{\partial z}dz$$

By Theorem of § 3,

$$\frac{\partial p}{\partial r} = \rho P$$

$$\frac{\partial p}{r\partial \phi} = \rho T \qquad \left|\ \because \delta s \text{ in the directions of } r, \phi, z \text{ is } \delta r, r\delta\phi, \delta z\right.$$

$$\frac{\partial p}{\partial z} = \rho Z$$

Therefore,

$$dp = \rho\, Pdr + \rho rTd\phi + \rho Zdz$$

$$\Rightarrow \quad dp = \rho\,(Pdz + rTd\phi + Zdz)$$

(3) In polar coordinates (r, θ, ϕ): Let (r, θ, ϕ) be the polar coordinates of a point P in the fluid. Let R, N, T be the components of acceleration in the direction of (r, θ, ϕ).

Since p is a function of (r, θ, ϕ), therefore, we have

$$dp = \frac{\partial p}{\partial r}dr + \frac{\partial p}{\partial \theta}d\theta + \frac{\partial p}{\partial \phi}d\phi$$

By Theorem of § 3,

$$\frac{\partial p}{\partial r} = \rho R, \quad \frac{\partial p}{r\partial \theta} = \rho N, \quad \frac{\partial p}{r\sin\theta\,\partial \phi} = \rho T$$

$\because$ δs in the directions of r, θ, ϕ is δr, $r\delta\theta$ and $r \sin\theta\ \delta\phi$

Therefore,

$$dp = \rho\ R\ dr + \rho\ r\ N\ d\theta + \rho\ r\ \sin\theta\ T\ d\phi$$

$$\Rightarrow dp = \rho\ (R\ dr + r\ N\ d\theta + r\ \sin\theta\ T\ d\phi)$$

This is called the differential equation of pressure in polar coordinates (r, θ, ϕ).

5. Conservative Field

If the work done by a force in displacing its point of application from a point A to a point B is independent of the path connecting A and B, i.e., remains the same whatsoever the path is followed from A to B, then we say that the force field is conservative.

Let X, Y, Z denote the components of the external forces per unit mass at any point (x, y, z) in the directions of x, y and z. Then, if the point (x, y, z) undergoes a displacement given by dx, dy, dz in the directions of x, y, z, then the work done by the forces during this displacement is $Xdx + Ydy + Zdz$.

If X, Y, Z are the differential coefficients of a function V with respect to x, y, z respectively, then, we may show that the work done by these forces on a particle in moving it from a point A to a point B depends only on the values of V for the initial and final positions of the particle and is independent of the path followed in moving from initial position A to final position B.

Such a system of forces in which the components X, Y, Z of the forces per unit mass are the differential coefficients of some function V of (x, y, z) such that $Xdx + Ydy + Zdz$ is a perfect differential, is known as a conservative system.

The function V is known as the potential function.

6. Condition of Equilibrium

The necessary and sufficient condition that must be satisfied by a given distribution of forces X, Y, Z so that the fluid may maintain equilibrium is that

$$X\left(\frac{\partial Y}{\partial z} - \frac{\partial Z}{\partial y}\right) + Y\left(\frac{\partial Z}{\partial x} - \frac{\partial X}{\partial z}\right) + Z\left(\frac{\partial X}{\partial y} - \frac{\partial Y}{\partial x}\right) = 0$$

Necessary Condition: Let p be the pressure at the point (x, y, z) in fluid. Let X, Y, Z be the components of the external force per unit mass of the fluid at the point (x, y, z) in the directions of x, y, z. Then, we have,

$$\left.\begin{aligned}\frac{\partial p}{\partial x} &= \rho X\\ \frac{\partial p}{\partial y} &= \rho Y\\ \frac{\partial p}{\partial z} &= \rho Z\end{aligned}\right\} \quad ...(1)$$

where ρ is the density of the fluid at the point (x, y, z).

Since p is a function of independent variables, x, y, z, therefore, we have

$$\frac{\partial^2 p}{\partial y \partial z} = \frac{\partial^2 p}{\partial z \partial y}, \frac{\partial^2 p}{\partial z \partial x} = \frac{\partial^2 p}{\partial x \partial z}, \frac{\partial^2 p}{\partial x \partial y} = \frac{\partial^2 p}{\partial y \partial x}$$

Now, $$\frac{\partial^2 p}{\partial y \partial z} = \frac{\partial^2 p}{\partial z \partial y}$$

$$\Rightarrow \quad \frac{\partial}{\partial y}\left(\frac{\partial p}{\partial z}\right) = \frac{\partial}{\partial z}\left(\frac{\partial p}{\partial y}\right) \quad \Rightarrow \frac{\partial}{\partial y}(\rho Z) = \frac{\partial}{\partial z}(\rho Y) \qquad | \text{ From (1)}$$

$$\Rightarrow \quad \rho\frac{\partial Z}{\partial y} + Z\frac{\partial \rho}{\partial y} = \rho\frac{\partial Y}{\partial z} + Y\frac{\partial \rho}{\partial z} \quad \Rightarrow \rho\left(\frac{\partial Y}{\partial z} - \frac{\partial Z}{\partial y}\right) = Z\frac{\partial \rho}{\partial y} - Y\frac{\partial \rho}{\partial z} \quad ...(2)$$

Similarly, we have

$$\rho\left(\frac{\partial Z}{\partial x} - \frac{\partial X}{\partial z}\right) = X\frac{\partial \rho}{\partial z} - Z\frac{\partial \rho}{\partial x} \quad ...(3)$$

$$\left| \text{ From } \frac{\partial^2 p}{\partial z \partial x} = \frac{\partial^2 p}{\partial x \partial z}\right.$$

and $$\rho\left(\frac{\partial X}{\partial y} - \frac{\partial Y}{\partial x}\right) = Y\frac{\partial \rho}{\partial x} - X\frac{\partial \rho}{\partial y} \quad ...(4)$$

$$\left| \text{ From } \frac{\partial^2 p}{\partial x \partial y} = \frac{\partial^2 p}{\partial y \partial x}\right.$$

Multiplying the equations (2), (3), and (4) by X, Y and Z respectively, and adding we get

$$X\left(\frac{\partial Y}{\partial z} - \frac{\partial Z}{\partial y}\right) + Y\left(\frac{\partial Z}{\partial x} - \frac{\partial X}{\partial z}\right) + Z\left(\frac{\partial X}{\partial y} - \frac{\partial Y}{\partial x}\right)$$

$$= \frac{1}{\rho}\left[XZ\frac{\partial \rho}{\partial y} - XY\frac{\partial \rho}{\partial z} + YX\frac{\partial \rho}{\partial z} - YZ\frac{\partial \rho}{\partial x} + ZY\frac{\partial \rho}{\partial x} - ZX\frac{\partial \rho}{\partial y}\right] = 0 \quad ...(5)$$

which is the required necessary condition of equilibrium to be satisfied by (X, Y, Z).

Sufficient Condition: Let us suppose that the condition (5) is satisfied. But (5) is the condition that $Xdx + Ydy + Zdz$ is an exact differential. If ρ is constant, then

$\rho\,(Xdx + Ydy + Zdz)$ is an exact differential, say, dp.

Therefore,

$$\rho\,(Xdx + Ydy + Zdz) = dp$$

$$\Rightarrow \quad \rho(Xdx + Ydy + Zdz) = \frac{\partial p}{\partial x}dx + \frac{\partial p}{\partial y}dy + \frac{\partial p}{\partial z}dz$$

This gives

$$\frac{\partial p}{\partial x} = \rho X$$

$$\frac{\partial p}{\partial y} = \rho Y$$

$$\frac{\partial p}{\partial z} = \rho Z$$

But these are the equations of equilibrium. Hence, the given condition is sufficient also.

7. Homogeneous Liquids

We know that the differential equation of pressure in Cartesian coordinates (x, y, z) is given by

$$dp = \rho\,(Xdx + Ydy + Zdz) \qquad \text{...(1)}$$

where X, Y, Z are the components of external forces per unit mass of the fluid at any point (x, y, z) in the directions of x, y and z and ρ is the density of the fluid.

If the fluid is homogeneous and incompressible, then ρ is constant. Now, since dp is a perfect differential, therefore (1) implies that $Xdx + Ydy + Zdz$ must be a perfect differential so that the equilibrium of the fluid may be maintained.

The other way round we can say that the force system must be conservative, i.e.,

$$Xdx + Ydy + Zdz = -\,dV \qquad \text{...(2)}$$

where V is the potential function

$$\Rightarrow \quad Xdx + Ydy + Zdz = -\left(\frac{\partial V}{\partial x}dx + \frac{\partial V}{\partial y}dy + \frac{\partial V}{\partial z}dz\right)$$

$$\Rightarrow \quad \left.\begin{aligned} X &= -\frac{\partial V}{\partial x} \\ Y &= -\frac{\partial V}{\partial y} \\ Z &= -\frac{\partial V}{\partial z} \end{aligned}\right\} \qquad \text{...(3)}$$

From (1) and (2), we have

$$dp = -\,\rho\, dV$$

$$\Rightarrow \quad d(p/\rho) = -\,dV \qquad |\because \rho \text{ is constant}$$

$\Rightarrow$ $p/\rho = -\,V + c$ where c is an arbitrary constant of integration/on integration

Note: when ρ = constant, i.e., liquid is homogeneous, then equations (2), (3) and (4) of § 6 become

$$\frac{\partial Y}{\partial z} = \frac{\partial Z}{\partial y}$$
$$\frac{\partial Z}{\partial x} = \frac{\partial X}{\partial z}$$
$$\frac{\partial X}{\partial y} = \frac{\partial Y}{\partial x}$$

which are evidently satisfied in this case by virtue of equations (3). [Refer § 7). It follows that the equilibrium of a homogeneous fluid under the action of such components of external forces per unit mass of the fluid at any point in the fluid is always possible.

8. Heterogeneous Fluids

In case of heterogeneous fluid, ρ is not constant. It is rather a function of independent variables *x, y, z*. Here, for the system of forces *X, Y, Z* to maintain the equilibrium of fluid, the necessary and sufficient condition that must be satisfied are as follows:

$$\frac{\partial}{\partial y}(\rho Z) = \frac{\partial}{\partial z}(\rho Y)$$
$$\frac{\partial}{\partial z}(\rho X) = \frac{\partial}{\partial x}(\rho Z) \qquad | \text{ Refer §6}$$
$$\frac{\partial}{\partial x}(\rho Y) = \frac{\partial}{\partial y}(\rho X)$$

9. Surface of Equal Pressure

The differential equation of pressure is $dp = \rho\,(X\,dx + Y\,dy + Z\,dz)$

where $p = p\,(x, y, z)$ is the pressure function at the point (x, y, z), ρ is the density of the fluid and *X, Y, Z* are the components of external forces per unit mass of the fluid.

If $\quad p$ = constant, then

$$dp = 0$$

$$\Rightarrow \quad Xdx + Ydy + Zdz = 0$$

Integration this, we get

$$\phi\,(x, y, z) = C$$

which represents a surface at every point of which the pressure is constant and is equal to C.

Such a surface is called surface of equal pressure.

For different values of C, we get different surfaces of equal pressure.

The external surface or free surface is obtained by making *p* equal to the pressure external to the fluid. If the external pressure is zero, then the equation of the free surface is given by

$$\phi\,(x, y, z) = 0$$

10. Line of Force

A line of force is a line (i.e., curve) such that the tangent at any point of it is in the direction of the resultant force at that point.

From Coordinate Geometry, we know that the direction cosines of the tangent are proportional to *dx*/*ds*, *dy*/*ds*, *dz*/*ds*. Also, the direction cosines of the line of action of the resultant external force at (x, y, z) are proportional to X, Y, Z.

Therefore, the differential equations of the line of force are:

$$\frac{\frac{dx}{ds}}{X} = \frac{\frac{dy}{ds}}{Y} = \frac{\frac{dz}{ds}}{Z}$$

$$\Rightarrow \qquad \frac{dx}{X} = \frac{dy}{Y} = \frac{dz}{Z}$$

11. Theorem

The surfaces of equal pressure are intersected orthogonally by the lines of force.

The equation of the surface of equal pressure is

$$p\,(x, y, z) = C \qquad \text{...(1)}$$

From analytical geometry, the direction cosines of the normal at any point (x, y, z) of this surface are proportional to $\frac{\partial p}{\partial x}, \frac{\partial p}{\partial y}, \frac{\partial p}{\partial z}$, i.e., proportional to

$\rho X, \rho Y, \rho Z$ $\quad \left| \because \frac{\partial p}{\partial x} = \rho X, \quad \frac{\partial p}{\partial y} = \rho Y, \quad \frac{\partial p}{\partial z} = \rho Z \right.$

i.e., proportional to X, Y, Z $\quad | \because \rho$ is constant

Hence the resultant force at any point is in the direction of the normal to the surface of equal pressure passing through that point.

But X, Y, Z are the direction ratios of the lines of force. Hence, the lines of force are parallel to normals.

Therefore, the surfaces of equal pressure are cut orthogonally by the lines of force.

12. Conservative Force System

If ρ is constant, i.e., the fluid is homogeneous, then the equation of pressure is given by

$$dp = \rho\,(X\,dx + Y\,dy + Z\,dz) \qquad \text{...(1)}$$

$\Rightarrow \rho\,(X\,dx + Y\,dy + Z\,dz)$ is an exact differential

$\Rightarrow X\,dx + Y\,dy + Z\,dz$ is an exact differential $\quad | \because \rho$ is constant

$\Rightarrow$ The force system is conservative.

Conversely: If the force system is conservative, then (1) may be written as

$$dp = -\rho\, dV \qquad ...(2)$$

where V is the potential function.

Since dp is a perfect differential, therefore, $\rho\, dV$ must be a perfect differential. Consequently, if the fluid is not homogeneous, i.e., ρ is not constant, then ρ must be a function of V and therefore ρ must be a function of p. Hence, when p = constant, then V = constant and also ρ = constant. It shows that surfaces of equal pressure are also the equi-potential surfaces and surfaces of equal density.

Note: When the force-system is not conservative, then these surfaces will not in general coincide.

13. Curves of Equal Pressure and Density

The equation of pressure is

$$dp = \rho\,(X\,dx + Y\,dy + Z\,dz) \qquad ...(1)$$

Let the fluid be heterogeneous and incompressible. Then, the density ρ will be a function of x, y, z.

For a surface of equal pressure,

$$p = \text{constant}$$

$$\Rightarrow dp = 0 \qquad ...(2)$$

For a surface of equal density,

$$\rho = \text{constant}$$

$$\Rightarrow d\rho = 0 \qquad ...(3)$$

From (1) and (2), we get

$$X\,dx + Y\,dy + Z\,dz = 0 \qquad ...(4)$$

Also, (3) gives

$$\frac{\partial \rho}{\partial x}dx + \frac{\partial \rho}{\partial y}dy + \frac{\partial \rho}{\partial z}dz = 0 \qquad ...(5)$$

(4) and (5) give

$$\frac{dx}{Z\frac{\partial \rho}{\partial y} - Y\frac{\partial \rho}{\partial z}} = \frac{dy}{X\frac{\partial \rho}{\partial z} - Z\frac{\partial \rho}{\partial x}} = \frac{dz}{Y\frac{\partial \rho}{\partial x} - X\frac{\partial \rho}{\partial y}} \qquad ...(6)$$

But from the conditions of equilibrium (Refer § 6), we have

$$Z\frac{\partial \rho}{\partial y} - Y\frac{\partial \rho}{\partial z} = \rho\left(\frac{\partial Y}{\partial z} - \frac{\partial Z}{\partial y}\right) \qquad ...(7)$$

$$X\frac{\partial \rho}{\partial z} - Y\frac{\partial \rho}{\partial x} = \rho\left(\frac{\partial Z}{\partial x} \quad \frac{\partial X}{\partial z}\right) \qquad ...(8)$$

$$Y\frac{\partial \rho}{\partial x} - X\frac{\partial \rho}{\partial y} = \rho\left(\frac{\partial X}{\partial y} - \frac{\partial Y}{\partial x}\right) \qquad ...(9)$$

In view of (7), (8) and (9), the equations (6) become

$$\frac{dx}{\frac{\partial Y}{\partial z}-\frac{\partial Z}{\partial y}}=\frac{dy}{\frac{\partial Z}{\partial x}-\frac{\partial X}{\partial z}}=\frac{dz}{\frac{\partial X}{\partial y}-\frac{\partial Y}{\partial x}} \quad \text{...(10)}$$

On integration, the equation (10) will yield two surfaces whose curve of intersection will be the curve of equal pressure and density.

ILLUSTRATIVE EXAMPLES

Example 1: A liquid of given volume V is at rest under the forces $X = -\mu x/a^2$, $Y = -\mu y/b^2$, $Z = -\mu z/c^2$. Find the pressure at any point of the liquid and the surfaces of equal pressure.

We have,

$$X = -\frac{\mu x}{a^2}$$

$$Y = -\frac{\mu y}{b^2}$$

$$Z = -\frac{\mu z}{c^2}$$

The differential equation of pressure is

$$dp = \rho\,(X\,dx + Y\,dy + Z\,dz)$$

$$\Rightarrow \quad dp = -\mu\rho\left(\frac{x}{a^2}dx+\frac{y}{b^2}dy+\frac{z}{c^2}dz\right)$$

Integration yields,

$$p = c-\frac{\mu\rho}{2}\left(\frac{x^2}{a^2}+\frac{y^2}{b^2}+\frac{z^2}{c^2}\right)$$

At the free surface, i.e., surface of zero pressure,

$$p = 0$$

$$\Rightarrow \quad \frac{x^2}{a^2}+\frac{y^2}{b^2}+\frac{z^2}{c^2}=\frac{2c}{\rho\mu}$$

This represents an ellipsoid. Let its volume be V. Then,

$$V = \frac{4}{3}\pi abc\left(\frac{2c}{\rho\mu}\right)^{3/2}$$

$$\therefore \quad c = \frac{\mu\rho}{2}\left(\frac{3V}{4\pi\, abc}\right)^{2/3}$$

$$\therefore \quad p = \frac{\mu\rho}{2}\left(\frac{3V}{4\pi\, abc}\right)^{2/3}-\frac{\mu\rho}{2}\left(\frac{x^2}{a^2}+\frac{y^2}{b^2}+\frac{z^2}{c^2}\right)$$

This gives the pressure at any point. Now, surfaces of equal pressure are given by

$$p = \text{constant} = K \text{ (say)}.$$

$$\therefore \quad K = C - \frac{\mu\rho}{2}\left(\frac{x^2}{a^2}+\frac{y^2}{b^2}+\frac{z^2}{c^2}\right)$$

$$\Rightarrow \quad \frac{\mu\rho}{2}\left(\frac{x^2}{a^2}+\frac{y^2}{b^2}+\frac{z^2}{c^2}\right) = c - K$$

$$\Rightarrow \quad \frac{x^2}{a^2}+\frac{y^2}{b^2}+\frac{z^2}{c^2} = \frac{2(c-K)}{\mu\rho} = \lambda^2 \text{ (say)}$$

which are clearly similar ellipsoids.

Example 2: A given volume of a heavy fluid is at rest under the action of a force to a fixed point varying as the distance from that point. Find the pressure at the point (a, b, c).

Sol. Take the fixed point O as origin. Let OX, OY, OZ be a set of rectangular axes through O. Let the components of the given force be given by $-\mu x$, $-\mu y$, $-\mu z$ along OX, OY, OZ respectively.

Since the fluid is given to be heavy, therefore, its weight will act vertically downwards.

Let X, Y, Z be the components of the external force per unit mass of the fluid parallel to the axes.

Then,

$$X = -\mu x$$
$$Y = -\mu y$$
$$Z = -\mu z - g$$

–g has been taken in RHS as the fluid is heavy

The differential equation of pressure is $dp = \rho\,(X\,dx + Y\,dy + Z\,dz)$

$\Rightarrow dp = \rho\,\{-\mu\,x\,dx - \mu\,y\,dy - (\mu z + g)\,dz\}$

Integration yields,

$$p = -\rho\mu\frac{x^2}{2} - \rho\mu\frac{y^2}{2} - \rho\mu\frac{z^2}{2} - \rho\,gz + c$$

where c is an arbitrary constant of integration.

$$\Rightarrow \quad p = c - \frac{\rho\mu}{2}\left(x^2+y^2+z^2+\frac{2}{\mu}gz\right)$$

At the free surface, $p = 0$

$$\therefore \quad 0 = c - \frac{\rho\mu}{2}\left(x^2+y^2+z^2+\frac{2}{\mu}gz\right) \Rightarrow \frac{2c}{\rho\mu} = x^2+y^2+z^2+\frac{2}{\mu}gz$$

$$\Rightarrow \quad \frac{g^2}{\mu^2}+\frac{2c}{\rho\mu} = x^2+y^2+\left(z+\frac{g}{\mu}\right)^2 \quad \Rightarrow x^2+y^2+\left(z+\frac{g}{\mu}\right)^2 = \frac{g^2}{\mu^2}+\frac{2c}{\rho\mu}$$

This shows that the free surface is a sphere. Let the radius of the free surface sphere be r. Then,

$$r^2 = \frac{g^2}{\mu^2} + \frac{2c}{\rho\mu}$$

$$\Rightarrow \qquad c = \frac{\rho\mu}{2}\left(r^2 - \frac{g^2}{\mu^2}\right)$$

$\therefore$ We have

$$p = \frac{\rho\mu}{2}\left(r^2 - \frac{g^2}{\mu^2}\right) - \frac{g\mu}{2}\left(x^2 + y^2 + z^2 + \frac{2}{\mu}gz\right)$$

At the point (a, b, c),

$$p = \frac{g\mu}{2}\left(r^2 - \frac{g^2}{\mu^2} - a^2 - b^2 - c^2 - \frac{2}{\mu}gc\right)$$

This gives the pressure at the point (a, b, c).

Example 3: Find the surfaces of equal pressure for a homogeneous fluid acted upon by two forces which very as the inverse square of the distances from two fixed points.

Prove that if the surface of no pressure be a sphere, the loci of points at which the pressure varies inversely as the distance from one of the centres of forces are also spheres.

Sol.

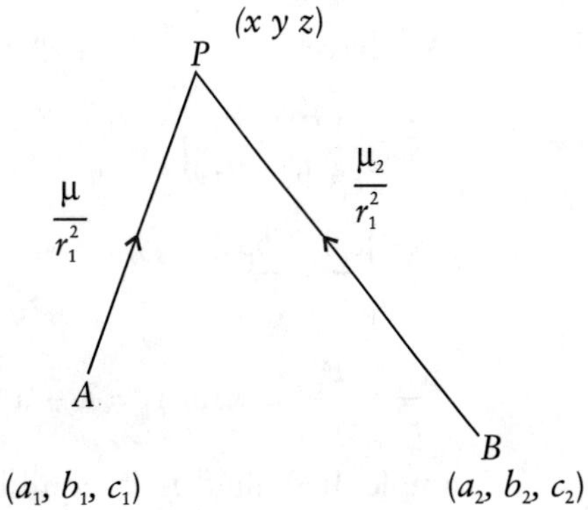

First Part: Let A (a_1, b_1, c_1) and B (a_2, b_2, c_2) be two fixed points. Let $P(x, y, z)$ be any point in the fluid.

Let $AP = r_1$ and $BP = r_2$. Then,

$$r_1^2 = (x - a_1)^2 + (y - b_1)^2 + (z - c_1)^2$$

and $\quad r_2^2 = (x - a_2)^2 + (y - b_2)^2 + (z - c_2)^2$

The differential equation of pressure at P is given by

$$dp = \rho\left(\frac{\mu_1}{r_1^2}dr_1 + \frac{\mu_2}{r_2^2}dr_2\right)$$

Integration yields,

$$p = c + \rho\left(\frac{\mu_1}{r_1} + \frac{\mu_2}{r_2}\right)$$

Surfaces of equal pressure are given by

$$p = \text{constant} = \lambda \text{ (say)}$$

$$\Rightarrow \quad \frac{\mu_1}{r_1} + \frac{\mu_2}{r_2} = \lambda$$

$$\Rightarrow \mu_1 r_2 + \mu_2 r_1 - \lambda r_1 r_2 = 0$$

Second Part: The surface of no pressure is given by $p = 0$

$$\Rightarrow \quad \rho\left(\frac{\mu_1}{r_1}+\frac{\mu_2}{r_2}\right)+c=0$$

This equation will represent a sphere when $c = 0$ as shown below.

$$\frac{\mu_1}{r_1}+\frac{\mu_2}{r_2}=0$$

$$\Rightarrow \mu_1 r_2 = -\mu_2 r_1$$

$$\Rightarrow \mu_1^2 r_2^2 = \mu_2^2 r_1^2$$

$$\Rightarrow \mu_1^2 [(x-a_2)^2+(y-b_2)^2+(z-c_2)^2] = \mu_2^2 [(x-a_1)^2+(y-b_1)^2+(z-c_1)^2]$$

which clearly represents a sphere.

Therefore, when the surface of equal pressure is a sphere, the pressure at any point of the liquid is given by

$$p=\rho\left(\frac{\mu_1}{r_1}+\frac{\mu_2}{r_2}\right) \quad |\because c=0 \text{ as shown above}$$

For points at which the pressure varies inversely as the distance from one of centres (say A) of forces,

$$p=\frac{\lambda}{r_1}$$

$\therefore$ We have

$$\frac{\lambda}{r_1}=\rho\left(\frac{\mu_1}{r_1}+\frac{\mu_2}{r_2}\right)$$

$$\Rightarrow \quad \frac{\lambda-\mu_1\rho}{r_1}=\frac{\rho\mu_2}{r_2}$$

$$\Rightarrow \quad \frac{A_1}{r_1}=\frac{B_1}{r_2}$$ where $A_1 = \lambda - \mu_1\rho$, $B_1 = \rho\mu_2$ which also represent spheres.

Example 4: A fluid is in equilibrium under a given system of forces. If $\rho_1 = \phi(x, y, z)$, $\rho_2 = \psi(x, y, z)$ be two possible values of the density at any point, show that the equations of the surfaces of equal pressure in either case are given by

$$\phi(x, y, z) + \lambda\,\psi(x, y, z) = 0, \text{ where } \lambda \text{ is an arbitrary parameter}$$

Sol. Let (X, Y, Z) be the given system of forces.

Let p_1 and p_2 be the pressures corresponding to the densities ρ_1 and ρ_2. Then, the differential equation of pressure gives

$$dp_1 = \rho_1 (X\,dx + Y\,dy + Z\,dz)$$

and $$dp_2 = \rho_2 (X\,dx + Y\,dy + Z\,dz)$$

$$\therefore \quad \frac{dp_1}{\rho_1} = \frac{dp_2}{\rho_2}$$

$$\Rightarrow \quad dp_1 = \frac{\rho_1}{\rho_2} dp_2$$

LHS is a perfect differential. For RHS to be a perfect differential, $\frac{\rho_1}{\rho_2}$ must be a function of p_2, say, $f(p_2)$.

If p_2 is constant, then,

$\frac{\rho_1}{\rho_2}$ is constant $= -\lambda$ (say)

$\therefore \quad \rho_1 + \lambda \rho_2 = 0$

$\Rightarrow \phi(x, y, z) + \lambda \psi(x, y, z) = 0$ $\quad | \because \rho_1 = \phi(x, y, z)$ and $\rho_2 = \psi(x, y, z)$

Example 5: Find the surfaces of equal pressure when a homogeneous liquid is subjected to a constant force in a fixed direction (gravity) and a force towards a fixed point and varying as the distance.

Sol. Let O be the fixed point. Let P be a point in the fluid at a distance r from O. μr is the force per unit mass towards O. Let g be the constant force per unit mass. Let the axis of Z be taken in the direction of g. Then, the equation of pressure is

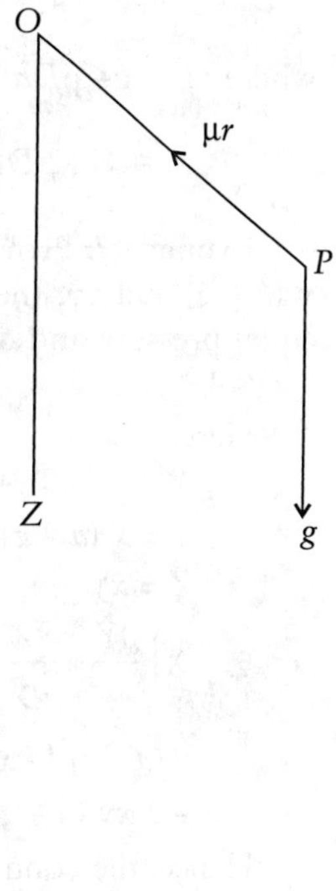

$$dp = \rho\,(g\,dz - \mu r\,dr)$$

| The negative sign before μr has been taken on account of the fact that it, being towards O, is opposite to the sense in which r increases.

Integration yields,

$$p = g\rho z - 1/2\ \mu\rho r^2 + c$$

where c is an arbitrary constant of integration.

$\Rightarrow p = g\rho z - 1/2\ \mu\rho\,(x^2 + y^2 + z^2) + c$

For surfaces of equal pressure,

$p = \text{constant} = \lambda$ (say)

$\therefore \quad g\rho z - 1/2\ \mu\rho\,(x^2 + y^2 + z^2) + c = \lambda$

$\Rightarrow \quad g\rho z - 1/2\ \mu\rho\,(x^2 + y^2 + z^2) = \lambda - c$

$\Rightarrow \quad g\rho z - 1/2\ \mu\rho\,(x^2 + y^2 + z^2) = \mu$ what $\mu = \lambda - c$ (constant)

This equation obviously represents concentric spheres for different values of μ.

Example 6: The particles of a sphere of homogeneous fluid of radius a are attracted to the centre with force inversely proportional to the distance from the centre. Prove that the pressure at a distance x from the surface (measured along a radius) is

$$C + D\,\rho\left(x + \frac{x^2}{2a} + \frac{x^3}{3a^2} + \ldots\right)$$

where C is a constant, *D* the force per unit mass at the surface of the sphere and ρ the constant density of the fluid.

Sol. Let *r* be the distance of a point in the fluid from the centre of the sphere and μ be the intensity of force. Then, the equation of pressure is

$$dp = -\rho\frac{\mu}{r}dr$$

Integration yields,

$p = -\rho\mu \log r + A$, where *A* is an arbitrary constant of integration.

Now, $\frac{\mu}{a} = D$

$\Rightarrow \quad \mu = Da$

$$\therefore\ p = -\rho Da \log r + A$$
$$= -\rho Da \log (a - x) + A$$
$$= -\rho Da \log a - \rho Da \log (1 - x/a) + A$$
$$= C + \rho Da\left(\frac{x}{a} + \frac{x^2}{2a^2} + \frac{x^3}{3a^3} + ...\right)$$

where $C = A - \rho Da \log a$ (constant)

$$= C + \rho D\left(x + \frac{x^2}{2a} + \frac{x^3}{3a^3} + ...\right)$$

Example 7: Prove that if the forces per unit mass at *x*, *y*, *z* parallel to the axes are $y(a - z)$, $x(a - z)$ and xy, the surfaces of equal pressures are hyperbolic paraboloids and the curves of equal pressure and density are rectangular hyperbolas.

Sol.

Here,

$$X = y(a - z)$$
$$Y = x(a - z)$$
$$Z = xy$$

$$\therefore\quad X\left(\frac{\partial Y}{\partial z} - \frac{\partial Z}{\partial y}\right) + Y\left(\frac{\partial Z}{\partial x} - \frac{\partial X}{\partial z}\right) + Z\left(\frac{\partial X}{\partial y} - \frac{\partial Y}{\partial x}\right)$$
$$= y(a - z)\{-x - x\} + x(a - z)(y + y) + xy\{(a - z) - (a - z)\}$$
$$= -2xy(a - z) + 2xy(a - z) = 0$$

Hence, the condition of equilibrium is satisfied.

Now, the surfaces of equal pressure are given by

p = constant

$\Rightarrow dp = 0 \qquad \Rightarrow \rho(Xdx + Ydy + Zdz) = 0$

$\Rightarrow Xdx + Ydy + Zdz = 0 \quad \Rightarrow y(a - z)\,dx + x(a - z)\,dy + xydz = 0$

Dividing throughout by $xy\,(a-z)$, we get

$$\frac{dx}{x}+\frac{dy}{y}+\frac{dz}{z}=0$$

Integrating, we get

$$\log x+\log y-\log(a-z)=\log c$$

where $\log c$ is an arbitrary constant of integration.

$$\Rightarrow\quad \log\frac{xy}{a-z}=\log c$$

$$\Rightarrow\quad \frac{xy}{a-z}=c$$

which are clearly hyperbolic paraboloids.

Again, the curves of equal pressure and density are given by

$$\frac{dx}{\dfrac{\partial Y}{\partial z}-\dfrac{\partial Z}{\partial y}}=\frac{dy}{\dfrac{\partial Z}{\partial x}-\dfrac{\partial X}{\partial z}}=\frac{dz}{\dfrac{\partial X}{\partial y}-\dfrac{\partial Y}{\partial x}}$$

$$\Rightarrow\quad \frac{dx}{-x-x}=\frac{dy}{y+y}=\frac{dz}{(a-z)-(a-z)}$$

$$\Rightarrow\quad \frac{dx}{-x}=\frac{dy}{y}=\frac{dz}{0}$$

The last member gives

$$dz=0$$

$$\Rightarrow\quad z=\text{constant}$$

Also, the first two fractions give

$$\frac{dx}{x}+\frac{dy}{y}=0$$

Integration gives,

$$\log x+\log y=\log c'$$

where $\log c'$ is an arbitrary constant of integration.

$$\Rightarrow\quad \log(xy)=\log c'$$

$$\Rightarrow\quad xy=c'\ \text{(constant)}$$

Thus, the curves of equal pressure and density are given by

$$xy=\text{constant},\ z=\text{constant}$$

which are clearly rectangular hyperbolas.

Aliter: We have

$$dp=\rho\,[Xdx+Ydy+Zdz]$$

$$=\rho\,[y\,(a-z)\,dx+x\,(a-z)\,dy+xydz]$$

$$=\rho\,[(a-z)\,(ydx+xdy)+xydz]$$

$$= \rho\,[(a-z)d(xy) - xyd\,(a-z)]$$

$$= \rho\left[\frac{(a-z)\,d(xy) - xy\,d(a-z)}{(a-z)^2}\right](a-z)^2$$

$$= \rho\,(a-z)^2\,d\left(\frac{xy}{a-z}\right) \qquad ...(1)$$

Since LHS is a perfect differential, therefore, RHS also must be a perfect differential which is possible only if

$$\rho\,(a-z)^2 = \text{constant} = c_1 \text{ (say)} \qquad ...(2)$$

Then, integrating (1), we get

$p = c_1\left(\dfrac{xy}{a-z}\right) + c_2$, where c_2 is an arbitrary constant of integration.

Now, the surfaces of equal pressure are given by

$$p = \text{constant} = c_3 \text{ (say)}$$

$$\Rightarrow \quad c_1\left(\frac{xy}{a-z}\right) + c_2 = c_3$$

$$\Rightarrow \quad \frac{xy}{a-z} = \frac{c_3 - c_2}{c_1} = \text{constant} = c_4 \text{ (say)} \qquad ...(3)$$

which are clearly hyperbolic paraboloids.

Again, when ρ = constant, then from (2), we get $(a-z)^2$ = constant

$$\Rightarrow \quad z = \text{constant} \qquad ...(4)$$

(3) and (4) give

$$xy = \text{constant} \qquad ...(5)$$

Hence, the curves of equal pressure and density are given by

$$xy = \text{constant},\ z = \text{constant}$$

which are clearly rectangular hyperbolas.

Example 8: If $X = y\,(y+z)$, $Y = z\,(z+x)$, $Z = y\,(y-x)$, show that the surfaces of equal pressure are the hyperbolic paraboloids

$$y\,(x+z) = c\,(y+z)$$

and the curves of equal pressure and density are given by the equations

$$y\,(x+z) = \text{constant and}$$

$$y + z = \text{constant}$$

Or

If $X = y\,(y+z)$, $Y = z\,(z+x)$, $Z = y\,(y-x)$, then find surfaces of equal pressure and the curves of equal pressure and density.

Sol. The equation of pressure is

$$dp = \rho\,(Xdx + Ydy + Zdz]$$

$$\Rightarrow \quad dp = \rho\,[y\,(y+x)\,dx + z\,(z+x)\,dy + y\,(y+x)\,dz]$$

For surfaces of equal pressure,

p = constant

$\Rightarrow dp = 0$

$\Rightarrow y(y+z)\,dx + z(z+x)\,dy + y(y-x)\,dz = 0$

$$\Rightarrow \frac{dx}{z+x}+\frac{zdy}{y(y+z)}+\frac{(y-x)dz}{(y+z)(z+x)} \quad | \text{ dividing throughout by } y\,(y+z)\,(z+x)$$

$$\Rightarrow \frac{dx}{z+x}+\frac{z\,dy}{y(y+z)}+\frac{(y+z)-(z+x)}{(y+z)(z+x)}dz=0$$

$$\Rightarrow \frac{dx}{z+x}+\frac{zdy}{y(y+z)}+\frac{dz}{z+x}-\frac{dz}{y+z}=0$$

$$\Rightarrow \frac{dx+dz}{z+x}+\frac{zdy-ydz}{y(y+z)}=0$$

$$\Rightarrow \frac{dx+dz}{x+z}+\frac{(y+z)dy-y(dy+dz)}{y(y+z)}=0$$

| adding and subtracting ydy in the numerator of the second term

$$\Rightarrow \frac{d(x+z)}{x+z}+\frac{dy}{y}-\frac{d(y+z)}{y+z}=0$$

Integration yields,

$$\log(x+z)+\log y-\log(y+z)=\log c$$

where $\log c$ is an arbitrary constant of integration.

$$\Rightarrow \log\frac{(x+z)y}{y+z}=\log c$$

$$\Rightarrow \frac{y(x+z)}{y+z}=c$$

$\Rightarrow y(x+z)=c(y+z)$

This equation represents surfaces of equal pressure for different values of c.

These surfaces are clearly hyperbolic paraboloids.

The curves of equal pressure and density are given by the differential equations

$$\frac{dx}{\dfrac{\partial Y}{\partial z}-\dfrac{\partial Z}{\partial y}}=\frac{dy}{\dfrac{\partial Z}{\partial x}-\dfrac{\partial X}{\partial z}}=\frac{dz}{\dfrac{\partial X}{\partial y}-\dfrac{\partial Y}{\partial x}}$$

$$\Rightarrow \frac{dx}{(2z+x)-(2y-x)}=\frac{dy}{-y-y}=\frac{dz}{2y+z-z}$$

$$\Rightarrow \frac{dx}{2x-2y+2z}=\frac{dy}{-2y}=\frac{dz}{2y} \Rightarrow \frac{dx}{x-y+z}=\frac{dy}{-y}=\frac{dz}{y}$$

Again, using the multipliers y, $x + z$, y respectively, we have

$$ydx + (x + z)\,dy + ydz$$
$$= y\,(x - y + z) - y\,(x + z) + y^2 = 0$$
$$\Rightarrow (ydx + xdy) + (zdy + ydz) = 0$$
$$\Rightarrow d\,(yx) + d\,(zy) = 0$$

Integrating, we get

$$xy + yz = \text{constant}$$
$$\Rightarrow y\,(x + z) = \text{constant}$$

Last two fractions give

$$dy + dz = 0$$

Integrating, we get

$$y + z = \text{constant}$$

Hence, the curves of equal pressure and density are given by

$$y\,(x + z) = \text{constant},\ y + z = \text{constant}$$

Example 9: Show that the force represented by $X = \mu\,(y^2 + yz + z^2)$, $Y = \mu\,(z^2 + zx + x^2)$, $Z = \mu\,(x^2 + xy + y^2)$ will keep a mass of liquid at rest, if density varies inversely as (distance)^2 from the plane $x + y + z = 0$. Show also that the curves of equal pressure and density will be circles.

Sol. Let d be the distance of the point (x, y, z) from the plane $x + y + z = 0$. Then,

$$d = \frac{x+y+z}{\sqrt{1+1+1}} = \frac{x+y+z}{\sqrt{3}} \qquad ...(1)$$

Let ρ be the density. Then,

$$\rho \propto \frac{1}{d^2}$$

$$\Rightarrow \rho = \frac{\lambda}{d^2} \qquad \Rightarrow \rho = \frac{3\lambda}{(x+y+z)^2}$$

$$\Rightarrow \rho = \frac{\lambda'}{(x+y+z)^2} \text{ where } \lambda' = 3\lambda \text{ (constant)} \qquad ...(2)$$

The conditions of equilibrium (Refer § 6) of the mass of liquid are

$$\frac{\partial}{\partial y}(\rho Z) = \frac{\partial}{\partial z}(\rho Y) \qquad ...(3)$$

$$\frac{\partial}{\partial x}(\rho Z) = \frac{\partial}{\partial z}(\rho X) \qquad ...(4)$$

and $$\frac{\partial}{\partial y}(\rho X) = \frac{\partial}{\partial x}(\rho Y) \qquad ...(5)$$

Now,

$$\frac{\partial}{\partial y}(\rho Z)=\frac{\partial}{\partial y}\left[\frac{\lambda'}{(x+y+z)^2}\mu(x^2+xy+y^2)\right]$$

$$=\lambda'\mu\left[\frac{(x+y+z)^2(x+2y)-(x^2+xy+y^2)2(x+y+z)}{(x+y+z)^4}\right]$$

$$=\frac{\lambda'\mu}{(x+y+z)^3}\left[(x+y+z)(x+2y)-2(x^2+xy+y^2)\right]$$

$$=\frac{\lambda'\mu}{(x+y+z)^3}[-x^2+2yz+xz+xy] \qquad ...(6)$$

and

$$\frac{\partial}{\partial z}(\rho Y)=\frac{\partial}{\partial z}\left[\frac{\lambda'}{(x+y+z)^2}\mu(x^2+zx+x^2)\right]$$

$$=\lambda'\mu\left[\frac{(x+y+z)^2(2z+x)-(z^2+zx+x^2)2(x+y+z)}{(x+y+z)^4}\right]$$

$$=\frac{\lambda'\mu}{(z+y+z)^3}\left[(x+y+z)(2z+x)-2(z^2+zx+x^2)\right]$$

$$=\frac{\lambda'\mu}{(z+y+z)^3}\left[-x^2+2yz+xz+xy\right] \qquad ...(7)$$

From (6) and (7), we have

$$\frac{\partial}{\partial y}(\rho Z)=\frac{\partial}{\partial z}(\rho Y)$$

Similarly, we can show that

$$\frac{\partial}{\partial x}(\rho Z)=\frac{\partial}{\partial z}(\rho X)$$

and

$$\frac{\partial}{\partial y}(\rho X)=\frac{\partial}{\partial x}(\rho Y)$$

Hence, the given forces will keep the mass of liquid at rest if ρ is given by (2).

Again, the curves of equal pressure and density are given by

$$\frac{dx}{\frac{\partial Y}{\partial z}-\frac{\partial Z}{\partial y}}=\frac{dy}{\frac{\partial Z}{\partial x}-\frac{\partial X}{\partial z}}=\frac{dz}{\frac{\partial X}{\partial y}-\frac{\partial Y}{\partial x}}$$

$$\Rightarrow \quad \frac{dx}{(x+2z)-(x+2y)}=\frac{dy}{(2x+y)-(y+2z)}=\frac{dz}{(2y+z)-(z+2x)}$$

$$\Rightarrow \quad \frac{dx}{2(z-y)} = \frac{dy}{2(x-z)} = \frac{dz}{2(y-x)} \quad \Rightarrow \frac{dx}{z-y} = \frac{dy}{x-z} = \frac{dz}{y-x}$$

$$= \frac{dx+dy+dz}{0} = \frac{xdx+ydy+zdz}{0}$$

$\Rightarrow \quad dx + dy + dz = 0$

and $\quad xdx + ydy + zdz = 0$

$\Rightarrow \quad x + y + z = \text{constant}$ | on integration ...(8)

and $\quad x^2 + y^2 + z^2 = \text{constant}$ ...(9)

Equations (8) and (9), taken together, give the curves of equal pressure and density. Equation (8) represents a plane while equation (9) represents a sphere.

Since the intersection of a sphere by a plane is a circle, therefore, the curves of equal pressure and density will be circles.

Aliter: The differential equation of pressure is

$$dp = \rho\,(X\,dx + Y\,dy + Z\,dz)$$

$$= \rho\mu\,[(y^2 + yz + z^2)\,dx + (z^2 + zx + z^2)\,dy + (z^2 + xy + y^2)\,dz]$$

$$= [\{(x + y + z)\,(y + z) - x\,(y + z) - yz\} + \{(x + y + z)\,(z + x) - y\,(z + x) - zx\}\,dy + \{(x + y + z)\,(x + y) - z\,(x + y) - xy\}dz]$$

$$= \rho\mu\,[(x + y + z)\,\{(y + z)\,dx + (z + x)\,dy + (x + y + z)\,dz\} - (yz + zx + xy)\,(dx + dy + dz)]$$

$$= \rho\mu\,[(x + y + z)\,d\,(xy + yz + zx) - (yz + zx + xy)\,(dx + dy + dz)]$$

$$= \rho\mu(x+y+z)^2\left[\frac{(x+y+z)d(xy+yz+zx)-(yz+zx+xy)d(x+y+z)}{(x+y+z)^2}\right]$$

$$= \rho\mu(x+y+z)^2\,d\left[\frac{yz+zx+xy}{x+y+z}\right] \quad ...(1)$$

The R.H.S. of equation (1) will be a perfect differential if

$$\rho\mu\,(x + y + z)^2 = \text{constant} = k \text{ (say)}$$

$$\Rightarrow \quad \rho = \frac{k}{\mu(x+y+z)^2}$$

$$= \frac{k}{3\mu\left(\dfrac{x+y+z}{\sqrt{3}}\right)^2}$$

$$= \frac{k}{3\mu d^2} \qquad \left|\because d = \frac{x+y+z}{\sqrt{3}}\right.$$

$$= \frac{\lambda}{d^2} \text{ where } \lambda = \frac{k}{3\mu} \text{ (constant)}$$

$\Rightarrow\ \rho\alpha\frac{1}{d^2}$, where d is the distance of the point (x, y, z) from the plane $x + y + z = 0$.

Hence, the mass of liquid will remain in equilibrium of $\rho\alpha\frac{1}{d^2}$; d specified above.

Again, the curves of equal pressure and density are given by ρ = constant and p = constant.

Now, $\quad \rho = \text{constant}$

$$\Rightarrow \quad x + y + z = \text{constant} \qquad ...(3)$$

and $\quad p = \text{constant}$

$$\Rightarrow \quad dp = 0$$

$$\Rightarrow \quad d\left(\frac{yz + zx + xy}{x + y + z}\right) = 0 \qquad | \text{ From (1)}$$

$$\Rightarrow \quad \frac{yz + zx + xy}{x + y + z} = \text{constant}$$

$$\Rightarrow \quad yz + zx + xy = \text{constant} \qquad |\text{From (3)} \qquad ...(4)$$

Hence, the curves of equal pressure and density are given by

$$x + y + z = \text{constant}$$

and $\quad yz + zx + xy = \text{constant}$

$$\Rightarrow \quad x + y + z = \text{constant}$$

and $\quad x^2 + y^2 + z^2 = \text{constant} \qquad | \because (x + y + z)^2 = x^2 + y^2 + z^2 + 2(xy + yz + zx)$

which represent a set of circles as explained in first method given above.

Example 10: A fluid rests in equilibrium in a field of force $X = y^2 + z^2 - xy - xz$, $Y = z^2 + x^2 - zy - xy$, $Z = x^2 + y^2 - xz - yz$. Show that the curves of equal pressure and density are a set of circles.

Sol. The differential equations of the curves of equal pressure and density are

$$\frac{dx}{\frac{\partial Y}{\partial z} - \frac{\partial Z}{\partial y}} = \frac{dy}{\frac{\partial Z}{\partial x} - \frac{\partial X}{\partial z}} = \frac{dz}{\frac{\partial X}{\partial y} - \frac{\partial Y}{\partial x}}$$

$$\Rightarrow \quad \frac{dx}{(2z - y) - (2y - z)} = \frac{dy}{(2x - z) - (2z - x)} = \frac{dz}{(2y - x) - (2x - y)}$$

$$\Rightarrow \quad \frac{dx}{y - z} = \frac{dy}{z - x} = \frac{dz}{x - y} = \frac{dx + dy + dz}{0} = \frac{x\,dx + y\,dy + z\,dz}{0}$$

$$\Rightarrow \quad dx + dy + dz = 0$$

and $\quad xdx + ydy + zdz = 0$

$$\Rightarrow \quad x + y + z = \text{constant} \qquad | \text{ on integration} \qquad ...(1)$$

and $\quad x^2 + y^2 + z^2 = \text{constant} \qquad ...(2)$

Equations (1) and (2), taken together, give the curves of equal pressure and density. Equation (1) represents a plane whereas equation (2) represents a sphere.

Since the intersection of a sphere by a plane is a circle, therefore, the curves of equal pressure and density are circles.

Example 11: A mass of fluid is at rest under the forces $X = (y + z)^2 - x^2$, $Y = (z + x)^2 - y^2$, $Z = (x + y)^2 - z^2$. Find the density and the surfaces of equal pressure. Also, show that the curves of equal pressure and density are a set of circles.

Sol. We have,

$$X\left(\frac{\partial Y}{\partial z}-\frac{\partial Z}{\partial y}\right)+Y\left(\frac{\partial Z}{\partial x}-\frac{\partial X}{\partial z}\right)+Z\left(\frac{\partial X}{\partial y}-\frac{\partial Y}{\partial x}\right)$$

$$= \{(y + z)^2 - x^2\} \{2 (y + z) - 2 (x + y)\} + \{(z + x)^2 - y^2\} \{2 (x + y) - 2 (y + z)\} + \{(x + y)^2 - z^2\} \{2 (y + z) - 2 (z + x)\}$$

$$= 2[\{(y + z)^2 - x^2\} (z - y) + \{(z + x)^2 - y^2\} (x - z) + \{(x + y)^2 - z^2\} (y + x)]$$

$$= 2(x + y + z) [(y + z - x) (z - y) + (z + x - y) (x - z) + (x + y - z) (y - x)]$$

$$= 2(x + y + z) [- \{x (z - y) + y (x - z) + z (y - x)\} - \{(y + z) (y - z) + (z + x) (z - x) + (x + y) (x - y)\}] = 0$$

Hence, the condition of equilibrium is satisfied. Hence, the fluid is in equilibrium.

We know that the equation of pressure is

$$dp = \rho (X\,dx + Y\,dy + Z\,dz)$$

$$= \rho [\{(y + z)^2 - x^2\} dx + \{(z + x)^2 - y^2\} dy + \{(x + y)^2 - z^3\} dz]$$

$$= \rho (x + y + z) \{(y + z - x) dx + (z + x - y) dy + (x + y - z) dz\}$$

$$= \rho (x + y + z) \{(ydx + xdy) + (zdx + xdz) + (zdy + ydz) - xdx - ydy - zdz\}$$

$$= \rho (x + y + z) \left\{d (xy) + d (zx) + d (yz) - d\left(\frac{x^2}{2}\right)-d\left(\frac{y^2}{2}\right)-d\left(\frac{z^2}{2}\right)\right\}$$

$$= \frac{-\rho(x+y+z)}{2}\, d (x^2 + y^2 + z^2 - 2xy - 2yz - 2zx) \quad \text{...(1)}$$

Since LHS is a perfect differential, therefore, RHS must also a perfect differential. This is possible only when

$\frac{-\rho(x+y+z)}{2}$ is a constant, say c.

Then, $\frac{-\rho(x+y+z)}{2} = c$

$$\Rightarrow \quad \rho = -\frac{2c}{x+y+z} \qquad \Rightarrow \rho \alpha \frac{1}{x+y+z} \quad \text{...(2)}$$

Also, (1) gives

$$dp = c\, d (x^2 + y^2 + z^2 - 2xy - 2yz - 2zx)$$

Integration yields,

$$p = c (x^2 + y^2 + z^2 - 2xy - 2yz - 2zx) + \lambda \quad \text{...(3)}$$

where λ is an arbitrary constant of integration.

For surface of equal pressure,

$$p = \text{constant}$$

$$\Rightarrow \quad x^2 + y^2 + z^2 - 2xy - 2yz - 2zx = \text{constant} \quad |\text{From (3)} \quad ...(4)$$

Equation (4) represents the equation of the surfaces of equal pressure.

Again, the curves of equal pressure and density are given by

$$p = \text{constant}$$

and $$\rho = \text{constant}$$

$$\Rightarrow \quad x^2 + y^2 + z^2 - 2xy - 2yz - 2zx = \text{constant} \quad |\text{From (4)} \quad ...(5)$$

and $$x + y + z = \text{constant} \quad |\text{From (2)} \quad ...(6)$$

We know that

$$(x + y + z)^2 = x^2 + y^2 + z^2 + 2(xy + yz + zx)$$

$$\Rightarrow \quad (x + y + z)^2 = 2(x^2 + y^2 + z^2) - \text{constant} \quad |\text{From (5)}$$

$$\Rightarrow \quad x^2 + y^2 + z^2 = \text{constant} \quad |\text{From (6)} \quad ...(7)$$

Hence, the curves of equal pressure and density and given by

$$x + y + z = \text{constant}$$

and $$x^2 + y^2 + z^2 = \text{constant}$$

which represent a set of circles as explained in Example 9.

Example 12: If the components parallel to the axes of the forces acting on an element of fluid at (x, y, z) be proportional to

$$y^2 + 2\lambda yz + z^2,\ z^2 + 2\mu zx + z^2,\ x^2 + 2\partial xy + y^2,$$

show that if equilibrium be possible, we must have $2\lambda = 2\mu = 2\partial = 1$

Sol. Let X, Y, Z be the components of forces per unit mass acting on an element of fluid at (x, y, z) parallel to the axes of the forces.

Then,

$$X = k(y^2 + 2\lambda yz + z^2)$$

$$Y = k(z^2 + 2\mu zx + x^2)$$

$$Z = k(x^2 + 2\partial xy + y^2)$$

The condition of equilibrium is

$$X\left(\frac{\partial Y}{\partial z} - \frac{\partial Z}{\partial y}\right) + Y\left(\frac{\partial Z}{\partial x} - \frac{\partial X}{\partial z}\right) + Z\left(\frac{\partial X}{\partial y} - \frac{\partial Y}{\partial x}\right) = 0$$

$$\Rightarrow \quad k^2\,[(y^2 + 2\lambda yz + z^2)\,\{2z + 2\mu z - 2y - 2\partial x\}$$
$$+ (z^2 + 2\mu zx + x^2)\,\{2x + 2\partial y - 2\lambda y - 2z\} +$$
$$(x^2 + 2\partial xy + z^2)\,\{2y + 2\lambda z - 2\mu z - 2x\}] = 0$$

$$\Rightarrow \quad (y^2 + 2\lambda yz + z^2)\,\{(z - y) + (\mu - \partial)\,x\} +$$
$$(z^2 + 2\mu zx + x^2)\,\{(x - z) + (\partial - \lambda)\,y\} +$$
$$(x^2 + 2\partial xy + z^2)\,\{(y - x) + (\lambda - \mu)\,z\} = 0$$

$$\Rightarrow \{(y^2 + yz + z^2) + (2\lambda - 1)\, yz\}\, \{(z - y) + (\mu - \partial)\, x\} +$$
$$\{(z^2 + zx + x^2) + (2\mu - 1)\, zx\}\, \{(x - z) + (\partial - \lambda)\, y\} +$$
$$\{(x^2 + xy + x^2) + (2\partial - 1)\, xy\}\, \{(y - x) + (\lambda - \mu)\, z\} = 0$$
$$\Rightarrow (z^2 - y^3 + x^3 - z^3 + y^3 - x^2) - [(2\lambda - 1)\, yz\, (y - z) + (2\mu - 1)\, zx\, (z - x) +$$
$$(2\partial - 1)\, xy\, (x - y)] + [(\mu - \partial)\, x\, (y^2 + z^2) +$$
$$(\partial - \lambda)\, y\, (z^2 + x^2) + (\lambda - \mu)\, z\, (x^2 + y^2)]$$
$$+ xyz\, [\mu - \partial + \partial - \lambda + \lambda - \mu]$$
$$+ 2xyz\, [\lambda(\mu - \partial) + \mu\, (\partial - \lambda) + \partial\, (\lambda - \mu)]$$
$$- xyz\, [\mu - \partial + \partial - \lambda + \lambda - \mu] = 0$$
$$\Rightarrow y^2z\, (-2\lambda + 1 + \lambda - \mu) +$$
$$yz^2\, (2\lambda - 1 + \partial - \lambda) +$$
$$z^2x\, (-2\mu + 1 + \mu - \partial) +$$
$$zx^2\, (2\mu - 1 + \lambda - \mu) +$$
$$x^2y\, (-2\partial - 1 + \partial - \lambda) +$$
$$xy^2\, (2\partial - 1 + \mu - \partial) = 0$$
$$\Rightarrow (\lambda + \mu - 1)\, (y^2z - x^2z) + (\mu + \partial - 1)\, (z^2x - xy^2) + (\partial + \lambda - 1)\, (x^2y - z^2y) = 0$$
$$\Rightarrow z(y^2 - x^2)\, (\lambda + \mu - 1) + x(z^2 - y^2)\, (\mu + \partial - 1) + y(x^2 - z^2)\, (\partial + \lambda - 1) = 0$$

Now, for this equation to be true for all values of x, y, z, we must have

$$\lambda + \mu - 1 = 0$$
$$\mu + \partial - 1 = 0$$
$$\partial + \lambda - 1 = 0$$

Solving these equations, we get

$$\lambda = \mu = \partial = 1/2$$
$$\Rightarrow \quad 2\lambda = 2\mu = 2\partial = 1$$

EXERCISE

1. If $X = y^2 + yz + z^2$, $Y = z^2 + zx + x^2$, $Z = x^2 + xy + y^2$, find the curves of equal pressure and density.

   Ans. A set of circles given by

   $$x^2 + y^2 + z^2 = \text{constant and } x + y + z = \text{constant}$$

2. A straight rod AB, every particle of which attracts with a force varying inversely as the square of the distance, is surrounded by a mass of homogeneous incompressible fluid. Show that the surfaces of equal pressure are ellipsoids of revolution.

3. In a solid sphere two spherical cavities, where radii are equal to half the radius of the solid sphere, are filled with liquid. The solid and liquid particles attract each other with forces which vary as the distance. Prove that the surfaces of equal pressure are spheres concentric with the solid sphere.

Chapter

5

Pressure on Plane Surfaces

1. Resultant Thrust or Resultant Fluid Pressure

We know that when a surface, plane or curved, is in contact with a fluid, then, the fluid exerts pressure on every element of it in a direction which is normal to the surface of the element. If the surface is a plane area, then, the pressure at any point of it is perpendicular to that plane area. The pressures at all the points on one side of the plane area constitute a system of like parallel forces whose magnitudes are known. All these parallel forces can be compounded into a single force acting at some definite point of the plane area, under consideration, whose magnitude is the arithmetical sum of all the parallel forces. This single force is called the resultant fluid pressure or resultant thrust or total thrust of the fluid on the plane surface.

Since its magnitude is the arithmetic sum of all the thrust on the various elements of the plane surface, therefore, the resultant fluid thrust on a plane area is termed as the *whole pressure*.

Thus, whole pressure may be defined as follows:

The whole pressure of a fluid on any surface in contact with it is the arithmetical sum of all the normal pressures exerted by the fluid on every element of the surface in contact with the fluid.

Note 1: If the surface under consideration is the curved surface, then, the thrusts at different elements of it will not form a system of parallel forces, i.e., they are not in the same direction. The whole pressure in this case also will be the arithmetical sum of all the thrusts at various elements of the curved surface.

However, this whole pressure, has no physical significance whatsoever and dose not solve any practical purpose.

Note 2: Students must clearly note the difference between whole pressure on plane surface and whole pressure on curved surface. In case of a plane surface, the whole pressure and the resultant pressure coincide. But in case of a curved surface, these two are entirely different.

Note 3: The method of finding out the resultant thrust on a curved surface will be taken into study in a chapter ahead.

2. Theorem

The whole pressure (or the resultant thrust) of a heavy homogeneous fluid on a plane area in contact with the fluid is equal to the product of the area of the plane surface and the pressure at its centre of gravity of the plane area.

Proof: Let the plane surface S in contact with the fluid be divided into an indefinitely large number of very small elementary areas $\alpha_1, \alpha_2, \alpha_3, ...$ whose depths from the free surface are $z_1, z_2, z_3, ...$ respectively.

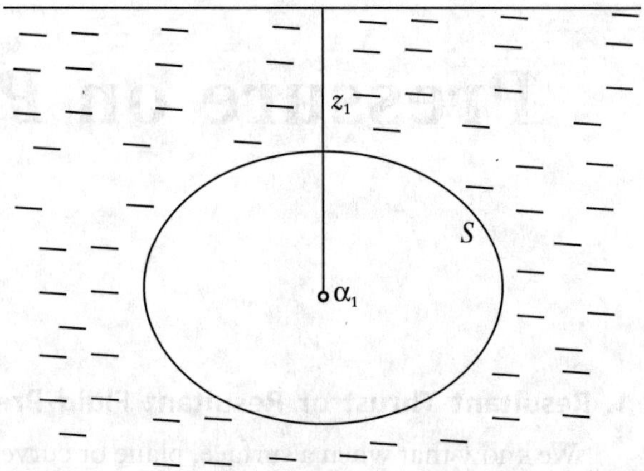

Since the elements are very small, the pressure at each element may be taken to be uniform. Therefore, pressures on these elements will be respectively.

$$w\alpha_1 z_1, w\alpha_2 z_2, w\alpha_3 z_3, ...$$

where w is the weight per unit volume of the fluid.

$\therefore$ Whole pressure on the plane area S

$=$ Arithmetical sum of the pressures on each element of the area S

$= w\alpha_1 z_1 + w\alpha_2 z_2 + w\alpha_3 z_3 + ...$

$= w(\alpha_1 z_1 + \alpha_2 z_2 + \alpha_3 z_3 + ...)$ ...(1)

But, from Statics, we know that if $\bar{z}$ be the depth of the centre of gravity of the given plane area, then,

$$\bar{z} = \frac{\alpha_1 z_1 + \alpha_2 z_2 + \alpha_3 z_3 + ...}{\alpha_1 + \alpha_2 + \alpha_3 + ...}$$

$$= \frac{\alpha_1 z_1 + \alpha_2 z_2 + \alpha_3 z_3 + ...}{S}$$

$$| \because S = \alpha_1 + \alpha_2 + \alpha_3 + ...$$

$$\Rightarrow \alpha_1 z_1 + \alpha_2 z_2 + \alpha_3 z_3 + ... = S\bar{z} \quad ...(2)$$

From (1) and (2), whole pressure on the plane area S

$= w\, S\, \bar{z}$

$= (w\, \bar{z})\, S$

$=$ Pressure at the C.G. × Area of plane surface

Note 1: The whole pressure (resultant thrust) on a plane surface depends only on the area of the surface and the position of the centroid of the plane surface.

Note 2: If a plane surface is immersed in a fluid, then the whole pressure (or whole thrust) on it is equal to $w\, S\, \bar{z}$, where,

w = weight per unit volume of the fluid

S = area of the plane surface

and $\bar{z}$ = depth of the centre of gravity of the plane area below the free surface (i.e., surface of zero pressure)

Note 3: The whole pressure on a plane area immersed in a fluid is equal to the product of weight per unit volume of the fluid, area of the plane surface immersed in the fluid and the depth of its C.G. below the free surface.

Note 4: The whole pressure of the fluid on a surface is equal to the weight of the column of fluid whose base is equal to the area of the given plane surface, and whose height is equal to the depth below the free surface of the fluid of the centre of gravity of the given plane surface in contact with the fluid.

Note 5: If the atmospheric pressure is taken into consideration, then pressure on the element α_1

$$= (\pi + w\, z_1)\, \alpha_1$$

and similarly for other elements.

Therefore, whole pressure

$$= \pi\, (\alpha_1 + \alpha_2 + \alpha_3 + \ldots.) + \pi\, (z_1\alpha_1 + z_2\alpha_2 + z_3\alpha_3 + \ldots.)$$

$$= \pi\, S + w\, S\, \bar{z}$$

$$= (\pi + w\, \bar{z})\, S$$

Let $\pi = wh_1$ where h_1 is the height of the effective surface above the free surface, then the whole pressure

$$= (wh_1 + w\bar{z})S$$

$$= w(h_1 + \bar{z})S$$

$$= w\, S\, \bar{z}$$

where $\bar{z} = h_1 + z$ is depth of the C.G. of the plane area below the effective surface.

3. Layers of Different Liquids

Let the liquid consist of layers of different immiscible n liquids of densities $\rho_1, \rho_2, \rho_3, \ldots, \rho_n$ and thickness $h_1, h_2, h_3, \ldots., h_n$ respectively beginning from the top where the pressure is zero. Let the plane area be entirely in contact with the lowest liquid. We divide the whole area into large number of elementary areas $\alpha_1, \alpha_2, \alpha_3, \ldots..$ Let $z_1, z_2, z_3, \ldots.$ be the depths of the elementary areas $\alpha_1, \alpha_2, \alpha_3, \ldots$ below the lowest surface of separation. Then, pressure on the element α_1

$$= g\rho_1 h_1\alpha_1 + g\rho_2 h_2\alpha_1 + g\rho_3 h_3\alpha_1 + \ldots\ldots$$

$$\ldots\ldots\ldots + g\rho_{n-1}h_{n-1}\alpha_1 + g\rho_n z_1\alpha_1$$

Similarly, pressure on the element α_2

$$= g\rho_1 h_1\alpha_2 + g\rho_2 h_2\alpha_2 + g\rho_3 h_3\alpha_2 + \ldots\ldots$$

$$\ldots\ldots\ldots + g\rho_{n-1}h_{n-1}\alpha_2 + g\rho_n z_2\alpha_2$$

and so on.

Therefore, whole pressure on the plane area

$$= g\rho_1 h_1\, (\alpha_1 + \alpha_2 + \alpha_3 + \ldots\ldots) +$$

$$g\rho_2 h_2\, (\alpha_1 + \alpha_2 + \alpha_3 + \ldots\ldots) +$$

$$g\rho_3 h_3\, (\alpha_1 + \alpha_2 + \alpha_3 + \ldots\ldots) +$$

- - - - - - - - - - - - - - - - - - -

- +

$$g\rho_{n-1}h_{n-1}(\alpha_1 + \alpha_2 + \alpha_3 + ...) +$$
$$g\rho_n(z_1\alpha_1 + z_2\alpha_2 + z_3\alpha_3 + ...)$$
$$= g\rho_1 h_1 S + g\rho_2 h_2 S + g\rho_3 h_3 S + ... + g\rho_{n-1}h_{n-1}S + g\rho_n S\bar{z} \qquad |\text{As in § 2}$$

where $\bar{z}$ is the depth of the C.G. of the plane area below the lowest surface of separation.

$$= (g\rho_1 h_1 + g\rho_2 h_2 + g\rho_3 h_3 + ... + g\rho_{n-1}h_{n-1} + g\rho_n\bar{z})S$$

Note: If the atmospheric pressure is taken into account, then the whole pressure π on the plane area

$$= (\pi + g\rho_1 h_1 + g\rho_2 h_2 + g\rho_3 h_3 + ... +$$
$$g\rho_{n-1}h_{n-1} + g\rho_n\bar{z})S$$

4. A Plane Area in Contact with more than one Fluid

If a plane area is in contact with more than one fluids, then, we divide the plane area into several parts such not each of these parts is in contact with one fluid only. Then, the pressure on each part is calculated by using the formula derived in § 3. The sum of these pressures gives the required whole pressure on the plane area.

5. Whole Thrust on a Horizontal Base

We know that the pressure at every point of a horizontal plane in contact with a fluid is the same. Therefore, the whole pressure on a horizontal base will be equal to the product of the pressure at any point of the base and the area of the base. It follows from here that the whole pressure of the fluid on the horizontal base of a vessel which contains it, does not depend on the shape of the vessel or upon the amount of liquid contained in it but only on the area of the base and the depth of the base below the effective surface.

Below are shown four vessels whose base areas and the heights are the same but the amounts of the liquids in them and their shapes are different. But the thrust on the base of each of the vessels is the same.

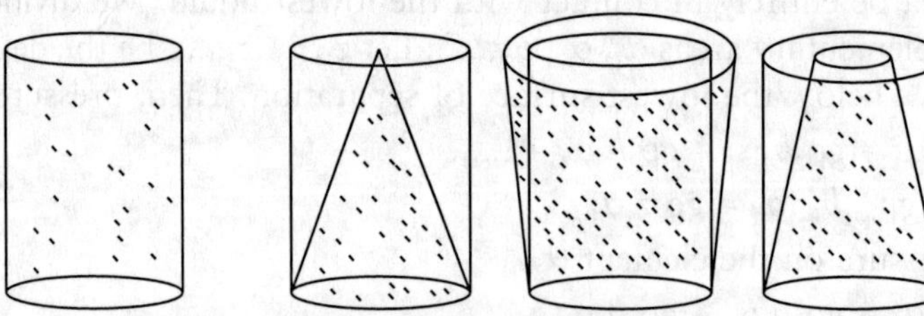

ILLUSTRATIVE EXAMPLES

Example 1: Find the whole pressure on a triangle, the depths of whose vertices are α, β and γ, the liquid being homogeneous and S being the area of the triangle.

Sol. Area of the triangle = S

Depth of C.G. of the triangle below the free surface.

$$= \frac{\alpha + \beta + \gamma}{3}$$

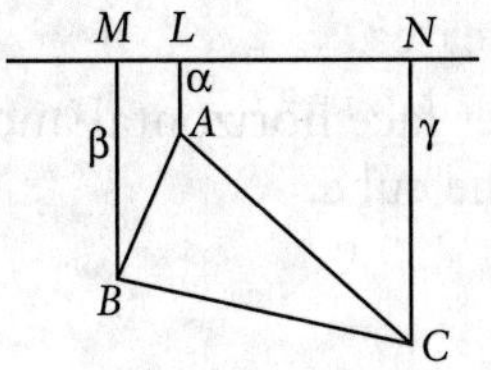

Let w be the weight per unit volume of the liquid.

Then, the required whole pressure

$$= wS\bar{z}$$

$$= wS\left(\frac{\alpha+\beta+\gamma}{3}\right)$$

Note 1: If one vertex A lies on the free surface, then $\alpha = 0$. Let β and γ be the depths of the vertices B and C respectively below the free surface. Then, the required whole pressure

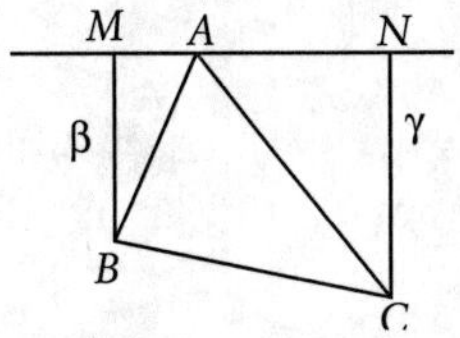

$$= wS\bar{z}$$

$$= wS\left(\frac{0+\beta+\gamma}{3}\right)$$

$$= wS\left(\frac{\beta+\gamma}{3}\right)$$

Note 2: If one vertex A lies in the free surface, then $\alpha = 0$. Let BC be horizontal. Then, depth of B below the free surface

= Depth of C below the free surface

= β (say)

Then, the required whole pressure

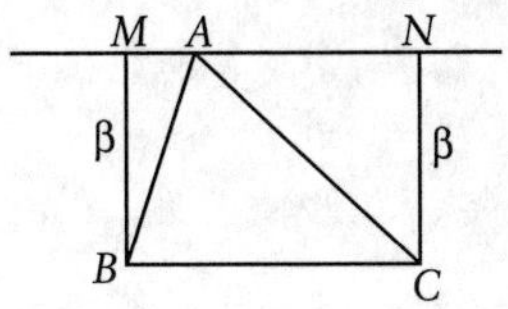

$$= wS\bar{z}$$

$$= wS\left(\frac{0+\beta+\beta}{3}\right)$$

$$= \frac{2}{3}wS\beta$$

Note 3: If side BC is on the free surface, then $\beta = 0 = \gamma$. Let the depth of the vertex A below the free surface be α. Then, the required whole pressure

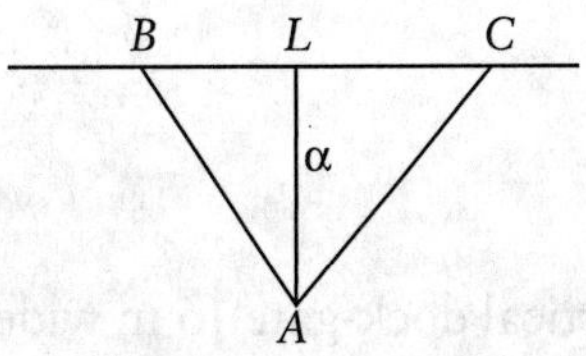

$$= wS\bar{z}$$

$$= wS\left(\frac{\alpha+0+0}{3}\right) \qquad = \frac{1}{3}wS\alpha$$

Example 2: A vessel containing some water rests on a horizontal table. If a person dips his hand in the water without touching the vessel, how is the pressure on the table affected?

Sol. Let w be the weight per unit volume of water. Let S be the area of the vessel in contact with the table. Let the water be filled to a height h in the vessel. Then, the whole pressure on the table

$$= wS\bar{z}$$

$$= wSh \qquad \text{...(1)}$$

When a person dips his hand in the water without touching the vessel, then w and S remain unaffected whereas the level of water in the vessel rises. Consequently, h increases.

It is therefore clear from (1) that the whole pressure on the table increases.

Example 3: A cube of 30 cms edge is suspended in water with its upper face horizontal and at a depth of 75 cms below the surface. Find the thrust on each face of the cube.

Sol.

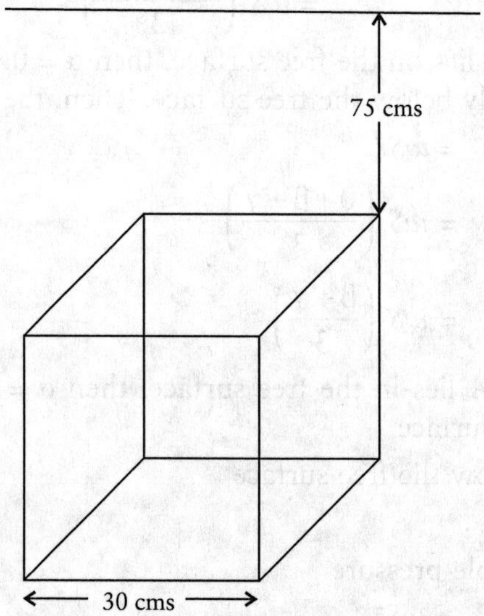

(1) Thrust on the upper face

$= wS\bar{z} = w\ (30 \times 30)\ 75 = 67500\ w = 67500$ grams wt

(2) Thrust on each vertical face

$= wS\bar{z} = w\ (30 \times 30)\ (75 + 30/2)$

$= w\ (30 \times 30)\ (90) = 81000\ w$

$= 81000$ grams wt

(3) Thrust on the lower face

$= wS\bar{z} = w\ (30 \times 30)\ (75 + 30)$

$= w\ (900)\ (105) = 94500\ w = 94500$ grams wt

Example 4: Find out the total thrust on one side of a rectangular vertical dock-gate 15 m wide, immersed in sea water to a depth of 10 m having given that 1 cm^3 of sea-water weighs one gm.

Sol. Area of the dock-gate immersed in sea water (S)

$= 15 \times 10$ sq m

$= 150$ sq m

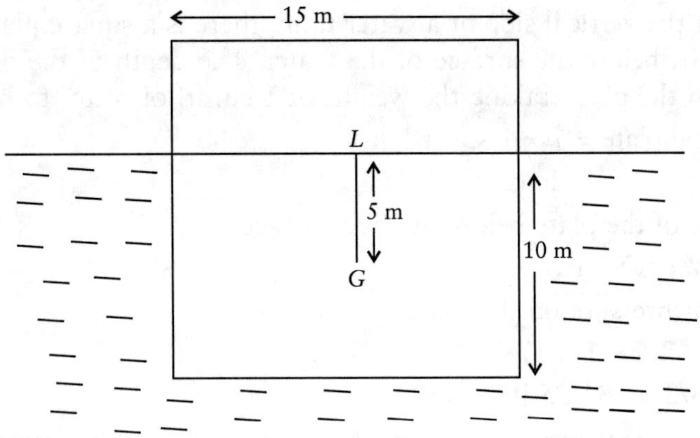

Depth of C.G. the area immersed below the free surface ($\bar{z}$) = 10/2 m = 5 m

∴ Required total thrust on one side of the dock-gate.

$= wS\bar{z}$

$= 100 \times 100 \times 150 \times 5$ gms wt

$= \dfrac{100\times100\times100\times150\times5}{1000}$ kg wt

= 750000 kg wt

Example 5: The width of a rectangular vertical dock-gate is 400 cm and on one side there is salt water (sp. gr. 1.026) to a depth of 240 cm. On the other side, there is fresh water. Find its depth if the thrusts on the two sides are equal.

Sol. Area immersed in salt water

$= 400 \times 240$ sq cm

Depth of C.G. of immersed area below the free surface

= 240/2 cm = 120 cm

∴ Whole pressure on one side of the dock-gate

$= wS\bar{z}$

$= 1.026\, g \times 400 \times 240 \times 120$ gms

Let the depth of the fresh water on the other side of the dock-gate be h cm.

∴ Whole pressure on the other side

$= wS\bar{z}$

$= 1 \times g \times (400 \times h) \times h/2$ gms

Since the thrusts on the two sides are equal, therefore

$$1 \times g \times (400 \times h) \times h/2 = 1.026 \times g \times 400 \times 240 \times 120$$

$\Rightarrow h/2 = 1.026 \times 240 \times 240$

$\Rightarrow h = 240\sqrt{1.026} \Rightarrow h = 240 \times 1.0129$

$\Rightarrow h = 243.096$ cm

Example 6: In the vertical side of a water tank, there is a square plate whose upper edge is horizontal and 8 ft. below the surface of the water. The depth of the plate is 1 foot. Find the resultant thrust on the plate, taking the weight of 1 cu. ft. of water to be 62.5 lbs.

Sol. Area of the plate = 1 × 1 sq. ft.

= 1 sq. ft.

Depth of C.G. of the plate below the free surface

= 8 + 1/2 = 17/2 ft.

∴ Resultant pressure on the plate

= $w S \bar{z}$ = 62.5 × 1 × 17/2

= 1062.5/2 = 531.25 lbs. wt.

Example 7: Find the thrust on a rectangular board, 3 ft. by 2 ft., immersed vertically in water with the smaller top edge horizontal and at a depth of 24.5 ft. below the surface of water, the atmospheric pressure being equal to a column of 34 ft. of water.

Sol.

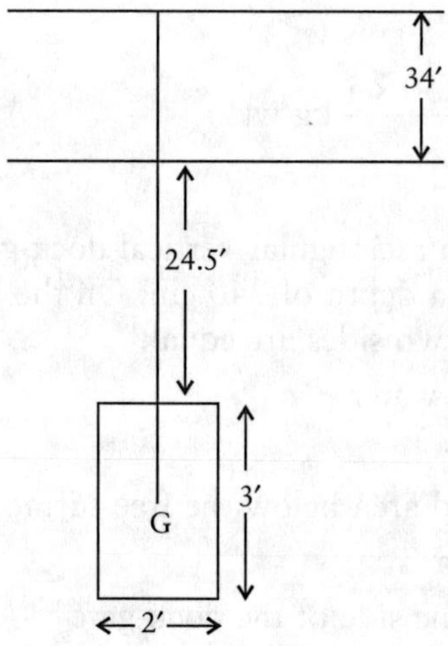

Area of the rectangular board

= 3 × 2 = 6 sq. ft.

Depth of C.G. of the board below the effective surface

= 34 + 24.5 + 3/2 = 34 + 24.5 + 1.5 = 60 ft.

∴ Required thrust

= $w S \bar{z}$ = 62.5 × 6 × 60 lbs. wt. = 22500 lbs. wt.

Example 8: A square plate, whose edge in 8 inches, is immersed in water, its upper edge being horizontal and at a depth of 12 inches below the surface of water. Find the thrust of the water on the surface of the plate when it is inclined at 45° to the horizon. The mass of a cubic foot of water is 64 lbs.

Sol. (1) Initial position of the plate

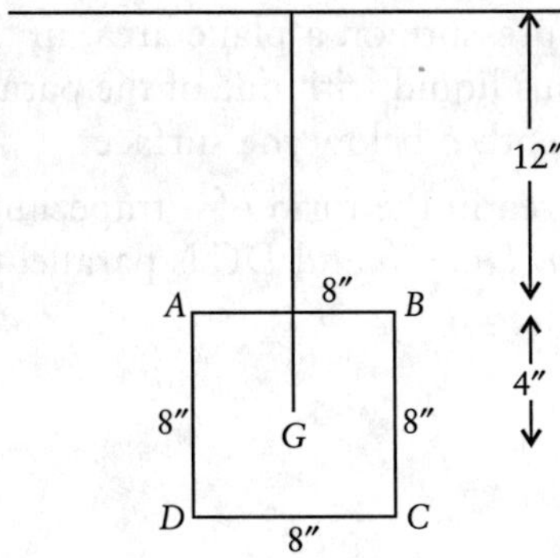

(2) Final position of the plate

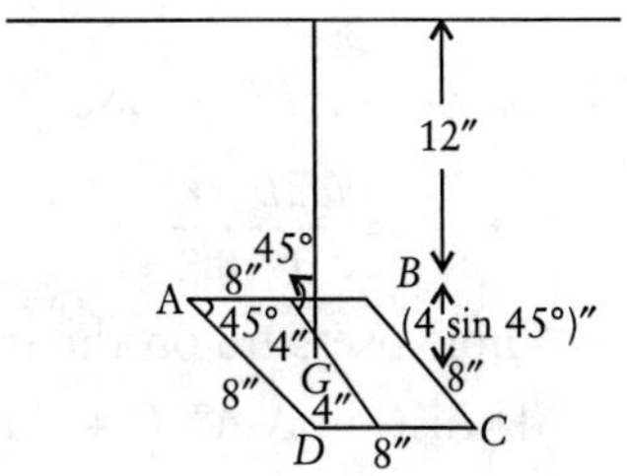

Area of the plate (S) $= 8 \times 8 = 64$ sq. inches

$$= \frac{64}{12 \times 12} \text{ sq.ft.}$$

$$= \frac{4}{9} \text{ sq.ft.}$$

$w = 64$ lbs.

Depth of C.G. of the plate ($\bar{z}$)

$= (12 + 4 \sin 45°)$ inches

$= (12 \times 4/\sqrt{2}\,)$ inches

$= (12 + 2\sqrt{2}\,)$ inches

$$= \frac{12 + 2\sqrt{2}}{12} \text{ ft.} = \left(1 + \frac{\sqrt{2}}{6}\right) \text{ ft.} = \left(1 + \frac{1}{3\sqrt{2}}\right) \text{ ft.}$$

$\therefore$ Required thrust

$= wS\bar{z}$

$$= 64 \times \frac{4}{9} \times \left(1 + \frac{1}{3\sqrt{2}}\right)$$

$$= 64 \times \frac{4}{9} \times \frac{3\sqrt{2} + 1}{3\sqrt{2}}$$

$$= 64 \times \frac{4}{9} \times \frac{3(1.4) + 1}{3(1.4)}$$

$$= 64 \times \frac{4}{9} \times \frac{5.2}{4.2} = 64 \times \frac{4}{9} \times \frac{26}{21}$$

$$= \frac{6656}{189} = 35.21 \text{ lbs. wt.}$$

Example 9: Find the whole pressure on a plane area, in the form of a trapezium immersed vertically in a heavy homogeneous liquid with one of the parallel sides of length a in the surface and the other of length b at a depth a below the surface.

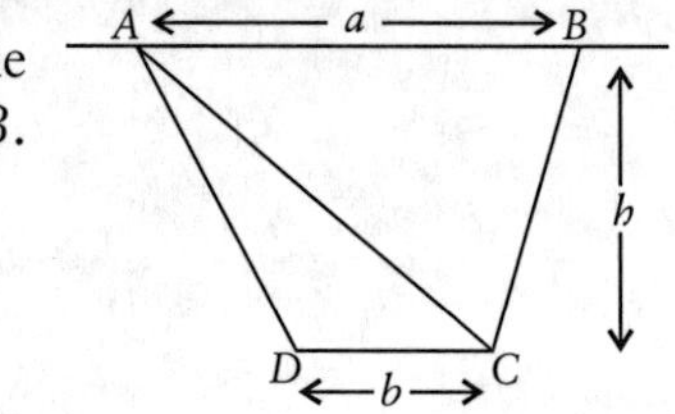

Sol. Let $ABCD$ be a plane area in the form of a trapezium. The side AB is in the surface. $AB = a$. $DC = b$ and DC is parallel to AB.

Distance between AB and $DC = h$

Join the diagonal AC.

Thrust on $\Delta\, ADC = wS\bar{z}$

$$= \frac{wbh}{2} \cdot \frac{2h}{3} \qquad = \frac{wbh^2}{3}$$

Thrust on $\Delta\, ABC = wS\bar{z}$

$$= \frac{wah}{2} \cdot \frac{h}{3} \qquad = \frac{wah^2}{6}$$

$\therefore$ Whole pressure on the trapezium $ABCD$

= Thrust on $\Delta\, ADC$ + thrust on $\Delta\, ABC$

$$= \frac{wbh^2}{3} + \frac{wah^2}{6}$$

$$= \frac{wh^2}{6}(2b + a)$$

Example 10: The sides of a cistern are vertical. Its base is a regular hexagon each side of which is 30 cm long. Find the depth if, when it is full of water. The thrust on each of its sides is the same as on its base.

Sol.

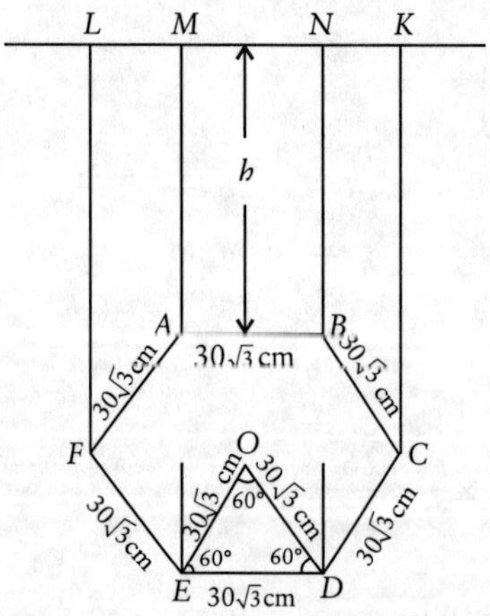

Let $ABCDEF$ be the regular hexagonal base of the cistern. O is the centre of the cistern. Let $ABNM$ be a side face of the cistern with height h.

Area of Δ *OED* = $\sqrt{3}/4\ (30\sqrt{3})^2 = 675\ \sqrt{3}$ cm²

$\therefore$ Area of hexagon *ABCDEF*

$= 6\ (675\sqrt{3})$

$= 4050\ \sqrt{3}$ cm²

Pressure on hexagon *ABCDEF* (P_1)

$= wS\bar{z} \qquad = w\ (4050\sqrt{3})\ h$

Pressure on a side face *ABNM* (P_2)

$= wS\bar{z}$

$= w\ (30\sqrt{3}\ h)\ h/2 \quad \Rightarrow = w\ 15\sqrt{3}\ h^2$

According to the question,

$P_1 = P_2$

$\Rightarrow \quad w\ (4050\sqrt{3})\ h = w\ 15\sqrt{3}\ h^2$

$\Rightarrow \quad h = 270$ cm

Example 11: An ellipse is placed with its minor axis on the surface of water and its plane axis as diameter. Find the total pressure of water on the portion of the area enclosed between the ellipse and circle.

Sol. *AB* is the minor axis of the ellipse lying on the surface of water.

Arc *ACB* is the arc of the circle described on *AB* as diameter.

Arc *ADB* is the arc of the ellipse.

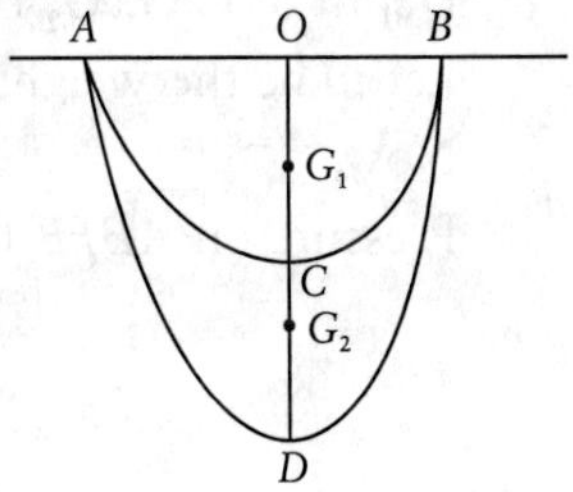

Let *AB* = 2*b* and *OD* = *a*.

Then, area of the semi-ellipse

$= 1/2\ \pi\ ab$

and, area of the semi-circle

$= 1/2\ \pi\ b^2$

Let G_1 be the centre of gravity of the semi-circle. Then,

$$OG_1 = \frac{4b}{3\pi}$$

Let G_2 be the centre of gravity of the semi-ellipse. Then,

$$OG_2 = \frac{4a}{3\pi}$$

Let *w* be the weight per unit volume of water. Then, pressure of water on the semi-ellipse

$= wS\bar{z}$

$$= w\frac{1}{2}\pi ab\frac{4a}{3\pi} \qquad = \frac{2}{3}a^2bw$$

and, pressure of water on the semi-circle

$= wS\bar{z}$

$$= w\frac{\pi b^2}{2}\cdot\frac{4b}{3\pi} \qquad = \frac{2}{3}b^3 w$$

∴ Required total pressure of water on the portion of the area enclosed between the ellipse and the circle.

= Pressure of water on the semi-ellipse

– Pressure of water on the semi-circle

$= 2/3\ a^2 b\ w - 2/3\ b^3 w$

$= 2/3\ b\ (a^2 - b^2)\ w$

Example 12: A square is placed with one side in the surface. Show how to draw horizontal line in the square dividing it into two portions, the thrusts on which are the same.

Let $ABCD$ be a square with side a whose one side AB is in the surface. Let us draw a horizontal line EF dividing the square $ABCD$ into two portions $ABFE$ and $EFCD$ such that

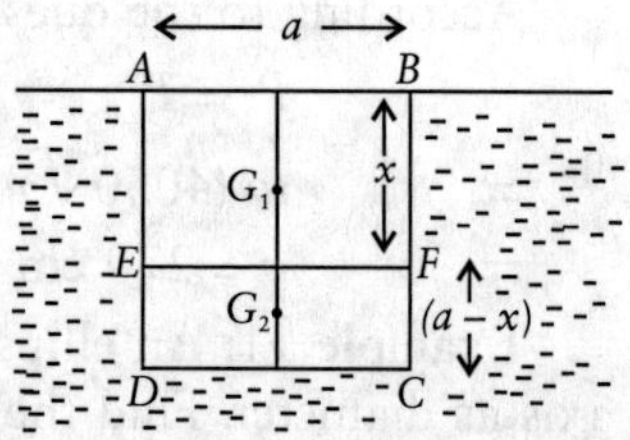

Pressure on $ABFE$ = Pressure on $EFCD$.

Let $\quad BF = x$

Then, $\quad FC = a - x$

Let G_1 and G_2 be the C.G. of the portions $ABFE$ and $EFCD$ respectively. Then, depth of C.G. G_1 below the free surface $= x/2$

Depth of C.G. G_2 below $EF = a–x/2$

Let w be the weight per unit volume of the liquid.

Now,

Pressure on $ABFE$ (P_1) $= wS\bar{z}$

$= wax \cdot x/2$

$= wax^2/2$

Pressure on $EFCD$ (P_2) $= wa(a-x)\left[x + \frac{a-x}{2}\right]$

$$= wa(a-x)\frac{(a+x)}{2}$$

$$= \frac{wa(a^2-x^2)}{2}$$

According to the question,

$P_1 = P_2$

$\Rightarrow \quad wa\frac{x^2}{2} = wa\frac{(a^2-x^2)}{2}$

$\Rightarrow \quad x^2 = a^2 - x^2 \quad \Rightarrow 2x^2 = a^2$

$\Rightarrow \quad x^2 = a^2/2 \quad \Rightarrow x = a/\sqrt{2}$

Example 13: A square lamina $ABCD$, which is immersed in water, has the side AB in the surface. Draw a line BE to a point E in CD such that the pressures on the two portions into which it divides the lamina may be equal.

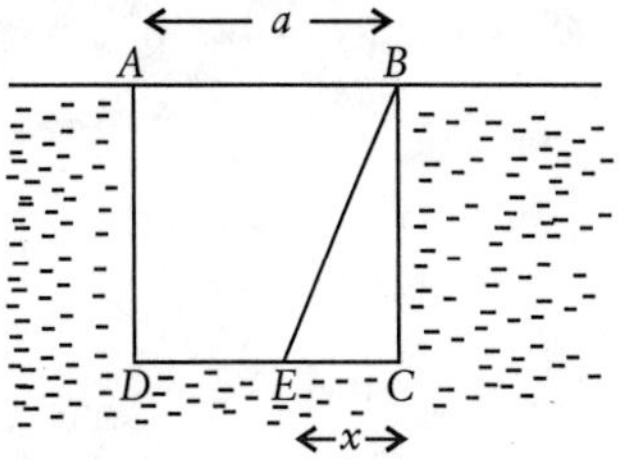

Sol. Let $ABCD$ be a square lamina with side a whose side AB is in the surface. Let $CE = x$.

Then,

Pressure on $\Delta\ BEC$ = Pressure on portion $ABED$

$\Rightarrow$ Pressure on $\Delta\ BEC$ = 1/2 Pressure on square $ABCD$

$$\Rightarrow \quad w \cdot \frac{ax}{2} \cdot \frac{2}{3}a = \frac{1}{2} \cdot w\,(a \cdot a)\frac{a}{2} \qquad \Big|\ \text{Pressure} = wS\bar{z}$$

where w is the weight per unit volume of water

$\Rightarrow \quad x = 3/4\ a$

$\Rightarrow \quad CE = 3/4\ CD \qquad \Rightarrow DE = CD - CE$

$\qquad = CD - 3/4\ CD \quad = CD/4$

$\therefore \qquad DE : EC = 1 : 3$

Example 14: Divide a square immersed vertically in a liquid, with one side in the surface by a straight line parallel to the diagonal, so that the thrusts on the two portions are equal.

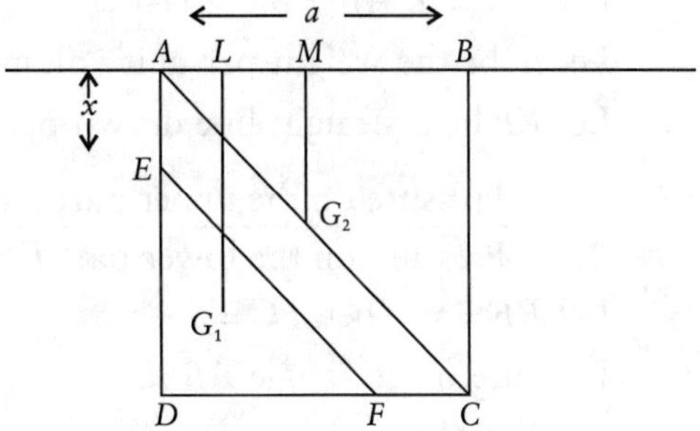

Sol. Let $ABCD$ be a square with side a whose one side AB is in the surface. Let EF be a line parallel to the diagonal AC of the square $ABCD$ such that $AE = x$.

According to the question.

Pressure on $\Delta\ DEF$ = Pressure on portion $AEFCB$

$\Rightarrow$ Pressure on $\Delta\ DEF$ = 1/2 Pressure on square $ABCD$

$$\Rightarrow \quad w\frac{(a-x)(a-x)}{2} \cdot \left(\frac{x+a+a}{3}\right)$$

$$1/2 \cdot w \quad a \cdot a \quad a/2 \qquad \Big|\ \text{Pressure} = wS\bar{z}$$

where w is weight per unit volume of the liquid and G_1 and G_2 are the C.G. of $\Delta\ DEF$ and square $ABCD$ respectively

$\Rightarrow 2\ (a - x)^2\ (x + 2a) = 3a^3$

$\Rightarrow 2x^3 - 6a^2x + a^3 = 0$

This equation gives the value of x.

Example 15: A rectangular lamina is immersed with one side in the free surface of a liquid. Show how to divide it into two parts by a straight line drawn parallel to this side, so that the pressures on the upper and lower parts are in the ratio 4 : 5.

Sol.

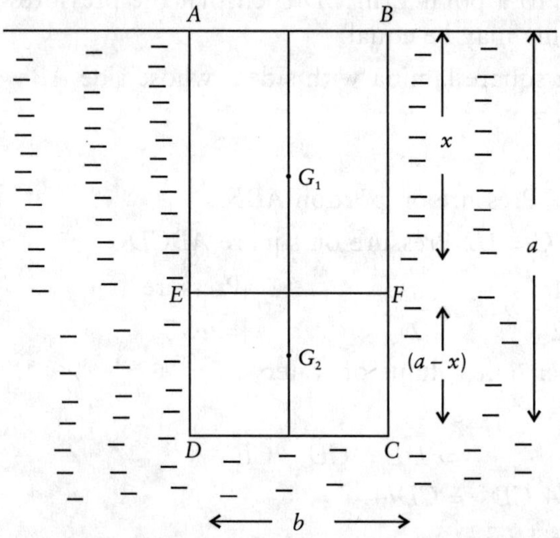

Let $ABCD$ be a rectangular lamina which is immersed with one side AB in the free surface of a liquid.

Let $AD = a$, $AB = b$

Let w be the weight per unit volume of the liquid.

Let EF be a straight line drawn parallel to AB such that

$$\frac{\text{Pressure on the upper part } ABFE}{\text{Pressure on the lower part } EFCD} = 4/5 \quad \text{...(1)}$$

Let $BF = x$. Then $FC = a - x$

Pressure on rectangle $ABFE$

$$= wS\bar{z}$$

$$= w\ (x\ b)\ x/2$$

$$= \frac{wbx^2}{2} \quad \text{...(2)}$$

Pressure on rectangle $EFCD$

$$= w(a-x)b\left[x + \frac{a-x}{2}\right]$$

$$= \frac{wb(a-x)(a+x)}{2}$$

$$= \frac{wb(a^2 - x^2)}{2} \quad \text{...(3)}$$

From (1), (2) and (3), we get

$$\frac{\frac{wbx^2}{2}}{\frac{wb(a^2-x^2)}{2}} = \frac{4}{5}$$

$\Rightarrow \quad \dfrac{x^2}{a^2-x^2} = \dfrac{4}{5} \qquad \Rightarrow 5x^2 = 4(a^2 - x^2)$

$\Rightarrow \quad 5x^2 = 4a^2 - 4x^2 \qquad \Rightarrow 9x^2 = 4a^2$

$\Rightarrow \quad x^2 = 4/9\ a^2 \qquad \Rightarrow x = 2/3\ a$

$\Rightarrow \quad a - x = a - 2/3\ a = a/3$

$\therefore \quad BC : FC = x : (a - x)$

$= 2a/3 : a/3$

$= 2 : 1$

Example 16: A triangle is immersed in a liquid with one vertex in the surface and the opposite side horizontal. Neglecting the atmospheric pressure, find the ratio in which the median through the vertex will be divided by a horizontal line which divides the triangle into two parts on which the pressures are equal.

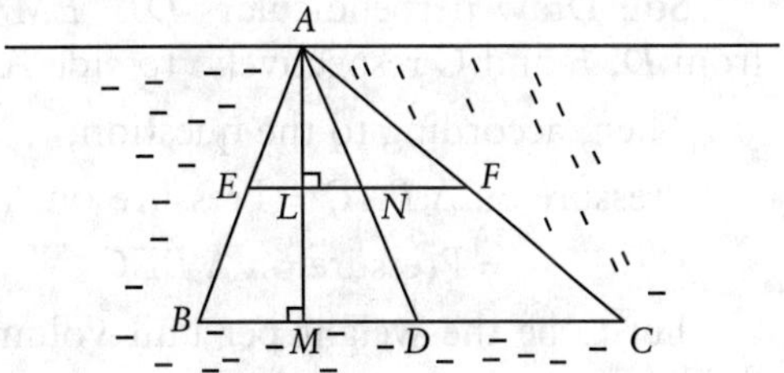

Sol. Let ABC be a triangle immersed in a liquid with one vertex A in the surface and the opposite side BC horizontal.

Let AD be the median through the vertex A. Let EF be a horizontal line which divides the triangle ABC into two parts AEF and $EFCB$ such that

Pressure on $\Delta\ AEF$ = Pressure on part $EFCB$

$\Rightarrow$ Pressure on $\Delta\ AEF$ = 1/2 Pressure on $\Delta\ ABC$ ...(1)

Let AM be the perpendicular drawn from vertex A to the side BC meeting EF in L. Let the median AD meet the line EF in N. Let w be the weight per unit volume of the liquid.

Pressure on $\Delta\ AEF = wS\bar{z}$

$= w\ (1/2\ EF.\ AL)\ (2/3\ AL)$

$= w/3\ EF\ .\ AL^2$ ...(2)

Pressure on $\Delta\ ABC = wS\bar{z}$

$= w\ (1/2\ BC.\ AM)\ (2/3\ AM)$

$= w/3\ BC\ .\ AM^2$ ...(3)

From (1), (2) and (3), we get

$w/3\ EF\ .\ AL^2 = 1/2\ w/3\ BC\ .\ AM^2$

$\Rightarrow \quad \dfrac{EF}{BC} = \dfrac{1}{2}\dfrac{AM^2}{AL^2}$ ...(4)

Now, $\Delta\ AEF \sim \Delta\ ABC$

$$\Rightarrow \quad \frac{EF}{BC} = \frac{AL}{AM} = \frac{AN}{AD} \quad \text{| By property of similar triangles} \qquad ...(5)$$

From (4) and (5), we get

$AL/AM = 1/2\ AM^2/AL^2$

$\Rightarrow \quad AL^3 = 1/2\ AM^3$

$\Rightarrow \quad AL^3/AM^3 = 1/2$

$\Rightarrow \quad AL/AM = 1/2^{1/3}$

$\Rightarrow \quad AN/AD = 1/2^{1/3}$ | From (5)

$\Rightarrow \quad AD = 2^{1/3}\ AN$ ...(6)

$ND = AD - AN$

$= 2^{1/3}\ AN - AN$ | From (6)

$(2^{1/3} - 1)\ AN$

$\Rightarrow \quad AN/AN = (2^{1/3} - 1)/1$

Example 17: The side *AB* of a triangle *ABC* is in the surface of a fluid and points *D, E* are taken in *AC*, such that the pressures on the triangles *BAD, BDE* and *BEC* are equal. Find *AD* : *DE* : *EC*.

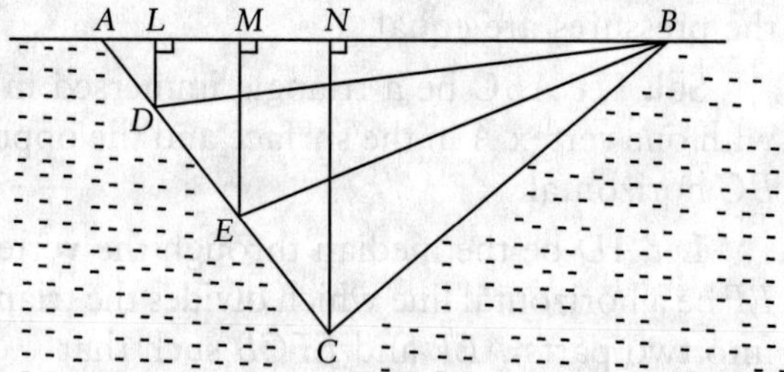

Sol. Draw perpendiculars *DL, EM* and *CN* from *D, E* and *C* respectively to side *AB*.

Then, according to the question,

Pressure on $\Delta\ BAD$ = Pressure on $\Delta\ BDE$

= Pressure on $\Delta\ BEC$ ...(1)

Let w be the weight per unit volume of the fluid. Then,

Pressure on $\Delta\ BAD = wS\bar{z}$

$= w\ .\ AB\ .\ DL/2\ .\ DL/3$

$= wAB\ .\ DL^2/6$ ...(2)

Pressure on $\Delta\ BAE = wS\bar{z}$

$= w\ .\ AB\ .\ EM/2\ .\ 1/3\ EM$

$= wAB\ .\ EM^2/6$ ...(3)

Pressure on $\Delta\ BAC = wS\bar{z}$

$= wAB\ .\ CN/2\ .\ CN/3$

$= wAB\ .\ CN^2/6$ ...(4)

From (1), we get

Pressure on $\Delta\ BAD$ = 1/2 Pressure on $\Delta\ BAE$

= 1/3 Pressure on $\Delta\ BAC$

$\Rightarrow \quad \dfrac{wAB \cdot DL^2}{6} = \dfrac{1}{2}\dfrac{wAB \cdot EM^2}{6} = \dfrac{1}{3}\dfrac{wAB \cdot CN^2}{6}$ | From (2), (3) and (4)

$\Rightarrow \quad DL^2 = 1/2\ EM^2 = 1/3\ CN^2$

$\Rightarrow \quad DL = EM/\sqrt{2} = CN/\sqrt{3}$

$\Rightarrow \quad DL/\sqrt{1} = EM/\sqrt{2} = CN/\sqrt{3}$ ...(5)

Now, $\Delta\ ALD \sim \Delta\ AME \sim \Delta\ ANC$

$\therefore\ AD/DL = AE/EM = AC/CN$ | $\because$ Corresponding sides of two similar triangles are proportional

From (5) and (6), we get

$AD/\sqrt{1} = AE/\sqrt{2} = AC/\sqrt{3}$

$\Rightarrow \quad AD/\sqrt{1} = AE - AD/\sqrt{2} - 1 = AC - AE/\sqrt{3} - \sqrt{2}$

$\Rightarrow \quad AD/\sqrt{1} = DE/\sqrt{2} - 1 = AC/\sqrt{3} - \sqrt{2}$

$\Rightarrow \quad AD : DE : EC = \sqrt{1} : \sqrt{2} - 1 : \sqrt{3} - \sqrt{2}$

$\Rightarrow \quad AD : DE : EC :: 1 : \sqrt{2} - 1 : \sqrt{3} - \sqrt{2}$

Example 18: An isosceles triangle is immersed vertically in a fluid with its vertex coincident with the surface of the fluid and its base horizontal. Determine how it must be divided by a line parallel to the base so that the pressures on the upper and lower parts may be respectively in the ratio $m : n$.

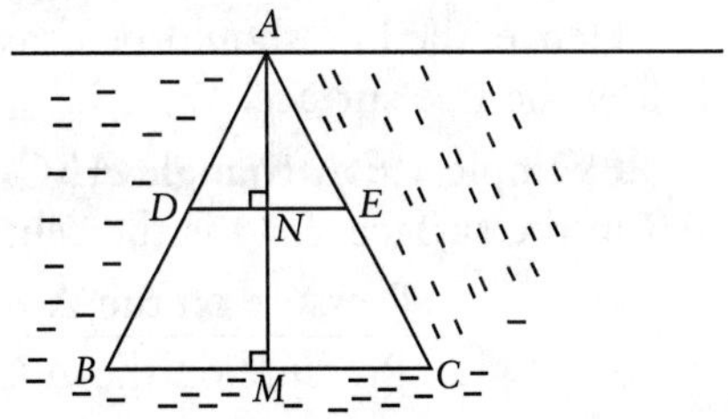

Sol. Let ABC be an isosceles triangle immersed vertically in a fluid with its vertex A coincident with the surface of the fluid and its base BC horizontal. Let DE be a line drawn parallel to the base BC such that

$$\frac{\text{Pressure on the upper part } ADE}{\text{Pressure on the lower part } DECB} = \frac{m}{n}$$

$$\Rightarrow \quad \frac{\text{Pressure on } DECB}{\text{Pressure on } ADE} = \frac{n}{m}$$

$$\Rightarrow \quad \frac{\text{Pressure on } DECB}{\text{Pressure on } ADE} + 1 = \frac{n}{m} + 1$$

$$\Rightarrow \quad \frac{\text{Pressure on } DECB + \text{Pressure on } ADE}{\text{Pressure on } ADE} = \frac{n+m}{m}$$

$$\Rightarrow \quad \frac{\text{Pressure on } \Delta ABC}{\text{Pressure on } \Delta ADE} = \frac{m+n}{m} \qquad \text{...(1)}$$

Draw perpendicular AM from A on BC intersecting DE at N. Then,

$AN \perp DE$ | By Geometry

Let w be the weight per unit volume of the fluid. Then,

Pressure on $\Delta\, ABC = wS\bar{z} = wBC\,.\,AM/2\,.\,2/3\,AM$

$$= w/3\ BC\,.\,AM^2 \qquad ...(2)$$

Pressure on $\Delta\, ADE = wS\bar{z}$

$$= wDE\,.\,AN/2\,.\,2/3\,AN$$

$$= w/3\ DE\,.\,AN^2 \qquad ...(3)$$

From (1), (2) and (3), we get

$$\frac{\frac{w}{3}BC\,.\,AM^2}{\frac{w}{3}DE\,.\,AN^2} = \frac{m+n}{m}$$

$$\Rightarrow \quad \frac{BC}{DE}\left(\frac{AM}{AN}\right)^2 = \frac{m+n}{m} \qquad ...(4)$$

Now, $\Delta\, ABC \sim \Delta\, ADE$

$BC/DE = AM/AN$ | By property of similar triangles ...(5)

From (4), and (5), we get

$$\left(\frac{AM}{AN}\right)^3 = \frac{m+n}{m} \quad \Rightarrow \left(\frac{AM}{AN}\right)^2 = \left(\frac{m+n}{m}\right)^{1/3}$$

$$\Rightarrow \quad \frac{AN}{AM} = \left(\frac{m}{m+n}\right)^{1/3} \quad \Rightarrow AN = \left(\frac{m}{m+n}\right)^{1/3} AM$$

Hence, the line should be drawn at a depth $(m/m+n)^{1/3}\, h$ where h is the depth of the base below the free surface.

Example 19: A triangle ABC is immersed in a liquid, its plane being vertical and the side AB in the surface. If O is the centre of the circumscribed circle of $\Delta\, ABC$, prove that

$$\frac{\text{Pressure on the } \Delta\, AOC}{\text{Pressure on the } \Delta\, OCB} = \frac{\sin 2B}{\sin 2A}$$

Sol.

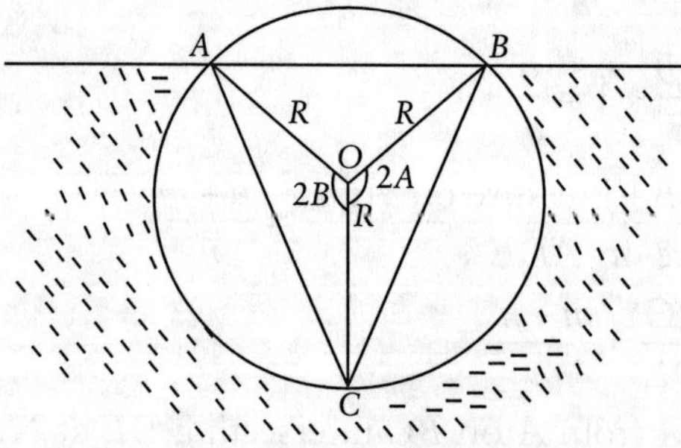

$\because$ O is the centre of the circumscribed circle of $\Delta\, ABC$

$\therefore \qquad \angle\, BOC = 2A$

and $\angle AOC = 2B$

Also, $OA = OB = OC = R$ (say)

Clearly, depth of C.G. of ΔAOC =

depth of C.G. of $\Delta BOC = h$ (say)

Area of $\Delta AOC = 1/2\ OA\ .\ OC \sin \angle AOC$

$= 1/2\ RR \sin 2B$

$= 1/2\ R^2 \sin 2B$

Area of $\Delta OCB = 1/2\ OB\ .\ OC \sin \angle BOC$

$= 1/2\ RR \sin 2A$

$= 1/2\ R^2 \sin 2A$

Let w be the weight per unit volume of the liquid.

Pressure on $\Delta AOC = wS\bar{z}$

$= w\ 1/2\ R^2 \sin 2B\ .\ h$

Pressure on $\Delta OCB = wS\bar{z}$

$= w\ 1/2\ R^2 \sin 2A\ .\ h$

$$\therefore \quad \frac{\text{Pressure on the } \Delta AOC}{\text{Pressure on the } \Delta OCB} = \frac{w\frac{1}{2}R^2 \sin 2B.h}{w\frac{1}{2}R^2 \sin 2A.h} = \frac{\sin 2B}{\sin 2A}$$

Example 20: A square of side a is dipped vertically in two liquids of densities ρ and σ with the upper side in the free surface. The depth of the upper liquid of density ρ is b ($< a$). Find the thrust on the square.

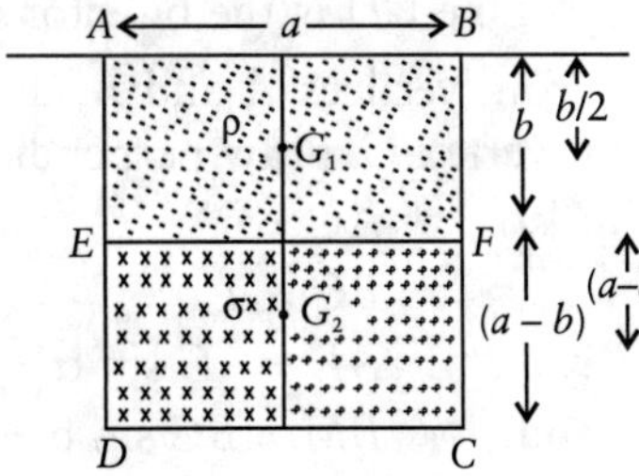

Sol. Let $ABCD$ be a square of side a dipped vertically in two liquids of densities ρ and σ ($\rho < \sigma$) with the upper side AB in the free surface.

$BF = b$

$\therefore \quad FC = a - b$

The portion $ABFE$ of the square $ABCD$ is in contact with the liquid of density ρ whereas the portion $EFCD$ of the square $ABCD$ is in contact with the liquid of density σ.

Thrust on the portion $ABFE$ (P_1)

$= wS\bar{z} = g\rho\ ab\ b/2 = 1/2\ g\rho\ ab^2$

Thrust on the portion $EF\ CD$ (P_2)

$= g\rho a(a - b)\, b + g\sigma a(a - b)\ (a - b/2) = ga(a - b)\ [\rho b + \sigma\ (a - b/2)]$

$\therefore$ Thrust on the square $ABCD$

= Thrust on the area $ABFE$ + Thrust on the area $EFCD$

$= 1/2\ g\rho\ ab^2 + ga(a - b)\ [\rho b + \sigma\ (a - b/2)] = 1/2\ g\rho\ ab\ (2a - b) + 1/2\ g\sigma a(a - b)^2$

Aliter: The thrust of the liquids on the square will remain the same even if we consider that one liquid of density ρ is in contact with the whole area *ABCD* and another liquid of density (σ – ρ) is in contact with the lower part *EFCD* only of the square *ABCD*.

Thrust due to the first liquid (P_1)

$= wS\bar{z} = g\rho\,(a\,a)\,a/2 = a^3\,g\rho/2$

Thrust due to the second liquid (P_2)

$= wS\bar{z} = g\,(\sigma - \rho)\,a(a - b)\,(a - b/2)$

$= ga\,(\sigma - \rho)\,(a - b)^2/2$

∴ Total thrust on the square $ABCD = P_1 + P_2$

$$= \frac{a^3 g\rho}{2} + g\frac{a(\sigma - \rho)(a - b)^2}{2}$$

$= 1/2\,a(a - b)^2\,g\sigma + 1/2\,ag\rho\,\{a^2 - (a - b)^2\}$

$= 1/2\,a(a - b)^2\,g\sigma + 1/2\,abg\rho\,(2a - b)$

Example 21: A triangle *ABC* is immersed in a liquid with the vertex C in the surface, and the sides *AC, BC* equally inclined to the surface. Show that the vertical through C divides the triangle into two others, the fluid pressures upon which are as:

$$b^3 + 3ab^2 : a^3 + 3a^2b$$

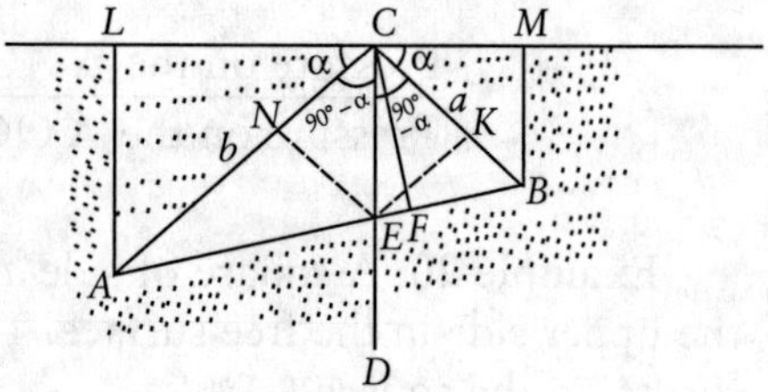

Sol. Let *ABC* be a triangle immersed in a liquid with the vertex C in the surface, and the sides *AC, BC* equally inclined (say at an angle α) to the surface. Let *CD* be the vertical through C intersecting the side *AB* in *E*. Then,

$\angle ACD = 90° - \alpha = \angle BCD$

⇒ Line *CD* is the bisector of $\angle ACB$.

From vertices *A* and *B*, draw perpendiculars *AL* and *BM* respectively to the free surface. From vertex C, draw perpendicular *CF* to the side *AB*.

Also, $AC = b$

$BC = a$

∴ $AL = AC \sin\alpha = b \sin\alpha$

and, $BM = BC \sin\alpha = a \sin\alpha$

By Geometry, *CD* divides the side *AB* in the ratio $b : a$

∴ $AE : BE = b : a$

Also, *AL, CD* and *BM* and parallel

The internal bisector of an angle of a triangle divides the opposite side in the ratio of the sides forming that angle

$$\therefore \quad CE = \frac{b(BM) + a(AL)}{b + a}$$

$$= \frac{b(a\sin\alpha) + a(b\sin\alpha)}{b + a} = \frac{2ab\sin\alpha}{a + b}$$

Depth of C.G. of Δ *ACE* below the free surface

(Depth of *A* + Depth of C + Depth of *E*)

$$= \frac{\text{below the free surface}}{3}$$

$$= \frac{b\sin\alpha + 0 + \dfrac{2ab\sin\alpha}{a+b}}{3} \qquad = \frac{b\sin\alpha + \dfrac{2ab\sin\alpha}{a+b}}{3}$$

Depth of C.G. of Δ BCE below the free surface

(Depth of B + Depth of C + Depth of E)

$$= \frac{\text{below the free surface}}{3}$$

$$= \frac{a\sin\alpha + 0 + \dfrac{2ab\sin\alpha}{a+b}}{3} \qquad = \frac{a\sin\alpha + \dfrac{2ab\sin\alpha}{a+b}}{3}$$

Fluid pressure on Δ ACE (P_1)

$= wS\bar{z}$

$$= w\frac{AE.CF}{2}\frac{1}{3}\left(b\sin\alpha + \frac{2ab\sin\alpha}{a+b}\right)$$

Fluid pressure on Δ BCE (P_2)

$= wS\bar{z}$

$$= w\frac{BE.CF}{2}\frac{1}{3}\left(a\sin\alpha + \frac{2ab\sin\alpha}{a+b}\right)$$

$\therefore$ Required ratio $= P_1/P_2$

$$= \frac{\dfrac{w.AE.CF}{2}\dfrac{1}{3}\left(b\sin\alpha + \dfrac{2ab\sin\alpha}{a+b}\right)}{\dfrac{w.BE.CF}{2}\dfrac{1}{3}\left(a\sin\alpha + \dfrac{2ab\sin\alpha}{a+b}\right)}$$

$$= \frac{AE}{BE}\left(\frac{b + \dfrac{2ab}{a+b}}{a + \dfrac{2ab}{a+b}}\right) \qquad = \frac{a}{b}\frac{b\left(1 + \dfrac{2a}{a+b}\right)}{a\left(1 + \dfrac{2b}{a+b}\right)}$$

$$= \frac{b^2(3a+b)}{a^2(a+3b)} \qquad = \frac{b^3 + 3ab^2}{a^3 + 3a^2b}$$

Aliter: Fluid pressure on Δ ACE (P_1) $= wS\bar{z}$

$$= \frac{w.\ AC.EN}{2}\left\{\frac{0 + b\sin\alpha + CE}{3}\right\} \qquad = \frac{w.\ AC.CE\cos\alpha}{2}\left(\frac{b\sin\alpha + CE}{3}\right)$$

Fluid pressure on ΔBCE (P_2)

$$= wS\bar{z} = \frac{w.\,BC.EK}{2}\left\{\frac{0 + a\sin\alpha + CE}{3}\right\} \qquad = \frac{w.\,BC.CE\cos\alpha}{2}\left(\frac{a\sin\alpha + CE}{3}\right)$$

$\therefore$ Required ratio = P_1/P_2

$$= \frac{\frac{w.\,AC.CE\cos\alpha}{2}\left(\frac{b\sin\alpha + CE}{3}\right)}{\frac{w.\,BC.CE\cos\alpha}{2}\left(\frac{a\sin\alpha + CE}{3}\right)} \qquad = \frac{b}{a}\frac{b\sin\alpha + CE}{a\sin\alpha + CE}$$

Now, by sine formula from the triangles ACE and BCE, we get

$$\frac{CE}{\sin A} = \frac{AE}{\cos\alpha}$$

and $$\frac{CE}{\sin B} = \frac{BE}{\cos\alpha}$$

$$\therefore \quad \frac{P_1}{P_2} = \frac{b}{a}\frac{b\sin\alpha + \frac{AE\sin A}{\cos\alpha}}{a\sin\alpha + \frac{BE\sin B}{\cos\alpha}}$$ | Putting the value of CE

$$= \frac{b}{a}\frac{b\sin\alpha\cos\alpha + AE\sin A}{a\sin\alpha\cos\alpha + BE\sin B}$$

$$= \frac{b}{a}\frac{b\sin 2\alpha + 2AE\sin A}{a\sin 2\alpha + 2BE\sin B}$$

Now, $AE : BE = b : a$

$$\Rightarrow \quad \frac{AE}{b} = \frac{BE}{a} = \frac{AE + BE}{b + a}$$

$$= \frac{AB}{a+b} = \frac{a}{a+b}$$

Also,

$$\angle C = 180° - 2\alpha$$

$\therefore$ $\sin C = \sin 2\alpha$

Hence, we get

$$\frac{P_1}{P_2} = \frac{b}{a}\frac{b\sin C + 2.\frac{bc}{a+b}\sin A}{a\sin C + 2\frac{ac}{a+b}\sin B}$$

$$= \frac{b}{a}\frac{b(a+b)\sin C + 2bc\sin A}{a(a+b)\sin C + 2ac\sin B} \qquad = \frac{b}{a}\left[\frac{b(a+b)c + 2bc.a}{a(a+b)c + 2ac.b}\right]$$ | by sine formula

$$= \frac{b}{a}\left[\frac{bc(3a+b)}{ac(a+3b)}\right]$$

$$= \frac{b^2(3a+b)}{a^2(a+3b)} = \frac{b^3+3ab^2}{a^3+3a^2b}$$

Example 22: A parallelogram is immersed in a homogeneous liquid with one side in the surface. Show how to draw horizontal lines dividing into n portions the thrusts on which are equal.

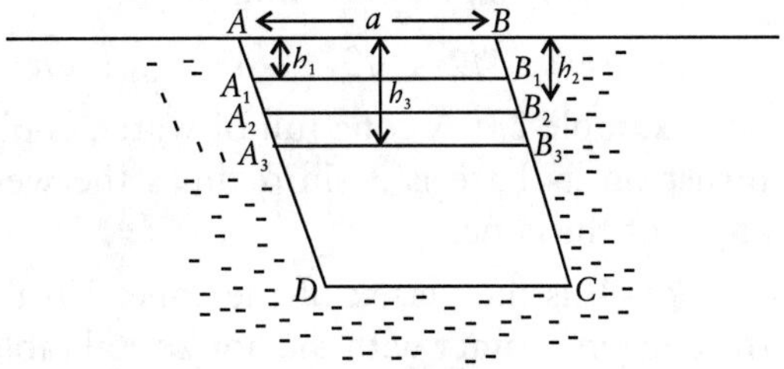

Sol. Let $ABCD$ be a parallelogram immersed in a homogeneous liquid with one side AB (= a), say, in the free surface. Let A_1B_1, A_2B_2, A_3B_3, ... be the horizontal lines dividing it into n portions the thrusts on which are equal. Let the respective depths of these lines below the free surface be h_1, h_2, h_3, ... respectively. Then,

area of the first strip = ah_1

area of the second strip = $a(h_2 - h_1)$

area of the third strip = $a(h_3 - h_2)$ and so on.

Also,

Depth of C.G. of the first strip below the free surface = $h_1/2$

Depth of C.G. of the second strip below the free surface = $(h_1 + h_2)/2$

Depth of C.G. of the third strip below the free surface = $(h_2 + h_3)/2$ and so on

Now,

Thrust on the first strip

$= wS\bar{z} \qquad = wah_1\, h_1/2 = 1/2\; awh_1^2$

Thrust on the second strip = $wS\bar{z}$

$= wa\,(h_2 - h_1)\,(h_1 + h_2)/2 \qquad = 1/2\; aw\,(h_2^2 - h_1^2)$

Thrust on the third strip

$= wS\bar{z}$

$= wa\,(h_3 - h_2)\,(h_2 + h_3)/2 \qquad = 1/2\; aw\,(h_3^2 - h_2^2)$ and so on.

According to the question, the thrusts on the strips are equal

$\therefore\; 1/2\; awh_1^2 = 1/2\; aw\,(h_2^2 - h_1^2) = 1/2\; aw\,(h_3^2 - h_2^2) = \ldots\ldots$

First two give

$h_1^2 = h_2^2 - h_1^2 \qquad \Rightarrow \qquad 2h_1^2 = h_2^2$

$\Rightarrow \quad h_1^2/h_2^2 = 1/2 \qquad \Rightarrow \qquad h_1/h_2 = 1/\sqrt{2}$

First and third give

$h_1^2 = h_3^2 - h_2^2 \qquad \Rightarrow h_1^2 = h_3^2 - 2h_1^2 \Rightarrow 3h_1^2 = h_3^2$

$\Rightarrow \quad h_1^2/h_3^2 = 1/3 \qquad \Rightarrow \qquad h_1/h_3 = 1/\sqrt{3}$

Similarly, we get

$h_1/h_4 = 1/\sqrt{4}$, $\quad h_1/h_5 = 1/\sqrt{5}$

$\vdots$

$h_1/h_n = 1/\sqrt{n}$

$\therefore \quad h_1 : h_2 : h_3 : h_4 : \ldots\ldots : h_n$

$= 1 : \sqrt{2} : \sqrt{3} : \sqrt{4} \;\ldots\ldots : \sqrt{n}$

Example 23: A cone full of water, is placed on its side on a horizontal table. Show that the thrust on its base is 3 sin α times the weight of the contained fluid, where 2α is the vertical angle of the cone.

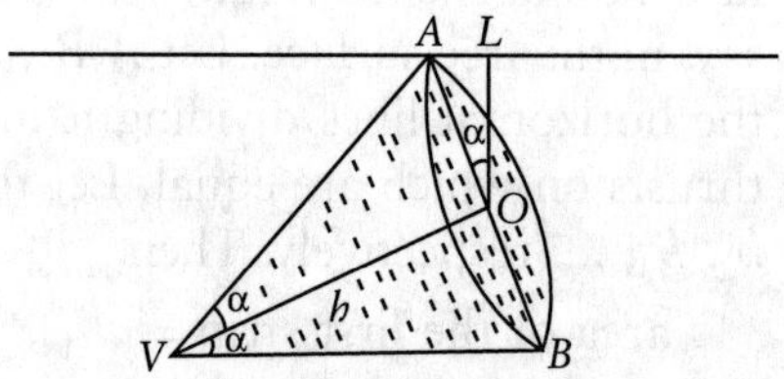

Sol. V is the vertex of the cone. VB is the generator of the cone in contact with the horizontal table. A is the highest point of the cone. The horizontal plane passing through A is the level of the free surface. Since 2α is the vertical angle of the cone, therefore, $\angle AOL = \alpha$.

Let h be the height of the cone.

Then, the radius of the base of the cone $= h \tan \alpha$

Also, depth of C.G. of the base of the cone below the free surface

$= OL \quad = AO \cos \alpha$

$= h \tan \alpha \cos \alpha \quad = h \sin \alpha$

$\therefore$ Thrust on the base of the cone

$= wS\bar{z} \quad = w\pi (h \tan \alpha)^2 \, h \sin \alpha$

$= w\pi h^2 \tan^2 \alpha \, h \sin \alpha = 3 \sin \alpha \, (1/3 \, \pi h^3 \tan^2 \alpha \, . \, w)$

$= 3 \sin \alpha$ (weight of the fluid contained in the cone).

Example 24: A closed hollow cone is just filled with liquid, and is placed with its vertex upwards and axis vertical. Divide its curved surface by a horizontal plane into two parts on which the whole pressures are equal.

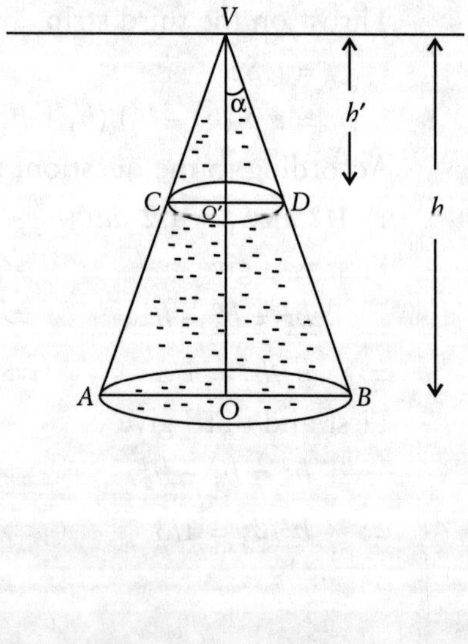

Sol. Let V be the vertex of the cone. Let O be the centre of the circular base. Let h be the height and α be the semi-vertical angle of the cone.

Then,

$OB = h \tan \alpha$

$VB = h \sec \alpha$

Let the plane CD divide the curved surface of the cone into two parts on which the whole pressures are equal. Let the vertical distance of this plane from the vertex V of the cone be h'. Then

$VO' = h'$

$O'D = h' \tan \alpha$, $VD = h' \sec \alpha$

Let G and G' be the C.G. of the curved surfaces of the cone (V, AB) and (V, CD) respectively. Then,

$VG = 2/3\ h$

$VG' = 2/3\ h'$

Curved surface of the cone (V, AB)

$= \pi h \sec\alpha\ h \tan\alpha = \pi h^2 \sec\pi \tan\alpha$

Curved surface of the cone (V, CD)

$= \pi h' \sec\alpha h' \tan\alpha = \pi h'^2 \sec\pi \tan\alpha$

Thrust on the curved surface of the cone (V, AB) $= wS\bar{z}$

$= w\pi h^2 \sec\alpha \sin\alpha\ (2/3\ h) = 2/3\ w\pi h^3 \sec\alpha \sin\alpha$

Thrust on the curved surface of the cone (V, CD) $= wS\bar{z}$

$= w\pi h'^2 \sec\alpha \tan\alpha\ (2/3\ h') = 2/3\ w\ \pi\ h'^3 \sec\alpha \tan\alpha$

According to the question,

Thrust on the curved surface of the cone (V, AB) = 2. Thrust on the curved surface of the cone (V, CD)

$\Rightarrow\quad 2/3\ w\ \pi\ h^3 \sec\pi \tan\alpha = 2\ (2/3\ w\ \pi\ h'^3 \sec\alpha \tan\alpha)$

$\Rightarrow\quad h^3 = 2h'^3$

$\Rightarrow\quad h'^3 = h^3/2$

$\Rightarrow\quad h' = h/2^{1/3}$

This gives the required depth of the plane CD from the vertex V of the closed hollow cone.

EXERCISE 1

1. Determine the total thrust on one side of a rectangular vertical dock-gate 50 ft. wide, immersed in salt water to a depth of 25 ft. having given that one cubic foot of salt water weighs 1026 ozs. (16 ozs. = 1 lb.)

   Ans. 1001953 1/8 lbs. wt.

2. The width of a rectangular vertical dock-gate is 40 ft. and on one side there is salt water (sp. gr. 1.026) to a depth of 24 ft. On the other side, there is fresh water. Find its depth if the thrusts on the two sides are equal.

   Ans. 24.3096 ft.

3. Taking the weight of a cubic foot of water as 1000 ozs. find the pressure of the water arising from its weight on a side of cistern 7 ft. wide and 8 ft. deep, if the cistern is filled with water.

   (given weight of 1 cu. ft. of water = 62.5 lbs.)

   Ans. 14000 lbs. wt.

   [**Hint:** Required pressure $= wS\bar{z}$

   $= 62.5 \times 7 \times 8 \times 8/2$ lbs. wt.

   $= 14000$ lbs. wt.]

4. In the vertical side of a water tank there is a square plate whose edge is horizontal and at a depth of 8 m. below the surface of the water. The depth of the plate is one m. Find the resultant pressure on the plate, taking the weight of 1 cm^3 of water to be 1 gm.

   Ans. 8500 kg. wt.

   [**Hint:** Resultant pressure = $wS\bar{z}$

   $= (100 \times 100 \times 100/1000) \times 1 \times (8+1/2)$

   $= 8500$ kg. wt.]

5. Find the thrust on a square board whose side is 1 foot when sunk in water to a depth of 20 feet, the board being horizontal and the atmospheric pressure at the surface being equivalent to a height of 30 inches of mercury.
   (sp. gr. of mercury being 13.59).

   Ans. 3373.4375 lbs. wt.

6. Find the thrust on a rectangular plate whose sides are a and b, the side a being horizontal and at a depth c below the surface and the plane of the rectangle being inclined at an angle θ to the vertical.

   Ans. $w\ (c + 1/2\ b \cos\theta)\ ab$ where w is the weight per unit volume of the liquid.

   [**Hint:**

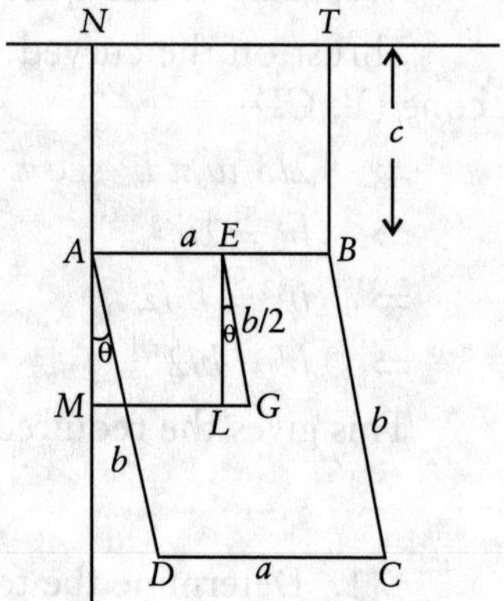

7. A square plate, whose edge is 80 cm, is immersed in water, its upper edge being horizontal at a depth of 120 cm below the surface of the water. Find the thrust of the water on the surface of the plate when it is inclined at 45° to the horizon; the mass of 1 cm^3 of water being 1 gm.

   Ans. $128\ (6 + \sqrt{2})$ kg. wt.

8. A rectangular vessel, one face of which is of height 100 cms and width 50 cms, is half filled with mercury (sp. gr. 13.5) and half with water. Find the thrust on this face.

   Ans. 36250 grams wt.

9. The resultant fluid pressure on a vertical circle of radius a is equal to twice the weight of sphere of the same liquid of radius a. The circle is now lowered through a depth equal to $2a$. Prove that the liquid pressure in the new position is 7/1 times the original one.

10. The lighter of two fluids, whose sp. gravities are as 2 : 3, rests on the heavier, to a depth of 4 cm. A square is immersed in a vertical position with one side in the upper surface. Determine the side of the square in order that the pressures on the portions in the two fluids may be equal.

    Ans. $4/3(1 + \sqrt{10})$ cm

11. One-third of a rectangular vessel is filled with mercury (sp. gr. 13.6) and the remainder with salt water (sp. gr. 1.04). If $3h$ be the depth of the vertical side and a be breadth, find the fluid thrust on a vertical side.

Ans. 10.96 gah^2.

12. A rectangular vessel contains three liquids which do not mix and whose sp. gravities are 1.0, 1.2 and 1.6, the thickness of which are 10, 9 and 5 cms respectively. Compare the total thrusts of the liquids on the different parts of a side of the vessel.

Ans. 40 : 87 : 92

13. The sides of a cistern are vertical. Its base is a horizontal regular hexagon each side of which is $\sqrt{3}$ metres long. Find the depth if, when it is full of water, the thrust on each of its sides in the same as on its base.

Ans. 9 m

14. An ellipse is placed with its major axis on the surface of water and its plane vertical. A circle (the auxiliary circle) is described on the major axis as diameter. Find the thrust of water on the portion of the area enclosed between the auxiliary circle and the ellipse.

Ans. $2/3a\,(a^2 - b^2)w$

15. Find the total thrust of water on a semi-circular area of radius 6 inches immersed in water with its diameter in the surface and the plane inclined to the vertical at an angle of 60°.

(12 inches = 1 foot)

Ans. $2\frac{29}{48}$ lbs. wt.

[**Hint:** Refer Example 8.

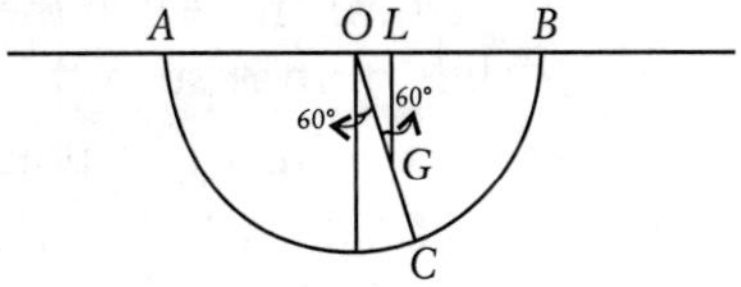

$OG = 4a/3\pi = 4 \times 6/3\pi = 8/\pi$ inches

$= 8/12\pi$ foot

$= 2/3\pi$ foot

$\bar{z} = LG = OG \cos 60°$

$= 2/3\pi \,.\, 1/2 = 1/3\pi$ foot

$S = 1/2\pi\, r^2 = 1/2\pi\, (1/2)^2$ | 6 inches = 1/2 foot

$= \pi/8$ sq. ft.

$\therefore$ Pressure required $= w\, S\, \bar{z}$

$$= w\frac{\pi}{8}\frac{1}{3\pi}$$

$= w/24$

$= 62.5/24$ lbs. wt.

$= 2\frac{29}{48}$ lbs. wt.]

16. A conical vessel filled with water stands with its plane base on a horizontal table. Prove that the thrust of the liquid on the base of the cone is three times the weight of the contained water. How do you explain the result?

[**Hint:** Thrust = $wS\bar{z}$

$= w\pi r^2 h = 3\,(1/3\pi r^2 hw)$]

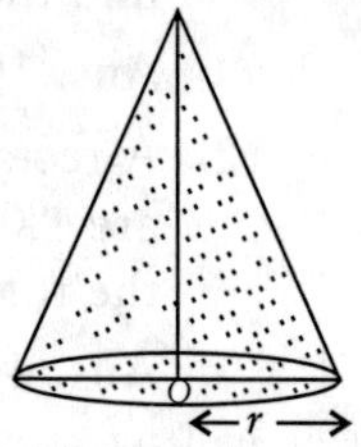

17. A quantity of water which just fills a cone, whose height is h and the radius of whose base is r, is poured into a cylinder, the radius of whose base is also r. Compare the thrust on the two bases, the axis of each vessel being vertical.

[**Hint:** $1/3\pi r^2 h = \pi r^2 H$

$\Rightarrow H = h/3$

$P_1 = w\pi r^2 h$ (cone)

$P_2 = w\pi r^2 h/3$ (cylinder)

$\therefore P_1 : P_2 = 3 : 1$]

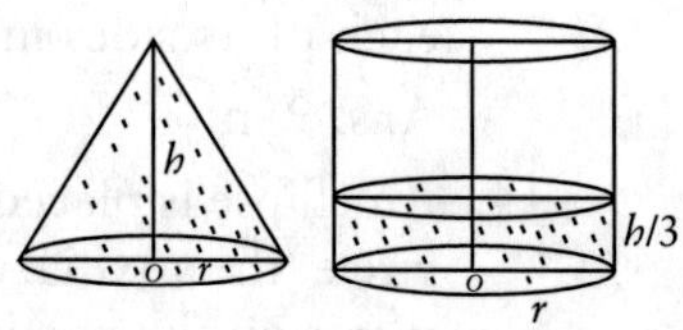

18. A rectangular vessel is full of water. Compare the fluid pressures on the lower and upper halves of a vertical side of the vessel.

Ans. 3 : 1.

19. $ABCD$ is a rectangle immersed in a homogeneous liquid with AB in the free surface and AD vertical. Show how to draw a straight line through A, dividing the rectangle into two portions equally pressed.

Ans. The dividing line is AE where E is on CD such that $EC = 1/h\ DC$

20. $ABCD$ is a rectangle, AB being equal to $2BC$. It is immersed in water with its plane vertical and AB in the surface. Show how to divide the area into two parts by a straight line from A such that the pressures on the two parts may be equal.

Ans. Join A to a point E in DC such that $DE : EC = 3 : 4$.

21. The side AB of a triangle ABC is in the surface of a fluid and a point D is taken in AC, such that the pressures on the triangles BAD and BDC are equal. Find the ratio $AD : AC$ and the ratio $AD : DC$.

Ans. $1 : \sqrt{2} : 1 : (\sqrt{2} - 1)$ or $(\sqrt{2} + 1) : 1$

[**Hint:** Refer Ex. 17. $AD : AC = 1 : \sqrt{2}$

$\Rightarrow AD/AC = 1/\sqrt{2}$

$\Rightarrow AC/AD = \sqrt{2}/1$

$\Rightarrow AC/AD - 1 = \sqrt{2} - 1/1$

$\Rightarrow AC - AD/AD = \sqrt{2} - 1/1$

$\Rightarrow DC/AD = \sqrt{2} - 1/1$

$\Rightarrow DC/AD = 1/\sqrt{2} + 1$

$\Rightarrow AD : DC = (\sqrt{2} + 1) : 1$]

22. The lighter of the two liquids of density ρ rests on the heavier of density σ, to a depth b. If square of side a is immersed in a vertical position with one side in the surface of the upper liquid and the thrusts on the two portions of the square in contact with the two liquids be equal, prove that

$$b\rho(3b - 2a) = \sigma(a - b)^2.$$

[**Hint:** Refer Ex. 20

$$P_1 = P_2$$

$$\Rightarrow \quad 1/2\, g\rho ab^2 = ga\,(a - b)\,\{\rho b + \sigma\,(a - b)/2\}$$

$$\Rightarrow \quad 1/2\, abg\rho\,\{b - 2\,(a - b)\} = 1/2\, g\sigma\,(a - b)^2\, a$$

$$\Rightarrow \quad b\rho\,(3b - 2a) = \sigma\,(a - b)^2]$$

23. A triangle ABC has its plane vertical and the side AB is in the surface of the liquid in which the triangle is immersed. Divide it by the straight lines drawn from B into n triangles on each of which the pressure shall be the same.

Ans. Distances of the points of division from B are in the ratio. $1 : \sqrt{2} : \sqrt{3} : \ldots : \sqrt{n}$

[**Hint:** Refer Ex. 22]

24. A parallelogram is immersed in water, one side being in the surface. Divide it into n parts by horizontal lines so that the thrusts on these parts may be equal. Show that the depths of the dividing lines are proportional to the square roots of the natural numbers.

25. A rectangle is immersed in a homogeneous liquid with one side in the surface. Show how to draw horizontal lines dividing it into n parts, the thrusts on which are equal.

Ans. $h_1 : h_2 : h_3 : \ldots : h_n = 1 : \sqrt{2} : \sqrt{3} : \ldots : \sqrt{n}$; $h_1, h_2, h_3, \ldots, h_n$ being the depths of the horizontal lines from the free surface.

26. A cone full of water is placed on its side on a horizontal table. If the thrust on the plane base is $1\frac{1}{2}$ times the weight of the contained liquid, show that the vertical angle of the cone is 60°.

[**Hint:** Refer Ex. 23.

$$\frac{\text{Thrust on the base}}{\text{Weight of the liquid contained}} = 3\sin\alpha = \frac{3}{2}$$

$$\Rightarrow \alpha = 30°$$

$$\Rightarrow 2\alpha = 60°]$$

27. A cone full of water is placed or its side on a horizontal table. Show that the thrust on the base is $3r/\sqrt{r^2 + h^2}$ times the weight of contained liquid, where h is the height and r is the radius of the base of the cone.

28. A conical vessel 10 cms high on a flat circular base of 5 cms radius, is filled with water. Calculate the vertical thrust on the base when the vertex is upwards.

Ans. 625 π gms. wt.

29. A rectangular plate is immersed vertically in a liquid with one side in the surface. Divide it into two parts by a line parallel to the surface such that the pressure on the upper half is double the pressure on the lower half.

Ans. The line should be drawn at a depth $\sqrt{\frac{2}{3}}$ b, where b is the vertical side of the rectangular plate.

30. A rectangular lamina is immersed vertically in a liquid with one side in the surface. Show how to draw a horizontal line dividing it into two parts such that the ratio of the thrusts on the upper and lower parts is 9 : 7.

Ans. The horizontal line should be drawn at a vertical depth $3b/4$ where b is the vertical side of the rectangular lamina.

HARDER ILLUSTRATIVE EXAMPLES

Example 1: The central board of a yacht is in the form of a trapezoid in which the parallel sides are 3 ft. and 5 ft. respectively, and the side perpendicular to these two is 4 ft., in length. Assuming that the last named side is parallel to the surface of the water at a depth of 1 foot and that the parallel sides are vertical, find the total thrust on the board.

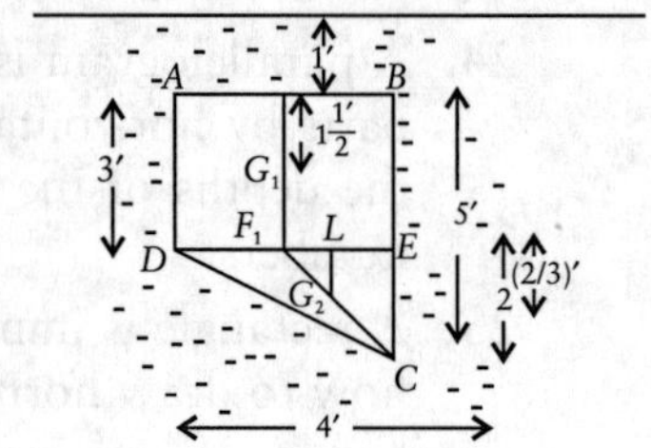

Sol. Let $ABCD$ be the central board of the yacht with side AB (= 4′) one foot below the free surface of the liquid and the sides AD (= 3′) and BC (= 5′) are vertical. From D, draw perpendicular DE to BC.

Let G_1 and G_2 be the C.G. of the rectangle $ABED$ and Δ DEC respectively. From G_2, draw perpendicular G_2L to DE. Then, by geometry,

$$G_2L = 1/3\ CE = 1/3\ (5 - 3) = \frac{2'}{3}$$

Depth of C.G. G_1 below the free surface of the liquid $= 1 + 1\frac{1}{2} = \frac{5'}{2}$

Depth of C.G. G_2 below the free surface of the liquid $= 1 + 3 + 2/3 = \frac{14'}{3}$

Let w be the weight per unit volume of water. Thrust on the rectangle $ABED$ (P_1)

$= wS\bar{z}$

$= w\ (4 \times 3)\ (1 + 1\frac{1}{2})$ $\quad = w\ 12\ .\ 5/2$

$= 30\ w$ $\quad = 30 \times 62.5$

$= 30 \times 125/2$ $\quad = 15 \times 125$ lbs. wt.

Thrust on the triangle DEC (P_2)

$= wS\bar{z}$

$= w\ (4 \times 2/2)\ 14/3$ $\quad = 56/3\ w$

$= 56/3 \times 62.5$ $\quad = 56/3 \times 125/2$

$= 28 \times 125/3$ lbs. wt.

∴ Total thrust on the board

= Thrust on the rectangle *ABED* + Thrust on the triangle *DEC*

$= P_1 + P_2 \qquad = 15 \times 125 + 28 \times 125/3$

$= 125\ (15 + 28/3) \qquad = 125\ (73/3)$

$= 9125/3 \qquad = 3041\frac{2}{3}$ lbs. wt.

Example 2: A parallelogram *ABCD* is immersed vertically in a homogeneous liquid with the side *AB* of length $2a$ in the surface. Points *F* and *E* are taken on *AB* and *DC* respectively such that $EC = x$ and $AF = y$. If *EF* divides *ABCD* into two parts on which the pressures are equal, show that $2x - y = a$.

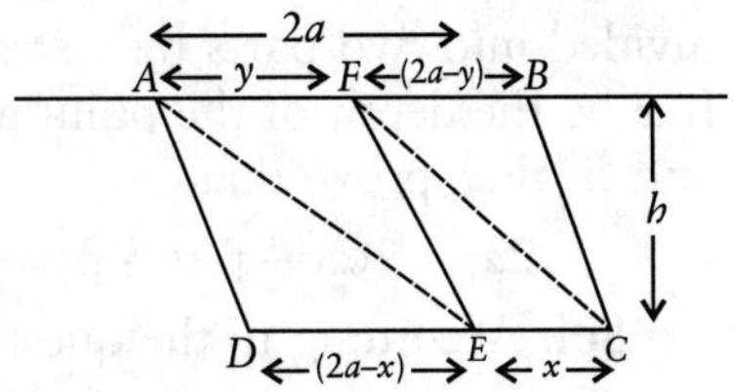

Sol. Let h be the height of the parallelogram taking *DC* as base, i.e., the vertical distance between sides *AB* and *DC* is h.

According to the question,

Pressure on part *AEFD* = Pressure on part *FBCE* ...(1)

Join *AE* and *FC*

Let w be the weight per unit volume of the liquid.

Pressure on Δ *ADE* $= wS\bar{z}$

$= w\ (2a - x)h/2\ .\ 2h/3 \qquad = w\ (2a - x)h^2/3$

Pressure on Δ *AFE* $= wS\bar{z} = w\ .\ y\ h/2\ .h/3 = wyh/6$

∴ Pressure on part *AEFD* = Pressure on Δ *ADE* + Pressure on Δ *AFE*

$$= \frac{w(2a - x)h^2}{3} + \frac{wyh^2}{6}$$

$$= \frac{wh}{}(4a - 2x + y) \qquad \text{...(2)}$$

Pressure on Δ *FEC* $= wS\bar{z}$

$$= \frac{wxh}{2}.\frac{2h}{3} \qquad = \frac{wxh^2}{3}$$

Pressure on Δ *FBC* $= wS\bar{z}$

$$= \frac{w(2a - y)h}{2}.\frac{h}{3} \qquad = \frac{w(2a - y)h^2}{6}$$

∴ Pressure on part *FBCE* = Pressure on Δ *FEC* + Pressure on Δ *FBC*

$$= \frac{wxh^2}{3} + \frac{w(2a - y)h^2}{6}$$

$$= \frac{wh^2}{6}(2x + 2a - y) \qquad \text{...(3)}$$

From (1), (2) and (3), we get

$$\frac{wh^2}{6}(4a-2x+y)=\frac{wh^2}{6}(2x+2a-y)$$

$\Rightarrow$ $4a - 2x + y = 2x + 2a - y$

$\Rightarrow$ $4x - 2y = 2a$

$\Rightarrow$ $2x - y = a$ | dividing throughout by 2

Example 3: A triangle, completely immersed in a liquid with the vertex C in its surface is divided into two parts by a straight line through A so that the thrusts on two parts are equal. If x be the depth of the point at which the dividing line cuts CB, and α, β be the depths of A and B, then prove that

$$2x^2 + 2\alpha x - \beta(\alpha + \beta) = 0$$

Sol. According to the question,

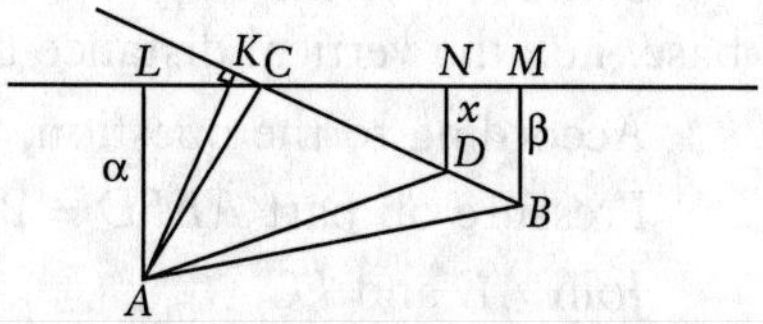

Pressure on $\Delta\, ACD$ = Pressure on $\Delta\, ABD$

$\Rightarrow$ Pressure on $\Delta\, ACD$ = 1/2 Pressure on $\Delta\, ABC$ ...(1)

Let w be the weight per unit volume of the liquid. Draw $AK \perp BC$.

Pressure on $\Delta\, ACD = wS\bar{z}$

$$= \frac{w\,CD.AK}{2}.\left(\frac{0+\alpha+x}{3}\right)$$

$$= \frac{w\,CD.AK(\alpha+x)}{6} \quad ...(2)$$

Pressure on $\Delta\, ABC = wS\bar{z}$

$$= \frac{w\,CB.AK}{2}.\left(\frac{0+\alpha+\beta}{3}\right)$$

$$= \frac{w\,CB.AK(\alpha+\beta)}{6} \quad ...(3)$$

From (1), (2) and (3), we get

$$\frac{w\,CD.AK(\alpha+x)}{6}=\frac{1}{2}\frac{w\,CB.AK(\alpha+\beta)}{6}$$

$\Rightarrow$ $2CD\,(\alpha + x) = CB\,(\alpha + \beta)$

$$\Rightarrow \quad \frac{CD}{CB}=\frac{\alpha+\beta}{2(\alpha+x)} \quad ...(4)$$

$\Delta\, CDN \sim \Delta\, CBM$

$$\therefore \quad \frac{CD}{CB}=\frac{DN}{BM}$$ | by property of similar triangles

$$\Rightarrow \quad \frac{CD}{CB}=\frac{x}{\beta} \quad ...(5)$$

From (4) and (5), we get

$$\frac{\alpha+\beta}{2(\alpha+x)} = \frac{x}{\beta}$$

$$\Rightarrow \quad 2x(\alpha + x) = \beta(\alpha + \beta)$$

$$\Rightarrow \quad 2x^2 + 2\alpha x - \beta(\alpha + \beta) = 0$$

Example 4: A parallelogram is immersed in a homogeneous fluid of density ρ, open to the atmospheric pressure π, with the side AB in the surface. E is a point on AB such that $AE = 2/3\ AB$. A straight line joins E to a point F on CD. Show that if the thrusts on $AEFD$ and $EBCF$ are equal, DF is given by

$$6\,DF(3\pi + 2\,g\rho b) = AB(6\pi + 5\,g\rho b),$$

where b is the depth of CD.

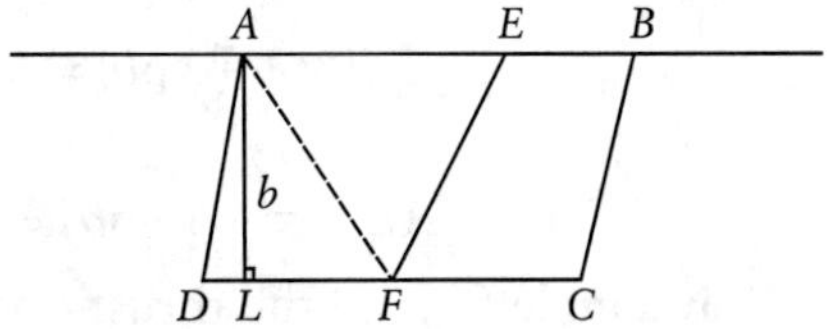

According to the question,

Thrust on $AEFD$ = Thrust on $EBCF$

$\Rightarrow$ Thrust on $AEFD$ = 1/2 Thrust on $ABCD$ ...(1)

b is the depth of CD

$\Rightarrow$ Height of the parallelogram is b.

Let $\bar{z}$ be the depth of C.G. of the area $AEFD$ from AB. Then,

$$\left\{\frac{(AE+DF)b}{2}\right\}\bar{z} = \left(\frac{AE.b}{2}\right).\frac{b}{3} + \left(\frac{DF.b}{2}\right).\frac{2b}{3}$$

$$\Rightarrow \quad (AE + DF)\,\bar{z} = (AE + 2DF)\,b/3$$

$$\Rightarrow \quad \bar{z} = \left(\frac{AE+2DF}{AE+DF}\right).\frac{b}{3}$$

$$\Rightarrow \quad \bar{z} = \left(\frac{\frac{2}{3}AB+2DF}{\frac{2}{3}AB+DF}\right)\frac{b}{3} \qquad ...(2)$$

Thrust on $AEFD$

$$= \left(\frac{AE+DF}{2}\right).b\,(\pi + g\rho\bar{z}) \qquad ...(3)$$

Thrust on $ABCD$

$$= AB\,.\,b\,(\pi + g\rho\,\frac{b}{2}) \qquad ...(4)$$

From (1), (3) and (4), we get

$$\left(\frac{AE+DF}{2}\right).b\,(\pi + g\rho\bar{z}) = \frac{1}{2}\,AB.\,b\,(\pi + g\rho\frac{b}{2})$$

$$\Rightarrow \quad (AE + DF)(\pi + g\rho\bar{z}) = AB\,.\,(\pi + g\rho\,\frac{b}{2})$$

$$\Rightarrow \quad \left(\frac{2}{3}AB+DF\right)\left\{\pi+g\rho\left(\frac{\frac{2}{3}AB+2DF}{\frac{2}{3}AB+DF}\right)\frac{b}{3}\right\}=AB(\pi+g\rho\frac{b}{2}) \quad | \text{ From (2)}$$

$$\Rightarrow \quad \left(\frac{2}{3}AB+DF\right)\pi+g\rho\left(\frac{2}{3}AB+2DF\right)\frac{b}{3}=AB\left(\pi+\frac{g\rho b}{2}\right)$$

$$\Rightarrow \quad DF(\pi+\frac{2}{3}g\rho b)=AB\left(\pi+\frac{g\rho b}{2}-\frac{2\pi}{3}-\frac{2}{9}g\rho b\right)$$

$$\Rightarrow \quad DF(\pi+\frac{2}{3}g\rho b)=AB\left(\frac{\pi}{3}+\frac{5}{18}g\rho b\right)$$

$$\Rightarrow \quad AB\ (6\pi + 5\ g\rho b) = 6\ DF\ (3\pi + 2\ g\rho b)$$

Example 5: Find the thrust on a vertical quadrilateral which has one side of length a in the surface, and the opposite side of length b parallel to it at a depth h.

If the fluid consists of a top layer of density ρ and thickness 1/2 h, and the rest of density σ, prove that the thrust on the quadrilateral is

$gh^2/48\ \{(7a + 11b)\ \rho + (a + 5b)\ \sigma\}$

Sol. (1)

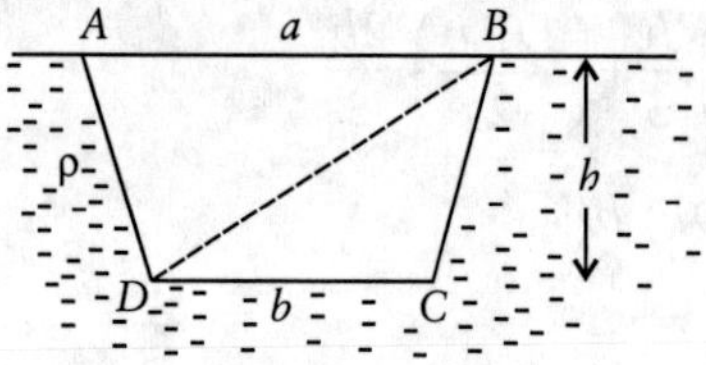

Join BD.

Thrust on quadrilateral $ABCD$

= Thrust on Δ ABD + Thrust on Δ BCD

$$= g\rho.\frac{ah}{2}.\frac{h}{3}+g\rho\frac{bh}{2}.\frac{2}{3}b$$

$$=\frac{g\rho h^2 a}{6}+\frac{g\rho h^2 b}{3} \qquad =\frac{g\rho h^2}{6}(a+2b)$$

(2)

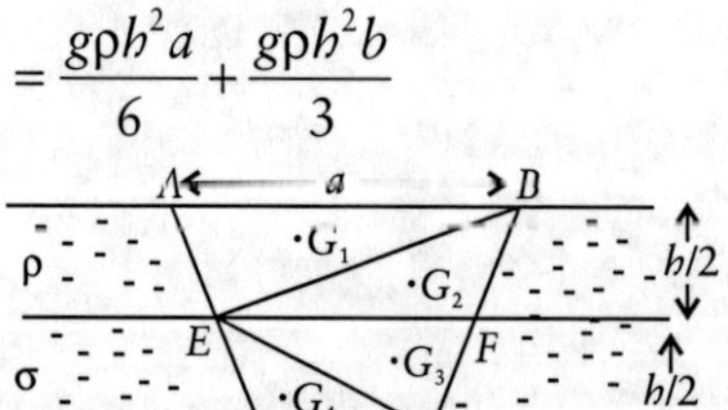

By geometry,

E and F are the mid-points of AD and BC respectively.

$\therefore \quad EF = 1/2\ (AB + DC) = 1/2\ (a + b)$

Join BE and CE.

Then the quadrilateral $ABCD$ is divided into four triangles. $\Delta\ ABE$, $\Delta\ BEF$, $\Delta\ CEF$ and $\Delta\ EDC$. Let the C.G. of these triangles be G_1, G_2, G_3 and G_4 respectively. Then,

$$\text{Depth of } G_1 \text{ below } AB = \frac{0+0+\frac{h}{2}}{3} = \frac{h}{6}$$

$$\text{Depth of } G_2 \text{ below } AB = \frac{0+\frac{h}{2}+\frac{h}{2}}{3} = \frac{h}{3}$$

$$\text{Depth of } G_3 \text{ below } EF = \frac{0+0+\frac{h}{2}}{3} = \frac{h}{6}$$

$$\text{Depth of } G_4 \text{ below } EF = \frac{0+\frac{h}{2}+\frac{h}{2}}{3} = \frac{h}{3}$$

Thrust on $\Delta\ ABE = wS\bar{z}$

$$= g\rho.\frac{a.\frac{h}{2}}{2} = \frac{h}{6} \qquad = \frac{g\rho a h^2}{24}$$

Thrust on $\Delta\ BEF = wS\bar{z}$

$$= g\rho.\frac{\frac{a+b}{2}.\frac{h}{2}}{2}.\frac{h}{3} \qquad = \frac{g\rho(a+b)h^2}{24}$$

Thrust on $\Delta\ CEF$

$$= \frac{\frac{1}{2}(a+b)\frac{h}{2}}{2}\left\{g\rho.\frac{h}{2}+g\sigma\frac{h}{6}\right\} \qquad = \frac{g\rho(a+b)h^2}{16}+\frac{g\sigma(a+b)h^2}{48}$$

Thrust on $\Delta\ EDC$

$$= \frac{b.\frac{h}{2}}{2}.\left\{g\rho\frac{h}{2}+g\sigma.\frac{h}{3}\right\} \qquad = \frac{g\rho b h^2}{8}+\frac{g\sigma b h^2}{12}$$

$\therefore$ Total thrust on the quadrilateral $ABCD$

= Thrust on $\Delta\ ABE$ + Thrust on $\Delta\ BEF$ + Thrust on $\Delta\ CEF$ + Thrust on $\Delta\ EDC$

$$= \frac{g\rho a h^2}{24}+\frac{g\rho(a+b)h^2}{24}+\frac{g\rho(a+b)h^2}{16}+\frac{g\sigma(a+b)h^2}{48}+\frac{g\rho b h^2}{8}+\frac{g\sigma b h^2}{12}$$

$$= \frac{g\rho h^2}{48}\{2a+2a+2b+3a+3b\}+\frac{g\sigma h^2}{48}\{a+b+4b\}$$

$$= \frac{gh^2}{48}\{\rho(7a+5b)+\sigma(a+5b)\}$$

Aliter to (2)

Join BD and HC

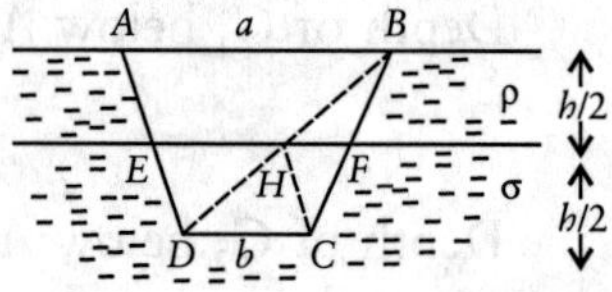

Thrust on $\Delta\, ABD$ = Thrust on $\Delta\, ABD$ in a liquid of density ρ + Thrust on $\Delta\, DHE$ in a liquid of density $(\sigma - \rho)$

$$= g\rho\frac{ah}{2}.\frac{h}{3}+\frac{\frac{a}{2}.\frac{h}{2}}{2}\frac{1}{3}\left(\frac{h}{2}\right)g(\sigma-\rho)$$

$$= \frac{g\rho a h^2}{6}+\frac{gah^2}{48}(\sigma-\rho) \qquad = \frac{gah^2}{48}(7\rho+\sigma) \qquad \text{...(1)}$$

Thrust on $\Delta\, BCD$ = Thrust $\Delta\, BCD$ in a liquid of density ρ + Thrust on the lower trapezium $HFCD$ in a liquid of density $(\sigma - \rho)$ ...(2)

Let $\bar{z}$ be the depth of C.G. of trapezium $HFCD$ below EF. Then,

$$\frac{b+\frac{b}{2}}{2}.\frac{h}{2}\bar{z} = \frac{b.\frac{h}{2}}{2}\frac{2}{3}\frac{h}{2}+\frac{\frac{b}{2}.\frac{h}{2}}{2}.\frac{1}{3}.\frac{h}{2} \qquad \mid \because HF = b/2 \text{ (by geometry)}$$

$$\Rightarrow \quad \frac{3bh}{8}\bar{z} = \frac{bh^2}{12}+\frac{bh^2}{48} \qquad \Rightarrow \quad \frac{3bh}{8}\bar{z} = \frac{5bh^2}{48}$$

$$\Rightarrow \quad \bar{z} = \frac{5h}{18} \qquad \text{...(3)}$$

From (2) and (3),

$$\text{Thrust on } \Delta\, BCD = \frac{bh}{2}\frac{2}{3}h\,g\rho+\frac{\left(b+\frac{b}{2}\right)\frac{h}{2}}{2}.\frac{5h}{18}.g(\sigma-\rho)$$

$$= \frac{bh^2 g\rho}{3}+\frac{3bh^2}{8}\frac{5}{18}.g(\sigma-\rho)$$

$$= \frac{bh^2 g\rho}{3}+\frac{5bh^2 g\sigma}{48}-\frac{5bh^2 g\rho}{48}$$

$$= \frac{11bh^2 g\rho}{48}+\frac{5bh^2 g\sigma}{48}$$

$$= \frac{bh^2 g}{48}(11\rho+5\sigma) \qquad \text{...(4)}$$

$\therefore$ Total thrust on the quadrilateral $ABCD$

= Thrust on $\Delta\, ABD$ + Thrust on $\Delta\, BCD$

$$= \frac{gah^2}{48}(7\rho + \sigma) + \frac{bh^2 g}{48}(11\rho + 5\sigma)$$

$$= \frac{gh^2}{48}[(7a + 11b)\rho + (a + 5b)\sigma]$$

Example 6: A rectangular area is immersed in a heavy liquid with two sides horizontal, and is divided by horizontal lines into strips on which the total thrusts are equal. Prove that if a, b, c are the depths of the consecutive strips,

$$a(a + b)(b - c) = c(b + c)(a - b)$$

Sol. Let $ABCD$ be a rectangular area immersed in a liquid with two sides AB and DC horizontal. Let $AB = CD = x$. Let h be the depth of the top of the upper must strip of the three consecutive strips below the free surface. Then,

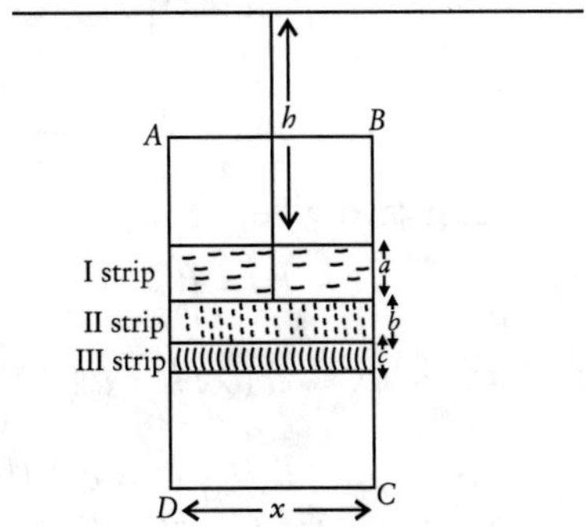

Thrust on the first strip $= wS\bar{z}$

$= g\rho\,(xa)\,(h + a/2)$

Thrust on the second strip $= g\rho\,(xb)\,(h + a + b/2)$

Thrust on the third strip $= g\rho\,(xc)\,(h + a + b + c/2)$

According to the question, these thrusts are equal

$\Rightarrow\ g\rho\,(xa)\,(h + a/2) = g\rho\,(xb)\,(h + a + b/2)$

$= g\rho\,(xc)\,(h + a + b + c/2)$

$\Rightarrow\ a(h + a/2) = b(h + a + b/2) = c(h + a + b + c/2)$

$$\Rightarrow \frac{h+\frac{a}{2}}{bc} = \frac{h+a+\frac{b}{2}}{ca} = \frac{h+a+b+\frac{c}{2}}{ab} \quad | \text{ dividing by } abc$$

$$= \frac{\left(h+c+\frac{b}{2}\right)-\left(h+\frac{a}{2}\right)}{ac-bc} = \frac{\left(h+a+b+\frac{c}{2}\right)-\left(h+a+\frac{b}{2}\right)}{ab-ca}$$

$$\Rightarrow \frac{\frac{1}{2}(a+b)}{c(a-b)} = \frac{\frac{1}{2}(b+c)}{a(b-c)}$$

$\Rightarrow\ a\,(a + b)\,(b - c) = c\,(b + c)\,(a - b)$

Aliter: Let the depth of C.G. of the middle strip be h

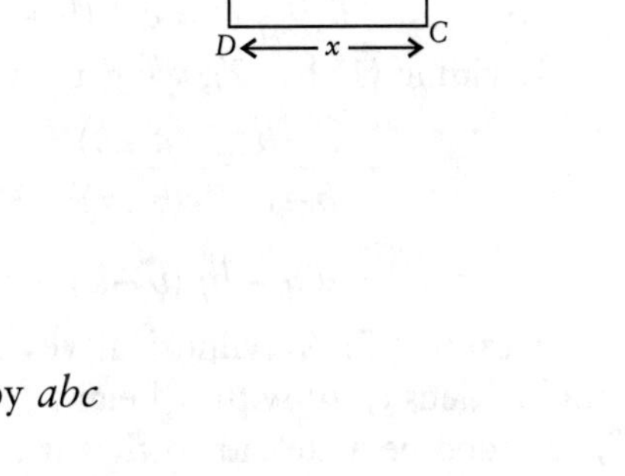

Then, depth of C.G. of the upper strip

$= h - (a + b/2)$

and, depth of C.G. of the lower strip

$= h + (b + c/2)$

Pressure on upper strip $(P_1) = w(ax)\left\{h - \frac{a+b}{2}\right\}$

Pressure on middle strip (P_2) = $w(bx)h$

Pressure on lower strip (P_3) = $w(cx)\left\{h-\frac{b+c}{2}\right\}$

According to the question,

$$P_1 = P_2 = P_3$$

First two give

$$P_1 = P_2$$

$$\Rightarrow \quad w(ax)\left\{h-\frac{a+b}{2}\right\} = w(bx)h$$

$$\Rightarrow \quad h(a-b) = \frac{a(a+b)}{2} \qquad ...(1)$$

Last two give

$$P_2 = P_3$$

$$\Rightarrow \quad w(bx)h = w(cx)\left\{h+\frac{b+c}{2}\right\}$$

$$\Rightarrow \quad h(b-c) = c/2\ (b+c) \qquad ...(2)$$

Dividing (1) by (2), we get

$$\frac{a-b}{b-c} = \frac{a(a+b)}{c(b+c)}$$

$$\Rightarrow \quad a(a+b)(b-c) = c(b+c)(a-b)$$

Example 7: A cylindrical vessel on a horizontal circular base of radius a, is filled with a liquid density w with a height h. If now a sphere of radius c and density greater than w is suspended be a thread so that it is completely immersed, prove that the increase of the whole pressure on the curved surface is

$$\frac{8\pi}{3a}wc^3\left(h+\frac{2c^3}{3a^2}\right)g$$

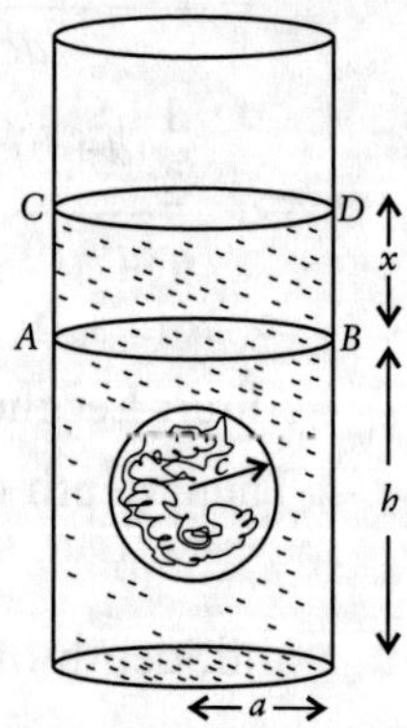

Sol. Let AB be the initial level of the liquid before the sphere is immersed. Let CD be the level of the liquid after the sphere is immersed. Let x be the increased height of the level when the level is raised. Then,

$$AC = BD = x$$

The volume of the liquid displaced by the sphere must be equal to the volume of the sphere.

$$\therefore \quad \pi a^2 x = \frac{4}{3}\pi c^3$$

$$\Rightarrow \quad x = \frac{4}{3}\frac{c^3}{a^2}$$

Pressure on the curved surface in the beginning $= wg.\ 2\pi ah.\ 1/2h$

$$= w\pi ah^2 g$$

Pressure on the curved surface after immersion

$$= wg2\pi a\ (h + x)\ .\ 1/2\ (h + x)$$
$$= w\pi a\ (h + x)^2\ g$$

$\therefore$ Increase of the whole pressure on the curved surface

$$= wga\ \{(h + x)^2 - h^2\}^2$$
$$= w\pi a\ (x^2 + 2hx)\ g$$
$$= w\pi a \left\{ \left(\frac{4c^3}{3a^2} \right)^2 + 2h \left(\frac{4c^3}{3a^2} \right) \right\} g \quad | \text{ From (1)}$$
$$= w\pi a \frac{4c^3}{3a^2} \left(\frac{4c^3}{3a^2} + 2h \right) g$$
$$= \frac{8\pi wc^3}{3a} \left(\frac{2c^3}{3a^2} + h \right) g$$

Example 8: A cylindrical tumbler, half filled with a liquid of density ρ, is filled up with a liquid of density ρ′ which does not mix with the former one. Show that the thrust on the base of the tumbler is to the whole pressure on its curved surface as

$$2r\ (\rho + \rho')\ \text{to}\ h\ (\rho + 3\rho')$$

where h is the height and r the radius of the base of the cylinder.

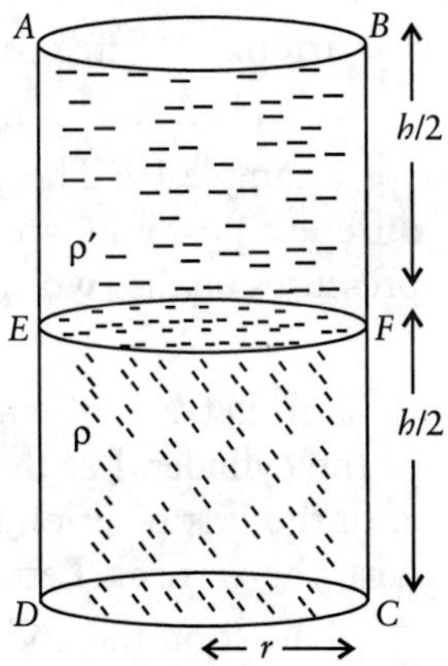

Sol. Let $ABCD$ be a cylindrical tumbler, half filled with a liquid of density ρ, which does not mix the former one. Clearly, each liquid stands to a height $h/2$.

Thrust on the base of the tumbler

$$= g\rho' \cdot h/2 \cdot \pi r^2 + g\rho \cdot h/2 \cdot \pi r^2$$
$$= \frac{g\pi r^2 h}{2}(\rho' + \rho) \quad \text{...(1)}$$

Whole pressure on the curved surface of the cylinder

= Whole pressure on the curved surface of the upper half + Whole pressure on the curved surface of the lower half

$$= g\rho'\ (2\pi r\ h/2)\ (1/2 \cdot h/2) + (2\pi r\ h/2)\ (g\rho'\ h/2 + g\rho \cdot 1/2 \cdot h/2)$$
$$= \frac{1}{4}\pi g\rho' rh^2 + \frac{\pi g\rho' rh^2}{2} + \frac{\pi g\rho rh^2}{4}$$
$$= 1/4\ \pi grh^2\ (3\rho' + \rho) \quad \text{...(2)}$$

From (1) and (2), we get

$$\frac{\text{Thrust on the base of the tumbler}}{\text{Whole pressure on the curved surface of the cylinder}}$$

$$= \frac{\dfrac{g\pi r^2 h(\rho'+\rho)}{2}}{\dfrac{\pi g r h^2(3\rho'+\rho)}{4}} = \frac{2r(\rho'+\rho)}{h(3\rho'+\rho)}$$

Example 9: A cylinder is filled with equal volumes of n different fluids which do not mix; the density of the highest is ρ, that of the next is 2ρ, and so on, that of the lowest being $n\rho$. Show that the whole pressures on the different portions of the curved surface of the cylinder are in the ratio $1^2 : 2^2 : 3^2 : .... : n^2$.

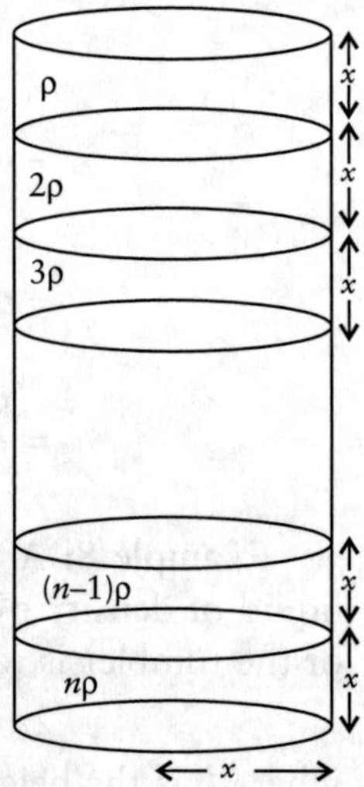

Sol. Let r be the radius of the base of the cylinder and let each liquid stand to a height x.

Whole pressure on the curved surface of the nth portion of the cylinder.

$= (g\rho \cdot x + g \cdot 2\pi \cdot x + g \cdot 3\rho \cdot x + ... + g\,(n-1)\,\rho \cdot x + gn\rho \cdot x/2)\; 2\pi rx$

$= 2\pi\, rx\; g\rho x\;\{1 + 2 + 3 + ... + (n-1) + n/2\}$

$$= 2\pi r\; g\rho x^2 \left\{\frac{n(n-1)}{2} + \frac{n}{2}\right\} \qquad = 2\pi r\; g\rho x^2\, \frac{n^2}{2}$$

$= \pi r g \rho x^2\, n^2$

Putting $n = 1, 2, 3, ..., n$, the whole pressures on the different portions of the curved surface of the cylinder are

$\pi rg\, \rho x^2\, 1^2,\ \pi rg\, \rho x^2\, 2^2,\ \pi rg\, \rho x^2\, 3^2,\ ....,\ \pi rg\, \rho x^2\, n^2$

Hence, their ratios are

$\pi rg\, \rho x^2\, 1^2 : \pi rg\, \rho x^2\, 2^2 : \pi rg\, \rho x^2\, 3^2 : .... : \pi rg\, \rho x^2\, n^2$

$= 1^2 : 2^2 : 3^2 : .... : n^2$

Example 10: The inclinations of the axis of a submerged solid cylinder to the vertical in two different positions are complementary to each other. If P and P^1 be the difference between the pressures on the two ends in the two cases, prove that the weight of the displaced fluid is equal to

$$(P^2 + P^1)^{1/2}$$

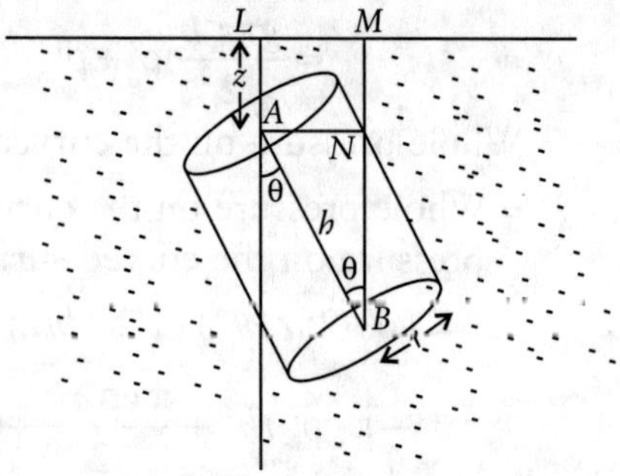

Sol. Let h be the height and r be the radius of the base of the cylinder. Let the depth of A from the free surface be z. In the first position let θ be the angle that the axis makes with the vertical. Let ρ be the density of the fluid.

Thrust on the end $A = g\rho\pi r^2\, z$

Thrust on the end $B = g\rho\, \pi r^2\, (z + h\cos\theta)$

$\therefore P =$ Difference between the pressures on the two ends A and B.

= Thrust on the end B – Thrust on the end A.

$= g\rho\pi x^2\, (z + h\cos\theta) - g\rho\pi x^2\, z$

$= g\rho\, \pi x^2\, h\cos\theta$ ...(1)

In the second position, the angle of inclination is $(90° - \theta)$.

$\therefore\quad P' = g\rho\, \pi r^2 h \cos(90° - \theta)$

$\Rightarrow\quad P' = g\rho\, \pi r^2 h \sin\theta$ ...(2)

From (1) and (2)

$$P^2 + P'^2$$

$$= (g\rho\ \pi r^2\ h \cos\theta) + (g\rho\ \pi r^2\ h \sin\theta) = (g\rho\ \pi r^2 h)^2$$

$$\Rightarrow \quad \sqrt{P^2 + P'^2} = g\rho\pi r^2 h = \text{Weight of the displaced fluid.}$$

Example 11: A hollow weightless cone of semi-vertical angle α and height h, is filled with liquid and freely hung from a point on the rim of the base. Show that the thrust of liquid on the base is

$$\frac{4\pi h^3 \tan^3\alpha \sin\alpha g\rho}{\sqrt{1+15\sin^2\alpha}}$$

where w is the weight per unit volume of the liquid.

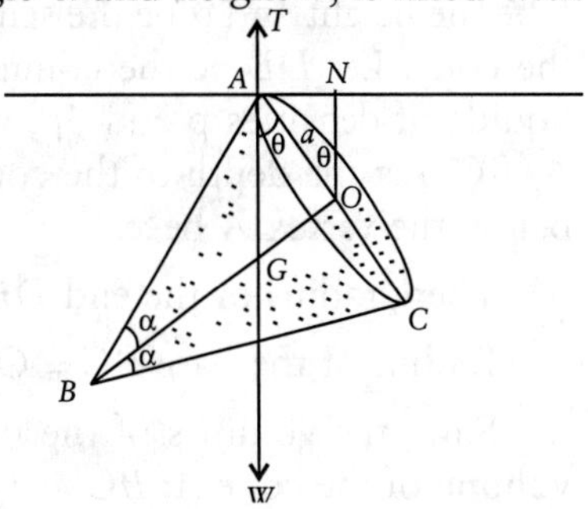

Sol. Let B be the vertex, α the semi-vertical angle and h be the height of the cone BAC.

Let the base radius of the cone be a.

Then, $\qquad \tan\alpha = a/h$ ...(1)

The cone is hung from a point A on the rim of the base. Let W be the weight of the liquid contained in the cone. Let T be the tension in the string.

Since these are only two forces W and T, therefore for the equilibrium of the system, T and W must act along the same line. Moreover, since W acts vertically downwards, therefore this line must be vertical. Hence, if G is the C.G. of the contained liquid, then AG must be vertical and $OG = h/4$.

Let $\angle OAG = \theta$. Then,

$$\tan\theta = \frac{OG}{OA} = \frac{\frac{h}{4}}{a} = \frac{h}{4a} = \frac{1}{4\tan\alpha} \qquad \text{| From (1)}$$

$$\therefore \quad \cos\theta = \frac{4\tan\alpha}{\sqrt{1+16\tan^2\alpha}}$$

$$= \frac{4\sin\alpha}{\sqrt{\cos^2\alpha + 16\sin^2\alpha}} = \frac{4\sin\alpha}{\sqrt{1+15\sin^2\alpha}} \qquad \text{...(2)}$$

Depth of O, the C.G. of the base, below A

$$= ON = a\cos\theta$$

∴ Thrust on the base

$$= g\rho\ .\ \pi a^2\ .\ a\cos\theta \qquad |\ P = wS\bar{z}$$

$$= g\rho\pi a^3 \cos\theta$$

$$= g\rho\ \ \pi h^3 \tan^3\alpha\ .\ \frac{4\sin\alpha}{\sqrt{1+15\sin^2\alpha}} \qquad \text{| From (1) and (2)}$$

$$= \frac{4\pi h^3 \tan^3\alpha \sin\alpha g\rho}{\sqrt{1+15\sin^2\alpha}}$$

Example 12: A hollow cone, whose axis is vertical and base downwards, is filled with equal volumes of two liquids whose densities are in the ratio 3 : 1. Show that the thrust on the base is $(3-\sqrt[3]{4})$ times as much as it is when the vessel is filled with the lighter liquid.

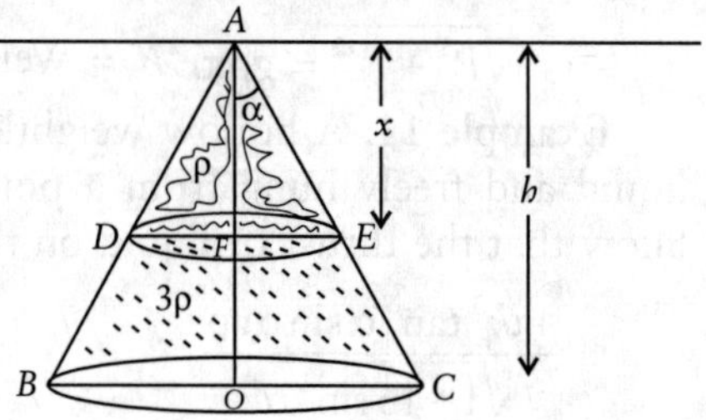

Sol. Let A, BC be a hollow cone, whose vertex is A, axis AO is vertical and base downwards. Let h be the height and α be the semi-vertical angle of the cone. Let DE be the common surface of the liquids of densities ρ and 3ρ, which fill the cone A, BC. Let the depth of the common surface DE below the vertex A be x.

Then, radius of the end $DE = FE = x \tan \alpha$

Radius of the base $BC = OC = h \tan \alpha$

Since the volumes of the two liquids are same, therefore volume of the cone A, DE = 1/2 volume of the cone A, BC

$$\Rightarrow \quad 1/3\ \pi\ (x \tan \alpha)^2\ x = 1/2\ .\ 1/3\ \pi\ (h \tan \alpha)^2\ h \Rightarrow x^3 = h^3/2$$

$$\Rightarrow \quad x = \frac{h}{(2)^{1/3}} \qquad ...(1)$$

Thrust on the base (P_1)

$$= \{g\rho\, x + g\ .\ 3\rho\ (h-x)\}\ \pi\ (h \tan \alpha)^2$$

$$= g\rho\ (3h - 2x)\ \pi\ (h \tan \alpha)^2$$

$$= g\rho\left\{3h - 2\frac{h}{(2)^{1/3}}\right\}\pi(h \tan \alpha)^2 \qquad |\text{From (1)}$$

$$= g\rho h\ (3 - 2^{2/3})\ \pi\ (h \tan \alpha)^2 \qquad ...(2)$$

Thrust on the base when the vessel is filled with the lighter liquid (P_2)

$$= g\rho h\ \pi\ (h \tan \alpha)^2 \qquad ...(3)$$

$$\therefore \quad \frac{P_1}{P_2} = \frac{g\rho h(3-2^{2/3})\pi(h\tan\alpha)^2}{g\rho h\pi(h\tan\alpha)^2}$$

$$= 3 - 2^{2/3} \quad \Rightarrow \quad = 3 - (4)^{1/3}$$

$$\Rightarrow \quad P_1 = (3-\sqrt[3]{4})P_2$$

Example 13: A thin hollow cone of negligible weight is filled with water, and is then suspended from a point in the rim and allowed gradually to take its position of equilibrium. Prove that if the vertical angle of the cone is $\cos^{-1} 2/3$, the surface of water will divide the generating line through the point of suspension in the ratio 2 : 1.

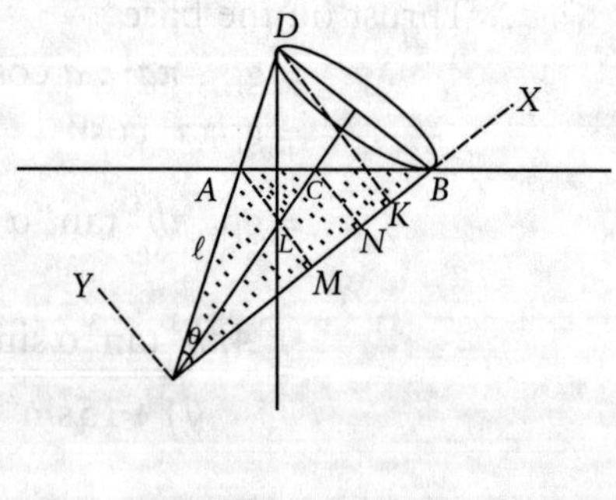

Sol. Let O, BD be a thin hollow cone of vertical angle θ where $\theta = \cos^{-1} 2/3$, i.e., $\cos \theta = 2/3$. Let AB be the surface of water. Let C be the mid-point of AB. Join OC. Take a point L on OC such that $OL = 3/4\ OC$. Then, point L is the C.G.

of the water contained. Let D be the point of suspension. For the equilibrium of the system, DL must be vertical, since the weight of water contained acts vertically downwards through L. Take O as origin, OB as X-axis and a line through O perpendicular to OB as Y-axis. Let $OB = a = OD$. Then, the coordinates of B are (a, O). Let OD be the generating line through the point of suspension.

Let $\quad OA = \ell$. Then,

$$A \to (OM, MA)$$

$$\Rightarrow \quad A \to (\ell \cos\theta, \ell \sin\theta)$$

$$\Rightarrow \quad A \to (2/3\ell, \sqrt{5}/3\ell) \qquad | \because \cos\theta = 2/3$$

Also, coordinates of D are

$$(OK, KD)$$

$$= (a\cos\theta, a\sin\theta)$$

$$= \left(\frac{2}{3}a, \frac{\sqrt{5}}{3}a\right)$$

$\because$ C is the mid-point of AB

$$\therefore C \to \left(\frac{\frac{2}{3}\ell + a}{2}, \frac{\frac{\sqrt{5}}{3}\ell + 0}{2}\right)$$

$$\Rightarrow C \to \left(\frac{3a+2\ell}{6}, \frac{\sqrt{5}}{6}\ell\right)$$

$\because$ L divides OC in the ratio 3 : 1

$$\therefore L \to \left\{\frac{3\frac{(3a+2\ell)}{6} + 1.0}{3+1}, \frac{3.\frac{\sqrt{5}\,\ell}{6} + 1.0}{3+1}\right\}$$

$$\Rightarrow L \to \left(\frac{(3a+2\ell}{8}, \frac{\sqrt{5}\,\ell}{8}\right)$$

$$\text{Slope of line } DL\ (m_1) = \frac{\frac{\sqrt{5}\,\ell}{8} - \frac{5}{3}a}{\frac{3a+2\ell}{8} - \frac{2a}{3}} = \frac{\sqrt{5}(3\ell - 8a)}{6\ell - 7a}$$

$$\text{Slope of line } AB\ (m_2) = \frac{0 - \frac{\sqrt{5}}{3}\ell}{a - \frac{2}{3}\ell} = \frac{\sqrt{5}\ell}{2\ell - 3a}$$

Since AB is horizontal and DL is vertical, therefore,

$$AB \perp DL$$

$$\therefore \quad m_1 m_2 = -1$$

$$\Rightarrow \quad \frac{\sqrt{5}(3\ell-8a)}{6\ell-7a}\cdot\frac{\sqrt{5}\ell}{2\ell-3a} = -1$$

$$\Rightarrow \quad 5\ell\,(3\ell - 8a) = -\,(6\ell - 7a)\,(2\ell - 3a)$$

$$\Rightarrow \quad 15\ell^2 - 40\,a\ell = -\,12\ell^2 + 18\,a\ell + 14\,a\ell - 21a^2 \quad \Rightarrow \quad 21a^2 - 72\,a\ell + 27\,\ell^2 = 0$$

$$\Rightarrow \quad 7a^2 - 24\,a\ell + 9\ell^2 = 0 \quad \Rightarrow \quad 7a^2 - 21a\ell - 3a\ell + 9\ell^2 = 0$$

$$\Rightarrow \quad 7a\,(a - 3\ell) - 3\ell\,(a - 3\ell) = 0 \quad \Rightarrow \quad (a - 3\ell)\,(7a - 3\ell) = 0$$

$$\Rightarrow \quad a - 3\ell = 0 \text{ or } 7a - 3\ell = 0 \quad \Rightarrow \quad a = 3\ell \text{ or } a = 3\ell/7$$

$a = 3\ell/7$ is inadmissible as $a > \ell$

$$\therefore \quad a = 3\ell$$

$$\Rightarrow \quad \ell = a/3 \quad \Rightarrow \quad OA = a/3$$

$$\therefore \quad AD = OD - OA$$

$$= a - a/3 = 2a/3$$

$$\therefore \quad OA : AD = a/3 : 2a/3 + 1 : 2$$

Example 14: A parallelogram is immersed in a fluid with a diagonal vertical, one extremity of which is in the surface of the liquid. Through this point lines are drawn dividing the parallelogram into three equal parts. Compare the pressures on these parts, and if P_2 be the pressure on the middle part and P_1, P_2, those on the other two, prove that $16\ P_2 = 11\ (P_1 + P_3)$.

Sol. Let $ABCD$ be a parallelogram, immersed in a fluid, with a diagonal AC vertical. Let the extremity A of the parallelogram $ABCD$ be in the surface of the liquid. Let the depths of B, C and D from the free surface be α, β and γ respectively.

Let the lines AE and AF be drawn through A which divide the parallelogram into three equal parts.

Hence, if $3S$ is the area of the parallelogram, then

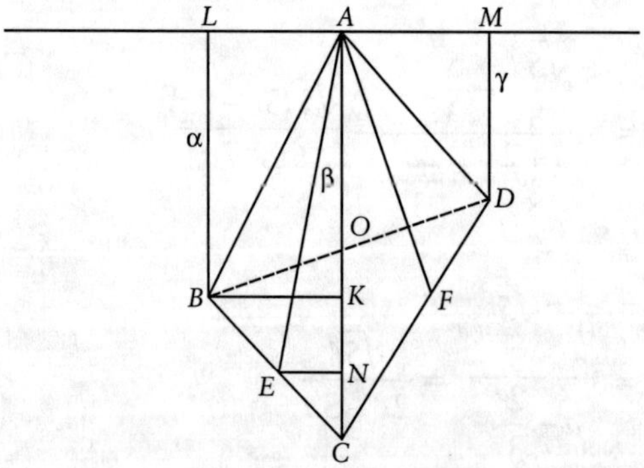

area of $\Delta\ ABE$ = area $AECF$ = area of $\Delta\ AFD = S$ ...(1)

Draw perpendiculars BK and EN from points B and E respectively on the diagonal AC.

By geometry,

area of $\Delta\, ABC$ = area of $\Delta\, ADC = 3S/2$ ...(2)

$\therefore$ Area of $\Delta\, AEC$ = Area of $\Delta\, ABC$ – Area of $\Delta\, ABE$

$= 3S/2 - S$ | From (1) and (2)

$= S/2$

$= 1/3\ (3S/2)$

$= 1/3$ Area of $\Delta\, ABC$ ...(3)

$$\Rightarrow \quad \frac{AC.EN}{2} = \frac{1}{3}.\frac{AC.BK}{2}$$

$\Rightarrow \quad EN = 1/3\ BK$ ...(4)

$\because \quad \Delta\, CNE \sim \Delta\, CKB$

$$\therefore \quad \frac{EN}{CK} = \frac{EN}{BK} \quad \text{| by property of similar triangles}$$

$$\Rightarrow \quad \frac{EN}{CK} = \frac{1}{3} \quad \text{| From (4)}$$

$\Rightarrow \quad CN = 1/3\ CK$ ...(5)

$\therefore \quad NK = CK - CN$

$= CK - 1/3\ CK$

$= 2/3\ CK$

$= 2/3\ (AC - AK) = 2/3\ (AC - LB)$

$= 2/3\ (\beta - \alpha)$ ...(6)

$\therefore$ Depth of E below the free surface

$= AN$

$= AK + KN$

$= \alpha + 2/3\ (\beta - \alpha)$ | From (6)

$= (\alpha + 2\beta)/3$ ...(7)

$\therefore$ Depth of C.G. of $\Delta\, ABC$ from the free surface

$$= \frac{0 + \alpha + \dfrac{\alpha + 2\beta}{3}}{3} = \frac{4\alpha + 2\beta}{9} \quad ...(8)$$

$\therefore$ Thrust on $\Delta\, ABE\ (P_1)$

$$= g\rho.S.\left(\frac{4\alpha + 2\beta}{9}\right) \quad ...(9)$$

Similarly, thrust on $\Delta\, ADF\ (P_3)$

$$= g\rho.S.\left(\frac{4\gamma + 2\beta}{9}\right) \quad ...(10)$$

$\therefore$ Thrust on Δ *AECF* (P_2)

$$= \text{Thrust on the parallelogram } ABCD - (P_1 + P_3)$$

$$= g\rho.3S.\frac{\beta}{2} - g\rho S\left(\frac{4\alpha+2\beta}{9}\right) - g\rho S\left(\frac{4\gamma+2\beta}{9}\right)$$ | From (9) and (10);

| Let the diagonals *AC* and *BD* of the parallelogram *ABCD* interect at O.
Then, depth of O (the C.G. of the parallelogram *ABCD*) below the free surface = β/2

$$= \frac{3g\rho S\beta}{2} - \frac{g\rho S}{9}(4\alpha+4\beta+4\gamma) \quad = \frac{3g\rho S\beta}{2} - \frac{4g\rho S}{9}(\alpha+\beta+\gamma)$$

$$= \frac{3g\rho S\beta}{2} - \frac{4g\rho S}{9}(\beta+\beta)$$ | $\because$ α + γ = β for the parallelogram *ABCD*

$$= \frac{3g\rho S\beta}{2} - \frac{8g\rho S\beta}{9} \quad = \frac{11}{18}g\rho S\beta \qquad \text{...(11)}$$

Now, $P_1 + P_3$

$$= \frac{8g\rho S\beta}{9}$$ | From above ...(12)

$$\therefore \quad \frac{P_2}{P_1+P_3} = \frac{\frac{11}{18}g\rho S\beta}{\frac{8g\rho S\beta}{9}} = \frac{11}{16}$$

$$\Rightarrow \quad 16\, P_2 = 11\,(P_1 + P_3)$$

Example 15: A hollow weightless hemisphere, filled with liquid is suspended from a point in the rim of its base. Show that the thrust on the plane base is to the weight of the contained liquid as $12 : \sqrt{73}$.

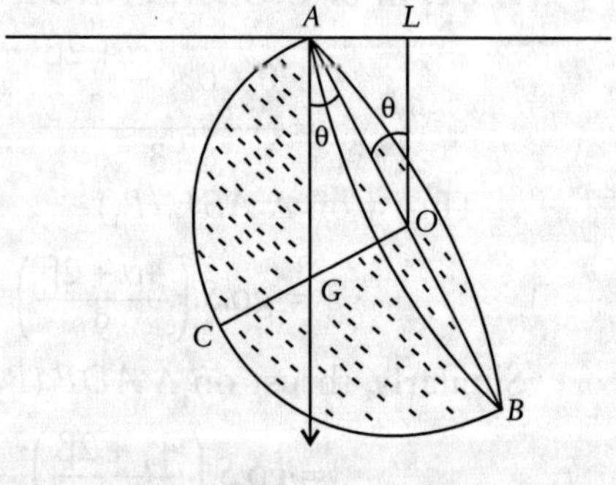

Sol. Let the hemisphere be suspended from a point *A* in the rim of its base. Let a be the radius of the hemisphere. Let *G* be the C.G. of the hemisphere full of liquid, i.e., solid hemisphere. Then

$$OG = 3a/8 \qquad \text{...(1)}$$

The hemisphere full of liquid is in equilibrium under the action of two forces: weight of the liquid acting at *G* vertically downwards and the tension at *A*.

Hence, for the equilibrium of the system, points A and G must be in the same vertical line. Let the base AB of the hemisphere be inclined to the vertical at an angle θ. Then,

$$\tan\theta = \frac{OG}{OA} = \frac{\frac{3a}{8}}{a} = \frac{3}{8} \qquad | \text{ From (1)} \qquad ...(2)$$

Let O be the centre of the circular base of the hemisphere. Then,

depth of O below the free surface

$$= OL$$

$$= a\cos\theta$$

∴ Thrust on the plane base of the hemisphere.

= Area of the base × Pressure at O

$$= \pi a^2 \,.\, g\rho \,.\, a\cos\theta$$

$$= \frac{8\pi g\rho a^3}{\sqrt{73}} \qquad | \text{ From (2)}$$

Weight of the contained liquid

$$= 2/3\ \pi a^3\, g\rho \qquad ...(4)$$

$$\therefore \frac{\text{Thrust on the plane base}}{\text{Weight of the contained liquid}}$$

$$= \frac{\frac{8\pi g\rho a^3}{\sqrt{73}}}{\frac{2}{3}\pi a^3 g\rho} = \frac{12}{\sqrt{73}}$$

Example 16: A cubical box is filled with two liquids of densities ρ and ρ′, the volume of each being the same. Find the pressure on the base and on one of the sides of the vessel.

Sol. Let a be the side of the cube.

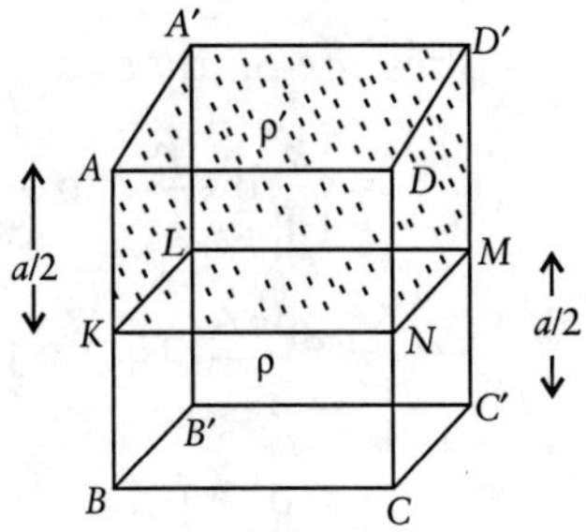

Since the volume of each liquid is the same, therefore, height of each liquid = $a/2$.

Let the densities of the lower and upper liquids be ρ and ρ′ (ρ > ρ′) respectively.

∴ Pressure on the base

$$= g\rho' \,.\, a^2 \,.\, a/2 + g\rho \,.\, a^2 \,.\, a/2$$

$$= \frac{a^3 g}{2}(\rho' + \rho) \qquad ...(1)$$

Pressure on the upper half side of the vessel (P_1)

$$= g\rho' \,.\, (a \,.\, a/2)\ 1/2\ (a/2) \qquad P = wS\bar{z}$$

$$= \frac{a^3 g\rho'}{8} \qquad ...(2)$$

Pressure on the upper half side of the vessel (P_2)

$$= \left(a.\frac{a}{2}\right)\left\{g\rho'.\frac{a}{2} + g\rho.\frac{1}{2}\frac{a}{2}\right\}$$

$$= \frac{a^2}{2}\left(\frac{g\rho' a}{2} + \frac{g\rho a}{4}\right)$$

$$= \frac{a^3}{8}g(2\rho' + \rho) \qquad ...(3)$$

$\therefore$ Pressure on one of the sides of the vessel

$= P_1 + P_2$

$$= \frac{a^3 g\rho'}{8} + \frac{a^3 g}{8}(2\rho' + \rho) \qquad = \frac{a^3 g}{8}(3\rho' + \rho) \qquad ...(4)$$

Example 17: A cubical box with vertical sides is filled with equal volumes of n different liquids, which do not mix, the density of the upper-must being ρ, that of the next 2 ρ, ..., and that of the lowest $n\rho$. Show that the thrust on the base is $(n + 1)$ times the thrust on that part of one of the sides which is in contact with the lowest liquid.

Sol. Thrust on the base (P_1)

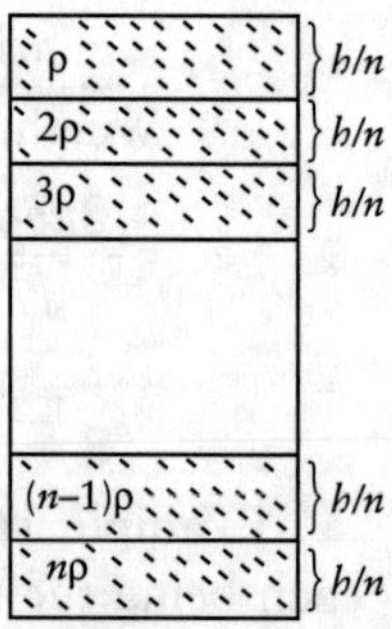

$$= \left(\frac{h}{n}.\rho + \frac{h}{n}.2\rho + \frac{h}{n}.3\rho + ... + \frac{h}{n}(n-1)\rho + \frac{h}{n}.n\rho\right)gA$$

where A is the area of the base

$$= \frac{gh\rho A}{n}(1 + 2 + 3 + ... + n) \qquad = \frac{gh\rho A}{n}\frac{n(n+1)}{2}$$

$$= \frac{gh\rho A(n+1)}{2} \qquad ...(1)$$

Pressure on that part of one of the sides which is in contact with the lowest liquid (P_2)

$$= \left\{\frac{h}{n}.\rho + \frac{h}{n}.2\rho + \frac{h}{n}..3\rho + ... + \frac{h}{n}(n-1)\rho + \frac{h}{2n}n\rho\right\}g\frac{A}{n}$$

$$= \frac{g\rho hA}{n^2}\left\{1 + 2 + 3 + ... + (n-1) + \frac{n}{2}\right\}$$

$$= \frac{g\rho hA}{n^2}\left\{\frac{n(n-1)}{2} + \frac{n}{2}\right\} \quad = \frac{g\rho hA}{2} \qquad ...(2)$$

From (1) and (2), we get

$$P_1 = (n + 1)P_2$$

Example 18: A vessel contains n different liquids resting in horizontal layers and of densities $\rho_1, \rho_2, ..., \rho_n$ starting the highest liquid. A triangle is held with its base in the upper surface of the highest liquid and with its vertex in the nth liquid. Prove that if Δ be the area of the triangle and $h_1, h_2, \ldots, h_n$ be the depths of the vertex below the upper surface of the 1st, 2nd, ..., n^{th} liquids respectively, the thrust on the triangle is

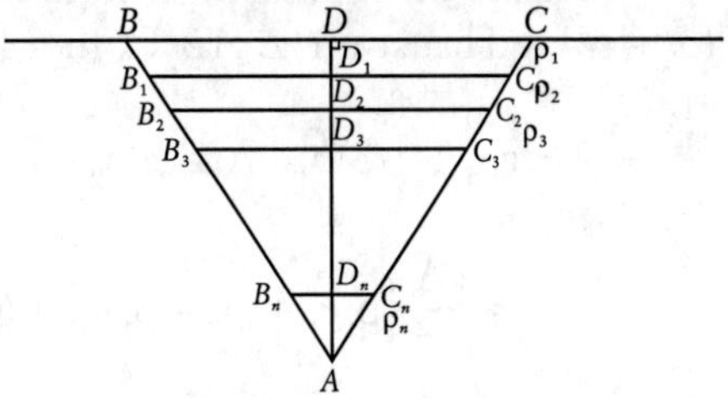

$$\frac{g\Delta}{3h_1^2}\{\rho_1(h_1^3 - h_2^3) + \rho_2(h_2^3 - h_3^3) + \ldots. + \rho_n h_n^3\}$$

Sol. Let ABC be the triangle immersed in the given n different liquids with its base BC in the upper surface of the liquid of density ρ_1 and the vertex A in the liquid of density ρ_n.

Draw perpendicular AD from A on BC.

Then,

$$\left.\begin{array}{l} AD = h_1 \\ AD_1 = h_2 \\ AD_2 = h_3 \\ \vdots \\ AD_n = h_n \end{array}\right\} \text{Given} \qquad \ldots(1)$$

Also, area of $\Delta\, ABC = D = BC\,.\,AD/2 = BC\,.\,h_1/2$

$$\Rightarrow BC = 2\Delta\,/h_1 \qquad \ldots(2)$$

From similar triangles ABC and $AB_1\,C_1$,

$$\frac{B_1C_1}{BC} = \frac{AD_1}{AD} = \frac{h_2}{h_1} \quad \Rightarrow \quad B_1C_1 = \frac{AD_1}{AD}\,.\,BC$$

$$= \frac{h_1}{h_2}\cdot\frac{2\Delta}{h_1} \qquad = \frac{2\Delta h_2}{h_1^2} \qquad \ldots(3)$$

Similarly,

$$B_2C_2 = \frac{2\Delta h_3}{h_1^2}\quad, \quad B_3C_3 = \frac{2\Delta h_4}{h_1^2}$$

$$\vdots \qquad \vdots$$

Areas of Δs ABC, ABC_1, AB_2C_2, AB_3C_3, ... are

$$\Delta, \frac{B_1C_1.AD_1}{2}, \frac{B_2C_2.AD_2}{2}, \frac{B_3C_3.AD_3}{2}, \ldots..$$

$$= \Delta, \frac{2\Delta h_2}{h_1^2}\cdot\frac{h_2}{2}, \frac{2\Delta h_3}{h_1^2}\cdot\frac{h_3}{2}, \frac{2\Delta h_4}{h_1^2}\cdot\frac{h_4}{2}, \ldots.$$

$$= \Delta, \frac{\Delta h_2^2}{h_1^2}, \frac{\Delta h_3^2}{h_1^2}, \frac{\Delta h_4^2}{h_1^2}, \ldots.$$

Now, thrust on ΔABC

= Thrust on ΔABC in a liquid of density ρ_1 + Thrust on ΔAB_1C_1 in a liquid of density $(\rho_2 - \rho_1)$ + Thrust on ΔAB_2C_2 in a liquid of density $(\rho_3 - \rho_1) - (\rho_2 - \rho_1)$, i.e., $\rho_3 - \rho_2$ and so on

$$= g\rho_1 \,.\, 1/3\ h_1 \,.\, \Delta + g(\rho_2 - \rho_1)\frac{1}{3}h_2.\frac{\Delta h_2^2}{h_1^2} + g(\rho_3 - \rho_2)\frac{1}{3}h_3\frac{\Delta h_3^2}{h_1^2} + \ldots.. + g(\rho_n - \rho_{n-1})\frac{1}{3}h_n\frac{\Delta h_n^2}{h_1^2}$$

$$= \frac{g\Delta}{3h_1^2}\{\rho_1 h_1^3 + (\rho_2 - \rho_1)h_2^3 + (\rho_3 - \rho_2)h_3^3 + \ldots. + (\rho_n - \rho_{n-1})h_n^3\}$$

$$= \frac{g\Delta}{3h_1^2}\{\rho_1(h_1^3 - h_2^3) + \rho_2(h_2^3 - h_3^3) + \ldots. + \rho_n h_n^3\}$$

Example 19: A semi-circle is immersed vertically in a liquid with the diameter in the surface. Show how to divide it into n sectors, such that the thrust on each in the same.

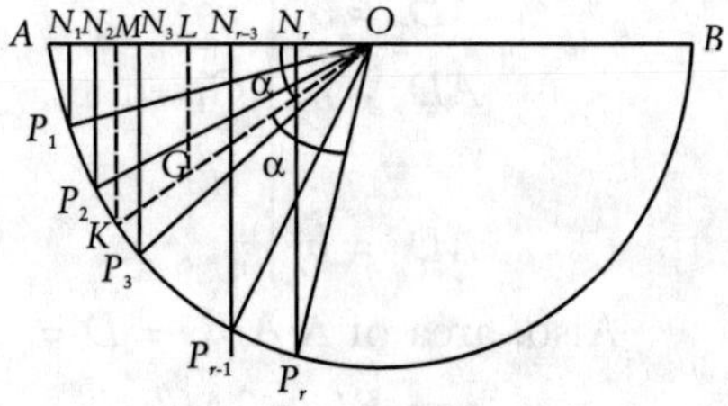

Sol. The diameter AB of the semi-circle is in the surface of the liquid. Let O be the centre of the semi-circular arc. Let the semi-circle be divided into n sectors AOP_1, P_1OP_2, P_2OP_3, ..., $P_{r-1}\ OP_r$ whose bounding radii are OP_1, OP_2, OP_3, ..., OP_r such that the thrust on each of them is the same. Therefore, there are r sectors in AOP_r. Let P_1N_1, P_2N_2, P_3N_3, ..., P_rN_r be the vertical heights of these r sectors respectively from the free surface.

Let $\angle AOP_r = 2\alpha$

Then, area of the sector $AOP_r = 1/2.\ a.\ a.\ 2\alpha = a^2\alpha$. Let G be the C.G. of sector AOP_r. Then, G lies on the bisector OK of $\angle AOP_r$, such that $OG = 2/3\ a \sin\alpha/\alpha$

$\therefore$ Depth of G below AB

$$= GL \qquad = OG\sin\alpha$$

$$= \frac{2}{3}a\frac{\sin\alpha}{\alpha}\sin\alpha \qquad = \frac{2}{3}a\frac{\sin^2\alpha}{\alpha}$$

$$\therefore \text{ Thrust on sector } AOP_r = g\rho\,.\,a^2\alpha\,.\frac{2}{3}a\frac{\sin^2\alpha}{\alpha} \qquad | P = wS\bar{z}$$

$$= 2/3\ a^3 \sin^2\alpha\ g\rho \qquad \ldots(1)$$

$$\text{Thrust on the semi-circle} = g\rho\,.\frac{1}{2}\pi a^2.\frac{4a}{3\pi} \qquad | P = wS\bar{z}$$

$$= 2/3\ a^3\ g\rho \qquad \ldots(2)$$

According the question,

$$\frac{\text{Thrust on sector } AOP_r}{\text{Thrust on semi-circle}} = \frac{r}{n}$$

$$\Rightarrow \quad \frac{\frac{2}{3}a^3 \sin^2 \alpha g\rho}{\frac{2}{3}a^3 g\rho} = \frac{r}{n}$$

$$\Rightarrow \quad \sin^2 \alpha = r/n$$

$$\therefore \quad \cos 2\alpha = 1 - 2\sin^2 \alpha = 1 - 2r/n$$

$$\therefore \quad AN_r = OA - ON_r$$

$$= a - a\cos 2\alpha \qquad = a - a(1 - 2r/n)$$

$$= 2ar/n$$

Putting $r = 1, 2, 3, \ldots, n$, we get

$$AN_1 = 2a/n \qquad AN_2 = 4a/n$$

$$AN_3 = 6a/n$$

$$\vdots \qquad \vdots$$

$$\therefore \quad N_1 N_2 = AN_2 - AN_1 = 2a/n \qquad N_2 N_3 = AN_3 - AN_2 = 2a/n$$

$$\vdots$$

Thus, $AN_1 = N_1 N_2 = N_2 N_3 = \ldots\ldots = 2a/n$

Hence, in order to get n sectors, the thrust on each of which is the same, divide the horizontal diameter AB into n equal parts, the length of each part being equal to $2a/n$. Through each point of division, draw perpendiculars to the diameter AB so as to meet the circumference. These points of intersection, when joined to O, will give the required n sectors.

EXERCISE 2

1. The lighter of two liquids where sp. gr. are 2 : 3 rests on the heavier, to a depth of 4″. A square is immersed in a vertical position with one side in the surface. Determine the side of the square in order that the thrusts on the portions in liquids may be equal.

   Ans. $4/3\,(1 + \sqrt{10})''$

2. A rectangle with its side a in the surface of the upper liquid and the side b vertical is in contact with two liquids of densities ρ and 2ρ. The area is divided by a horizontal line so that the pressures on the two parts are equal. Prove that the horizontal line divides the side in the ratio $(2\sqrt{6} + 1) : 8$.

3. Find the thrust on a vertical quadrilateral which has one side of length 50 cm in the surface, and the opposite side of length 90 cm parallel to it at a depth 40 cm.

   If the fluid consists of a top layer of density 1.1 gm/cm^3 and thickness 20 cm, and the rest of density 1.2 gm/cm^3, then find the resultant thrust on the quadrilateral.

   Ans. 67819800 C.G.S. Units

4. A rectangular box is half filled with water and remaining half with oil. If the oil be half as heavy as water, show that the total thrust on a side of the box is one fourth greater than it would be if the box were filled with oil only.

5. A triangle is immersed in a fluid with one of its sides in the surface. Find the position of a point within the triangle, such that if it be joined to the angular points, the triangle will be divided into three others, the fluid pressures upon which are equal.

   Ans. The point lies in the line from the vertex bisecting the base and at a depth $1/\sqrt{3}$ times the depth of the vertex.

6. A hollow weightless cylinder filled with water is suspended freely from a point on the rim of its plane end. If the height of the cylinder be twice the radius of plane circular ends, show that the thrusts on the ends are in the ratio 3 : 1.

   [**Hint:** Refer Ex. 10.

   Thrust on end O $(P_1) = g\rho\pi r^2 \,.\, OL = g\rho\pi r^2 r \sin 45°$

   $$= \frac{g\rho\,\pi r^3}{\sqrt{2}}$$

   Thrust on end O' $(P_2) = g\rho\pi r^2 \,.\, O'M$

   $$= g\rho\pi r^2\,(O'N + NM)$$
   $$= g\rho\pi r^2\,(r\cos 45° + AG)$$
   $$= g\rho\pi r^2\,(r/\sqrt{2} + r\sqrt{2})$$
   $$= \frac{3g\rho\,\pi r^3}{\sqrt{2}}$$

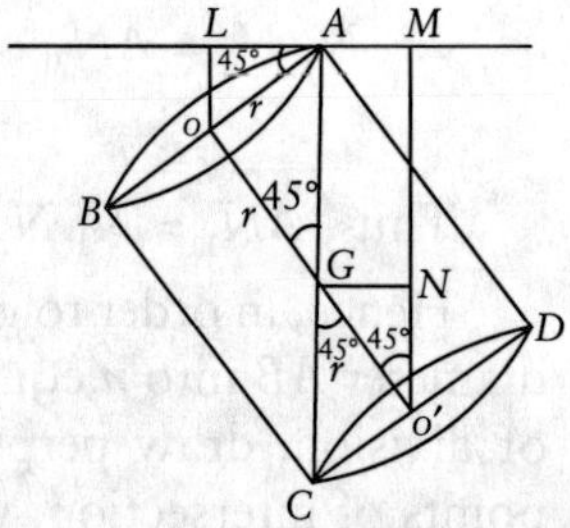

   $$\therefore \quad \frac{P_2}{P_1} = \frac{1}{3}]$$

7. A hollow weightless cone of vertical angle 2α, is filled with a fluid and suspended freely from a point on the rim of its base. Show that the thrust on the base is to the weight of the fluid the cone would contain in the ratio

   $$12\sin^2\alpha : \quad \cos\alpha\sqrt{1+15\sin^2\alpha}$$

   [**Hint:** Refer Ex. 11

   $$\text{Required ratio} = \frac{4\pi h^3\tan^3\alpha\sin\alpha g\rho}{\sqrt{1+15\sin^2\alpha}} : \frac{1}{3}\pi h^3\tan^2\alpha g\rho$$

   $$= \frac{12\sin^2\alpha}{\cos\alpha\sqrt{1+15\sin^2\alpha}}\Bigg]$$

8. A hollow weightless hemisphere, filled with liquid is suspended freely from a point in the rim of its base. Prove that the whole pressure on the curved surface and that on the base are in the ratio 19 : 8.

   [**Hint:** Refer Ex. 15.

   $OG_1 = a/2$, where G_1 is the C.G. of the curved surface of the hemispherical shell

   Depth of G_1 below the free surface

   $$= OA\cos\theta + OG_1\sin\theta$$
   $$= a\cos\theta + a/2\sin\theta = a/2\,(2\cos\theta + \sin\theta)$$

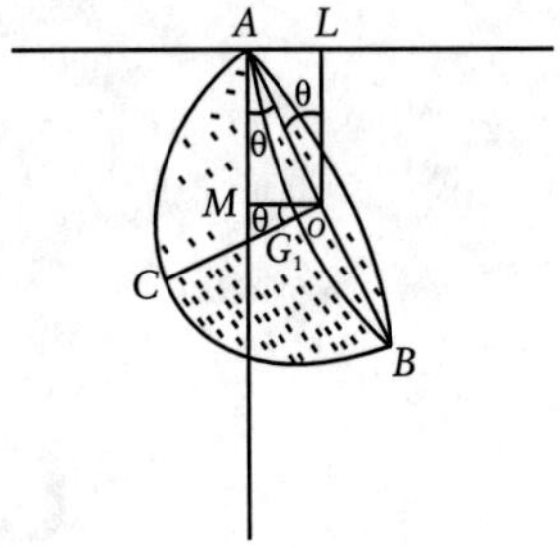

where $\tan\theta = 3/8$

Whole pressure on the curved surface

$= wS\bar{z}$

$= g\rho \,.\, 2\pi a^2 \, a/2\,(2\cos\theta + \sin\theta)$

$= g\rho\pi\, a^3\,(2\cos\theta + \sin\theta)$

Whole pressure on the base $= wS\bar{z} = g\rho \,.\, \pi a^2 \,.\, a\cos\theta = g\rho\pi\, a^3 \cos\theta$, where $\tan\theta = 3/8$

$$\therefore \text{ Required ratio } = \frac{g\rho\pi a^3(2\cos\theta + \sin\theta)}{g\rho\pi a^3\cos\theta} = 2 + \tan\theta = 2 + \frac{3}{8} = \frac{19}{8}$$]

Chapter
6

Centre of Pressure

1. Centre of Pressure

We know that if a plane area is immersed in a liquid, then the pressure at any point of it is normal to the plane surface and is proportional to the depth of the point from the free surface. Thus the pressures at all points on one side of the plane area form a system of like parallel forces which can be compounded into one single force acting at some definite point of the plane of the area. This single force is called the resultant fluid pressure or fluid thrust whose magnitude is the arithmetical sum of the magnitudes of these forces. The definite point of the plane area where this resultant fluid pressure acts is known as the centre of pressure (abbreviated as C.P.).

Definition: The centre of pressure of a plane immersed in a liquid is that point of the area at which the resultant fluid pressure on one side of the area acts. We know that the intensity of pressure at any point in heavy homogeneous liquid varies as the depth of that point below the free surface. Hence, it increases with the depth of the point under consideration below the free surface. If the plane area under consideration is horizontal, then the intensity of pressure at every point of it will be the same. As a result, the resultant thrust will act through the centre of gravity (C.G.) of the area. Hence in this case the C.P. of the plane area will coincide with its C.G.

However, if the plane area is not horizontal, then the pressure at any point in the lower part will be greater than the pressure at any point in the upper part. This is true for every point in the lower part with reference to every point in the upper part. Hence, the centre of pressure of a plane area, which is not horizontally immersed in a liquid, will be below the C.G. of the plane area.

The above discussion establishes the following proposition:

> "The depth of the centre of pressure of a plane area, which is not immersed horizontally in a liquid, is greater than the depth of centre of gravity (centroid) of that plane area."

Aliter: Let a plane area, whose whole area is S, be immersed in a liquid such that it is not horizontal after immersion. Divide the plane area into a large number of small elementary areas α_1, α_2, α_3, ... whose respective depths below the free surface are z_1, z_2, z_3, ... respectively. Let w be the weight per unit volume of the liquid. Then the thrusts on the various elements α_1, α_2, α_3, ... are $w\alpha_1 z_1$, $w\alpha_2 z_2$, $w\alpha_3 z_3$, ... respectively. Then, depth of C.P.

$$= \frac{(w\alpha_1 z_1)z_1 + (w\alpha_2 z_2)z_2 + (w\alpha_3 z_3)z_3 + \ldots\ldots}{w\alpha_1 z_1 + w\alpha_2 z_2 + w\alpha_3 z_3 + \ldots\ldots} = \frac{\Sigma\alpha_1 z_1^2}{\Sigma\alpha_1 z_1}$$

and, depth of C.G.

$$= \frac{w'\alpha_1 z_1 + w'\alpha_2 z_2 + w'\alpha_3 z_3 + \ldots\ldots}{w'\alpha_1 + w'\alpha_2 + w'\alpha_3 + \ldots\ldots}$$

where $w' = g\sigma$, σ being the surface density the plane area immersed.

$$= \frac{\Sigma\alpha_1 z_1}{\Sigma\alpha_1}$$

Now,

depth of C.P. – depth of C.G.

$$= \frac{\Sigma\alpha_1 z_1^2}{\Sigma\alpha_1 z_1} - \frac{\Sigma\alpha_1 z_1}{\Sigma z_1}$$

$$= \frac{\Sigma\alpha_1 z_1^2 \Sigma\alpha_1 - (\Sigma\alpha_1 z_1)^2}{\Sigma\alpha_1 z_1 \Sigma\alpha_1}$$

$$= \frac{\alpha_1\alpha_2(z_1 - z_2)^2 + \alpha_1\alpha_3(z_1 - z_3)^2 + \ldots\ldots\ldots\ldots\ldots}{\Sigma\alpha_1 z_1 \Sigma\alpha_1}$$

which is clearly a positive quantity (> 0)

| The square of a real number, positive or negative, is positive.

$\Rightarrow$ Depth of C.P. > Depth of C.G.

Hence, the required result is established.

2. Method of Finding the Centre of Pressure (C.P.)

Let a plane area S be immersed in a liquid with its plane vertical. Take the line of intersection of this plane area with the free surface as X-axis and a line perpendicular to it lying in the plane of the area as Y-axis.

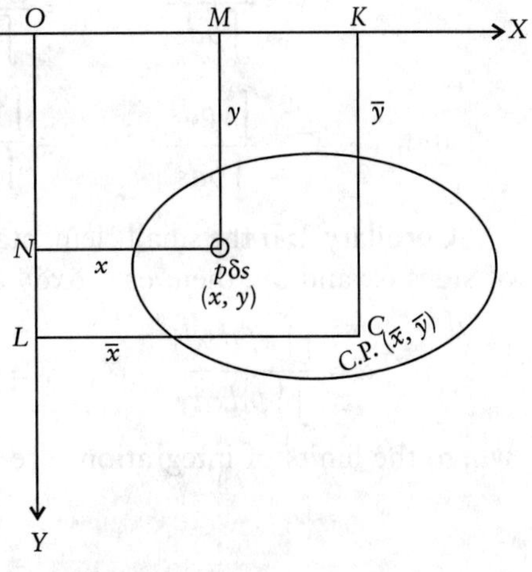

Divide the plane area into a large number of small elementary areas and consider one such elementary onto δs surrounding the point $P(x, y)$, which is the position of its C.P. Let p be the intensity of pressure at the point P. Then, the pressure on the element δs is $p\delta s$. Its moment about the axis of $Y = xp\delta s$. Hence, the sum of the moments of the pressures on all the elements of the type δs of the plane area

$\Sigma\ xp\delta s.$

Again, total pressure on the whole plane area = $\Sigma p\delta s$. Let $(\bar{x}, \bar{y})$ be the coordinates of the required centre of pressure C. Then, the total pressure will act at $(\bar{x}, \bar{y})$.

Now, taking moments about Y-axis, we get

$$\bar{x}\,\Sigma p\delta s = \Sigma xp\delta s$$

| The moment of the total pressure on the whole plane area about Y-axis is equal to the algebraic sum of the moments of pressures on each elementary plane area about Y-axis [from statics]

$$\Rightarrow \qquad \bar{x} = \frac{\Sigma xp\delta s}{\Sigma p\delta s}$$

Similarly, by equating moments about X-axis, we get

$$\bar{y}\,\Sigma p\delta s = \Sigma xp\delta s$$

$$\Rightarrow \qquad \bar{y} = \frac{\Sigma xp\delta s}{\Sigma p\delta s}$$

If the process of summation is carried out by using the integration technique, then we have,

$$\bar{x} = \frac{\int xpds}{\int pds}, \quad \bar{y} = \frac{\int ypds}{\int pds}$$

$\therefore$ The C.P. is

$$\left(\frac{\int xpds}{\int pds}, \frac{\int ypds}{\int pds}\right)$$

where the limits of integration are so chosen such that the whole plane area is covered.

Corollary 1: If the liquid, under consideration, is homogeneous of density ρ (say), then the pressure p at $P\ (x, y)$ is given by $g\rho x$ and then the have,

$$\bar{x} = \frac{\int xpds}{\int pds} = \frac{\int xg\rho xds}{\int g\rho xds} = \frac{\int x^2 ds}{\int xds}$$

$$\text{and,} \quad \bar{y} = \frac{\int ypds}{\int pds} = \frac{\int yg\rho xds}{\int g\rho xds} = \frac{\int xyds}{\int xds}$$

Corollary 2: If the small elementary area surrounding the point $P\ (x, y)$ be taken as a rectangle of sides δx and δy, then, $\delta s = \delta x\delta y$ and then the coordinates $(\bar{x}, \bar{y})$ of the C.P. C are given by

$$\bar{x} = \frac{\iint xpdxdy}{\iint pdxdy}, \quad \bar{y} = \frac{\iint ypdxdy}{\iint pdxdy}$$

where the limits of integrations are so chosen such that the entire plane area is covered.

3. Independence of the Position of C.P.

If a plane area, immersed in a homogeneous liquid at rest, is rotated about its line of intersection with the free surface of the liquid, then the position of centre of pressure of the plane area relative to the plane area does not change.

OR

The position of the centre of pressure of a plane area is independent of the inclination of plane area to the vertical.

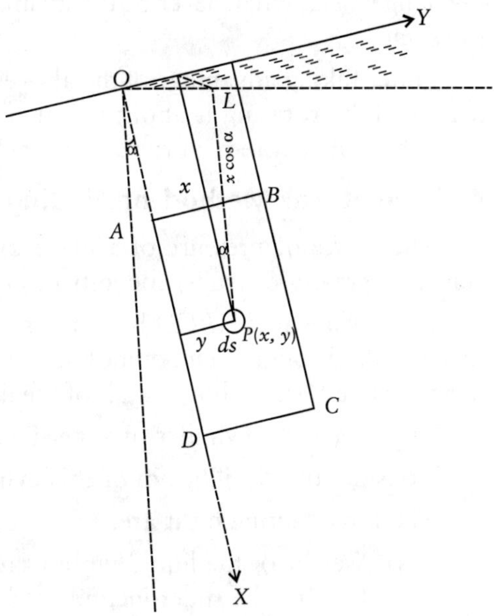

Let *ABCD* be the plane area immersed in liquid. Let α be the inclination of plane area to the vertical. Take the line of intersection of this plane area with the free surface of the liquid as Y-axis and a line perpendicular to it lying in the plane of the area as X-axis. Take any point *P* (x, *y*) in the plane area *ABCD*. Surrounding *P*, consider an element of area *ds*. Form *P*, draw perpendicular *PL* to the free surface. Then,

$$PL = x \cos \alpha$$

Let p be the intensity of pressure at *P*. Then,

$$p = g\rho x \cos \alpha$$

where ρ is the density of the liquid.

From § 2 above, if $(\bar{x}, \bar{y})$ are the coordinates of C.P., then

$$\bar{x} = \frac{\int xpds}{\int pds}$$

and, $$\bar{y} = \frac{\int ypds}{\int pds}$$

Substituting $p = g\rho x \cos \alpha$ in these, we get

$$\bar{x} = \frac{\int xg\rho x \cos \alpha ds}{\int g\rho x \cos \alpha ds} = \frac{\int x^2 ds}{\int xds}$$

and, $$\bar{y} = \frac{\int yg\rho x \cos \alpha ds}{\int g\rho x \cos \alpha ds}$$

$$= \frac{\int xyds}{\int xds}$$

We see that these values of $\bar{x}$ and $\bar{y}$ do not involve α. Hence, they are independent of α.

This shows that the position of the C.P. of the plane area relative to the plane area remains unchanged, i.e., whatsoever be the inclination of the plane area to the vertical, C.P. $(\bar{x}, \bar{y})$ remains the same.

Note: Taking advantage of the above assertion, we may make of the plane of the area vertical, if needed, by rotating it about its line of intersection with the free/effective surface as in each case, the C.P. will remain the same (unchanged).

4. Geometrical Method of Finding the Centre of Pressure

The centre of pressure of a plane area lies vertically below the centre of gravity of super-incumbent liquid.

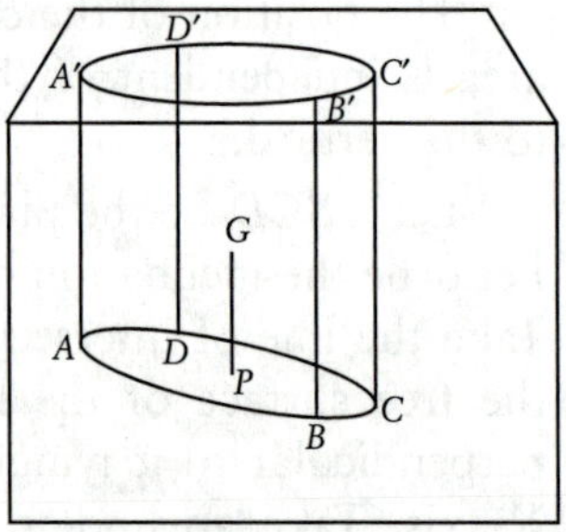

Let a plane area $ABCD$ be immersed (not in a vertical position) in a liquid. Through each point of its boundary, draw vertical lines to meet the free surface (horizontal) of the liquid in the curve $A'B'C'D'$.

Thus, we get a cylinder of super-incumbent liquid.

Consider the equilibrium of this cylinder.

The forces acting on it are:

(i) Weight of the liquid within this cylinder, acting vertically downwards through G, the C.G. of the super-incumbent liquid.

(ii) Resultant vertical thrust on the plane area $ABCD$ acting through P, the centre of pressure of this area.

(iii) The thrust of the surrounding liquid on the curved surface of the cylinder, acting horizontally.

[There is no thrust on the plane end $A'B'C'D'$, it being in the free surface of the liquid.]

For the sake of vertical equilibrium of this cylinder, resolving all these forces vertically, we find that

Vertical thrust on the plane area $ABCD$ passing through its centre of pressure P is equal in magnitude but opposite in direction to the weight of the liquid in the cylinder acting vertically downwards through G, the C.G. of the super-incumbent liquid.

Hence, line GP must be vertical, i.e., the centre of pressure of a plane area lies vertically below the centre of gravity of the super-incumbent liquid.

Note: If the plane area is vertical, then there will be no super-incumbent liquid. In such a situation, we may rotate the plane about its line of intersection with the free surface so that it becomes inclined. Then, we may find the position of C.P. by §4 which will be the same when the plane area is vertical. [Note that the position of the C.P. of a plane area relative to the plane area remains unchanged, whatsoever be the inclination of plane area to be vertical.]

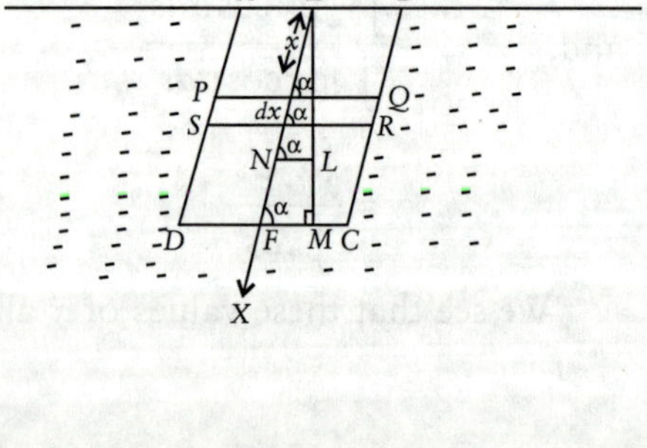

5. Centre of Pressure in some Standard Cases

(1) *Centre of pressure of a parallelogram immersed in a homogeneous liquid with one side in the free surface.*

Let $ABCD$ be a parallelogram immersed in a homogeneous liquid with one side AB in the free surface. If the parallelogram $ABCD$ is not vertical, then it may be made vertical by its rotation about its line of intersection AB with the free surface. These is no loss of generality in doing so as the position of C.P. is not affected (Refer §3).

Let E and F be the mid-points of sides AB and DC respectively. Take line EF as X-axis. Let line EF make an angle α to the horizontal.

Let $AD = a$ and $AB = b$.

Divide the parallelogram $ABCD$ into a large number of elementary strips parallel to the free surface. Let $PQRS$ be one of such strips which is at a distance x from the origin E in the free surface along EF. Let dx be the width of this strip along EF.

Since the strip $PQRS$ is very thin, therefore, the pressure at each point of the strip may be taken to be the same. Hence, the whole pressure on it acts at the mid-point which lies on EF. This is also true for any such other strip. Hence, C.P. of the parallelogram will lie on the line EF, i.e., X-axis.

$\therefore\ \bar{y} = 0$ where $(\bar{x}, \bar{y})$ are the coordinates of C.P.

p = intensity of pressure at any point of the strip $PQRS$

$= g\rho x \sin\alpha$

where ρ is the density of the liquid.

dS = area of the strip = $b\ dx \sin\alpha$

$$\therefore\quad \bar{x} = \frac{\int xpds}{\int pds}$$

$$= \frac{\int_0^a xg\rho x \sin\alpha b dx \sin\alpha}{\int_0^a g\rho x \sin\alpha b dx \sin\alpha}$$

$$= \frac{\int_0^a x^2 dx}{\int_0^a x\, dx} = \frac{\left(\frac{x^3}{3}\right)_0^a}{\left(\frac{x^2}{2}\right)_0^a} = \frac{\frac{a^3}{3}}{\frac{a^2}{2}} = \frac{2}{3}a = \frac{2}{3}EF$$

Hence the C.P. is along EF at a distance 2/3 EF from E.

Note 1: If $ABCD$ is a rectangle, then $\alpha = \pi/2$ and the result is the same.

Note 2: If $ABCD$ is a square, then $\alpha = \pi/2$ and $a = b$. Hence, the result is the same.

Note 3: By geometry, depth of C.P. of the parallelogram $ABCD$ below the free surface.

= 2/3 depth of DC below the free surface.

Let the C.P. of parallelogram $ABCD$ be N. Then,

$EN/EF = 2/3$

$\because\ \Delta ENL \sim \Delta EFM$

$\therefore$ $EN/EF = EL/EM = 2/3$ | by property of similar triangles

$\Rightarrow$ $EL = 2/3\ EM$

Note 4: By geometry, depth of C.G. of the parallelogram $ABCD$ below AB

$= 1/2\ EM$ | Refer Note 3

Depth of C.P. of the parallelogram $ABCD$ below AB

$$= \frac{1+1}{2+1}\ EM \quad = 2/3\ EM$$

(Note 4 can be treated as a trick to remember the formula for C.P.)

(2) *Centre of pressure of a triangular area immersed in a homogeneous liquid with its vertex in the surface and base horizontal.*

Let ABC be a triangular area immersed in a homogeneous liquid with its vertex A in the surface and base BC horizontal. Let AD be the median of triangular area ABC. Take line AD as X-axis. Let median AD make an angle α with the horizon.

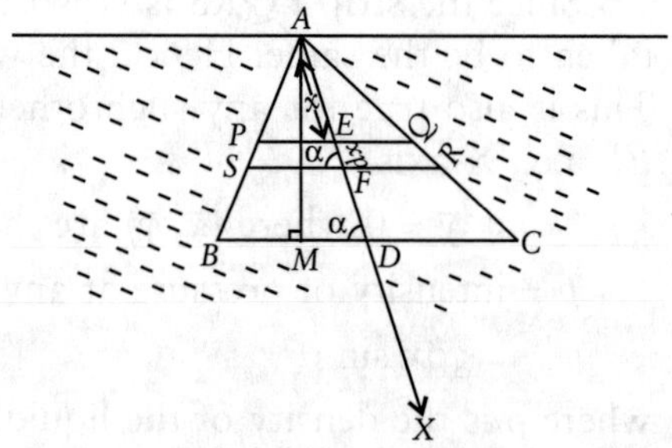

By symmetry, the C.P. will lie on the median AD, i.e., X-axis | Refer §5 (1)

$\therefore\ \bar{y} = 0$ where $(\bar{x}, \bar{y})$ are the coordinates of C.P.

Consider an elementary strip $PQRS$ parallel to the base BC of $\Delta\ ABC$ at a distance x along the median AD from A.

Let dx be the width of this strip along AD.

Let $BC = a$ and $AD = r$

$\therefore$ $\Delta\ APQ \sim \Delta\ ABC$

$\therefore$ $PQ/BC = AE/AD$ | by property of similar triangles

$\Rightarrow$ $PQ/a = x/r$ $\Rightarrow$ $PQ = ax/r$

Area of the strip $PQRS$ (ds)

$= PQ\ .\ EF \sin\alpha \quad = ax/r\ dx \sin\alpha$

Depth of the strip from the free surface $= x \sin\alpha$

$\therefore$ p = Intensity of pressure at any point of the strip $PQRS$

$= g\rho x \sin\alpha$

where ρ is the density of the liquid.

$$\therefore \quad \bar{x} = \frac{\int xp\,ds}{\int p\,ds} = \frac{\int_0^r xg\rho x \sin\alpha \frac{ax}{r} dx \sin\alpha}{\int_0^r g\rho x \sin\alpha \frac{ax}{r} dx \sin\alpha}$$

$$= \frac{\int_0^r x^3 dx}{\int_0^r x^2 dx} = \frac{\left(\frac{x^4}{4}\right)_0^r}{\left(\frac{x^3}{3}\right)_0^r} = \frac{\frac{r^4}{4}}{\frac{r^3}{3}} = \frac{3}{4} r \quad = \frac{3}{4} AD$$

Hence the C.P. of $\Delta\ ABC$ lies on the median AD at a distance 3/4 AD from A.

Note 1: Depth of C.P. $\Delta\ ABC$ below the free surface = 3/4 depth of base BC below the free surface | Refer §5 (1)

Note 2: Depth of C.G. of $\Delta\ ABC$ below the free surface = 2/3 depth of base BC below the free surface | Refer §5 (1)

Note 3: Depth of C.G. of $\Delta\ ABC$ below the free surface = 2/3 AM | Refer §5 (1)

Depth of C.P. of $\Delta\ ABC$ below the free surface $= \dfrac{2+1}{3+1}\ AM$ = 3/4 AM | Refer §5 (1)

(Note 3 can be treated as a tact to remember formula for C.P.)

(3) *Centre of Pressure of a triangular area immersed in a homogeneous liquid with one side in the surface.*

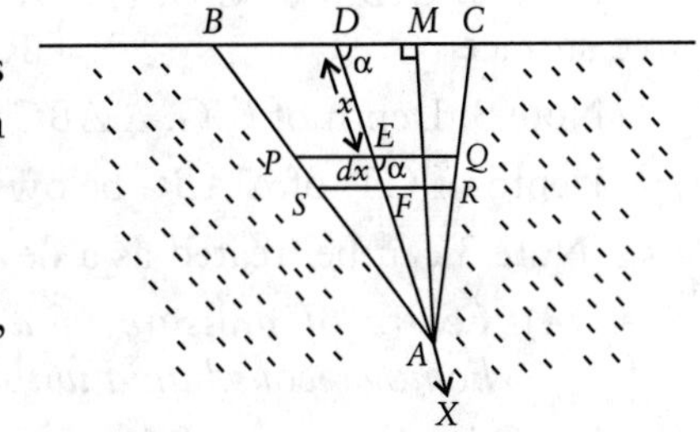

Let ABC be a triangular area immersed in a homogeneous liquid with one side BC in the surface. Let AD be the median of the triangular area ABC. Take line DA as X-axis.

Let median AD make an angle α with the horizon.

By symmetry, the C.P. will lie on the median DA, i.e., X-axis | Refer §5 (1)

$\therefore\ \bar{y} = 0$ where $(\bar{x}, \bar{y})$ are the coordinates of C.P.

Consider an elementary strip $PQRS$ parallel to the base BC of $\Delta\ ABC$ at a distance x along the median DA from D. Let dx be the width of the strip along DA.

Let $BC = a$ and $DA = r$.

$\because\quad \Delta\ APQ \sim \Delta\ ABC$

$\therefore\quad PQ/BC = AE/AD$ | by property of similarity of triangles

$\Rightarrow\quad PQ/a = AD - DE/AD$

$\Rightarrow\quad PQ/a = (r - x)/r$

$\Rightarrow\quad PQ = a\ (r - x)/r$

Area of the strip $PQRS\ (ds) = PQ\ .\ EF \sin\alpha$

$$= \frac{a(r-x)}{r} dx \sin\alpha$$

Depth of the strip from the free surface = $x \sin\alpha$

$\therefore\quad p$ = Intensity of pressure at any point of the strip $PQRS$

$= g\rho x \sin\alpha$

where ρ is the density of the liquid.

$$\therefore\quad \bar{x} = \frac{\int xpds}{\int pds} = \frac{\int_0^r x.g\rho x \sin\alpha\ \dfrac{a(r-x)}{r} dx \sin\alpha}{\int_0^r g\rho x \sin\alpha\ \dfrac{a(r-x)}{r} dx \sin\alpha}$$

$$= \frac{\int_0^r (r-x)x^2 dx}{\int_0^r (r-x)x\, dx} = \frac{\left(\frac{rx^3}{3} - \frac{x^4}{4}\right)_0^r}{\left(\frac{rx^2}{2} - \frac{x^3}{3}\right)_0^r} = \frac{r}{2} = \frac{1}{2} DA$$

Hence the C.P. of Δ ABC lies on the median DA at a distance 1/2 DA from D, i.e., C.P. is the middle point of DA.

Note 1: Depth of C.P. Δ ABC below the free surface = 1/2 Depth of vertex A below the free surface

| Refer §5 (1)

Note 2: Depth of C.G. Δ ABC below the free surface = 1/3 Depth of vertex A below the free surface | Refer §5 (1)

Note 3: Depth of C.G. Δ ABC below the free surface = 1/3 AM | Refer §5 (1)

Depth of C.P. of Δ ABC below the free surface = (1+1)/(3+1) AM = 1/2 AM

(Note 3 can be treated as a device to remember the formula for C.P.)

(4) *Centre of pressure of a vertical circular area of radius a wholly immersed in a homogeneous liquid with its centre at a depth h below the free surface.*

Let O be the centre of the circle at a depth h below the free surface. Then, $OE = h$. Take O as the origin, vertical line OX through O as X-axis and the horizontal line OY through O as Y-axis. Let a be the radius of circular area. Then, the equation of the circle is

$$x^2 + y^2 = a^2 \quad ...(1)$$

By symmetry, C.P. lies on the vertical diameter AB.

$\therefore$ $\bar{y} = 0$ where $(\bar{x}, \bar{y})$ are the coordinates of C.P. (P_1)

Consider an elementary horizontal strip $PQRS$ at a distance x from O and of thickness dx. Take a point Q on the circle whose coordinates are (x, y)

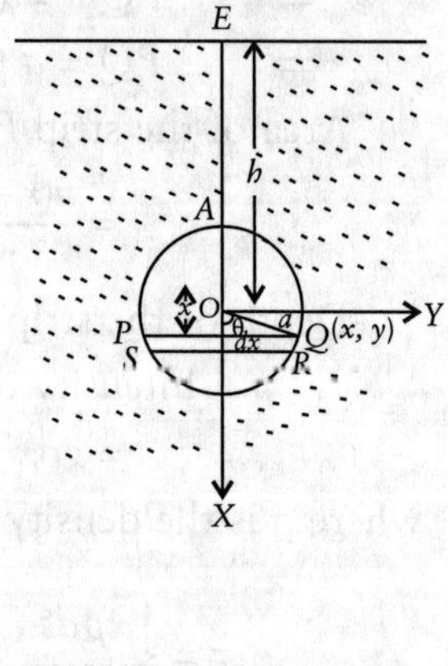

ds = area of the elementary strip $PQRS$

$= 2y dx$

p = intensity of pressure at any point of the strip $PQRS$

$= g\rho\,(h + x)$

where ρ is the density of the liquid.

$$\therefore \quad \bar{x} = \frac{\int xp\,ds}{\int p\,ds} = \frac{\int_{-a}^{a} xg\rho(h+x)2y\,dx}{\int_{-a}^{a} g\rho(h+x)2y\,dx}$$

$$= \frac{\int_{-a}^{a} xy(h+x)dx}{\int_{-a}^{a} y(h+x)dx}$$

Put $x = a\cos\theta$

$y = a\sin\theta$ | parametric equations of the circle

Then, $dx = -a\sin\theta\, d\theta$

$$= \frac{\int_{\pi}^{0} a\cos\theta . a\sin\theta(h + a\cos\theta)(-a\sin\theta\, d\theta)}{\int_{\pi}^{0} a\sin\theta(h + a\cos\theta)(-a\sin\theta\, d\theta)}$$

$$= \frac{a\int_{0}^{\pi}\cos\theta\sin^2\theta(h + a\cos\theta)d\theta}{\int_{0}^{\pi}\sin^2\theta(h + a\cos\theta)d\theta} = \frac{ha\int_{0}^{\pi}\cos\theta\sin^2\theta d\theta + a^2\int_{0}^{\pi}\cos^2\theta\sin^2\theta d\theta}{h\int_{0}^{\pi}\sin^2\theta d\theta + a\int_{0}^{\pi}\sin^2\theta\cos\theta d\theta}$$

$$= \frac{a^2\int_{0}^{\pi}\cos^2\theta\sin^2\theta d\theta}{h\int_{0}^{\pi}\sin^2\theta d\theta}$$ | by a property of definite integrals

$$= \frac{2a^2\int_{0}^{\pi/2}\cos^2\theta\sin^2\theta d\theta}{2h\int_{0}^{\pi}\sin^2\theta d\theta}$$ | by a property of definite integrals

$$= \frac{a^2\int_{0}^{\pi/2}\sin^2\theta\cos^2\theta d\theta}{h\int_{0}^{\pi/2}\sin^2\theta d\theta} = \frac{\frac{a^2\sqrt{\frac{3}{2}}\sqrt{\frac{3}{2}}}{2\sqrt{3}}}{h.\frac{1}{2}.\frac{\pi}{2}}$$

$$= \frac{\frac{a^2.\frac{1}{2}\sqrt{\pi}.\frac{1}{2}.\sqrt{\pi}}{4}}{h\frac{\pi}{4}} = \frac{a^2}{4h}$$

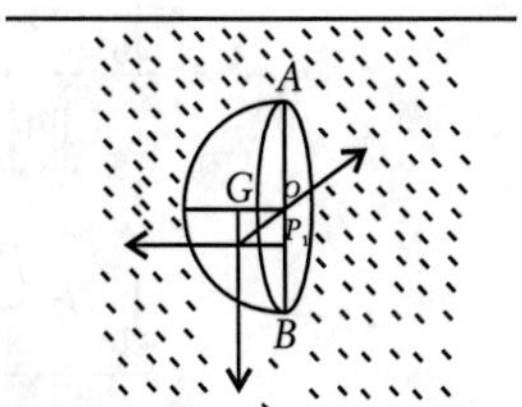

Hence, the depth of the C.P. (P_1) from the free surface

$h + a^2/4h$

Aliter 1: Constant a hemisphere on the circle $x^2 + y^2 = a^2$ as base and consider the equilibrium of the liquid contained in it.

This hemisphere is in equilibrium under the action of the following forces.

(1) Weight of the liquid $A = 2/3\ \pi a^3 g\rho$ contained in it acting through its C.G. G where $OG = 3/8a$ and ρ is the density of the liquid.

(2) Thrust on the circle $= \pi a^2 g\rho h$ acting through the centre of pressure P_1 which evidently lies on AB due to symmetry.

(3) The resultant thrust on the curved surface will essentially pass through O for the equilibrium of the system. | by statics

Taking moments about O, we get

$$\frac{2}{3}\pi a^3 g\rho \times OG = \pi a^2 g\rho h \times OP_1$$

$$OP_1 = \frac{2}{3}\times\frac{3}{8}\frac{a^2}{h} = \frac{a^2}{4h}$$

Aliter 2: By double integration

Let the radius of the circle be a. Let O be the centre of the circle at a depth h below the free surface. Take O as pole and OB as initial line. Let P be a point of the circular area with coordinates (r, θ).

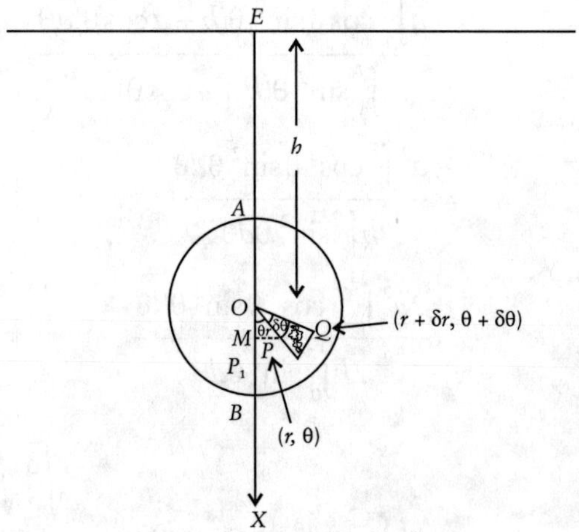

Let Q $(r + \delta r, \theta + \delta\theta)$ be a neighbouring point.

Then, elementary area = $rd\theta dr$

p = Intensity of pressure at any point of the elementary area

$= g\rho\ (h + r\cos\theta)$

By symmetry C.P. will lie on the initial line below O.

Let $\overline{x}$ be the depth of C.P. below O. Then,

$$\overline{x} = \frac{\int xpds}{\int pds}$$

$$= \frac{2\int_{\theta=0}^{\pi}\int_{r=0}^{a} xg\rho(h + r\cos\theta)r\,d\theta\,dr}{2\int_{\theta=0}^{\pi}\int_{r=0}^{a} g\rho(h + r\cos\theta)r\,d\theta\,dr}$$

$$= \frac{\int_0^{\pi}\int_0^{a} r^2(h + r\cos\theta)\cos\theta d\theta\,dr}{\int_0^{\pi}\int_0^{a} r(h + r\cos\theta)d\theta\,dr} \qquad \Big|\ \because x = r\cos\theta$$

$$= \frac{\int_0^{\pi}(h\frac{a^3}{3} + \frac{a^4}{4}\cos^2\theta)d\theta}{\int_0^{\pi}(h\frac{a^2}{2} + \frac{a^3}{3}\cos\theta)d\theta}$$

$$= \frac{\frac{a^4}{4}.\frac{\pi}{2}}{h\frac{a^2}{2}\pi} = a^2/4h$$

Corollary 1: *When the plane of the circle is not vertical.*

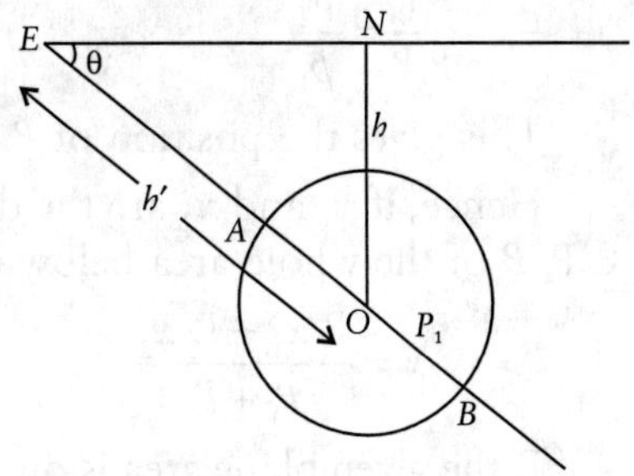

If the plane of the circle is not vertical, then also by §3, the distance of the C.P. from the centre will be $a^2/4h$ but here h will stand for the slant distance of the centre from the horizontal free surface measured along the inclined plane of the circle. However, if h denotes the vertical depth of O from the free surface, then the distance of the centre O measured along the inclined plane will be $h' = h/\sin\theta$ and consequently the distance of C.P. from the centre will be

$$\frac{a^2}{4h'} = \frac{a^2 \sin\theta}{4h}$$

Corollary 2: When the circular area is just immersed

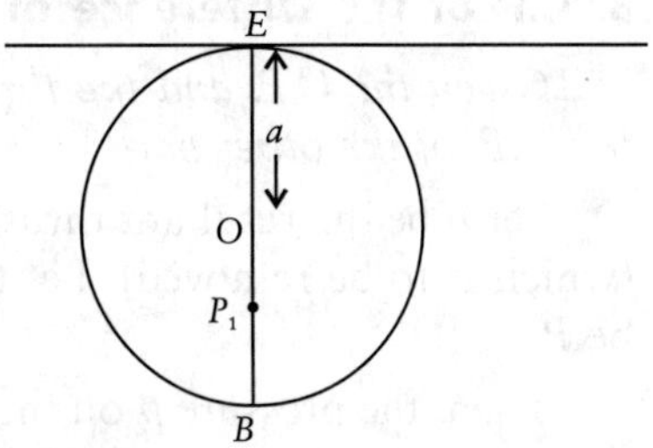

If the circular area is just immersed in a homogeneous liquid vertically, then

$h = a$. Therefore,

$$OP_1 = \frac{a^2}{4h} = \frac{a^2}{4a} = \frac{a}{4}$$

Hence, the depth of the C.P. of the circular area from the free surface

$$= EP$$
$$= EQ + OP_1$$
$$= a + a/4 = 5a/4$$

6. Analogy in the Methods of Determination of C.P. and C.G.

We have seen that the C.P. of lamina of any shape is the point through which the resultant of a system of parallel forces which are actually fluid thrusts acting on infinitesimally small elementary areas into which the given lamina is subdivided. It therefore follows that the methods of determination of C.P. are analogous to the methods that are used in the determination of C.G. This is named as analogy in the methods of determination of C.P. and C.G.

7. C.P. of the Sum of Two Areas

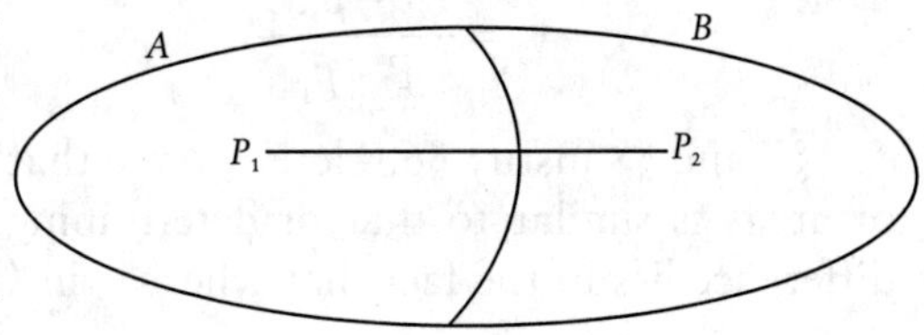

Given the position of the C.P. and the thrust on each of the two parts into which a plane area is divided, to find the position of C.P. of the whole area.

Let a plane area be divided into two parts A and B. Let P_1 and P_2 be the C.P. of the parts A and B respectively. Let p_1 and p_2 be the thrusts on the parts A and B respectively. Now thrust p_1 acts at P_1 and thrust p_2 acts at P_2. We are to find where the total thrust $(p_1 + p_2)$ acts. Since this total pressure is the resultant of thrusts p_1 and p_2, therefore, by statics, $(p_1 + p_2)$ will act at a point P lying on the straight line $P_1 P_2$ joining P_1 and P_2 such that

$$p_1 \times P_1 P = p_2 \times P_2 P$$

$$\Rightarrow \quad \frac{P_1P}{P_2P} = \frac{p_2}{p_1}$$

This gives the position of P.

Hence, if x_1 and x_2 are the depths of P_1 and P_2 below the free surface, then the depth ($\bar{x}$) of C.P. P of the whole area below the free surface is given by

$$\bar{x} = \frac{P_1 x_1 + P_2 x_2}{P_1 + P_2}$$

If the given plane area is divided into more than two portions on which the resultant thrusts are P_1, P_2, P_3, ..., the depths of whose centres of pressure below the free surface are x_1, x_2, x_3, ... respectively, then the depth ($\bar{x}$) of C.P. P of the whole area below the free surface is given by

$$\bar{x} = \frac{P_1 x_1 + P_2 x_2 + P_3 x_3 + ...}{P_1 + P_2 + P_3 + ...}$$

8. C.P of the Difference of Two Areas

Given the C.P. and the thrust on the whole area and on one part, to find the position of the C.P. of the other part

Let p be the resultant thrust on the whole area and p_1 be the resultant thrust on one part A (which is to be removed). Let the C.P. of whole plane area be P and that of the removed part be P_1.

Then, the pressure p on the whole area is the resultant of the pressure p_1 on the part A and the pressure $(p - p_1)$ on the other part B. Let it act at P_2. Then, be statics, $P_1 \, P \, P_2$ is a straight such that

$$p_1 \, . \, PP_1 = (p - p_1) \, PP_2$$

$$\Rightarrow \quad \frac{PP_2}{PP_1} = \frac{p_1}{(p - p_1)}$$

This gives the position of P_2, the required C.P. Hence if x and x_1 are the depths of P and P_1 below the free surface, then the depth x_2 of the C.P. of the remaining part below the free surface is given by

$$x_2 = \frac{Px - p_1 x_1}{P - p_1}$$

§7 and §8 justify §6. Here, we see that the method of determining C.P. of sum or difference of areas is similar to that of determining C.G. of sum or difference of two areas. The only difference lies in the fact that whereas in C.G. weight is involved, in C.P., thrust is taken.

ILLUSTRATIVE EXAMPLES

Example 1: Prove that the centre of pressure of triangle ABC immersed in a homogeneous liquid with the side BC in the surface, coincides with the centre of two equal forces each $1/6 \, w\Delta\alpha$ acting at the middle points of AB and AC, where α is the vertical depth of A below BC, Δ the area of the triangle and w the specific weight of the liquid. Also prove that the resultant of these two forces is equal to the whole pressure on the triangle.

Sol. Let E and F be the middle points of AB and AC respectively. Let D be the middle point of BC. Then, AD is a median of the $\Delta\, ABC$. Then, centre of pressure of triangle ABC is P such that $DP = PA$, i.e., P is the mid-point of AD. |Refer §5 (3)

Now, centre of two equal forces each equal $1/6\ w\Delta\alpha$ will be the mid-point of EF. |By statics.

By geometry, the mid-point of EF is P.

Hence, the C.P. of $\Delta\, ABC$ coincides with the centre of two equal forces each equal to $1/6\ w\Delta\alpha$.

Also, resultant of these two forces (P_1)

$= 1/6\ w\Delta\alpha + 1/6\ w\Delta\alpha$ |The forces being like parallel (By statics)

$= 1/3\ w\Delta\alpha$

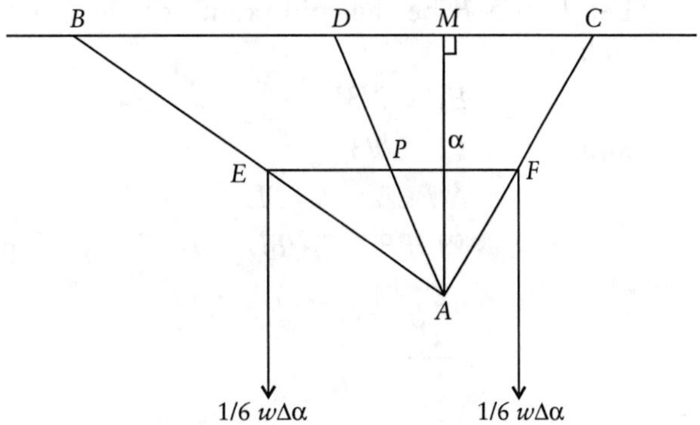

and, whole pressure on the triangle (P_2)

$= wS\,\bar{z}$

$= w\Delta\ \alpha/3$

$= 1/3\ w\Delta\alpha$

Clearly, $P_1 = P_2$

Example 2: A rectangle is immersed vertically in a heavy homogeneous liquid with two of its sides horizontal and at depths a and b below the surface. Find the depth of the centre of pressure.

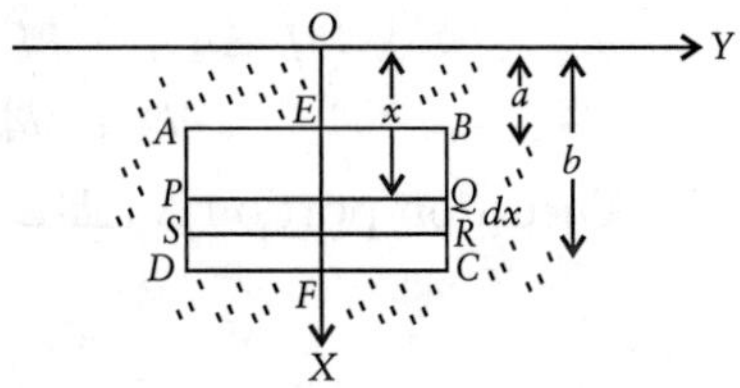

Sol. Let $ABCD$ be a rectangle immersed vertically in a heavy homogeneous liquid of density ρ with its two sides AB and DC horizontal and at depths a and b below the free surface. Let $AB = d$. Consider an elementary strip $PQRS$ of the rectangle at a depth x below the free surface and of thickness dx. Then, elementary area $\delta s = d\ \delta x$.

p = Intensity of pressure at any point of this strip.

$= g\rho x$

Let E and F be the mid-points of AB and DC respectively. Then, by symmetry, C.P. will lie on EF.

Take O, the point of intersection of line EF with free surface, as origin, OEF as X-axis and a line through O, perpendicular to OEF as Y-axis. Let $(\bar{x}, \bar{y})$ be the coordinates of C.P. referred to these axes. Then,

$$\bar{y} = 0$$

$$\bar{x} = \frac{\int xp\,ds}{\int p\,ds} \qquad = \frac{\int_a^b xg\rho x\,d.dx}{\int_a^b g\rho x\,d.dx}$$

$$= \frac{\int_a^b x^2\,dx}{\int_a^b x\,dx} = \frac{\left(\frac{x^3}{3}\right)_a^b}{\left(\frac{x^2}{2}\right)_a^b} \quad \Rightarrow \quad = \frac{\frac{b^3 - a^3}{3}}{\frac{b^2 - a^2}{2}} = \frac{2}{3}\left(\frac{a^2 + ab + b^2}{a + b}\right)$$

Example 3: *P* is the C.P. of the rectangle *ABCD*, the side *AB* being in the surface. Prove that the line through *A* and *P* divides the area into two portions, the pressures on which are equal.

Sol. Let $AB = a$ and $AD = b$. Let *AP* be produced to meet the side *DC* of the rectangle in *L*.

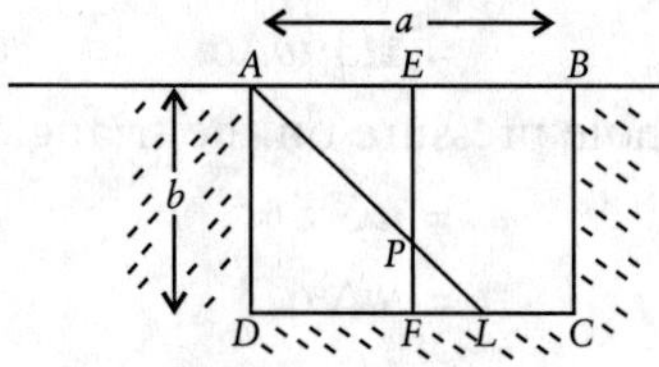

Let *E* and *F* be the mid-points of the sides *AB* and *DC*. Then,

$$EP = 2/3b$$

and $$PF = b/3$$

$\because$ $\Delta\, PEA \sim \Delta\, PFL$

$\because$ $PE/PF = EA/FL$ | by property of similar triangles

$$\Rightarrow \quad \frac{\frac{2}{3}b}{\frac{1}{3}b} = \frac{\frac{a}{2}}{FL}$$

$$\Rightarrow \quad FL = a/4$$

$$\therefore \quad DL = DF + FL$$

$$= a/2 + a/4 = 3a/4$$

Thrust on portion $\Delta\, ADL\ (P_1) = wS\,\bar{z}$

$$= w\frac{DL\,.\,AD}{2}\,.\,\frac{2}{3}AD$$

$$= \frac{w}{3}\,.\,DL\,.\,AD^2$$

$$= \frac{w}{3}\,.\,\frac{3a}{4}\,.\,b^2$$

$$= \frac{wa}{4}\,b^2 \qquad \text{...(1)}$$

Thrust on rectangle $ABCD\ (P_2) = wS\bar{z}$

$$= wAB\,.\,AD\,\frac{AD}{2}$$

$$\frac{wAB\,.\,AD^2}{2}$$

$$\frac{wab^2}{2} \qquad \text{...(2)}$$

∴ Thrust on area *AE BCL* (P_3)

= Thrust on rectangle *ABCD*–Thrust on Δ *ADL*

$$= P_2 - P_1 \qquad = \frac{wab^2}{2} - \frac{wab^2}{4}$$

$$= \frac{wab^2}{4} \qquad ...(3)$$

From (1) and (3), we get

$P_1 = P_3$

⇒ Thrust on portion Δ *ADL* = Thrust on portion *AEBCL*.

Hence, the thrusts on the two portions are equal.

Example 4: Prove that the horizontal line through the centre of pressure of a rectangle, immersed in a liquid with one side in the surface, divides the rectangle in two parts, the fluid pressures on which are in the ratio 4 : 5.

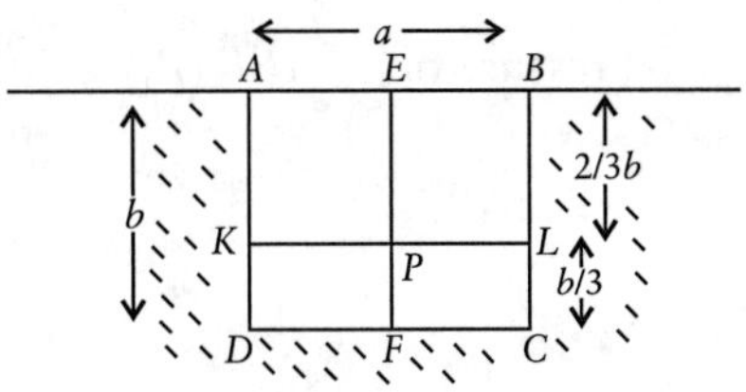

Sol. Let *ABCD* be a rectangle immersed in a liquid with one side *AB* in the surface. Let *AB* = *a* and *AD* = *b*. Let *E* and *F* be the mid-points of sides *AB* and *DC* respectively. Let *P* be the C.P. of the rectangle *ABCD*. Then, *P* will lie on *EF* (by symmetry) such that *EP* = 2/3 *b* and *PF* = *b*/3. |Refer §5(1)

Let *KL* be the horizontal line through *P*. Then, it divides the rectangle *ABCD* in two parts *ABLK* and *KLCD*.

Fluid pressure on part *ABLK* (P_1) = $wS\bar{z}$

$= w\,(a\,.\,2/3b)\;1/2\,(2/3b)$

$= 2/9\; wab^2 \qquad ...(1)$

Fluid pressure on part *KLCD* = $wS\bar{z}$

$= w\,(a\,.\,b/3)\,(2/3b + 1/2\,.\,b/3) = w\,.\,ab/3\;5b/6$

$= 5wab^2/18 \qquad ...(2)$

$$\therefore \qquad = \frac{P_1}{P_2} = \frac{\frac{2}{9}wab^2}{\frac{5}{18}wab^2} = \frac{4}{5} \qquad \Rightarrow \; P_1 : P_2 = 4:5$$

Example 5: A triangle is wholly immersed in a liquid with its base on the surface. Show that a horizontal straight line drawn through the centre of pressure of the triangle divides it into two parts, the pressures on which are equal.

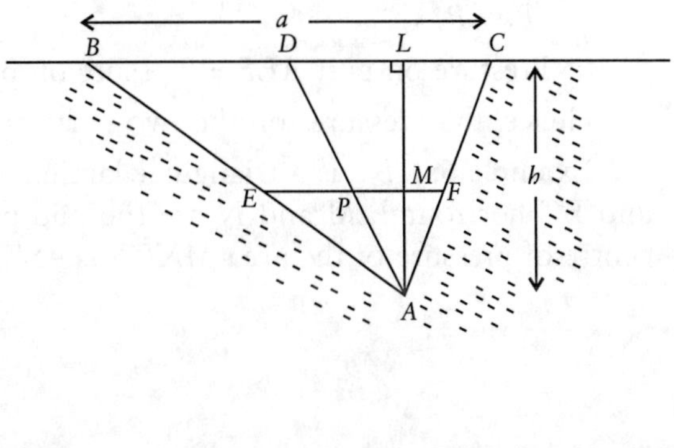

Sol. Let *ABC* be a triangle wholly immersed in a liquid with its base *BC* in the surface. Let *D* be the mid-point of *BC*. Then, *AD* is a median of Δ *ABC*. By symmetry, the centre of pressure *P* of the Δ *ABC* will lie on the median *AD* such that *DP* = 1/2 *DA*. |Refer § 5(3)

Draw perpendicular AL from A on base BC to meet the free surface at L. Let EF be a horizontal line drawn through P which intersects line AL at M. Then, it divides the triangle ABC into two parts $\Delta\, AEF$ and area $EFCB$. Also, $AL = 2\,AM$

| Refer § 5(3) Note 3 ...(1)

Let $BC = a$ and $AL = h$

$\because\ \Delta\, AEF \sim \Delta\, ABC$

$\therefore\ EF/BC = AM/AL$ | By property of similar triangles

$\Rightarrow\ EF/BC = 1/2$ | From (1)

$\Rightarrow\ EF = 1/2\ BC = 1/2\ a$ ...(2)

Depth of C.G. of $\Delta\, AEF$ below EF

$= 1/3\,.\,h/2 = h/6$

Depth of C.G. of $\Delta\, ABC$ below $BC = h/3$

$$\text{Pressure on } \Delta\, AEF\ (P_1) = w.\left(\frac{EF\,.\frac{h}{2}}{2}\right)\left(\frac{h}{2}+\frac{h}{6}\right)$$

$$= \frac{w\frac{1}{2}a\,.\frac{h}{2}\left(\frac{2h}{3}\right)}{2}$$

$$= \frac{wah^2}{12} \qquad ...(3)$$

Pressure on $\Delta\, ABC\ (P_2) = wS\,\bar{z}$

$= w.\ BC\,.\,h/2\,.\,h/3$

$= wah/2\,.\,h/3$

$= wah^2/6$ ...(4)

$\therefore$ Pressure on area $EFCB\ (P_3) = P_2 - P_1$

$$= \frac{wah^2}{6} - \frac{wah^2}{12}$$

$$= \frac{wah^2}{12} \qquad ...(3)$$

From (1) and (3)

$P_1 = P_3$

$\Rightarrow$ Pressure on part AEF = Pressure on part $EFCB$.

Hence, the pressures on the two parts are equal.

Example 6: ABC is a triangular lamina vertically immersed in a fluid with A in the surface and BC horizontal. M and N are the mid-points of AB and AC. Prove that the depth of the centre of pressure of the area $MNCB$ is $45/56d$, where d is the depth of B.

Sol. Since MN is a line joining the mid-points of the sides AB and AC of $\Delta\, ABC$, therefore,

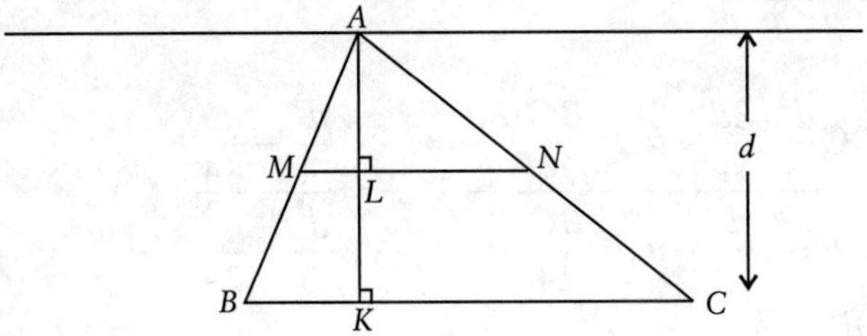

$$MN \,||\, BC$$

and $MN = 1/2\, BC$ ...(1)

| By geometry (Mid-Point Theorem)

$\because\ \Delta\, AMN \sim \Delta\, ABC$

$\therefore\ AL/AK = MN/BC$ | by property of similar triangles

$= 1/2$

$\Rightarrow\ AL/d = 1/2$

$\Rightarrow\ AL = 1/2\, d$ ...(2)

Pressure on $\Delta\, AMN\ (P_1) = wS\bar{z}$

$$= \frac{w.MN\,.\,AL}{2}\,.\,\frac{2}{3}AL \qquad = \frac{w.MN\,.\,AL^2}{3}$$

$$= \frac{w.MN\,.\,\frac{d^2}{4}}{3} \qquad \text{| From (2)}$$

$$= \frac{wd^2}{12}MN \qquad = \frac{wd^2}{12}\,.\,\frac{1}{2}BC$$

$$= \frac{wd^2BC}{24} \qquad \text{...(3)}$$

Depth of C.G. of $\Delta\, AMN = 3/4\, AL = 3/4\ 1/2\, d = 3d/8$ ...(4)

Pressure or $\Delta\, ABC\ (P) = wS\bar{z}$

$$= \frac{wBC\,.\,AK}{2}\,.\,\frac{2}{3}AK$$

$$= \frac{wBCd^2}{3} \qquad \text{...(5)}$$

Depth of C.G. of $\Delta\, ABC\ (x) = 3/4\, AK = 3d/4$ ...(6)

Let x_2 be the depth of the centre of pressure of the area $MNCB$. Then,

$$\bar{x} = \frac{Px - P_1x_1}{P - P_1} \qquad \text{| Refer §8}$$

$$\overline{x} = \frac{Px - P_1x_1}{P - P_1}$$

$$= \frac{wBC\dfrac{d^2}{3}.\dfrac{3d}{4} - wd^2\dfrac{BC}{24}.\dfrac{3d}{8}}{wBC\dfrac{d^2}{3} - wd^2\dfrac{BC}{24}} \qquad = \frac{\dfrac{d^3}{4} - \dfrac{d^3}{64}}{\dfrac{d^2}{3} - \dfrac{d^2}{24}}$$

$$= \frac{\dfrac{15d^3}{64}}{\dfrac{7d^2}{24}} \qquad = \frac{15d^3}{64} \cdot \frac{24}{7d^2}$$

$$= \frac{45d}{56}$$

Example 7: Prove that the depth of the centre of pressure of a trapezium immersed in water with the side a in the surface and the parallel side b at a depth h below the surface is

$$\left(\frac{a+3b}{a+2b}\right)\frac{h}{2}$$

Sol. Let $ABCD$ be a trapezium immersed in water with the side AB (= a) in the surface and the parallel side DC (= b) at a depth h below the free surface.

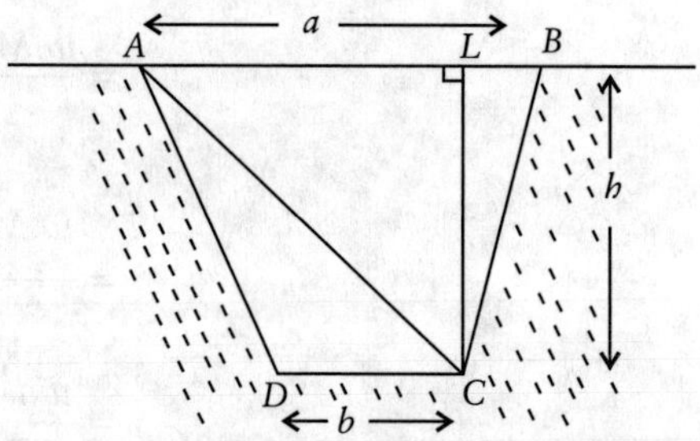

Draw perpendicular CL from C on the side AB to meet the side AB in L.

Then, $CL = h$

Join AC.

Then, AC divides the trapezium $ABCD$ into two parts $\Delta\ ADC$ and $\Delta\ ABC$.

Pressure on $\Delta\ ADC\ (P_1) = wS\,\overline{z}$

$$= w.\frac{bh}{2}.\frac{2}{3}h$$

$$= \frac{1}{3}wbh^2$$

Depth of C.P. of $\Delta\ ADC$ below the free surface $(\overline{x}_1) = 3/4h$ | Refer § 5(2)

Pressure on $\Delta\ ABC\ (P_2) = wS\,\overline{z}$

$$= w.\frac{ah}{2}.\frac{h}{3}$$

$$= \frac{wah^2}{6}$$

Depth of C.P. of Δ ABC below the free surface $(x_2) = h/2$ |Refer § 5(3)

$\therefore$ Depth of C.P. of the trapezium $ABCD$ below the free surface $(\bar{x})$

$$= \frac{P_1x_1 + P_2x_2}{P_1 + P_2} \quad | \text{ Refer § 7}$$

$$= \frac{\frac{1}{3}wbh^2\,\frac{3}{4}h + \frac{wah^2}{6}.\frac{h}{2}}{\frac{1}{3}wbh^2 + \frac{wah^2}{6}} = \frac{\frac{bh^3}{4} + \frac{ah^3}{12}}{\frac{bh^2}{3} + \frac{ah^2}{6}}$$

$$= \frac{\frac{3bh^3 + ah^3}{12}}{\frac{2bh^3 + ah^2}{6}} = \frac{(a+3b)h}{(a+2b)2}$$

Example 8: A quadrilateral $ABCD$, with sides AB, CD parallel (and DA, BC equal) is immersed in a homogeneous liquid with AB in the free surface. Show that the centre of pressure will be the intersection of the diagonals if

$$AB = \sqrt{3}\,CD$$

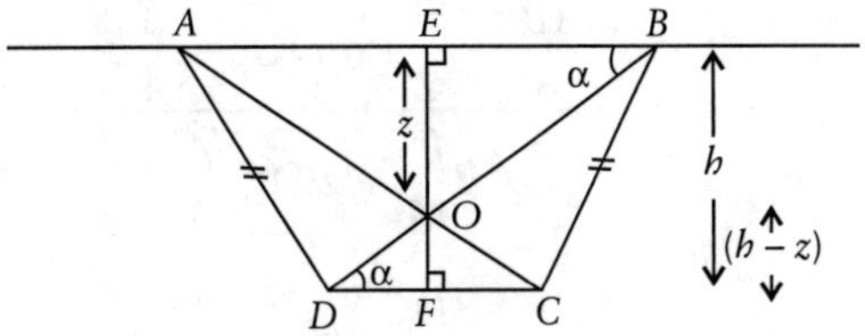

Sol. Let O be the point of intersection of the diagonals AC and BD of the quadrilateral $ABCD$.

Let h be the depth of DC below the free surface AB.

Through O, draw a line EOF perpendicular to the parallel sides AB and DC.

Let the depth of O below the free surface be z. Then,

$$OE = z$$

$$\therefore \quad OF = EF - OE = h - z$$

$$\because \quad \Delta\, PAB \sim PCD$$

$$\therefore \quad \frac{AB}{CD} = \frac{OE}{OF} \quad | \text{ by property of similar triangles}$$

$$\Rightarrow \quad \frac{AB}{CD} = \frac{z}{h-z} \quad \Rightarrow \quad \frac{CD}{AB} = \frac{h-z}{z}$$

$$\Rightarrow \quad \frac{CD}{AB} = \frac{h}{z} - 1 \quad \Rightarrow \quad \frac{CD}{AB} + 1 = \frac{h}{z}$$

$$\Rightarrow \quad \frac{CD + AB}{AB} = \frac{h}{z} \quad \Rightarrow \quad z = h\,.\,\frac{AB}{CD + AB} \quad ...(1)$$

Thrust on $\Delta\, ABD\ (P_1) = wS\,\bar{z}$

$$= \frac{wAB.h}{2} \cdot \frac{h}{3} \qquad = \frac{wABh^2}{6}$$

Depth of C.P. of $\Delta\, ABD\ (x_1) = h/2$ |Refer § 5(3) ...(2)

Thrust on $\Delta\, BCD\ (P_2) = wS\,\bar{z}$

$$= \frac{wDC.h}{2} \cdot \frac{2}{3}h$$

$$= \frac{wDC.h^2}{3} \qquad ...(3)$$

Depth of C.P. of $\Delta\, BCD\ (x_2) = 3/4$ |Refer § 5(2) ...(4)

$\therefore$ Depth of C.P. of the trapezium $ABCD$ below the free surface.

$$= \frac{P_1x_1 + P_2x_2}{P_1 + P_2}$$ |Refer § 7 since area of trapezium $ABCD$ = area of $\Delta\, ABD$ + area of $\Delta\, BCD$

$$= \frac{\frac{wABh^2}{6} \cdot \frac{h}{2} + wDC\frac{h^2}{3} \cdot \frac{3}{4}h}{wAB\frac{h^2}{6} + wDC\frac{h^2}{3}} = \frac{\frac{ABh^3}{12} + DC\frac{h^3}{4}}{\frac{ABh^2}{6} + DC\frac{h^2}{3}}$$

$$= \frac{\left(\frac{AB + 3DC}{12}\right)h^3}{\left(\frac{AB + 2DC}{6}\right)h^2} = \frac{(AB + 3DC)h}{2(AB + 2DC)} \qquad ...(5)$$

If the centre of pressure is the intersection of the diagonals, then the C.P. is O whose depth from the free surface is

$$z = \frac{hAB}{CD + AB}$$

$\therefore\quad z = \bar{x}$

$$\Rightarrow \quad \frac{hAB}{CD + AB} = \frac{(AB + 3DC)h}{2(AB + 2DC)}$$

$\Rightarrow\ 2AB\ (AB + 2DC) = (AB + CD)\ (AB + 3DC)$

$\Rightarrow\ 2AB^2 + 4AB\,.\,DC = AB^2 + 3AB\,.\,DC + DC\,.\,AB + 3DC^2$

$\Rightarrow\ AB^2 = 3DC^2$

$\Rightarrow\ AB = \sqrt{3}\,DC$

Aliter: Refer Ex. 7

Depth of C.P. of trapezium $ABCD$

$$= \frac{(a+3b)h}{2(a+2b)} \text{ where } a = AB,\ b = DC \quad \text{...(1)}$$

Now, $\tan\alpha = \frac{OE}{EB} = \frac{OF}{FD}$

$\Rightarrow \frac{OE}{OF} = \frac{EB}{FD}$

$\Rightarrow \frac{OF}{h-OE} = \frac{\frac{1}{2}AB}{\frac{1}{2}DC}$ | by geometry

$\Rightarrow \frac{OE}{h-OE} = \frac{a}{b}$

$\Rightarrow b\ OE = ah - a\ OE$

$\Rightarrow (a + b)\ OE = ah$

$\Rightarrow OE = ah/a + b$ ...(2)

According to the question,

$\frac{ah}{a+b} = \frac{(a+3b)h}{2(a+2b)}$ | $\because$ O is the C.P. of the trapeziums $ABCD$

$\Rightarrow 2a(a + 2h) = (a + b)(a + 3b) \Rightarrow 2a^2 + 4ah = a^2 + 3ab + ab + 3b^2$

$\Rightarrow a^2 = 3b^2 \Rightarrow a = \sqrt{3}\, b \Rightarrow AB = \sqrt{3}\, DC$

Example 9: ABC is a triangular lamina immersed vertically in water with C in the surface and AB horizontal. Show how to divide the area by a horizontal line PQ into two portions, the thrusts on which are equal, P and Q being points on AC and BC respectively.

If h be the length of the perpendicular from C on AB, prove that the height above AB of the centre of pressure of the area $APQB$ in the above case is

$\frac{1}{8}h(3\sqrt[3]{4} - 4)$

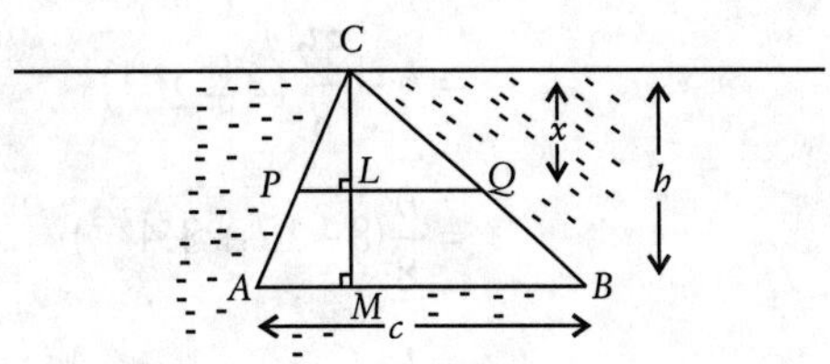

Sol. Let the depth of PQ below C be x. Let $AB = c$

$\because \Delta CPQ \sim \Delta CAB$

$\therefore PQ/AB = CL/CM$ | by property of similar triangles

$\Rightarrow PQ/c = x/h$

$\Rightarrow PQ = cx/h$ ...(1)

$\because$ Thrust on ΔCPQ = Thrust on portion $PQBA$

$\therefore$ Thrust on ΔCAB = 2 Thrust on ΔCPQ

$\Rightarrow w.\left(\frac{AB.h}{2}\right)\left(\frac{2}{3}h\right) = 2.w\left(\frac{PQ.x}{2}\right)\frac{2}{3}x$ | $P = wS\bar{z}$

$\Rightarrow \quad h^2 AB = 2x^2 PQ$

$\Rightarrow \quad h^2 c = 2x^2 \, cx/h$

$\Rightarrow \quad h^3 = 2x^3$

$\Rightarrow \quad x^3 = h^3/2$

$\Rightarrow \quad x = h/(2)^{1/3}$

Hence the horizontal line PQ should be drawn at a depth $h/(2)^{1/3}$, where h is the depth of AB below the free surface.

Let the thrust on $\Delta\ CPQ$ be P.

Let the thrust on $\Delta\ CAB$ be $2P$. | Since the thrusts on the two parts made by PQ are equal

Also, let x_1 and x_2 be the depths of C.P.'s of $\Delta\ CAB$ and $\Delta\ CPQ$ below the free surface. Then,

$$x_1 = 3/4\ h$$

$$x_2 = 3/4\ x = \frac{3}{4}\frac{h}{\sqrt[3]{2}}$$

$\therefore$ The depth of C.P. of the area $APQB$ below the free surface $(\bar{x})$

$$\frac{2Px_1 - Px_2}{2P - P} \qquad \text{| Refer §8}$$

$$= 2x_1 - x_2$$

$$= 2\,.\,3/4h - \frac{3}{4}\frac{h}{\sqrt[3]{2}}$$

$$= 3/4\ h\left(2 - \frac{1}{\sqrt[3]{2}}\right)$$

$\therefore$ Required height above AB of the centre of pressure of the area $APQB = h - \bar{x}$

$$= h - \frac{3h}{4}\left(2 - \frac{1}{\sqrt[3]{2}}\right) \qquad = h - \frac{3h}{8}\left(4 - \frac{2}{\sqrt[3]{2}}\right)$$

$$= h - \frac{3h}{8}(4 - 2^{2/3}) \qquad = h - \frac{3h}{8}(4 - 4^{1/3})$$

$$= \frac{h}{8}(8 - 12 + 3.4^{2/3}) \qquad = \frac{h}{8}(3.4^{1/3} - 4)$$

$$= \frac{h}{8}(3\sqrt[3]{4} - 4)$$

Example 10: A square lamina of side a, has a portion of it in the form of the inscribed circle removed from it and the remaining figure is immersed vertically in water with one side of the square in the surface. Prove that the depth of the C.P. is $= \dfrac{a}{24}\cdot\dfrac{64 - 15\pi}{4 - \pi}$

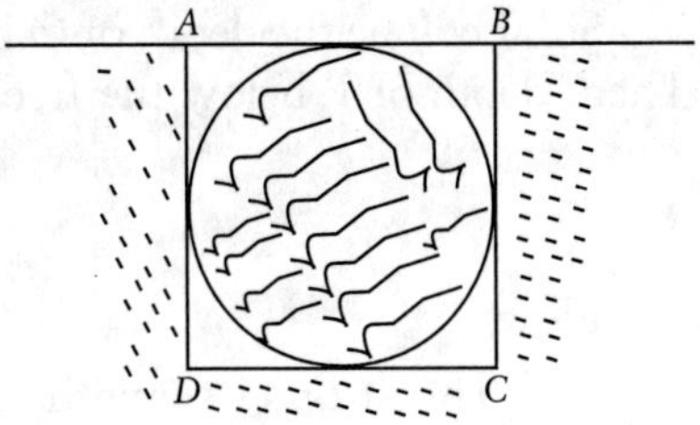

Sol. Let $ABCD$ be a square lamina of side a from which a portion in the form of the inscribed circle (shown shaded) has been removed and the remaining figure is immersed vertically in water with one side AB in the surface.

Radius of the inscribed circle = 1/2 side of the square = $a/2$

Now,

Thrust on the square $ABCD$ (P)

$= wS\,\bar{z}$

$= wa^2 a/2 = wa^3/2$

Depth of C.P. of the square $ABCD$ below the free surface $(x) = 2/3a$

Thrust on the inscribed circle (P_1)

$= wS\,\bar{z}$

$= w\pi\,(a/2)^2\,a/2$

$= w\pi a^3/8$

Depth of C.P. of the inscribed circle (x_1)

$$= \frac{a}{2} + \frac{\left(\frac{a}{2}\right)^2}{4\left(\frac{a}{2}\right)} \qquad | \text{ Refer § 5(4)}$$

$$= \frac{a}{2} + \frac{\frac{a^2}{4}}{2a} \qquad = \frac{a}{2} + \frac{a}{8} = \frac{5a}{8}$$

$\therefore$ The depth of C.P. of the remaining figure

$$= \frac{Px - P_1x_1}{P - P} \qquad | \text{ Refer § 8}$$

$$= \frac{\frac{wa^3}{2}\cdot\frac{2}{3}a - \frac{w\pi a^3}{8}\cdot\frac{5a}{8}}{\frac{wa^3}{2} - \frac{w\pi a^3}{8}}$$

$$= \frac{\frac{2}{3}a - \frac{5a}{32}\pi a}{1 - \frac{\pi}{4}} \qquad = \frac{a}{24}\left(\frac{64 - 15\pi}{4 - \pi}\right)$$

Example 11: A plane lamina consists of a circular disc (radius a and centre O) from which a circular portion (radius $a/2$ and centre P) has been cut. The lamina is completely immersed in a homogeneous fluid with its plane vertical and P vertically below O. If OP is equal to $1/2a$ and the centre of pressure of the lamina is at O, prove that O is at a depth $11/8a$ below the surface of the liquid.

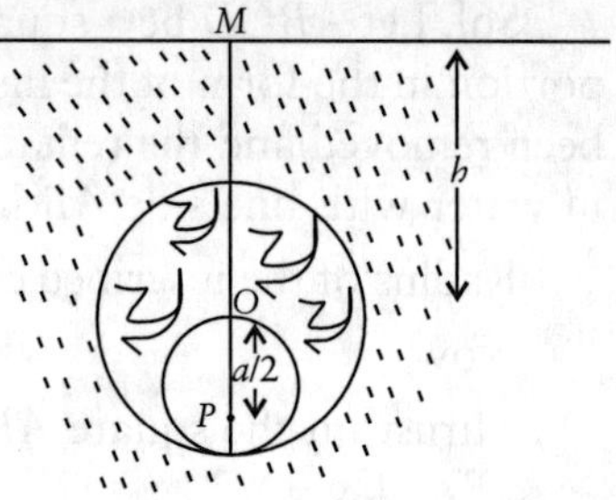

Sol. Let h be the depth of O below the surface of the liquid. Then, depth of P below the free surface = MP

$$= OM + OP$$
$$= h + a/2$$

Now,

P_1 = Thrust on circle with centre $O = wS\,\bar{z}$

$$= w\pi a^2 \,.\, h$$

x_1 = Depth of C.P. of the circle with centre O below O

$= a^2/4h$ | Refer § 5(4)

P_2 = Thrust on circle with centre P

$= wS\,\bar{z}$

$= w\pi\,(a/2)^2 \,.\, (h + a/2)$

x_2 = Depth of C.P. of the circle with centre P below O

$$= \frac{a}{2} + \frac{\left(\frac{a}{2}\right)^2}{4\left(h + \frac{a}{2}\right)}$$ | Refer § 5(4)

If the C.P. of the remainder area (shown shaded) is at O, then depth of C.P. of the remainder area below O is zero.

$$\therefore \quad \frac{P_1 x_1 - P_2 x_2}{P_1 - P_2} = 0$$ | Refer § 8

$$\Rightarrow \quad P_1 x_1 - P_2 x_2 = 0$$

$$\Rightarrow \quad w\pi a^2 h \frac{a^2}{4h} - w\pi\left(\frac{a}{2}\right)^2\left(h + \frac{a}{2}\right)\left\{\frac{a}{2} + \frac{\left(\frac{a}{2}\right)^2}{4\left(h + \frac{a}{2}\right)}\right\} = 0$$

$$\Rightarrow \quad \left\{a^2 - \left(h + \frac{a}{2}\right)\left(\frac{a}{2} + \frac{a^2}{16\left(h + \frac{a}{2}\right)}\right)\right\} = 0 \quad \Rightarrow \quad a^2 - \frac{a^2}{16} - \left(h + \frac{a}{2}\right)\frac{a}{2} = 0$$

$$\Rightarrow \quad a^2 - \frac{a^2}{16} - \frac{a^2}{4} - \frac{ha}{2} = 0 \quad \Rightarrow \quad \frac{11a^2}{16} - \frac{ha}{2} = 0$$

$$\Rightarrow \quad \frac{11a}{16} - \frac{h}{2} = 0 \quad \Rightarrow \quad h = \frac{11a}{8}$$

Hence, O is at a depth $11a/8$ below the surface of the fluid.

Example 12: A circular area is immersed in a liquid with its plane vertical and its centre at a depth h. If the area is lowered slowly so that the centre is at a depth $f(t)$ at time t, show that the velocity of the centre of pressure at that instant is

$$f(t)\left[1-\frac{a^2}{4\{f(t)\}^2}\right], \text{ where } a \text{ is the radius.}$$

Sol. Depth of centre of the area at time $t = h = f(t)$

Therefore, depth of C.P. of the circular area at any time t is given by

$$x = h + a^2/4h \qquad | \text{ Refer §5(4)}$$
$$= f(t) + a^2/4f(t)$$

Hence, the velocity of the centre of pressure at any time t is given by

$$\dot{x} = \frac{dx}{dt}$$

$$= f'(t) - \frac{a^2}{4\{f(t)\}^2} f'(t) \qquad = f'(t)\left[1-\frac{a^2}{4\{f(t)\}^2}\right]$$

Example 13: A flat circular plate of radius a, lies on a plane inclined at 30° to the horizontal and is subjected to water pressure on one face. The centre of pressure is at a distance $(a/16)$ from the geometric centre. Show that the geometric centre is at a depth $2a$ below the free surface of the water.

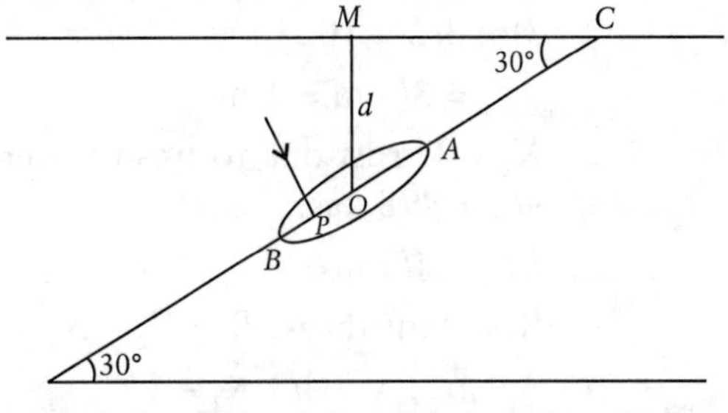

Sol. Let O be the geometric centre of the flat circular plate. Let the depth of O below the free surface be d. Then, $OM = d$. Let AB be the diameter of the plate along the line of greatest slope. Produce BA to meet the free surface at C. Then,

$$OC = d \text{ cosec } 30° = 2d \qquad ...(1)$$

Now let the plane of circular plate be rotated so that it becomes vertical and its centre O be at a depth $CO = 2d$ below the free surface. Then, the depth of the C.P. of the circular plate below the centre O

$$= \frac{a^2}{4h} \qquad | \text{Refer § 5(4)}$$

$$\Rightarrow \quad OP = \frac{a^2}{4OC}$$

$$\Rightarrow \quad \frac{a}{16} = \frac{a^2}{4(2d)} \qquad | \text{ From (1) and } OP = a/16 \text{ [given]}$$

$$\Rightarrow d = 2a$$

Hence, the geometric centre is at a depth $2a$ below the free surface of the water.

EXERCISE 1

1. A rectangular plate 3 metre long and 1 metre wide is immersed in water vertically in such a way that its 3 metre side is parallel to the water surface and is 1 metre below it. Find the position of centre of pressure.

   Ans. 14/9 m below the free surface on the line joining the mid-points of the sides of the plate parallel to the water surface.

2. In the vertical side of a vessel containing water, there is a square trap-door, opening freely outwards about a hinge in its upper edge which is horizontal. The length of a side of the door is 3 cm and water rises in the vessel upto the level of the hinge. Find the least force in grammes that will keep the trap-door closed.

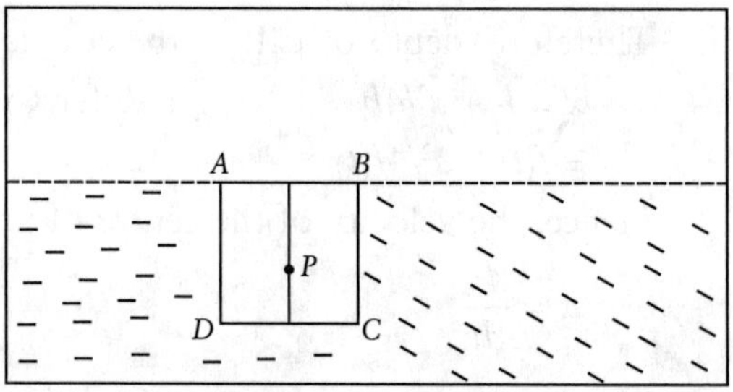

   Ans. 9 grammes weight

3. The gate of a dock is 3 m wide and 6 m deep and it has the pressure due to 4 m of fresh water on one side, 3 m of sea water on the other side. Find the magnitude and position of the resultant pressure on the surface of the gate.

   (Fresh water weighs 998 kg. per cu. m. and sea-water weighs 1025 kg. per cu. m.)

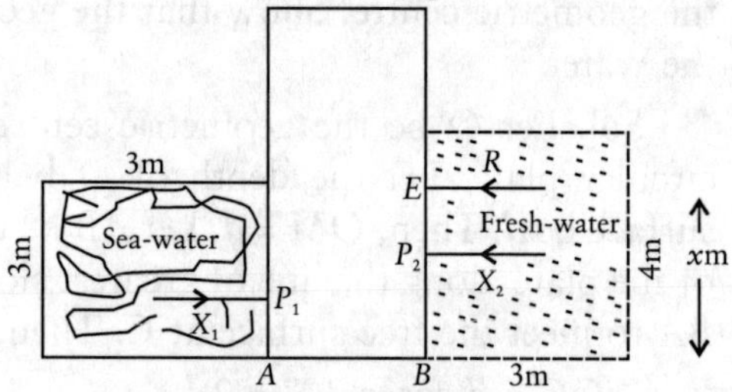

   [**Hint:** X_1 = Thrust due to sea-water = $(3 \times 3) \times 3/2 \times 1025$ kgs.

   $AP_1 = 3/3$ m = 1 m

   X_2 = Thrust due to fresh water = $(4 \times 3) \times 4/2 \times 998$ kgs.

   $BP_2 = 4/3$ m

   Resultant thrust $R = X_2 - X_1 = 10114.5$ kgs.

   Again, $Rx = 4/3\, X_2 - 1X_1$ | Taking moments about A

   $\Rightarrow \quad x = 2.7$ m (approx.)]

4. A rectangle is immersed vertically in a liquid whose density varies as the depth below the free surface. If the horizontal sides of the rectangle are at depths a and b below the surface, then show that the depth of the C.P. is

$$\frac{3}{4}\frac{(a+b)(a^2+b^2)}{(a^2+ab+b^2)}$$

   [**Hint:** Refer Ex. 2

$$\rho = \lambda x$$

$$\therefore \quad \bar{x} = \frac{\int_a^b xg\lambda x.x.d.dx}{\int_a^b g\lambda x.x.d.dx}$$

$$= \frac{\int_a^b x^3 dx}{\int_a^b x^2 dx} = \frac{\left(\frac{x^4}{4}\right)_a^b}{\left(\frac{x^3}{3}\right)_a^b}$$

$$= \frac{\frac{b^4 - a^4}{4}}{\frac{b^3 - a^3}{3}} = \frac{3}{4}\frac{(a+b)(a^2+b^2)}{(a^2+ab+b^2)}\Bigg]$$

5. A rectangular area of sides a and b has the side of length a vertical and the centre of the rectangle at a depth h below the surface of water in which it is immersed. Prove that the C.P. is at a depth $a^2/12h$ below the centre.

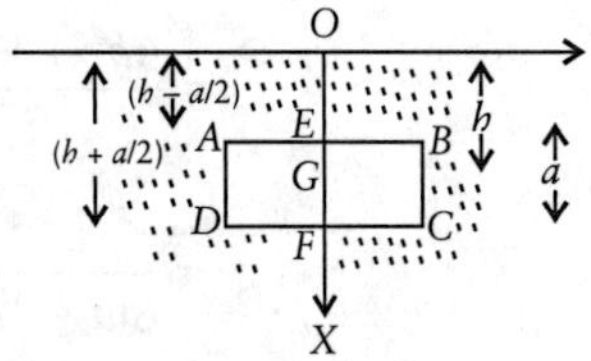

[**Hint:** Refer Ex. 2

Replace a by $(h - a/2)$ and b by $(h + a/2)$. Then,

Depth of C.P. below the free surface $(\overline{x})$

$$= \frac{2}{3}\frac{\left(h-\frac{a}{2}\right)^2 + \left(h-\frac{a}{2}\right)\left(h+\frac{a}{2}\right) + \left(h+\frac{a}{2}\right)^2}{\left(h-\frac{a}{2}\right) + \left(h+\frac{a}{2}\right)}$$

$$= \frac{3h^2 + \frac{a^2}{4}}{3h} = h + \frac{a^2}{12h}$$

$\therefore$ Depth of C.P. below the centre

$$= \left(h + \frac{a^2}{12h}\right) - h = \frac{a^2}{12h}\Bigg]$$

6. A rectangle is immersed vertically in water with two sides horizontal and at depths h and $a + h$ respectively below the effective surface. Prove that the distance of C.P. from the upper side is $\frac{a}{h}\cdot\frac{3h+2a}{2h+a}$. Show also that the C.P. is below the C.G. of the area and as the depth increases, C.P. approaches to C.G. but never coincides with it.

[**Hint:** Refer Ex. 2.

Replace a by h and b by $(a + h)$. Then,

Depth of C.P. below the free surface $(\overline{x})$

$$= \frac{2}{3}\frac{h^2 + h(a+h) + (a+h)^2}{h + (a+h)}$$

$$= \frac{2}{3}\frac{a^2 + 3h^2 + 3ah}{a + 2h}$$

$\therefore$ Depth of C.P. from the upper side = $\bar{x} - h$

$$= \frac{2}{3}\frac{a^2 + 3h^2 + 3ah}{a+2h} - h \qquad = \frac{2a^2 + 6h^2 + 6ah - 3ah - 6h^2}{3(a+2h)}$$

$$= \frac{2a^2 + 3ah}{3(a+2h)} \qquad = \frac{a(2a+3ah)}{3(a+2h)}$$

Depth of C.G. $(\bar{z}) = \frac{h+(a+h)}{2} = \frac{a+2h}{2}$

$$\therefore \quad \bar{x} - \bar{z} = \frac{2}{3}\frac{a^2 + 3h^2 + 3ah}{a+2h} - \frac{a+2h}{2}$$

$$= \frac{4a^2 + 12h^2 + 12ah - 3(a^2 + 4ah + 4h^2)}{6(a+2h)}$$

$$= \frac{a^2}{6(a+2h)} \qquad (>0) \text{ as } a > 0,\ h > 0$$

$\Rightarrow \quad \bar{x} > \bar{z}$

$\Rightarrow$ C.P. is below the C.G. of the area.

As the depth h increases, $\bar{x} - \bar{z}$ decreases and $\to 0$ when $h \to \infty$.

It shows that C.P. approaches to C.G. as h increases but never coincides with it.]

7. A rhombus is immersed in a liquid with a vertex in the surface and the diagonal through the vertex vertical. Prove that the centre of pressure divides the diagonal in the ratio 7 : 5.

[**Hint:** By symmetry, the C.P. of rhombus will lie at some point P on AC. Depth of C.P. P of rhombus $ABCD$ will be the same as depth of C.P. of $\Delta\, ABC$ or $\Delta\, ACD$.

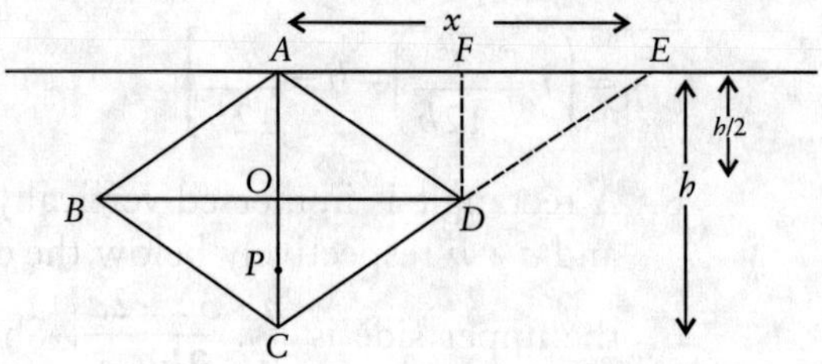

P = Pressure on $\Delta\, ACE = w\,(xh/2)\,h/3 = w/6\; xh^2$

$\bar{x}$ = Depth of C.P. of $\Delta\, ACE = 1/2h$

P_1 = Pressure on $\Delta\, ADE = w\,(x\,.\,h/2/2)\,h/2/3 = wxh^2/24$

x_1 = Depth of C.P. of $\Delta\, ADE = 1/2\,(h/2) = h/4$

$\therefore$ Depth of C.P. of $\Delta\, ACD$ (x_2)

$$= \frac{P\bar{x} - P_1x_1}{P - P_1} = \frac{7}{12}h$$

$$\Rightarrow \quad AP = \frac{7}{12}h$$

$\therefore\ PC = h - 7/12\ h = 5/12\ h$

$\therefore\ AP : PC = 7 : 5]$

8. Find the centre of pressure of a square lamina immersed in a fluid with one vertex in the surface and the diagonal vertical.

   Ans. C.P. divides the vertical diagonal in the ratio 7 : 5.

9. Determine the depth of the centre of pressure of a circular area whose plane is inclined at an angle θ to the horizontal.

   [**Hint:** Refer Ex. 13.

$$OC = d \operatorname{cosec}\theta$$

$$OP = \frac{a^2}{4OC} = \frac{a^2}{4d \operatorname{cosec}\theta}$$

$$\therefore \quad CP = OC + OP = d \operatorname{cosec}\theta + \frac{a^2}{4d \operatorname{cosec}\theta}$$

$\therefore$ Depth of C.P. of the circular area below the free surface = $CP \sin\theta$

$$= \left(d \operatorname{cosec}\theta + \frac{a^2}{4d \operatorname{cosec}\theta}\right)\sin\theta$$

$$= d + \frac{a^2 \sin^2\theta}{4d}$$

9. Effect of Further Immersion

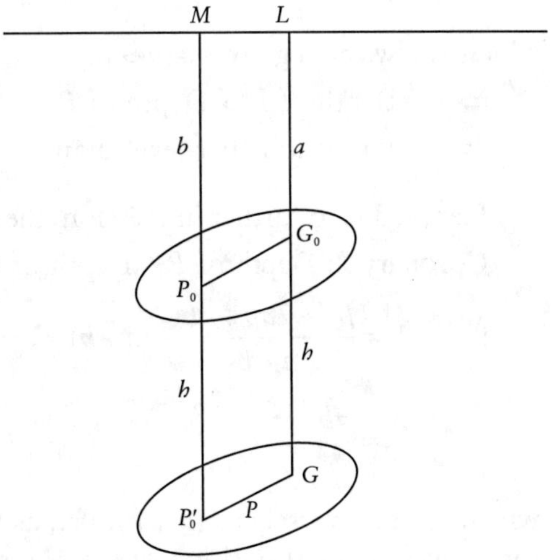

A plane area is immersed in a homogeneous liquid such that the depths of its centre of gravity and centre of pressure are a and b respectively. If the whole area is now lowered, without any rotation, to a further depth h, to find the new position of the centre of pressure of the area.

Let a plane area be immersed in a homogeneous, liquid with its centre of gravity G and centre of pressure P at depths a and b respectively below the free surface.

Let its area be S. Then,

the thrust on this area = $wS\,\bar{z}$

$= g\rho Sa$

where ρ is the density of the liquid. This thrust acts at P_0.

Now this plane area is lowered further in its own plane without any rotation, to a depth h. Let the new positions of the points G_0 and P_0 be G and P_0' respectively. Then, the depths of G_0 and P_0 below the free surface are now $a + h$ and $b + h$ respectively.

Due to this further immersion, the pressure at every point of S increases by an amount $g\rho h$ and this extra pressure is uniform on the whole area. Therefore, the resultant increased thrust due to this further immersion is $g\rho h\, S$ acting at G.

Thus in the new position of the plane area, there are two thrusts.

(i) $g\rho h\, S$ acting at G.

and (ii) $g\rho a\, S$ acting at P_0'

Since these two thrusts are like parallel, therefore, they can be compounded into a simple force acting at P, a point on the line $P_0' G$, such that

$$g\rho a\, S \times P_0'P = g\rho h\, S \times GP \qquad | \text{ By statics}$$

$$\Rightarrow \quad \frac{P_0'P}{GP} = \frac{h}{a}$$

If $\bar{z}$ is the depth of P below the free surface, then by taking moments about the free surface, we get

$$(g\rho a\, S + g\rho h\, S)\, \bar{z} = g\rho a\, S\,(b + h) + g\rho h\, S\,(a + h)$$

$$\Rightarrow \quad \bar{z} = \frac{h^2 + 2ah + ab}{a + h}$$

Corollary 1: Depth of P – Depth of P_0'

$$= \frac{h^2 + 2ah + ab}{a + h} - (b + h)$$

$$= -\frac{b - a}{a + h} h$$

which is always negative as $b > a$.

$\Rightarrow$ Depth of P_0' > Depth of P

$\Rightarrow$ P is at a higher level than P_0'

Hence, due to further immersion, the centre of pressure is raised through the distance $\frac{b-a}{a+h}h$

Corollary 2: Depth of P – Depth of G

$$= \frac{h^2 + 2ah + ab}{a + h} - (a + h)$$

$$= \frac{ab - a^2}{a + h}$$

which varies inversely as $(a + h)$, i.e., as the depth of the C.G. [Note that a and b are constants for a given original position of the plane area in the liquid]

It follows that greater the value of h, smaller is the difference between the depths of P and G. i.e., the centre of pressure P and the centre of gravity G come closer as h is increased and when h is infinite, P and G coincide.

Corollary 3: If the centre of pressure of the given plane area be known when the atmospheric pressure is not taken into account, then its position can be determined when the atmospheric pressure is taken into account for if π is the atmospheric pressure, then its effect is equivalent to supposing that the given area has been further lowered through a depth $\pi/g\rho$, because a head of $\pi/g\pi$ of that liquid above its surface would produce the same pressure π as due to atmosphere.

Corollary 4: The above theorem (corollary 3) is also true if instead of the area being lowered further through a distance h, a head (height) h of the same liquid has been superimposed on the surface of the liquid.

ILLUSTRATIVE EXAMPLES

Example 1: An equilateral triangle, each of where sides is $180\sqrt{3}$ cm long, is immersed vertically in water with its side in the surface which is open to the air. If the water barometer stands at 1020 cm, find the depth of centre of pressure of the triangle.

Sol. Side of the equilateral triangle = $180\sqrt{3}$ cm

∴ Height of the equilateral triangle

$$= 180\sqrt{3}\sin 60°$$

$$= 180\sqrt{3}\,\frac{\sqrt{3}}{2}$$

$$= 270 \text{ cm}$$

∴ a = Depth of C.G. = 1/3 × 270 = 90 cm

b = Depth of C.P. = 1/2 × 270° = 135 cm

Also, h = 1020 cm

∴ Depth of the centre of pressure of the triangle from the free surface

$$= \frac{h^2 + 2ah + ab}{a + h}$$

$$= \frac{(1020)^2 + (90)(1020) + (90)(135)}{90 - 1020}$$

$$= \frac{1040400 + 183600 + 12150}{1110}$$

$$= \frac{1236150}{1110} = \frac{123615}{111} = \frac{41205}{37}\text{cm}$$

∴ Depth of the centre of pressure of the triangle from the side in the surface.

$$= \frac{41205}{37} - 1020 = \frac{41205 - 37740}{37} = \frac{3465}{37} = 93\frac{24}{37}\text{cm}$$

Example 2: A plane area of any shape is completely submersed in a fluid of uniform density. Its plane is vertical. The depth of immersion is varied without rotating the area in its own plane. Show that the product of

(i) the depth of the centre of gravity below the surface, and

(ii) the depth of its centre of pressure below its centre of gravity is constant, the atmospheric pressure is to be neglected.

Sol. Refer figure of § 9

Depth of the centre of gravity below the surface = $a + h$

Depth of centre of pressure below centre of gravity

$$= \frac{ab - a^2}{a + h} \quad | \text{ Refer corollary 2 of § 9}$$

$\therefore$ Required product

$$= (a + h)\left(\frac{ab - a^2}{a + h}\right)$$

$$= ab - a^2$$

which is constant because it is independent of h.

Example 3: A square lamina is just immersed vertically in water and is then lowered through a depth h. If a is the length of the edge of the square, prove the distance of the centre of pressure from the centre of the square is $\frac{a^2}{6a + 12h}$.

Sol. Let $ABCD$ be the given square lamina, the length of whose edge is a, immersed vertically in water such that its edge AB is in the free surface. Then, the depth of C.G. (G_0) of the square $ABCD$ below the free surface = $EG_0 = a/2$

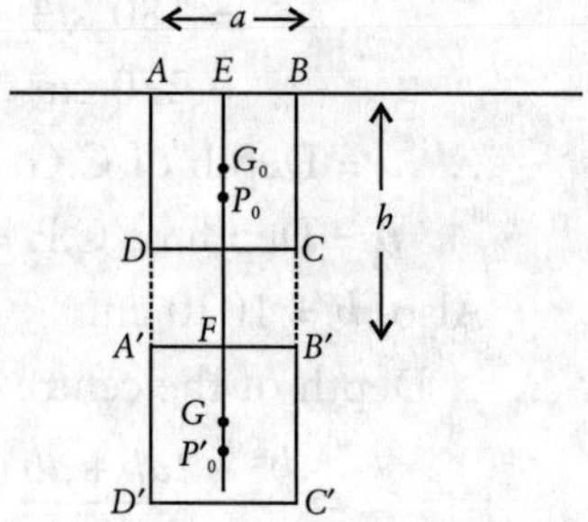

and, the depth of C.P. (P_0) of the square $ABCD$ below the free surface = $EP_0 = 2a/3$.

Now let the square be lowered through a depth h. Let $A'B'C'D'$ be the new position of the square $ABCD$. Then, the depth of the new C.G. (G) below the free surface = $EG = h + a/2$

and, the depth of the new C.P. (P'_0) below the free surface

$$= EP'_0 = \frac{h^2 + 2\left(\frac{a}{2}\right)h + \left(\frac{a}{2}\right)\left(\frac{2a}{3}\right)}{\frac{a}{2} + h} \quad | \text{ Refer § 9 for } \bar{z}$$

$$= \frac{h^2 + ah + \frac{a^2}{3}}{\frac{a}{2} + h}$$

$$= \frac{2(3h^2 + 3ah + a^2)}{3(a + 2h)}$$

$\therefore$ Distance of the new C.P. (P'_0) from the centre of the square

$$(G) = GP'_0$$

$$= EP'_0 - EG$$

$$= \frac{2(3h^2 + 3ah + a^2)}{3(a+2h)} - \left(h + \frac{a}{2}\right)$$

$$= \frac{12h^2 + 12ah + 4a^2 - 3(a+2h)(a+2h)}{6(a+2h)}$$

$$= \frac{12h^2 + 12ah + 4a^2 - 3a^2 - 12h^2 - 12ah}{6a + 12h}$$

$$= \frac{a^2}{6a + 12h}$$

Example 4: A triangle of height h is immersed in a liquid with the base horizontal and vertex in the surface. If the atmospheric pressure is equivalent to a head of H feet of the liquid, prove that the centre of pressure is raised to a height

$$\frac{hH}{4(2h + 3H)}$$

in the plane of the triangle.

Sol. Let $\Delta\, ABC$, whose height is h, be immersed in a liquid with its base BC horizontal and vertex A in the surface. When the atmospheric pressure is neglected, the depths of its C.G. and C.P. below the free surface are 2/3 h and 3/4 h respectively. When the atmospheric pressure, which is equivalent to a head of H feet of the liquid, is taken into consideration, then the effect of this atmospheric pressure is equivalent to supposing that the triangle ABC has been lowered further through the distance h in its own plane. In this case,

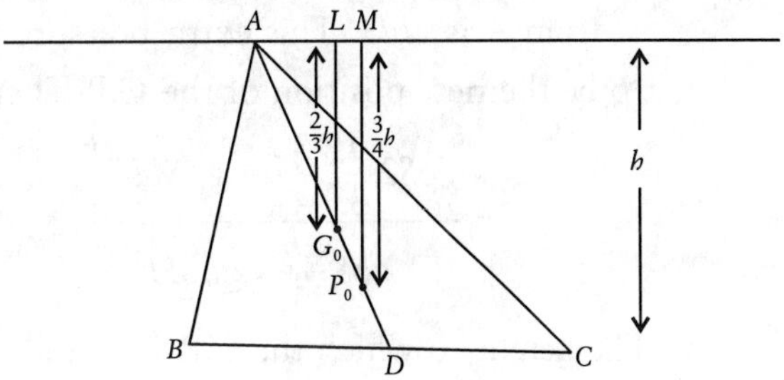

The depth of C.P. of the $\Delta\, ABC$ below the effective surface

$$= \frac{H^2 + 2\left(\frac{2h}{3}\right)(H) + \left(\frac{2h}{3}\right)\left(\frac{3h}{4}\right)}{\frac{2}{3}h + H} \qquad \text{| Refer § 9 for } \bar{z}$$

$$= \frac{H^2 + \frac{4}{3}hH + \frac{h^2}{2}}{\frac{1}{3}(2h + 3H)} = \frac{6H^2 + 8hH + 3h^2}{2(2h + 3H)}$$

$\therefore$ Depth of C.P. of the ΔABC below A

$$= \frac{6H^4 + 8hH + 3h^2}{2(2h + 3H)} - H$$

$$= \frac{4hH + 3h^2}{2(2h + 3H)}$$

$\therefore$ The height through which the C.P. is raised

$$= \frac{3h}{4} - \frac{4hH + 3h^2}{2(2h + 3H)}$$

$$= \frac{3h(2h + 3H) - 2(4hH + 3h^2)}{4(2h + 3H)}$$

$$= \frac{hH}{4(2h + 3H)}$$

Aliter: Let ρ be the density of the liquid. Let S be the area of ΔABC. When the atmospheric pressure is taken into account, then the thrusts acting on the triangle ABC are

(i) The whole thrust $g\rho S2h/3$ acting at P'_0 whose depth from A is $3/4h$.

(ii) The increased thrust $g\rho SH$ on account of the atmospheric acting at G whose depth from A is $2h/3$. This extra pressure is uniform on the whole triangle.

Let ρ be the new position of the C.P. Then, its depth from A

$$= \frac{g\rho S2\frac{h}{3} \cdot \frac{3}{4}h + g\rho SH\frac{2h}{3}}{g\rho S\frac{2h}{3} + g\rho SH} = \frac{3h^2 + 4hH}{2(2h + 3H)}$$

$\therefore$ The height to which the C.P. is raised

$$= \frac{3h}{4} - \frac{3h^2 + 4hH}{2(2h + 3H)}$$

$$= \frac{3h(2h + 3H) - 2(3h^2 + 4hH)}{4(2h + 3H)}$$

$$= \frac{hH)}{4(2h + 3H)}$$

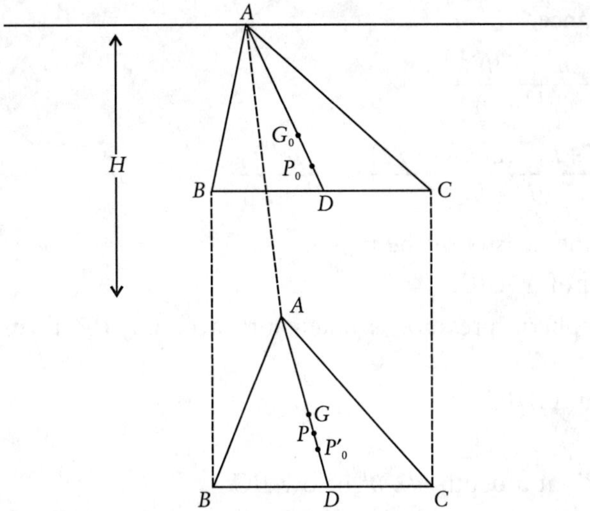

Example 5: A triangle has its base in the surface of a liquid and its vertex downwards. If the atmospheric pressure is equivalent to a height h of liquid, prove that the centre of pressure will be higher by a distance $\frac{1}{2}\frac{hh'}{h+h'}$ than it is when the atmospheric pressure is neglected, where h' is the depth of the C.G. of the triangle from the surface of the liquid.

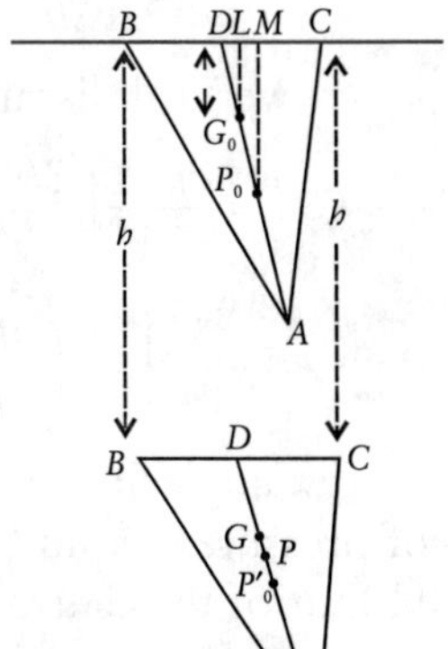

Sol. Let a triangle ABC have its base BC in the surface of a liquid and vertex A downwards when the atmospheric pressure is neglected, then the depth of its C.G. below the free surface is h'.

Therefore, depth of the vertex A below the free surface = $3h'$

$\therefore$ Depth of C.P. of $\Delta\, ABC$ below the free surface

$$= \frac{1}{2}(3h') = \frac{3}{2}h'$$

Now, if the atmospheric pressure, which is equivalent to head h of the liquid, is taken into account, then its effect is equivalent to supposing that the $\Delta\, ABC$ has been further lowered through a distance h in its own plane. In this case, the depth of C.P. of the $\Delta\, ABC$ below the effective surface

$$= \frac{h^2 + 2hh' + h' \cdot \left(\frac{3}{2}h'\right)}{h' + h} \qquad \text{| Refer § 9 for } \bar{z}$$

$$= \frac{2h^2 + 4hh' + 3h'^2}{2(h' + h)}$$

$\therefore$ Depth of C.P. of ABC below BC

$$= \frac{2h^2 + 4hh' + 3h'^2}{2(h' + h)} - h \qquad = \frac{2hh' + 3h'^2}{2(h' + h)}$$

$\therefore$ Required distance

$$= \frac{3h'}{2} - \frac{2hh' + 3h'^2}{2(h'+h)}$$

$$= \frac{3h'(h'+h) - (2hh' + 3h'^2)}{2(h'+h)} = \frac{hh'}{2(h+h')}$$

Aliter: Let ρ be the density of the liquid.

Let S be the area of $\Delta\, ABC$.

When the atmospheric pressure is taken into account, the thrusts acting on the triangle ABC are

(i) Pressure on $\Delta\, ABC$

$= wSh'$

acting at P'_0 at a depth 3/2 h' below BC

(ii) Additional pressure wSh due to the atmospheric pressure acting at G at a depth h' below BC

Let P be the new position of C.P. Then, its depth from BC

$$= \frac{wSh' - h' \quad wSh \quad h'}{wSh' \quad wSh} = \left(\frac{3h' \quad 2h}{h' \quad h}\right)\frac{\;'\;}{\;}$$

$\therefore$ Vertical distance between P and P_0'

$$= \frac{3h'}{2} - \left(\frac{3h' + 2h}{h' + h}\right)\frac{h'}{2}$$

$$= \frac{h'}{2}\left\{3 - \left(\frac{3h' + 2h}{h' + h}\right)\right\} = \frac{hh'}{2(h+h')}$$

Example 6: When the depth of the liquid is increased by an amount a, the depth of the centre of pressure is found to increase by y, and when instead, the depth of the liquid is increased by b, that of the centre of pressure is found to increase by z. Show that the depth of the centre of gravity of the area in the original state of the liquid is

$$= \frac{ab(b - a + y - z)}{az - by}$$

Sol. Refer figure of § 9

Let x and d be the depths of C.G. and C.P. of the area in the original state of the liquid.

When the depth of the liquid is increased by an amount a, then the depth of the centre of pressure is found to increase by y. | Given

$\therefore$ From $\bar{z} = \dfrac{h^2 + 2ah + ab}{a + h}$, we get

$$d+y=\frac{a^2+2xa+xd}{x+a}$$

$\therefore \quad (d + y)(x + a) = a^2 + 2xa + xd$

$\Rightarrow \quad xy + ay + ad = a^2 + 2ax \qquad ...(1)$

Again when the depth of the liquid is increased by an amount b, then the depth of the centre of pressure is found to increase by z. | Given

$\therefore$ From $\bar{z}=\dfrac{h^2+2ah+ab}{a+h}$, we get

$$d+z=\frac{b^2+2xb+xd}{x+b}$$

$\Rightarrow \quad (d + z)(x + b) = b^2 + 2xb + xd$

$\Rightarrow \quad xz + bz + bd = b^2 + 2bx \qquad ...(2)$

Multiplying (1) by b and (2) by a and then subtracting, we get

$x(by - az) + aby - abz = a^2b - ab^2$

$\Rightarrow \quad x(by - az) = a^2b - ab^2 - aby + abz$

$\Rightarrow \quad x(by - az) = ab\,(a - b - y + z)$

$\Rightarrow \quad x=\dfrac{ab(b-a+y-z)}{az-by}$

EXERCISE 2

1. (a) A triangle is wholly immersed in a liquid with its C.G. at a depth 2 cm and C.P. at a depth 3 cm. If a triangle be lowered through a distance 2 cm, find the depth of the C.P. in the new position.

   Ans. $4\frac{1}{2}$ cm.

   [**Hint:** a = 2 cm, b = 3 cm, h = 2 m]

   (b) An equilateral triangle, each of whose sides is $6\sqrt{3}$ feet long, is immersed vertically in water with its side in the surface which is open to the air. If the water barometer stands 34 feet, find the depth of the centre of pressure of the triangle.

   Ans. $3\frac{9}{74}$ feet.

2. A triangle is immersed in a liquid with the base horizontal and vertex in the surface. If the atmospheric pressure is equivalent to a head of H feet of the liquid, find through what height the centre of pressure of the triangle is raised in the plane of the triangle?

   Ans. $\dfrac{6H^2+8hH+3h^2}{2(2h+3H)}$, where h is the height of the triangle.

   [**Hint:** Refer Ex. 4]

3. A vertical circular area is immersed in water. If the depth of its centre is doubled, prove that the distance between its centre and centre of pressure will be halved.

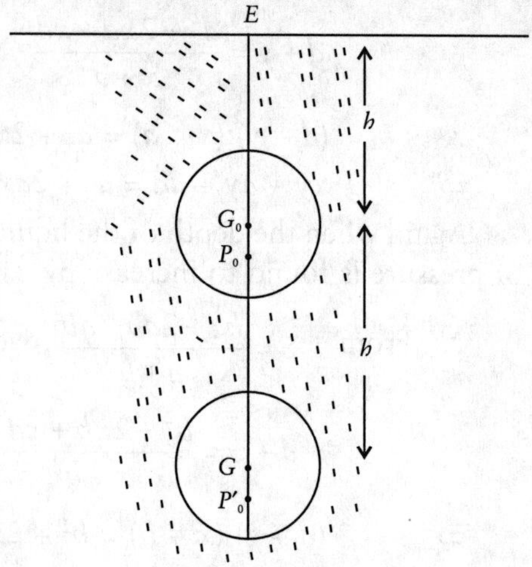

[Hint:

$$G_oP_o = \frac{a^2}{4h} \qquad \Big| \text{Refer § 5(4)}$$

$$EP'_o = \frac{h^2 + 2hh + h\left\{h + \dfrac{a^2}{4h}\right\}}{h+h}$$

| Here, 'a' = h, 'b' = $h + a^2/4h$, 'h' = h
| Refer § 9 for $\bar{z}$

$$= \frac{h^2 + 2h^2 + h^2 + \dfrac{a^2}{4}}{2h} = 2h + \frac{a^2}{8h}$$

$$\therefore\ GP'_0 = EP'_0 - EG$$

$$= \left(2h + \frac{a^2}{8h}\right) - 2h = \frac{a^2}{8h} = \frac{1}{2}\left(\frac{a^2}{4h}\right) = \frac{1}{2}G_oP_o\ ]$$

10. To Find the Depth of C.P. of a Triangle in Terms of Depths of its Three Vertices

Case I: When one vertex A is in the free surface.

Let ABC be a triangle, immersed in a liquid, with one vertex A in the free surface and vertices B and C at depths b and c respectively from the free surface. Produce CB to meet the free surface in D.

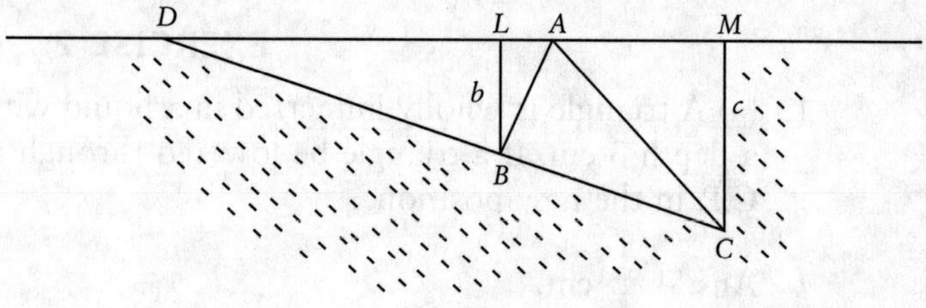

Let ρ be the density of the liquid. Now,

P_1 = thrust on $\Delta\, ACD$

$$= g\rho \cdot \frac{c}{3} \frac{AD \cdot c}{2} \qquad \Big| \text{'}P = wS\bar{z}\text{'}$$

$$= \frac{g\rho c^2 AD}{6}$$

z_1 = Depth of C.P. of $\Delta\, ACD$ below the free surface = $c/2$

P_2 = Thrust on $\Delta\, ABD$

$$= g\rho \cdot \frac{b}{3} \cdot \frac{AD \cdot b}{2} \qquad \Big| \text{'}P = wS\bar{z}\text{'}$$

$$= \frac{g\rho b^2 AD}{6}$$

z_2 = Depth of C.P. of $\Delta\, ABD$ below the free surface = $b/2$

Let $\bar{z}$ be the depth of C.P. of ΔABC below the free surface. Then,

$$\bar{z} = \frac{P_1 z_1 - P_2 z_2}{P_1 - P_2}$$

$$= \frac{g\rho c^2 \frac{AD}{6} \cdot \frac{c}{2} - g\rho b^2 \frac{AD}{6} \cdot \frac{b}{2}}{g\rho c^2 \frac{AD}{6} - g\rho b^2 \frac{AD}{6}}$$

$$= \frac{1}{2} \frac{c^3 - b^3}{c^2 - b^2}$$

$$= \frac{1}{2} \frac{(c-b)(b^2 + c^2 + bc)}{(c-b)(c+b)}$$

$$= \frac{1}{2} \frac{b^2 + c^2 + bc}{b+c}$$

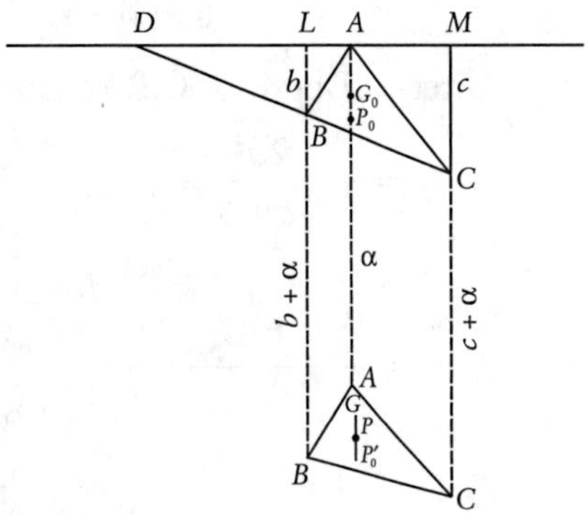

Case II: When the depths of the vertices, A, B and C from the free surface are α, β and γ respectively.

Let the triangle of Case I be now lowered further through a depth α without any rotation. Then, the depths of A, B and C from the free surface will be α, $b + \alpha$ and $c + \alpha$ respectively.

Due to this additional depth, an extra pressure $g\rho\alpha S$ will act at the centre of gravity G, where S is the area of the triangle.

In this position, two like parallel thrusts are acting on the triangle ABC.

(i) Thrust $g\rho.S.\ b + c/3$ acting at P'_0 whose depth from A is $\frac{1}{2}\frac{b^2 + c^2 + bc}{b+c}$ |Refer case I

(ii) Thrust $g\rho\alpha S$ acting at G whose depth from A is $b + c/3$

$\therefore$ Depth of the final C.P. (P) below the horizontal line through A

$$= \frac{g\rho S\left(\frac{b+c}{3}\right) \cdot \frac{1}{2}\left(\frac{b^2 + c^2 + bc}{b+c}\right) + g\rho\alpha S\left(\frac{b+c}{3}\right)}{g\rho S\left(\frac{b+c}{3}\right) + g\rho\alpha S}$$ | Refer § 7

$$= \frac{1}{2} \frac{b^2 + c^2 + bc + 2\alpha(b+c)}{b + c + 3\alpha}$$

$\therefore$ Depth of P below the free surface

$$= \alpha + \frac{1}{2} \frac{b^2 + c^2 + bc + 2\alpha(b+c)}{b + c + 3\alpha}$$

If the depths of B and C from the free surface are β and γ respectively, then

$\beta = b + \alpha$

and $\gamma = c + \alpha$

$\therefore$ $b = \beta - \alpha$

and $c = \gamma - \alpha$

$\therefore$ Depth of P below the free surface

$$= \alpha + \frac{1}{2}\frac{(\beta-\alpha)^2 + (\gamma-\alpha)^2 + (\beta-\alpha)(\gamma-\alpha) + 2\alpha(\beta-\alpha+\gamma-\alpha)}{\beta-\alpha+\gamma-\alpha+3\alpha}$$

$$= \alpha + \frac{1}{2}\frac{\beta^2+\gamma^2+\beta\gamma-\alpha\beta-\gamma\alpha-\alpha^2}{\alpha+\beta+\gamma}$$

$$= \frac{\alpha^2+\beta^2+\gamma^2+\alpha\beta+\beta\gamma+\gamma\alpha}{2(\alpha+\beta+\gamma)}$$

Aliter 1: Depth of C.P. of $\Delta\, ABC$ in this new position

$$= \frac{h^2+2ah+ab}{a+h} \qquad | \text{ Refer § 9}$$

$$= \frac{\alpha^2 + 2\frac{b+c}{3}.\alpha + \frac{b+c}{3}.\frac{1}{2}\frac{b^2+c^2+bc}{b+c}}{\frac{b+c}{3}+\alpha}$$

| Here, 'a' = $b + c/3$; 'h' = α, 'b' = $b^2 + c^2 + bc/2(b + c)$ by case I

$$= \frac{6\alpha^2+4\alpha(b+c)+(b^2+c^2+bc)}{2(b+c)+6\alpha}$$

$$= \frac{6\alpha^2+4\alpha(\beta-\alpha+\gamma-\alpha)+[(\beta-\alpha)^2+(\gamma-\alpha)^2+(\beta-\alpha)(\gamma-\alpha)]}{2(\beta-\alpha+\gamma-\alpha)+6\alpha}$$

$$= \frac{6\alpha^2+4\alpha(\beta+\gamma-2\alpha)+(\beta^2+\alpha^2-2\alpha\beta+\gamma^2+\alpha^2-2\gamma\alpha+\beta\gamma-\alpha\beta-\gamma\alpha+\alpha^2)}{2(\alpha+\beta+\gamma)}$$

$$= \frac{6\alpha^2+4\alpha\beta+4\gamma\alpha-8\alpha^2+3\alpha^2+\beta^2+\gamma^2-3\alpha\beta-3\gamma\alpha+\beta\gamma}{2(\alpha+\beta+\gamma)}$$

$$= \frac{\alpha^2+\beta^2+\gamma^2+\alpha\beta+\beta\gamma+\gamma\alpha}{2(\alpha+\beta+\gamma)}$$

Aliter 2: Let ABC be a triangle immersed in a liquid such that the depths of its vertices A, B and C from the free surface are α, β and γ respectively. Draw a horizontal line BD through B which divides the triangle ABC into two triangles ABD and BCD.

Consider an elementary strip $PQRS$ of $\Delta\, ABD$ at a depth z below the free surface and of breadth dz.

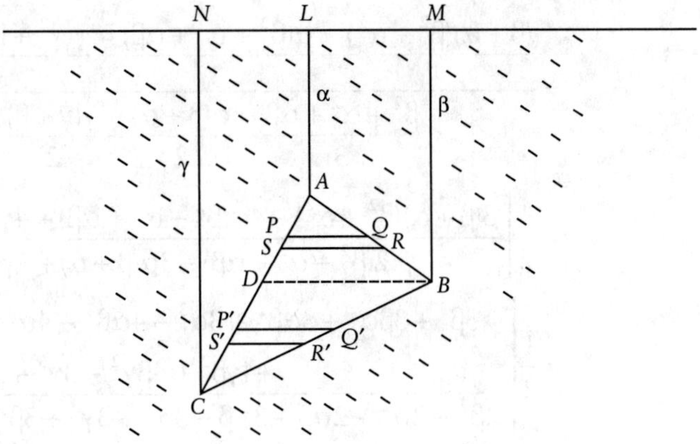

$\because \quad \Delta\, APQ \sim \Delta\, ADB$

$$\therefore \quad \frac{PQ}{DB} = \frac{z-\alpha}{\beta-\alpha}$$

| by property of similar triangles

$$\Rightarrow \quad PQ = \frac{z-\alpha}{\beta-\alpha} DB$$

dS = Area of the strip $PQRS$

$= PQ\, dz$

$$= \frac{z-\alpha}{\beta-\alpha} DB\, dz$$

p = intensity of pressure at any point on the strip $PQ = g\rho z$

$$\therefore \text{ Thrust on strip } PQ = \frac{z-\alpha}{\beta-\alpha} DB\, dz\, g\rho.z \qquad \text{...(1)}$$

Similarly, the thrust on strip $P'Q'R'S'$ of $\Delta\, BCD$ at a depth z

$$= \frac{\gamma-z}{\gamma-\beta} DB\, dz.g\rho.z \qquad \text{...(2)}$$

Let $\bar{z}$ be the depth of the C.P. of $\Delta\, ABC$ from the free surface. Then,

$$\bar{z} = \frac{\int_\alpha^\beta z\frac{z-\alpha}{\beta-\alpha} DB.dz.g\rho z + \int_\beta^\gamma z\frac{\gamma-z}{\gamma-\beta} DB.dz.g\rho z}{\int_\alpha^\beta \frac{z-\alpha}{\beta-\alpha} DB.dz.g\rho z + \int_\beta^\gamma z\frac{\gamma-z}{\gamma-\beta} DB.dz.g\rho z}$$

$$= \frac{\frac{1}{\beta-\alpha}\int_\alpha^\beta z^2(z-\alpha)dz + \frac{1}{\gamma-\beta}\int_\beta^\gamma z^2(\gamma-z)dz}{\frac{1}{\beta-\alpha}\int_\alpha^\beta z(z-\alpha)dz + \frac{1}{\gamma-\beta}\int_\beta^\gamma z(\gamma-z)dz}$$

$$= \frac{\frac{1}{\beta-\alpha}\left[\frac{z^4}{4} - \alpha\frac{z^3}{3}\right]_\alpha^\beta + \frac{1}{\gamma-\beta}\left(\gamma\frac{z^3}{3} - \frac{z^4}{4}\right)_\beta^\gamma}{\frac{1}{\beta-\alpha}\left(\frac{z^3}{3} - \alpha\frac{z^2}{2}\right)_\alpha^\beta + \frac{1}{\gamma-\beta}\left(\gamma\frac{z^2}{2} - \frac{z^3}{3}\right)_\beta^\gamma}$$

$$= \frac{\frac{1}{\beta-\alpha}\left[\frac{\beta^4-\alpha^4}{4} - \frac{\alpha}{3}(\beta^3-\alpha^3)\right] + \frac{1}{\gamma-\beta}\left[\frac{\gamma^4-\gamma\beta^3}{3} - \frac{\gamma^4-\beta^4}{4}\right]}{\frac{1}{\beta-\alpha}\left(\frac{\beta^3-\alpha^3}{3} - \alpha\frac{\beta^2-\alpha^2}{2}\right) + \frac{1}{\gamma-\beta}\left(\frac{\gamma^3-\gamma\beta^2}{2} - \frac{\gamma^3-\beta^3}{3}\right)}$$

$$= \frac{\frac{(\beta+\alpha)(\beta^2+\alpha^2)}{4} - \frac{\alpha(\beta^2+\alpha^2+\alpha\beta)}{3} + \frac{\gamma(\gamma^2+\beta^2+\beta\gamma)}{3} - \frac{(\gamma+\beta)(\gamma^2+\beta^2)}{4}}{\frac{\beta^2+\alpha^2+\alpha\beta}{3} - \frac{\alpha(\beta+\alpha)}{2} + \frac{\gamma(\gamma+\beta)}{2} - \frac{(\gamma^2+\beta^2+\beta\gamma)}{3}}$$

$$= \frac{1}{2}\left[\frac{3(\beta+\alpha)(\beta^2+\alpha^2) - 4\alpha(\beta^2+\alpha^2+\alpha\beta) + 4\gamma(\gamma^2+\beta^2+\beta\gamma) - 3(\gamma+\beta)(\gamma^2+\beta^2)}{2(\beta^2+\alpha^2+\alpha\beta) - 3\alpha(\beta+\alpha) + 3\gamma(\gamma+\beta) - 2(\gamma^2+\beta^2+\beta\gamma)}\right]$$

$$= \frac{1}{2}\left[\frac{\begin{array}{l}3\beta^3 + 3\beta\alpha^2 + 3\alpha\beta^2 + 3\alpha^3 - 4\alpha\beta^2 - 4\alpha^3 - 4\alpha^2\beta + 4\gamma^3 \\ \qquad + 4\gamma\beta^2 + 4\beta\gamma^2 - 3\gamma^3 - 3\gamma\beta^2 - 3\beta\gamma^2 - 3\beta^3\end{array}}{2\beta^2 + 2\alpha^2 + 2\alpha\beta - 3\alpha\beta - 3\alpha^2 + 3\gamma^2 + 3\beta\gamma - 2\gamma^2 - 2\beta^2 - 2\beta\gamma}\right]$$

$$= \frac{1}{2}\left[\frac{\gamma^3 + \gamma\beta^2 + \beta\gamma^2 - \alpha^2\beta - \alpha\beta^2 - \alpha^3}{-\alpha^2 - \alpha\beta + \beta\gamma + \gamma^2}\right]$$

$$= \frac{1}{2}\left[\frac{\gamma^3 - \alpha^3 + \gamma\beta^2 - \alpha\beta^2 + \beta\gamma^2 - \alpha^2\beta}{\gamma^2 - \alpha^2 + \beta\gamma - \alpha\beta}\right]$$

$$= \left[\frac{(\gamma-\alpha)(\gamma^2+\alpha^2+\gamma\alpha) + \beta^2(\gamma-\alpha) + \beta(\gamma-\alpha)(\gamma+\alpha)}{(\gamma-\alpha)(\gamma+\alpha) + \beta(\gamma-\alpha)}\right]$$

$$= \frac{1}{2}\left\{\frac{\gamma^2+\alpha^2+\gamma\alpha+\beta^2+\beta(\gamma+\alpha)}{\gamma+\alpha+\beta}\right\}$$

$$= \frac{1}{2}\frac{\alpha^2+\beta^2+\gamma^2+\alpha\beta+\beta\gamma+\gamma\alpha}{\alpha+\beta+\gamma}$$

Corollary 1: The above expression can be written as

$$\bar{z} = \frac{\frac{\alpha+\beta}{2}\cdot\frac{\alpha+\beta}{2} + \frac{\beta+\gamma}{2}\cdot\frac{\beta+\gamma}{2} + \frac{\gamma+\alpha}{2}\cdot\frac{\gamma+\alpha}{2}}{\frac{\alpha+\beta}{2} + \frac{\beta+\alpha}{2} + \frac{\gamma+\alpha}{2}}$$

which shows that the centre of pressure of the triangle coincides with the centre of three parallel forces acting at the middle points of the sides and proportional (in magnitude) to the corresponding depths of the middle points.

Corollary 2: The above expression can also be written as

$$\bar{z} = \frac{\alpha(2\alpha+\beta+\gamma) + \beta(\alpha+2\beta+\gamma) + \gamma(\alpha+\beta+2\gamma)}{(2\alpha+\beta+\gamma) + (\alpha+2\beta+\gamma) + (\alpha+\beta+2\gamma)}$$

which shows that the centre of pressure of the triangle coincides with the centre of parallel forces acting at vertices A, B, C and proportional (in magnitude) to $(2\alpha + \beta + \gamma)$, $(\alpha + 2\beta + \gamma)$, $(\alpha + \beta + 2\gamma)$ respectively.

ILLUSTRATIVE EXAMPLES

Example 1: An equilateral triangle *ABC* of altitude *h* has one vertex fixed at a depth $2h$ below the surface of a liquid. From the position in which *BC* is horizontal above *A*. The triangle is turned till *BC* is again horizontal but below *A*, the plane of the triangle always remaining vertical. Show that the shift of the centre of pressure relative to the triangle is $h/16$.

Sol. Let *ABC* be an equilateral triangle whose vertex *A* is fixed at a depth $2h$ below the free surface of a liquid. Since the altitude of Δ ABC is *h* and its base *BC* is horizontal above *A*, therefore, the depths of the vertices *A, B, C* of the Δ *ABC* in this position are $2h$, *h*, *h* respectively.

∴ Depth of C.P. of Δ *ABC* below the free surface (AP_1)

$$= \frac{\alpha^2 + \beta^2 + \gamma^2 + \alpha\beta + \beta\gamma + \gamma\alpha}{2(\alpha + \beta + \gamma)}$$

$$\frac{(2h)^2 + (h)^2 + (h)^2 + (2h)(h) + (h)(h) + (h)(2h)}{2(2h + h + h)} \qquad \text{| Here, } \alpha = 2h,\ \beta = h,\ \gamma = h$$

$$= \frac{4h^2 + h^2 + h^2 + 2h^2 + h^2 + 2h^2}{8h} = 11h/8$$

∴ Depth of C.P. from $A = 2h - 11h/8 = 5h/8$

Again in the second position, let $AB'C'$ be the new position of the triangle. Then, the depths of the vertices *A*, B', C' of the Δ $AB'C'$ in this position are $2h$, $3h$, $3h$ respectively.

∴ Depth of the new C.P. of Δ $AB'C'$ below the free surface in this case.

$$= \frac{(2h)^2 + (3h)^2 + (3h)^2 + (2h)(3h) + (3h) + (3h) + (3h)(2h)}{2(2h + 3h + 3h)}$$

| Refer §10, Case II for $\bar{z}$; Here, $\alpha = 2h$, $\beta = 3h$, $\gamma = 3h$

$$= \frac{4h^2 + 9h^2 + 9h^2 + 6h^2 + 9h^2 + 6h^2}{16h}$$

$$= \frac{43h^2}{16h} \qquad = \frac{43}{16}h$$

∴ Depth of C.P. of Δ $AB'C'$ below *A*

$43h/16 - 2h = 11h/16$

∴ The shift of the centre of pressure relative to the triangle (P_1P_2)

$= AP_2 - AP_1 \qquad = 11h/16 - 5h/8$

$= h/16$

Example 2: A quadrilateral is immersed vertically having two sides of lengths $2a$, *a* parallel to the surface at depths *h*, $2h$ respectively.

Show that the depth of the centre of pressure is $3h/2$.

Sol. Let *ABCD* be a quadrilateral immersed vertically having two sides of lengths $2a$, *a* parallel to the surface at depths *h*, $2h$ respectively.

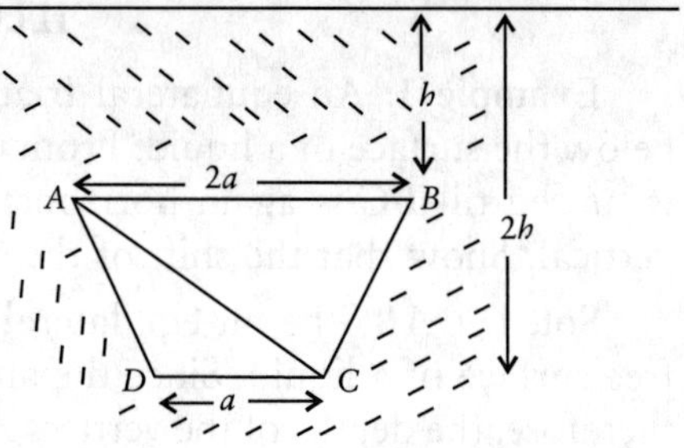

Join AC. Then,

Area of the quadrilateral $ABCD$

= Area of the $\Delta\, ADC$ + Area of the $\Delta\, ABC$.

Now,

P_1 = Thrust on the $\Delta\, ADC$

$$= g\rho\left(\frac{ah}{2}\right)\left(\frac{h+2h+2h}{3}\right) \qquad | \; P = wS\,\bar{z}$$

where ρ is the density of the liquid

$$= \frac{5g\rho a h^2}{6}$$

z_1 = Depth of C.P. of the $\Delta\, ADC$ below the free surface

$$= \frac{h^2+(2h)^2+(2h)^2+h(2h)+(2h)(2h)+(2h)h}{2(h+2h+2h)}$$

$17h/10$ $\qquad$ |Refer §10 case II for $\bar{z}$

P_2 = Thrust on $\Delta\, ABC$

$$= g\rho\left(\frac{2ah}{2}\right)\left(\frac{h+h+2h}{3}\right) \qquad | \; P = wS\,\bar{z}$$

$$= \frac{4g\rho a h^2}{3}$$

z_2 = Depth of C.P. of the $\Delta\, ABC$ below the free surface

$$= \frac{h^2+h^2+(2h)^2+h.h+h(2h)+(2h)h}{2(h+h+2h)}$$

$= 11h/8$ $\qquad$ |Refer §10 case II for $\bar{z}$

Let $\bar{z}$ be the depth of the quadrilateral $ABCD$ below the free surface. Then,

$$\bar{z} = \frac{P_1 z_1 + P_2 z_2}{P_1 + P_2} \qquad | \text{ Refer § 7}$$

$$= \frac{\frac{5g\rho a h^2}{6}\cdot\frac{17}{10}h + \frac{4g\rho a h^2}{3}\cdot\frac{11h}{8}}{\frac{5g\rho a h^2}{6} + \frac{4g\rho a h^2}{3}}$$

$$= \frac{\frac{17}{12}+\frac{11}{6}}{\frac{5}{6}+\frac{4}{3}}h \qquad = \frac{\frac{39}{12}}{\frac{13}{6}}h = \frac{3}{2}h$$

Example 3: Find the centre of pressure of a square lamina immersed in a fluid, with one vertex in the surface and the diagonal vertical. Show that the centre of pressure divides the diagonal in the ratio 7 : 5.

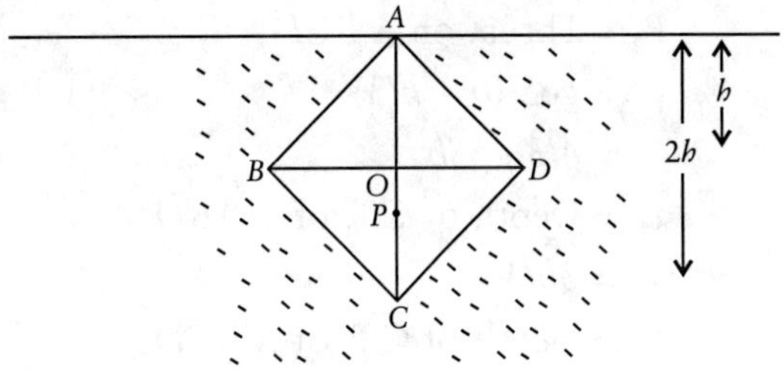

Sol. Let $ABCD$ be a square lamina immersed in a liquid, with one vertex A in the surface and the diagonal AC vertical. By symmetry, the centre of pressure of the square $ABCD$ will lie on AC and its depth from A will be the same as the depth of C.P. of ABC from A.

Let $AC = 2h$

Then, depth of B = depth of D below the free surface = h

∴ Depth of C.P. of $\Delta\ ABC$ from A below the free surface

$$= \frac{(0)^2 + (h)^2 + (2h)^2 + (0)(h) + (h)(2h) + (2h)(0)}{2(0 + h + 2h)}$$

$= 7h/6$ | Refer § 10 case II for $\bar{z}$

Let P be the C.P. of the square $ABCD$. Then,

$AP = 7h/6$

∴ $PC = AC - AP = 2h - 7h/6 = 5h/6$

∴ $AP : PC = 7h/6 : 5h/6 = 7 : 5$

Example 4: A parallelogram $ABCD$ is immersed in a liquid with A in the surface, and BD horizontal. Prove that the centre of pressure P lies on AC such that

$AP : AC = 7 : 2$

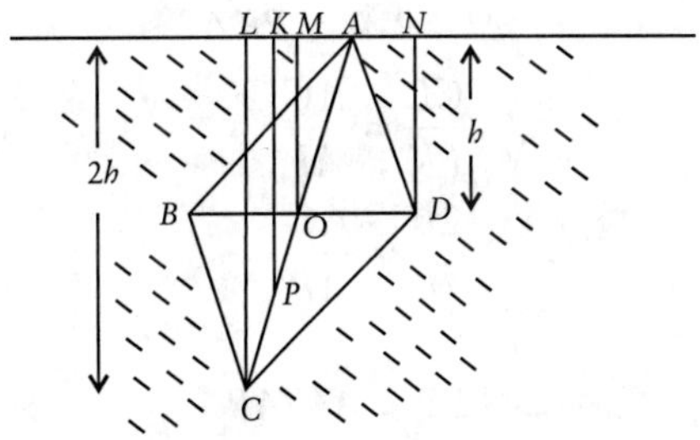

Sol. $ABCD$ is a parallelogram immersed in a liquid with A in the surface, and BD horizontal. Let O be the point of intersection of the diagonals AC and BD of the parallelogram $ABCD$. Then,

Area of parallelogram $ABCD$.

= Area of $\Delta\ ABC$ + Area of $\Delta\ ADC$.

Since the diagonal of a parallelogram bisect each other, therefore, AO is a median of $\Delta\ ABD$ and CO is a median of $\Delta\ BCD$; O being the middle point of AC and BD both.

The C.P. of $\Delta\ ABD$ will lie on the median AO. The C.P. of $\Delta\ BCD$ will lie on the median CO. Therefore, C.P. of the parallelogram $ABCD$ will lie on AC.

Let it be P.

Let area of $\Delta\ ABD$ = area of $\Delta\ BCD = S$

P_1 = Thrust on ABD

$= g\rho S\ 2/3\ h$ | $P = wS\ \bar{z}$

$= 2/3\ g\rho Sh$

where ρ is the density of the liquid and h is the depth of BD below the free surface.

P_2 = Thrust on Δ *BCD*

$= g\rho S\,(h + h/3)$ | $P = wS\bar{z}$

$= 4/3\ g\rho Sh$

z_1 = Depth of C.P. of Δ *ABD*

$= 3h/4$

z_2 = Depth of C.P. of Δ *BCD*

$$= \frac{h^2 + h^2 + (2h)^2 + (h)(h) + (h)(2h) + (2h)(h)}{2(h + h + 2h)}$$

$= 11h/8$ | Refer § 10 case II for $\bar{z}$

Let $\bar{z}$ be the depth of C.P. (*P*) of the parallelogram *ABCD* below the free surface. Then,

$$\bar{z} = \frac{P_1 z_1 + P_2 z_2}{P_1 + P_2} \quad \text{| Refer § 7}$$

$$= \frac{\left(\frac{2}{3} g\rho Sh\right)\left(\frac{3h}{4}\right) + \left(\frac{4}{3} g\rho Sh\right)\left(\frac{11h}{8}\right)}{\frac{2}{3} g\rho Sh + \frac{4}{3} g\rho Sh}$$

$$= \frac{\frac{1}{2} + \frac{11}{6}}{2} = \frac{7h}{6}$$

$\because$ Δ*CLA* ~ Δ *PKA*

$$\therefore \quad \frac{CL}{PK} = \frac{AC}{AP} \quad \text{| By property of similar triangles}$$

$$\therefore \quad \frac{2h}{\frac{7h}{6}} = \frac{AC}{AP}$$

$\Rightarrow 12/7 = AC/AP \quad \Rightarrow \quad AP/AC = 7/12 \quad \Rightarrow \quad AP : AC = 7 : 12$

Example 5: A quadrilateral in immersed in water with two angular points in the surface and the other two at depths *a* and *b*. If *x* and *y* are the depths below the surface of its centres of gravity and pressure respectively, show that $6xy + ab = 3x\,(a + b)$

Sol. Let *ABCD* be a quadrilateral immersed in water with two angular points *A* and *B* in the surface and the other two, i.e., *C* and *D* at depths *a* and *b* below the surface.

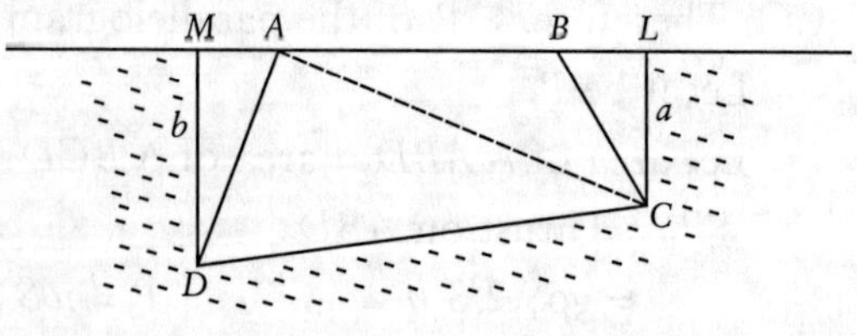

Join *AC*.

Then, area *ABCD* = area *ABC* + area *ADC*.

Let S_1 and S_2 be the areas of Δ *ABC* and Δ *ADC* respectively.

Depth of C.G. of Δ *ABC* = *a*/3

Depth of C.G. of Δ *ADC* = *a* + *b*/3

Let x be the depth of the C.G. of the quadrilateral $ABCD$ below the free surface. Then,

$$(S_1 + S_2)x = S_1 . \frac{a}{3} + S_2 . \frac{a+b}{3}$$

$$= (S_1 + S_2)\frac{a}{3} + S_2 \frac{b}{3}$$

$\Rightarrow \quad 3 (S_1 + S_2)x = (S_2 + S_2)\, a + S_2 b$

$\Rightarrow \quad (3x - a)\,(S_2 + S_2) = S_2 b$

$\Rightarrow \quad 3x - a/b = S_2 / S_2 + S_2$ ...(1)

Depth of C.P. of $\Delta\, ABC = a/2$

Depth of C.P. of $\Delta\, ADC \quad = \dfrac{b^2 + a^2 + ab}{2(a+b)}$ | Refer § 10 case I for $\bar{z}$

Thrust on $\Delta\, ABC \quad = g\rho S_1 \dfrac{a}{3}$ | $P = wS\,\bar{z}$

Thrust on $\Delta\, ADC \quad = g\rho S_2 \dfrac{a+b}{3}$ | $P = wS\,\bar{z}$

Let y be the depth of C.P. of quadrilateral $ABCD$ below the free surface. Then,

$$g\rho(S_1 + S_2)xy = g\rho S_1 . \frac{a}{3} . \frac{a}{2} + g\rho S_2 \frac{a+b}{3} . \frac{a^2 + b^2 + ab}{2(a+b)}$$

$\Rightarrow \quad 6(S_1 + S_2)xy = a^2(S_1 + S_2) + S_2 b(a+b)$

$\Rightarrow \quad 6xy = a^2 + \dfrac{S_2}{S_1 + S_2} b(a+b)$

$= a^2 + \dfrac{3x - a}{b} b(a+b)$ | From (1)

$= a^2 + (3x - a)\,(a + b) = 3x(a + b) - ab$

$\Rightarrow \quad 6xy + ab = 3x(a + b)$

Example 6: Show that the depth of the centre of pressure of a rhombus totally immersed with one diagonal vertical and its centre at a depth h is $h + \dfrac{a^2}{24h}$, where a is the length of the vertical diagonal.

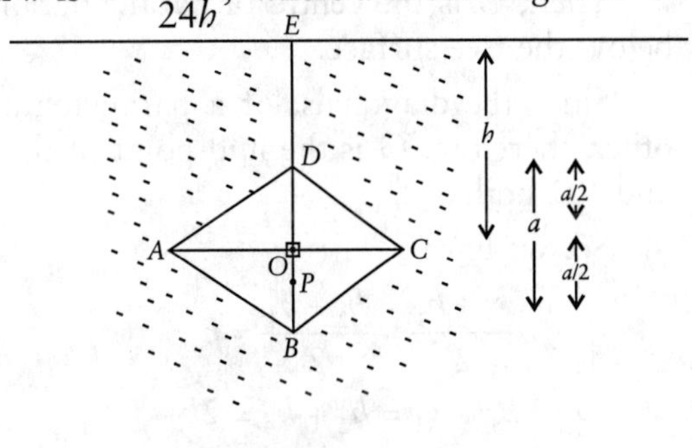

Sol. Let $ABCD$ be a rhombus totally immersed in a liquid with one diagonal DB vertical and its centre O at a depth h below the free surface. [Note that the centre O is the point of intersection of the diagonals AC and DB of the rhombus $ABCD$.]

Since a is the length of the vertical diagonal DB, therefore,

$DO = OB = a/2$ | by geometry

Since the rhombus $ABCD$ is symmetrical about the vertical diagonal DB, therefore, the C.P. of the rhombus will lie on DB. Also, the depth of C.P. of rhombus $ABCD$ will be the same as the depth of C.P. of $\Delta\ ABD$.

The depths of the vertices A, B and D of $\Delta\ ABD$ below the free surface are h, $h + a/2$ and $h - a/2$ respectively.

$\therefore$ Depth of C.P. of $\Delta\ ABD$ below the free surface

$$= \frac{\alpha^2 + \beta^2 + \gamma^2 + \alpha\beta + \beta\gamma + \gamma\alpha}{2(\alpha + \beta + \gamma)}$$

$$= \frac{h + \left(h + \frac{a}{2}\right)^2 + \left(h - \frac{a}{2}\right)^2 + h\left(h + \frac{a}{2}\right) + \left(h + \frac{a}{2}\right)\left(h - \frac{a}{2}\right) + \left(h - \frac{a}{2}\right)h}{2\left(h + h + \frac{a}{2} + h - \frac{a}{2}\right)}$$

$$= \frac{h^2 + h^2 + \frac{a^2}{4} + ah + h^2 + \frac{a^2}{4} - ah + h^2 + \frac{ah}{2} + h^2 - \frac{a^2}{4} + h^2 - \frac{ah}{2}}{6h}$$

$$= \frac{6h^2 + \frac{a^2}{4}}{6h} \qquad = h + \frac{a^2}{24}$$

Let P be the C.P. of the rhombus $ABCD$.

Then, depth of P below the free surface $= h + a^2/24h$.

Example 7: A parallelogram has its corners at depths h_1, h_2, h_3, h_4 below the surface of a liquid, and its centre at a depth h. Show that the depth of its centre of pressure is

$$\frac{h_1^2 + h_2^2 + h_3^2 + h_4^2 + 8h^2}{12h}$$

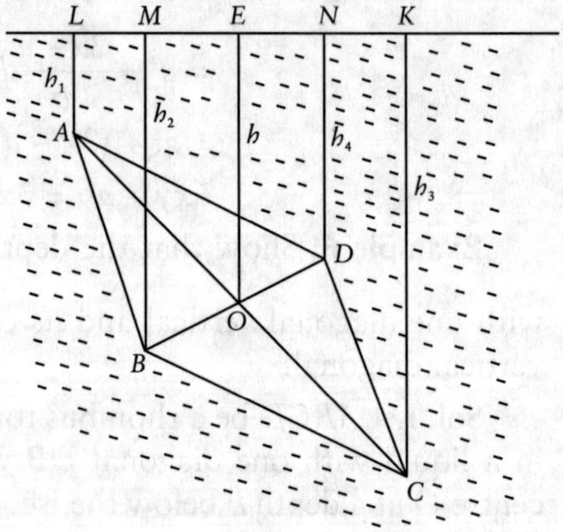

Sol. Let $ABCD$ be a parallelogram immersed in a liquid with its corners A, B, C and D at depth h_1, h_2, h_3 and h_4 respectively below the free surface. Join AC and BD.

Let the diagonals AC and BD intersect at O.

Then, O is the centre of parallelogram at a depth h below the free surface.

Since the diagonals, of a parallelogram bisect each other, therefore, O is the mid-point of the diagonals AC and BD both.

So, we have

$$\frac{h_1 + h_3}{2} = \frac{h_2 + h_4}{2} = h$$

$$\Rightarrow \quad h_1 + h_3 = h_2 + h_4 = 2h \qquad \text{...(1)}$$

Parallelogram $ABCD = \Delta\, ABD + \Delta\, ACD$ ...(2)

Let the area of $\Delta\, ABD = S$

Then, area of $\Delta\, ACD = S$

P_1 = Thrust on $\Delta\, ABD$

$$= g\rho\,.\,S\,.\left(\frac{h_1+h_2+h_4}{3}\right) \qquad | \; P = wS\,\bar{z} \qquad = g\rho S\left(\frac{h_1+2h}{3}\right) \qquad | \text{ From (1)}$$

where ρ is the density of the liquid.

z_1 = Depth of C.P. of $\Delta\, ABD$ below the free surface

$$\frac{h_1^2+h_2^2+h_4^2+h_1h_2+h_2h_4+h_4h_1}{2(h_1+h_2+h_4)} \qquad | \text{ Refer § 10 case II for } \bar{z}$$

$$\frac{h_1^2+h_2^2+h_4^2+h_1(h_2+h_4)+h_2h_4}{2(h_1+h_2+h_4)}$$

$$\frac{h_1^2+h_2^2+h_4^2+h_1(2h)+h_2h_4}{2(h_1+2h)} \qquad | \text{ From (1)}$$

P_2 = Thrust on $\Delta\, ACD$

$$= g\rho S\left(\frac{h_2+h_3+h_4}{3}\right) \qquad | \; P = wS\,\bar{z}$$

$$= g\rho S\left(\frac{h_3+2h}{3}\right) \qquad | \text{ From (1)}$$

z_2 = Depth of C.P. of $\Delta\, ACD$ below the free surface

$$= \frac{h_1^2+h_3^2+h_4^2+h_2h_3+h_3h_4+h_4h_2}{2(h_2+h_3+h_4)} \qquad | \text{ Refer § 10 case II for } \bar{z}$$

$$= \frac{h_2^2+h_3^2+h_4^2+h_3(h_2+h_4)+h_4h_2}{2(h_2+h_3+h_4)}$$

$$= \frac{h_2^2+h_3^2+h_4^2+h_3(2h)+h_4h_2}{2(2h+h_3)} \qquad | \text{ From (1)}$$

Let $\bar{z}$ be the depth of the parallelogram $ABCD$ below the free surface. Then,

$$\bar{z} = \frac{P_1z_1+P_2z_2}{P_1+P_2} \qquad | \text{ Refer § 7}$$

$$= g\rho S\left(\frac{h_1+2h}{3}\right).\left[\frac{h_1^2+h_2^2+h_4^2+2hh_1+h_2h_4}{2(h_1+2h)}\right] + g\rho S\left(\frac{h_3+2h}{3}\right).\left[\frac{h_2^2+h_3^2+h_4^2+2hh_3+h_2h_4}{2(2h+h_3)}\right]$$

$$g\rho S\left(\frac{h_1+2h}{3}\right) + g\rho S\left(\frac{h_3+2h}{3}\right)$$

$$= \frac{1}{2}\frac{\left(h_1^2+h_2^2+h_4^2+2hh_1+h_2h_4\right)+\left(h_2^2+h_3^2+h_4^2+2hh_3+h_2h_4\right)}{(h_1+2h)+(h_3+2h)}$$

$$= \frac{1}{2}\frac{\left(h_1^2+h_2^2+h_3^2+h_4^2\right)+\left(h_4^2+h_2^2+2h_2h_4\right)+\left(2hh_1+2hh_3\right)}{h_1+h_3+4h}$$

$$= \frac{1}{2}\frac{\left(h_1^2+h_2^2+h_3^2+h_4^2\right)+\left(h_2+h_4\right)^2+2h\left(h_1+h_3\right)}{2h+4h} \quad | \text{ From (1)}$$

$$= \frac{1}{2}\frac{\left(h_1^2+h_2^2+h_3^2+h_4^2\right)+(2h)^2+2h(2h)}{6h} \quad | \text{ From (1)}$$

$$= \frac{h_1^2+h_2^2+h_3^2+h_4^2+8h^2}{12h}$$

Example 8: A parallelogram whose plane is vertical and centre at depth h below the surface is totally immersed. Show that if a and b are the lengths of the projections of its sides on a vertical line, then the depth of its centre of pressure will be $h + a^2 + b^2/12h$.

Sol. Refer Ex. 7

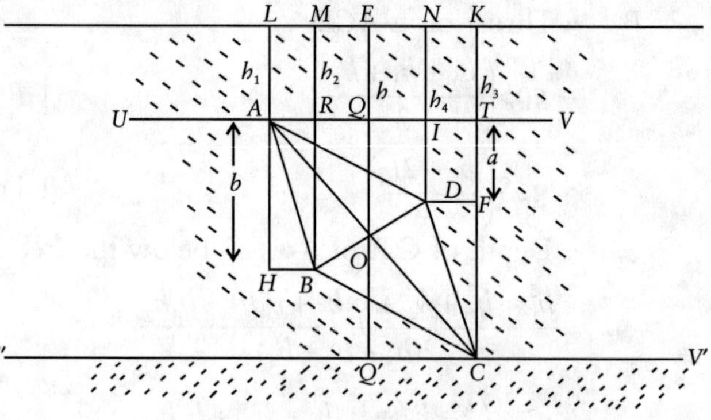

Through A, the uppermost vertex of the parallelogram $ABCD$, draw a horizontal line UV meeting the vertical lines through B and D in R and I respectively. $OE = h$ | Given

$TF = a$

$AH = b$

From figure, since O is the mid-point of BD, therefore,

$$OQ = \frac{TF+RB}{2} = \frac{TF+AH}{2} = \frac{a+b}{2}$$

Let h_1, h_2, h_3, and h_4 be the depths of the vertices of the parallelogram $ABCD$ below the free surface. Then,

$h_1 = AL$

$= QE \qquad = OE - OQ \qquad = h - (a + b)/2$

$h_2 = MB$

$= MR + RB \qquad = LA + AH$

$= h - (a + b)/2 + b \quad = h - (a - b)/2$

$h_3 = KC$

$= EQ' \qquad = EO + OQ'$

$= EO + OQ \qquad$ | by geometry

$= h + (a + b)/2$

$h_4 = ND$

$= NI + ID \qquad = LA + TF$

$= h - (a + b)/2 + a \quad = h + (a - b)/2$

$\therefore$ Depth of C.P. of parallelogram $ABCD$ below the free surface

$$\frac{h_1^2 + h_2^2 + h_3^2 + h_4^2 + \ h^2}{12} \qquad | \text{ From Ex. 7}$$

$$= \frac{\left(h - \dfrac{a+b}{2}\right)^2 + \left(h - \dfrac{a-b}{2}\right)^2 + \left(h + \dfrac{a+b}{2}\right)^2 + \left(h + \dfrac{a-b}{2}\right)^2 + 8h^2}{12h}$$

$$= \frac{\begin{array}{c} h^2 + \left(\dfrac{a+b}{2}\right)^2 - h(a-b) + h^2 + \left(\dfrac{a-b}{2}\right)^2 - h(a-b) \\ +h^2 + \left(\dfrac{a+b}{2}\right)^2 + h(a-b) + h^2 + \left(\dfrac{a-b}{2}\right)^2 + h(a-b) + 8h^2 \end{array}}{12h}$$

$$= \frac{4h^2 + a^2 + b^2 + 8h^2}{12h}$$

$$= \frac{12h^2 + a^2 + b^2}{12h} \qquad = h + \frac{a^2 + b^2}{12h}$$

Example 9: A square whose side is $2a$ is completely immersed in a homogeneous liquid in a vertical plane with its centre at a depth d. Prove that the centre of pressure is at a distance $a^2/3d$ below the centre of the square, whatever be the inclination of the side of the square to the vertical.

Sol.

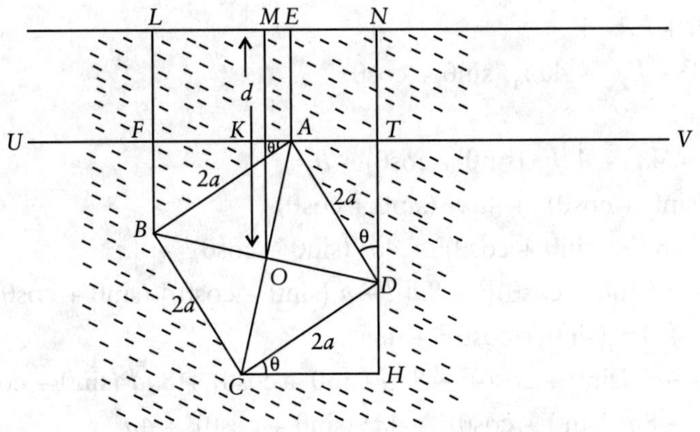

Let $ABCD$ be a square of side $2a$ completely immersed in a homogeneous liquid in a vertical plane with its centre O at a depth d. Then, $OM = d$.

Let $AE = h_1$. Let θ be the inclination of side AD of the square to the vertical.

Depth of B below the free surface (h_2)

$$= BL \qquad = BF + FL$$
$$= 2a \sin\theta + AE \qquad = 2a \sin\theta + h_1$$

Depth of C below the free surface (h_3)

$$= HN \qquad = HD + DT + TN = HD + DT + AE$$
$$= 2a \sin\theta + 2a \cos\theta + h_1 \qquad = 2a (\sin\theta + \cos\theta) + h_1$$

Depth of D below the free surface (h_4)

$$= DN \qquad = DT + TN$$
$$= DT + AE \qquad = 2a \cos\theta + h_1$$

$\because$ O is the mid-point of BD

$$\therefore \quad OM = \frac{BL + DN}{2}$$

$$\Rightarrow \quad d = \frac{h_2 + h_4}{2}$$

$$\Rightarrow \quad d = 1/2 \, [(2a \sin\theta + h_1) + (2a \cos\theta + h_1)]$$
$$= a(\sin\theta + \cos\theta) + h_1$$
$$\Rightarrow \quad h_1 = d - a\,(\sin\theta + \cos\theta) \qquad ...(1)$$

Now, $h_2^2 + h_4^2$

$$= (2a \sin\theta + h_1)^2 + (2a \cos\theta + h_1)^2$$
$$= 2h_1^2 + 4a^2 + 4ah_1 (\sin\theta + \cos\theta)$$
$$\Rightarrow \quad h_1^2 + h_2^2 + h_4^2 = 3h_1^2 + 4a^2 + 4ah_1 (\sin\theta + \cos\theta) \qquad ...(2)$$
$$h_3^2 = [2a (\sin\theta + \cos\theta) + h_1]^2$$
$$= 4a^2 (\sin\theta + \cos\theta)^2 + h_1^2 + 4ah_1 (\sin\theta + \cos\theta) \qquad ...(3)$$

From (2) and (3),

$$h_1^2 + h_2^2 + h_3^2 + h_4^2 = 3h_1^2 + 4a^2 + 4ah_1 (\sin\theta + \cos\theta) + h_1^2$$
$$+ 4a^2 (\sin\theta + \cos\theta)^2 + 4ah_1 (\sin\theta + \cos\theta)$$
$$= 4h_1^2 + 8ah_1 (\sin\theta + \cos\theta)^2 + 4a^2 (\sin\theta + \cos\theta)^2 + 4a^2$$
$$= 4[d - a (\sin\theta + \cos\theta)]^2 + 8a[d - a (\sin\theta + \cos\theta)] (\sin\theta + \cos\theta)$$
$$+ 4a^2 (\sin\theta + \cos\theta)^2 + 4a^2$$
$$= 4d^2 + 4a^2 (\sin\theta + \cos\theta)^2 - 8ad (\sin\theta + \cos\theta) + 8ad (\sin\theta + \cos\theta)$$
$$- 8a^2 (\sin\theta + \cos\theta)^2 + 4a^2 (\sin\theta + \cos\theta)^2 + 4a^2$$
$$= 4d^2 + 4a^2 \qquad ...(4)$$

∴ Depth of centre of pressure of the square (which is essentially a parallelogram) below the free surface

$$= \frac{h_1^2 + h_2^2 + h_3^2 + h_4^2 + 8d^2}{12d} \qquad | \text{ Refer Ex. 7}$$

$$= \frac{4d^2 + 4a^2 + 8d^2}{12d} \qquad | \text{ From (4)}$$

$$= \frac{12d^2 + 4a^2}{12d} \qquad = d + \frac{a^2}{3d}$$

∴ Depth of C.P. below the centre of the square

$$= \left(d + \frac{a^2}{3d}\right) - d \qquad = \frac{a^2}{3d}$$

Example 10: If a quadrilateral area be entirely immersed in water, and α, β, γ, δ be the depths of its four corners, and h that of its centre of gravity, show that the depth of its centre of pressure is

$$\frac{1}{2}(\alpha + \beta + \gamma + \delta) - \left[\frac{\beta\gamma + \gamma\alpha + \alpha\beta + \alpha\delta + \beta\delta + \gamma\delta}{6h}\right]$$

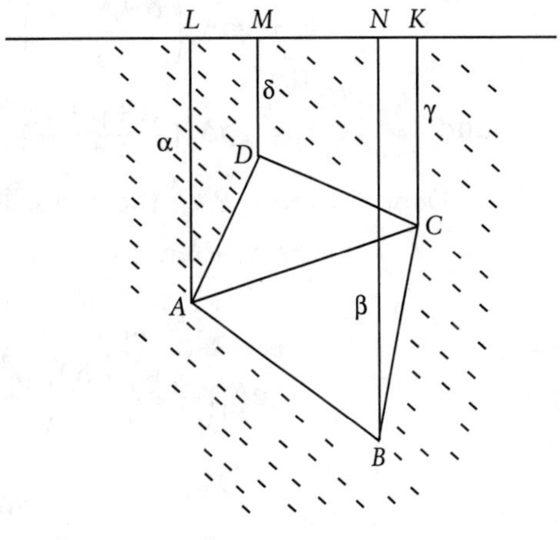

Sol. Let $ABCD$ be a quadrilateral area entirely immersed in water, the depths of whose four corners A, B, C, D below the free surface be α, β, γ, δ respectively.

Join AC. Then, area $ABCD$ = area ADC + area ABC

Let Δ_1 = area ADC

Let Δ_2 = area ABC

Let z_1, z_2, be the depths of C.G. of Δ_1 and Δ_2 respectively below the free surface. Then,

$$z_1 = \frac{\alpha + \gamma + \delta}{3}$$

and $$z_2 = \frac{\alpha + \beta + \gamma}{3}$$

Let h be the depth of the centre of gravity of the quadrilateral $ABCD$ below the free surface. Then,

$$h = \frac{\Delta_1 z_1 + \Delta_2 z_2}{\Delta_1 + \Delta_2} \qquad | \text{ By statics} \qquad ...(1)$$

$$\Rightarrow \quad (\Delta_1 + \Delta_2)\, h = \Delta_1 z_1 + \Delta_2 z_2$$

$$\Rightarrow \quad \Delta_1 (h - z_1) = \Delta_2 (z_2 - h)$$

$$\Rightarrow \quad \frac{\Delta_1}{\Delta_2} = \frac{z_2 - h}{h - z_1}$$

$$= \frac{\frac{1}{3}(\alpha+\beta+\gamma)-h}{h-\left(\frac{\alpha+\gamma+\delta}{3}\right)}$$

$$= \frac{\alpha+\beta+\gamma-3h}{3h-\alpha-\gamma-\delta} \qquad ...(2)$$

Let x_1 and x_2 be the depths of the centres of pressure of the triangle ADC and ABC respectively below the free surface. Then,

$$x_1 = \frac{\alpha^2+\gamma^2+\delta^2+\alpha\gamma+\gamma\delta+\delta\alpha}{2(\alpha+\gamma+\delta)}$$ | Refer § 10 case II for $\bar{z}$

and $$x_2 = \frac{\alpha^2+\beta^2+\gamma^2+\alpha\beta+\beta\gamma+\gamma\alpha}{2(\alpha+\beta+\gamma)}$$

Let P_1 and P_2 be the pressures on triangles ADC and ABC respectively. Then,

$$P_1 = g\rho\Delta_1\left(\frac{\alpha+\gamma+\delta}{3}\right)$$ | $P = wS\bar{z}$

and $$P_2 = g\rho\Delta_2\left(\frac{\alpha+\beta+\gamma}{3}\right)$$

∴ Depth of the C.P. of the quadrilateral $ABCD$ below the free surface

$$= \frac{P_1x_1+P_2x_2}{P_1+P_2}$$ | Refer § 7

$$= \frac{g\rho\Delta_1\left(\frac{\alpha+\gamma+\delta}{3}\right)\left(\frac{\alpha^2+\gamma^2+\delta^2+\alpha\gamma+\gamma\delta+\delta\alpha}{2(\alpha+\gamma+\delta)}\right) + g\rho\Delta_2\left(\frac{\alpha+\beta+\gamma}{3}\right)\left(\frac{\alpha^2+\beta^2+\gamma^2+\alpha\beta+\beta\gamma+\gamma\alpha}{2(\alpha+\beta+\gamma)}\right)}{g\rho\Delta_1\left(\frac{\alpha+\gamma+\delta}{3}\right)+g\rho\Delta_2\left(\frac{\alpha+\beta+\gamma}{3}\right)}$$

$$= \frac{1}{2}\,\frac{(\alpha^2+\gamma^2+\delta^2+\alpha\gamma+\gamma\delta+\delta\alpha)\frac{\Delta_1}{\Delta_2}+(\alpha^2+\beta^2+\gamma^2+\alpha\beta+\beta\gamma+\gamma\alpha)}{\frac{\Delta_1}{\Delta_2}(\alpha+\gamma+\delta)+(\alpha+\beta+\gamma)}$$

$$= \frac{1}{2}\,\frac{(\alpha^2+\gamma^2+\delta^2+\alpha\gamma+\gamma\delta+\delta\alpha)\left(\frac{\alpha+\beta+\gamma-3h}{3h-\alpha-\gamma-\delta}\right)+(\alpha^2+\beta^2+\gamma^2+\alpha\beta+\beta\gamma+\gamma\alpha)}{\left(\frac{\alpha+\beta+\gamma-3h}{3h-\alpha-\gamma-\delta}\right)(\alpha+\gamma+\delta)+(\alpha+\beta+\gamma)}$$

$$= \frac{1}{2}(\alpha + \beta + \gamma + \delta) - \left[\frac{\beta\gamma + \gamma\alpha + \alpha\beta + \alpha\delta + \beta\delta + \gamma\delta}{6h}\right] \quad | \text{ after simplification}$$

Example 11: A regular hexagon of side *a* is immersed in water with one side in the surface. Show that the depth of its centre of pressure to that if its centre of gravity is 23 : 18.

Sol. Let *ABCDEF* be a regular hexagon of side a immersed in water with one side *AB* in the free surface. Let *G* be the C.G. (essentially the centre) of the hexagon. Join all the vertices to G. Then, the whole hexagon is divided into six equilateral triangles of equal areas.

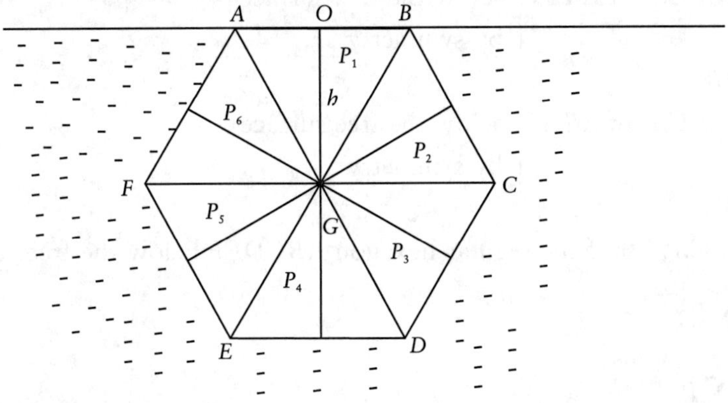

Let the area of each triangle be *S* and the depth of *G* below the free surface be *h*.

P_1 = Thrust on $\Delta\, ABG$

$= g\rho.S.h/3 \qquad | \; P = wS\bar{z}$

P_2 = Thrust on $\Delta\, BCG$

$= g\rho.S.2h/3 = 2P_1 \qquad | \; P = wS\bar{z}$

P_3 = Thrust on $\Delta\, CDG$

$= g\rho.S.(h + h/3) = g\rho.S.4h/3 \qquad | \; P = wS\bar{z}$

$= 4P_1$

P_4 = Thrust on $\Delta\, DEG$

$= g\rho.S.(h + 2h/3) = g\rho.S.5h/3 \qquad | \; P = wS\bar{z}$

$= 5P_1$

P_5 = Thrust on $\Delta\, EFG$

$= g\rho.S.(h + h/3) = g\rho.S.4h/3 \qquad | \; P = wS\bar{z}$

$= 4P_1$

P_6 = Thrust on $\Delta\, FAG$

$= g\rho.S.2h/3 \qquad | \; P = wS\bar{z}$

$= 2P_1$

z_1 = Depth of C.P. of $\Delta\, ABG$ below the free surface = $1/2h$

z_2 = Depth of C.P. of $\Delta\, BCG$ below the free surface = $3/4h$

z_3 = Depth of C.P. of $\Delta\, CDG$ below the free surface

$$= \frac{h^2 + h^2 + (2h)^2 + h^2 + h(2h) + (2h)h}{2(h + h + 2h)} = 11/8h$$ | Refer §10 case II for $\bar{z}$

z_4 = Depth of C.P. of Δ *DEF* below the free surface

$$= \frac{h^2 + (2h)^2 + (2h)^2 + h + (2h) + (2h) + (2h) + 2h(h)}{2(h + 2h + 2h)}$$ | Refer §10 case II for $\bar{z}$

= 17/10*h*

z_5 = Depth of C.P. of Δ *EFG* below the free surface

= z_3 | by symmetry

= 11*h*/8

z_6 = Depth of C.P. of Δ *FAG* below the free surface

= z_2 | by symmetry

= 3*h*/4

∴ Depth of C.P. (P) of the regular hexagon *ABCDEF* below the free surface

$$= \frac{\sum_{i=1}^{6} P_i z_i}{\sum_{i=1}^{6} P_i}$$

$$= \frac{(P_1)\left(\frac{1}{2}h\right) + (2P_1)\left(\frac{3}{4}h\right) + (4P_1)\left(\frac{11}{8}h\right) + (5P_1)\left(\frac{17}{10}h\right) + (4P_1)\left(\frac{11h}{8}\right) + (2P_1)\left(\frac{3h}{4}\right)}{P_1 + 2P_1 + 4P_1 + 5P_1 + 4P_1 + 2P_1}$$

⇒ $OP = 23/18h$ | after simplification

where O is the middle point of the side *AF*.

Also, *OG* = *h*, where *G* is the C.G. of the regular hexagon.

∴ $OP/OG = 23/18$

⇒ $OP : OG = 23 : 18$

EXERCISE 3

1. A triangle of height 1 cm is immersed vertically in a liquid with its base parallel to the surface and at a depth of 3 cm. If the vertex be downwards, find the depth of the centre of pressure of the triangle.

   Ans. 67/20 cm

   [**Hint:** Here, α = 3 cm, β = 3 cm, γ = 4 cm

   | Refer §10 case II for $\bar{z}$]

2. A plane triangle (height 30 cm) is immersed in a liquid, its base horizontal and vertex uppermost. Find the difference of level between the C.P. and C.G. of the triangle, given the vertex is 10 cm below the surface.

   Ans. 5/3 cm

[**Hint:** Here, $\alpha = 10$ cm, $\beta = 40$ cm, $\gamma = 40$ cm

$\therefore \bar{z} = 95/3$ cm | Refer §10 case II for $\bar{z}$

$$\text{Depth of C.G.} = \frac{10+40+40}{3} = 30 \text{ cm}$$

$\therefore$ Difference $= 95/3 - 30 = 5/3$ cm]

3. A plane triangular area is immersed in a liquid of uniform density with its plane vertical. One side horizontal and the opposite corner downwards. Its vertical altitude is h and the horizontal side is at a depth h below the effective surface. Show that its centre of pressure is at a depth $11/8h$ below the surface.

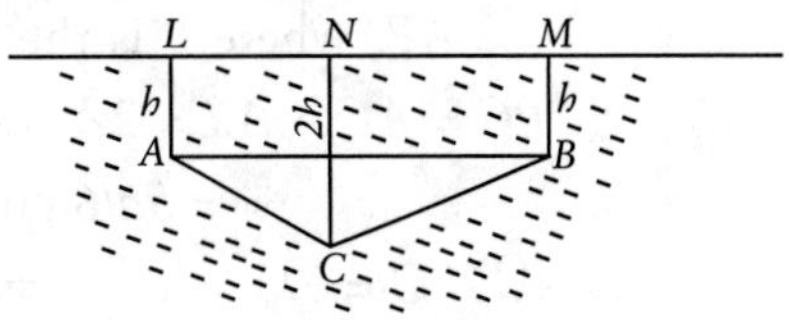

[**Hint:** Here, $\alpha = h$, $\beta = h$, $\gamma = 2h$ | Refer §10 case II for $\bar{z}$]

4. Show that the centre of pressure of a triangular lamina, the depths of whose angular points are a, b, c is at a depth

$$= \frac{a^2+b^2+c^2-ab-bc-ca}{6(a+b+c)}$$ below the centre of gravity of the lamina.

[**Hint:** Required depth

= Depth of C.P. – Depth of C.G.

$$= \frac{a^2+b^2+c^2+ab+bc+ca}{2(a+b+c)} - \frac{a+b+c}{3}\Bigg]$$

5. Find the centre of pressure of an isosceles triangle immersed with its plane vertical and base horizontal as for below the surface as its vertex.

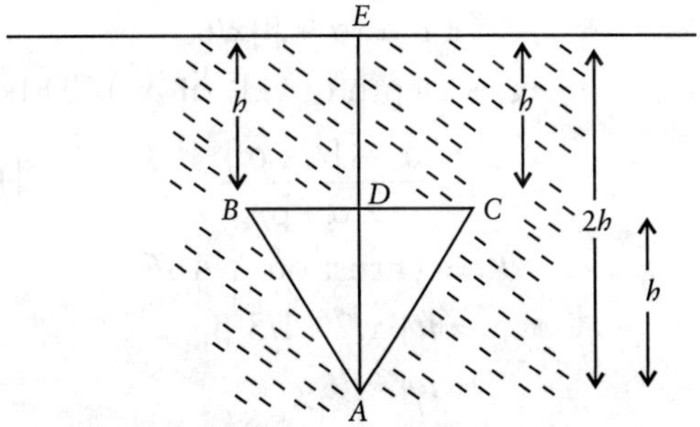

Ans. 11/8 h, where h is the altitude of the triangle.

[**Hint:** Here, $\alpha = 2h$, $\beta = h$, $\gamma = h$
| Refer §10 case II for $\bar{z}$]

6. A rhombus is immersed in a liquid with a vertex in the surface and the diagonal through the vertex vertical. Prove that the centre of pressure divides the diagonal in the ratio 7 : 5.

[**Hint:** Refer Ex. 3.

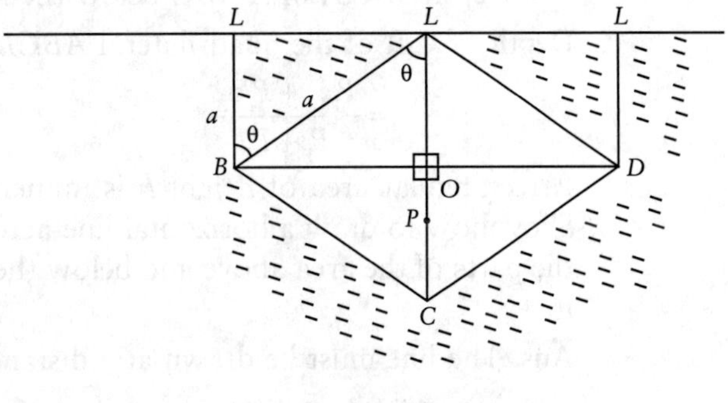

Aliter: Depth of C.P. of rhombus $ABCD$
= Depth of C.P. of $\Delta\, ABC$ below the free surface.

Let $AB = a$ and $\angle BAC = \theta$

Then, $BL = DM = a \cos\theta$

$$AC = 2\,a \cos\theta$$

$\therefore\ \alpha = 0,\ \beta = a\cos\theta,\ \gamma = 2a\cos\theta$

$\therefore\ \bar{z} = 7a/6 \cos\theta$ | Refer §10 case II for $\bar{z}$

$= AP$, where P is the C.P. of the rhombus $ABCD$.

$\therefore\ PC = AC - AP = 2AO - AP = 2a\cos\theta - 7/6a\cos\theta$

$$= 5a/6 \cos\theta$$

$\therefore\ AP/PC = 7/5 \quad \Rightarrow \quad AP : PC = 7 : 5$]

7. A lamina in the shape of a quadrilateral $ABCD$ has the side CD in the surface, and the sides AD, BC vertical and of lengths α and β respectively. Prove that the depth of the centre of pressure is $\dfrac{1}{2}\dfrac{\alpha^4-\beta^4}{\alpha^3-\beta^3}$ *or* $\dfrac{1}{2}\dfrac{(\alpha+\beta)(\alpha^2+\beta^2)}{\alpha^2+\alpha\beta+\beta^2}$.

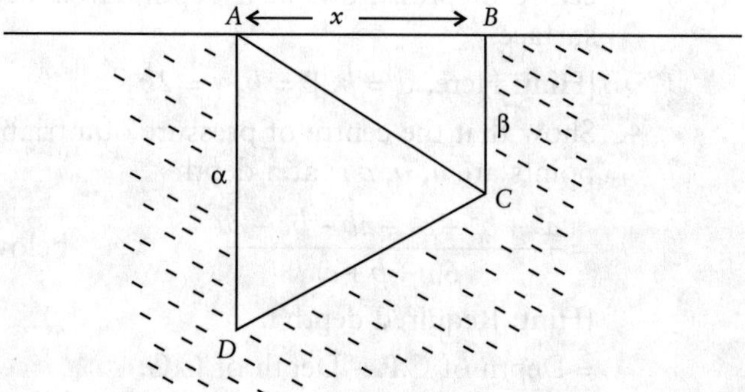

[**Hint:**

P_1 = Thrust on $\Delta\,ACD$

$= w\,.\,\alpha x/2\,.\,\alpha + \beta/3$

$= w\alpha\,(\alpha + \beta)x/6$

z_1 = Depth of C.P. of $\Delta\,ACD$ below the free surface

$$= \frac{\alpha^2+\beta^2+\alpha\beta}{2(\alpha+\beta)}$$ | Refer §10 case II for $\bar{z}$

P_2 = Thrust on $\Delta\,ABC$

$= w\beta x/2\,.\,1/3\,\beta$

$= w\beta^2 x/6$

z_2 = Depth of C.P. of $\Delta\,ABC$ below the free surface = $\beta/2$

$\therefore$ Depth of C.P. of the quadrilateral $ABCD$ below the free surface ($\bar{z}$)

$$= \frac{P_1 z_1 + P_2 z_2}{P_1 + P_2}$$ | Refer § 7]

8. A rectangular area of height h is immersed in a liquid with one side in the surface. Show how to draw a horizontal line across the area so that the centres of pressure of the parts of the area above and below the line of division shall be equally distant from it.

Ans. The line must be drawn at a distance x from the upper side where x is given by

$x = 1/2h\,(\sqrt{5} - 1)$.

9. The centre of pressure of a parallelogram completely immersed in water is in one of its diagonals. Prove that the other diagonal is horizontal.
10. A square is immersed with its diagonal vertical and its lowest point as deep again as its highest point. Find the depth of the centre of pressure.

    Ans. 7/6 h where $2h$ is the length of the vertical diagonal.
11. A trapezium $ABCD$ is immersed with the side AB in the surface of water and the sides $AD = a$, $BC = b$ are vertical. Prove that the vertical line through the centre of pressure divides AB in the ratio $a^2 + 2ab + 3b^2 : 3a^2 + 2ab + b^2$ and the depth of the centre of pressure is $\frac{1}{2}\frac{(a+b)(a^2+b^2)}{a^2+ab+b^2}$
12. A regular hexagon is immersed in water with one side in the surface. Find the C.P. of the upper half.

    Ans. 7/10 h where h is the depth of the centre of the regular hexagon from the free surface, i.e., C.P. divides OG in the ration 7 : 3. Where O is the mid-point of the side in the surface and G is the C.G. of the regular hexagon.

    [**Hint:** Refer Ex. 11]

11. Illustrative Examples (on determination of centre of pressure by using integration technique)

Example 1: A rectangle is immersed vertically in water with two sides horizontal and at depths h and $a + h$ respectively below the effective surface. Prove that the distance of the C.P. from the upper side is

$$\frac{a}{3}\frac{3h+2a}{2h+a}$$

Show also that the C.P. is below the C.G. of the area, and as the depth increases, approaches but never coincides with it.

Sol. Let a rectangle $ABCD$ be immersed vertically in water with two sides AB and DC horizontal and at depths h and $a + h$ respectively below the effective surface. Then, $MB = h$ and $MC = a + h$.

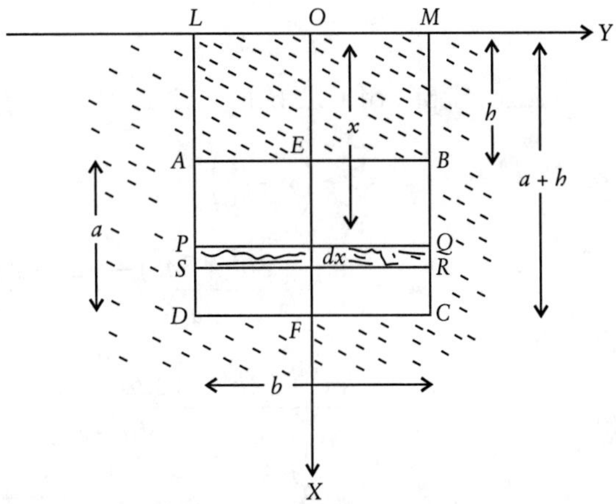

Let the vertical lines drawn through A and B meet the free surface in L and M respectively. Let O be the mid-point of LM. Take O as origin and a vertical line OEF through O as X-axis and a horizontal line OM through O as Y-axis.

By symmetry, C.P. of the rectangle $ABCD$ will lie on the vertical line EF (X-axis) where E and F are the mid-points of AB and DC respectively.

Let $(\bar{x}, \bar{y})$ be the coordinates of C.P. of the rectangle $ABCD$ referred to these axes.

Then, $\bar{y} = 0$

Divide the rectangular area $ABCD$ into horizontal strips and consider one such strip $PQRS$ at a depth x below the free surface and of thickness dx.

Let $AB = b$ and ρ be the density of water.

Then,

ds = Area of the strip PQRS

$= b\,dx$

p = Intensity of pressure at any point of the strip $PQRS = g\rho x$

Hence, the depth of C.P. ($\bar{x}$) of the rectangle $ABCD$ below the free surface.

$$= \frac{\int xp\,ds}{\int p\,ds} \qquad = \frac{\int_h^{a+h} xg\rho x\,b\,dx}{\int_h^{a+h} g\rho x\,b\,dx}$$

$$= \frac{\int_h^{a+h} x^2\,dx}{\int_h^{a+h} x\,dx} \qquad = \frac{\left(\dfrac{x^3}{3}\right)_h^{a+h}}{\left(\dfrac{x^2}{2}\right)_h^{a+h}}$$

$$= \frac{2}{3}\left[\frac{(a+h)^3 - h^3}{(a+h)^2 - h^2}\right] \qquad = \frac{2}{3}\left[\frac{(a+h-h)\{(a+h)^2 + h^2 + (a+h)h\}}{(a+h-h)(a+h+h)}\right]$$

$$= \frac{2}{3}\,\frac{a^2 + 3h^2 + 3ah}{a+2h}$$

$\therefore$ Distance of the C.P. from the upper side

$$= \frac{2}{3}\,\frac{a^2 + 3h^2 + 3ah}{a+2h} - h \qquad = \frac{2a^2 + 6h^2 + 6ah - 3ah - 6h^2}{3(a+2h)}$$

$$= \frac{2a^2 + 3ah}{3(a+2h)} \qquad = \frac{a}{3}\,\frac{2a+3h}{a+2h}$$

Also, depth of C.P. below C.G.

$$= \bar{x} - \left(h + \frac{a}{2}\right) \qquad = \frac{2}{3}\,\frac{a^2 + 3h^2 + 3ah}{a+2h} - \frac{2h+a}{2}$$

$$= \frac{4(a^2 + 3h^2 + 3ah) - 3(2h+a)^2}{6(a+2h)}$$

$$= \frac{4a^2 + 12h^2 + 12ah - 12h^2 - 3a^2 - 12ah}{6(a+2h)}$$

$$= \frac{a^2}{6(a+2h)} > 0 \text{ as } a \text{ and } h \text{ are both positive.}$$

Hence, C.P. is below the C.G. of the area.

We see that h increases, the Depth of C.P. below C.G. decreases, i.e., C.P. approaches C.G. However, in actuality it never vanishes until h is in finite, i.e., C.P. will never coincide with C.G.

Example 2: A semi-circular area of radius a is immersed vertically with its diameter horizontal at a depth b. If the circumference be below the centre, prove that the depth of the centre of pressure below the free surface is

$$\frac{1}{4}\frac{3\pi(a^2+4b^2)+32ab}{4a+3\pi b}$$

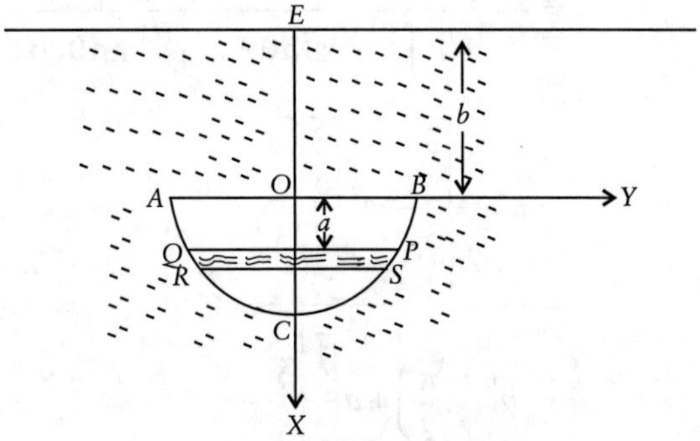

Sol. Let ACB be a semi-circular area of radius a immersed vertically with its bounding diameter AOB horizontal at a depth b from the free surface

Let O be the mid-point of AB. Take O as origin and OX, OY as axes as shown in the figure. Then, the equation of the circle is $x^2 + y^2 = a^2$. By symmetry, it is clear that C.P. lies on OX. Hence, $\bar{y} = 0$ where $(\bar{x}, \bar{y})$ are the coordinates of C.P. of the semi-circular area ACB.

Consider an elementary strip $PQRS$ of thickness dx and at a depth x below the boundary diameter AOB. Let the coordinates of P be (x, y). Then,

dS = Area of the strip $PQRS$

$= 2y\, dx$

p = Intensity of pressure at any point of the strip $PQRS$

$= g\rho\,(b + x)$

where ρ is the density of the liquid.

Hence, the depth of C.P. of the semi-circular area ACB below the bounding diameter AOB

$$= \bar{x} = \frac{\int xp\,dS}{\int p\,dS}$$

$$= \frac{\int_0^a xg\rho(b+x)2y\,dx}{\int_0^a g\rho(b+x)2y\,dx}$$

$$= \frac{\int_0^a x(b+x)\sqrt{a^2-x^2}\,dx}{\int_0^a (b+x)\sqrt{a^2-x^2}\,dx} \qquad \Big|\ \because x^2+y^2=a^2$$

$$= \frac{b\int_0^a x\sqrt{a^2-x^2}\,dx + \int_0^a x^2\sqrt{a^2-x^2}\,dx}{b\int_0^a \sqrt{a^2-x^2}\,dx + \int_0^a x\sqrt{a^2-x^2}\,dx}$$

Put $x = a\sin\theta$

$\because dx = a\cos d\theta$

$$= \frac{b\int_0^{\pi/2} a\sin\theta a\cos\theta a\cos\theta d\theta + \int_0^{\pi/2} a^2\sin^2\theta a\cos\theta a\cos\theta d\theta}{b\int_0^{\pi/2} a\cos\theta a\cos\theta d\theta + \int_0^{\pi/2} a\sin\theta a\cos\theta a\cos\theta d\theta}$$

$$= \frac{a^3 b\int_0^{\pi/2}\sin\theta\cos^2\theta d\theta + a^4\int_0^{\pi/2}\sin^2\theta\cos^2\theta d\theta}{ba^2\int_0^{\pi/2}\cos^2 d\theta + a^3\int_0^{\pi/2}\sin\theta\cos^2\theta d\theta}$$

$$= \frac{ab\dfrac{\Gamma 1\,\Gamma\frac{3}{2}}{2\Gamma\frac{5}{2}} + a^2\dfrac{\Gamma\frac{3}{2}\Gamma\frac{3}{2}}{2\Gamma 3}}{b\left(\dfrac{1}{2}\cdot\dfrac{\pi}{2}\right) + a\dfrac{\Gamma 1\,\Gamma\frac{3}{2}}{2\Gamma\frac{5}{2}}} = \frac{\dfrac{ab.\frac{1}{2}\sqrt{\pi}}{2.\frac{3}{2}.\frac{1}{2}.\sqrt{\pi}} + \dfrac{a^2.\frac{1}{2}\sqrt{\pi}\frac{1}{2}\sqrt{\pi}}{2.2.1}}{\dfrac{b\pi}{4} + a\dfrac{\frac{1}{2}\sqrt{\pi}}{2.\frac{3}{2}.\frac{1}{2}\sqrt{\pi}}}$$

$$= \frac{\dfrac{ab}{3} + \dfrac{a^2\pi}{16}}{\dfrac{b\pi}{4} + \dfrac{a}{3}} = \frac{\dfrac{16ab + 3a^2\pi}{48}}{\dfrac{3b\pi + 4a}{12}}$$

$$= \frac{16ab + 3a^2\pi}{4(3b\pi + 4a)}$$

$\therefore$ Depth of the C.P. below the free surface

$$= \frac{16ab + 3a^2\pi}{4(3b\pi + 4a)} + b$$

$$= \frac{16ab + 3a^2\pi + 12b^2\pi + 16ab}{4(3b\pi + 4a)}$$

$$= \frac{3\pi(a^2 + 4b^2) + 32ab}{4(3b\pi + 4a)}$$

Example 3: Prove that the depth of the centre of pressure of the area bounded by two concentric semi-circles with their common bounding diameter in the surface is

$\dfrac{3\pi}{16}\cdot\dfrac{a^4 - b^4}{a^3 - b^3}$, a and b being their radii.

Sol. Let $A'DB'$ and ACB be two concentric semi-circles of radii a and b respectively with their common bounding diameter $B'A'$ and BA in the surface. Let P_1 and P_2 be the fluid pressures on the semi-circles of radii a and b respectively. Let their centres of pressure be at depths z_1 and z_2 below the free surface.

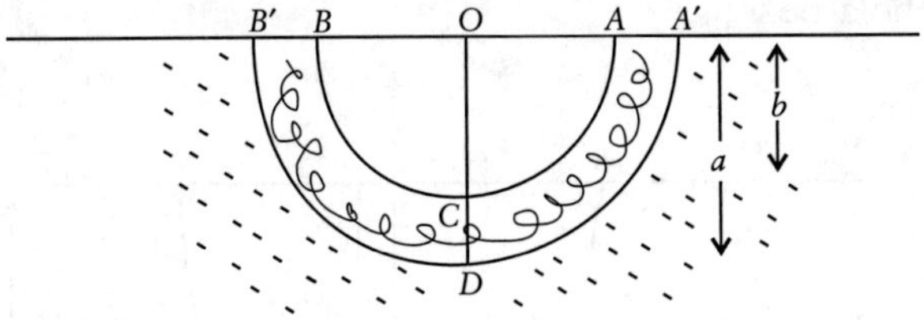

Also, the depths of centres of gravity of these semi-circles below the free surface are $4a/3\pi$ and $4b/3\pi$ respectively.

P_1 = Thrust on semi-circle $A'DB'$ of radius a

$$= g\rho\frac{1}{2}\pi a^2 \cdot \frac{4a}{3\pi} \qquad | \; P = wS\,\bar{z}$$

$$= \frac{2}{3}a^3 g\rho$$

Similarly, P_2 = Thrust on semi-circle ACB of radius b

$$= 2/3\; b^3 g\rho$$

$$z_1 = 3/16\; a\pi \qquad | \text{ Refer Ex. 2 in case } b = 0$$

$$z_2 = 3/16\; b\pi$$

Let $\bar{x}$ be the required depth of C.P. of the area bounded by two concentric semi-circles. Then,

$$\bar{x} = \frac{P_1 z_1 - P_2 z_2}{P_1 - P_2} \qquad | \text{ Refer § 8}$$

$$= \frac{\frac{2}{3}a^3 g\rho \cdot \frac{3}{16}a\pi - \frac{2}{3}b^3 g\rho \frac{3}{16}b\pi}{\frac{2}{3}a^3 g\rho - \frac{2}{3}b^3 g\rho}$$

$$= \frac{3\pi(a^4 - b^4)}{16(a^3 - b^3)}$$

$$= \frac{3\pi(a^2 - b^2)(a^2 + b^2)}{16(a - b)(a^2 + ab + b^2)}$$

$$= \frac{3\pi(a + b)(a^2 + b^2)}{16(a^2 + ab + b^2)}$$

Example 4: A segment of a parabola cut off by a double ordinate at a distance h from the vertex is immersed with this ordinate in the surface of a liquid. Find the resultant thrust on it. Find also the position of the point where its line of action meets the lamina. Neglect the atmospheric pressure.

Sol. Let O be the vertex of the parabola. Take O as origin, a vertical line through O as X-axis and a line through O, perpendicular to OX as Y-axis. Referred to these axes, let the equation of the parabola be

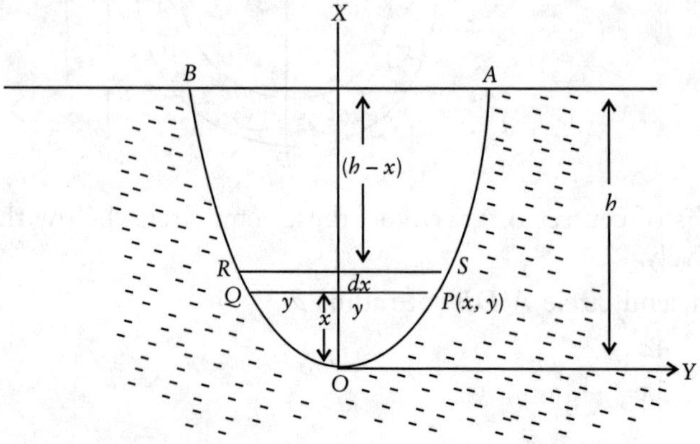

$$y^2 = 4ax \qquad ...(1)$$

Let AB be a double ordinate which cuts off a segment AOB such that AB is at a distance h from the vertex O. AB is in the surface.

By symmetry, C.P. lies on OX.

$$\therefore \qquad \bar{y} = 0$$

where $(\bar{x}, \bar{y})$ are the coordinates of C.P.

Consider an elementary strip PQRS of thickness dx parallel to AB at a height x above O. Let the coordinates of P be (x, y). Then,

dS = Area of the strip PQRS

$= 2y\, dx$

p = Intensity of pressure at any point of the strip PQRS

$= g\rho\,(h - x)$

$\therefore$ Pressure on the strip PQRS

$= 2g\rho\,(h - x)y\,dx$

Hence, the whole pressure

$$= 2\int_0^h g\rho(h-x)y\,dx \qquad = 2g\rho\int_0^h (h-x)2\sqrt{ax}\,dx$$

$$= 4g\rho\sqrt{a}\int_0^h (h-x)\sqrt{x}\,dx \qquad = 4g\rho\sqrt{a}\left[\frac{2}{3}hx^{3/2} - \frac{2}{5}x^{5/2}\right]_0^h$$

$$= 4g\rho\sqrt{a}\,h^{5/2}\left(\frac{2}{3} - \frac{2}{5}\right) \qquad = 4g\rho\sqrt{a}\,h^{5/2}\,\frac{4}{15}$$

$$= \frac{16}{15}g\rho h^2\sqrt{ah}$$

Let $\bar{x}$ be the height of C.P. of the parabolic area AOB above the vertex O, then,

$$\bar{x} = \frac{\int xp dS}{\int p dS} = \frac{\int_0^h xg\rho(h-x)2y\,dx}{\int_0^h g\rho(h-x)2y\,dx}$$

$$= \frac{\int_0^h x(h-x)4\sqrt{ax}\,dx}{\int_0^h (h-x)4\sqrt{ax}\,dx} \qquad | \text{ From (1)}$$

$$= \frac{\int_0^h (xh-x^2)\sqrt{x}\,dx}{\int_0^h (h-x)\sqrt{x}\,dx} = \frac{\int_0^h (hx^{3/2}-x^{5/2})dx}{\int_0^h (hx^{1/2}-x^{3/2})dx}$$

$$= \frac{\left(\frac{2}{5}hx^{5/2}-\frac{2}{7}x^{7/2}\right)_0^h}{\left(\frac{2}{3}hx^{3/2}-\frac{2}{7}x^{5/2}\right)_0^h} = \frac{h\left(\frac{2}{5}-\frac{2}{7}\right)}{\left(\frac{2}{3}-\frac{2}{5}\right)} = h\,.\frac{\frac{2}{35}}{\frac{2}{15}} = \frac{15h}{35}$$

$$= \frac{3h}{7}$$

$\therefore$ Depth of C.P. below the free surface

$= h - 3h/7 \qquad = 4h/7$

Example 5: An ellipse is completely immersed with its minor axis horizontal and at a depth h. Find the position of the centre of pressure.

Sol. Take the centre C of the ellipse as origin. Take the major axis and the minor axis of the ellipse as X-axis and Y-axis respectively.

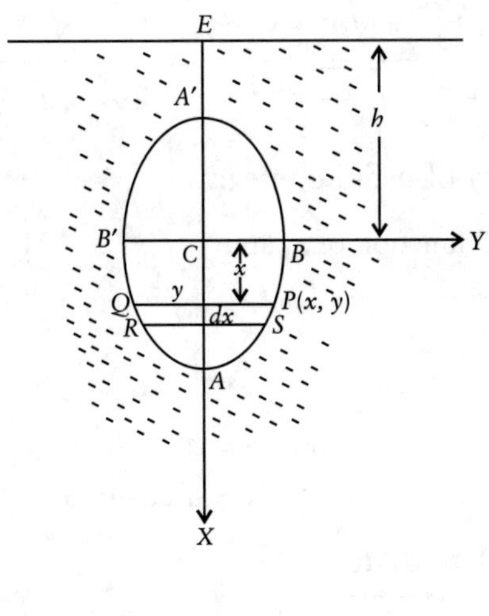

Let $AA' = 2a$ and $BB' = 2b$.

Then, the equation of the ellipse is

$$\frac{x^2}{a^2}+\frac{y^2}{b^2}=1$$

$$\Rightarrow \quad y=\frac{b}{a}\sqrt{a^2-x^2} \qquad \text{...(1)}$$

Let the minor axis BB' be at a depth h below the free surface. By symmetry, C.P. of the ellipse will lie on the major axis.

$$\therefore \quad \bar{y}=0$$

where $(\bar{x}, \bar{y})$ are the coordinates of the C.P. of the ellipse referred to the axes CX and CY.

Consider an elementary strip $PQRS$ of the ellipse at a depth x below CY. Let its width be dx. Let $P \to (x, y)$. Then,

$$dS = \text{Area of the strip}$$
$$= 2y\, dx$$
$$= 2\frac{b}{a}\sqrt{a^2-x^2}\, dx \qquad | \text{ From (1)}$$

p = Intensity of pressure at any point of the strip $PQRS$

$= g\rho\,(h + x)$

where ρ is the density of the liquid.

Let $\bar{x}$ be the depth of C.P. of the ellipse below the centre C. Then,

$$\bar{x}=\frac{\int xp\,dS}{\int p\,dS} = \frac{\int_{-a}^{a} xg\rho(h+x)\,2\frac{b}{a}\sqrt{a^2-x^2}\,dx}{\int_{-a}^{a} g\rho(h+x)\,2\frac{b}{a}\sqrt{a^2-x^2}\,dx}$$

$$=\frac{h\int_{-a}^{a} x\sqrt{a^2-x^2}\,dx+h\int_{-a}^{a} x^2\sqrt{a^2-x^2}\,dx}{h\int_{-a}^{a}\sqrt{a^2-x^2}\,dx+\int_{-a}^{a} x\sqrt{a^2-x^2}\,dx}=\frac{2\int_0^a x^2\sqrt{a^2-b^2}\,dx}{2h\int_0^a\sqrt{a^2-b^2}\,dx}$$

| by property of definite integrals

$$\left[\int_{-a}^{a} f(x)dx = 0, \text{if } f(x) \text{ is on odd function of } x, \text{ and } \int_{-a}^{a} f(x)dx = 2\int_0^a f(x) \text{ is } f(x) \text{ is an even function of } x.\right]$$

$$=\frac{\int_0^a x^2\sqrt{a^2-x^2}\,dx}{h\int_0^a\sqrt{a^2-x^2}\,dx}$$

Put $a = a\sin\theta$

$\therefore\ dx = a\cos\theta\, d\theta$

$$=\frac{\int_0^{\pi/2} a^2\sin^2\theta.a\cos\theta\; a\cos\theta d\theta}{h\int_0^{\pi/2} a\cos\theta\; a\cos\theta d\theta}$$

$$= \frac{a^2 \int_0^{\pi/2} \sin^2 \theta \cos^2 \theta d\theta}{h \int_0^{\pi/2} \cos^2 \theta d\theta} = \frac{a^2 \left. \Gamma\frac{3}{2} \Gamma\frac{3}{2} \right/ 2\Gamma 3}{h.\frac{1}{2}.\frac{\pi}{2}} = \frac{\frac{a^2 . \frac{1}{2}\sqrt{\pi} . \frac{1}{2}\sqrt{\pi}}{2.2}}{\frac{h\pi}{4}}$$

$= a^2/4h$

Note: Depth of C.P. below the free surface = $h + a^2/4h$

Example 6: A uniform elliptic lamina, whose axes are $2a$ and $2b$, is half immersed in water, the axis $2b$ being in the surface. Find the centre of pressure.

Sol. Take the centre C of the ellipse as origin, the major axis as X-axis and the minor axis as Y-axis,

$CA = a,\ CB = b$

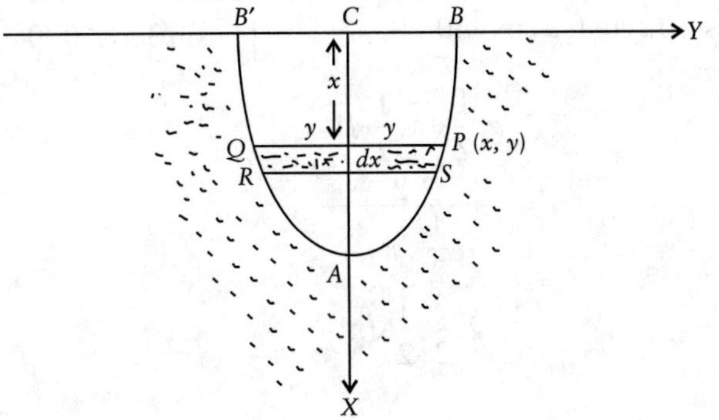

The equation of the ellipse is

$$\frac{x^2}{a^2} + \frac{y^2}{b^2} = 1$$

$$y = \frac{b}{a}\sqrt{a^2 - x^2} \qquad \text{...(1)}$$

By symmetry, the C.P. will lie on X-axis.

$\therefore\ \bar{y} = 0$

where $(\bar{x}, \bar{y})$ are the coordinates of C.P. referred to coordinates axes CX and CY.

Consider an elementary strip $PQRS$ of thickness dx and at a depth x below the free surface. Let $P \rightarrow (x, y)$. Then,

dS = Area of the elementary strip $PQRS$

$= 2y\, dx$

$= 2\frac{b}{a}\sqrt{a^2 - x^2}\ dx$

p = Intensity of pressure at any point of the strip

$= g\rho x$

where ρ is the density of water.

$$\therefore \quad \bar{x} = \frac{\int xpdS}{\int pdS} = \frac{\int_0^a x.g\rho x\, 2\frac{b}{a}\sqrt{a^2-x^2}\,dx}{\int_0^a g\rho x\, 2\frac{b}{a}\sqrt{a^2-x^2}\,dx}$$

$$= \frac{\int_0^a x^2\sqrt{a^2-x^2}\,dx}{\int_0^a x\sqrt{a^2-x^2}\,dx}$$

Put $x = a\sin\theta$

$\therefore\ dx = a\cos\theta\, d\theta$

$$= \frac{\int_0^{\pi/2} a^2\sin^2\theta\, a\cos\theta\, a\cos\theta d\theta}{\int_0^{\pi/2} a\sin\theta\, a\cos\theta\, a\cos\theta d\theta} = \frac{a\int_0^{\pi/2}\sin^2\theta\cos^2\theta d\theta}{\int_0^{\pi/2}\sin\theta\cos^2\theta d\theta}$$

$$= \frac{a\dfrac{\Gamma\frac{3}{2}\,\Gamma\frac{3}{2}}{2\Gamma 3}}{\dfrac{\Gamma 1\,\Gamma\frac{3}{2}}{2\Gamma\frac{5}{2}}} = \frac{\dfrac{a.\frac{1}{2}\sqrt{\pi}.\frac{1}{2}\sqrt{\pi}}{2.2}}{\dfrac{\frac{1}{2}\sqrt{\pi}}{2.\frac{3}{2}.\frac{1}{2}\sqrt{\pi}}}$$

$$= \frac{\frac{a\pi}{16}}{\frac{1}{3}} = \frac{3\pi a}{16}$$

Example 7: A semi-ellipse bounded by its minor axis is just immersed in a liquid, the density of which varies as the depth. If the minor axis be in the surface, find the eccentricity in order that the focus may be the centre of pressure.

Sol. Refer Ex. 6.

Here, $p = g\rho x = g\lambda x x \quad = g\lambda x^2$

$dS \quad = 2y\, dx = 2\frac{b}{a}\sqrt{a^2-x^2}\, dx$

$$\therefore \quad \bar{x} = \frac{\int xpdS}{\int pdS}$$

$$= \frac{\int_0^a x.g\lambda x^2.2\frac{b}{a}\sqrt{a^2-x^2}\,dx}{\int_0^a g\lambda x^2.2\frac{b}{a}\sqrt{a^2-x^2}\,dx} = \frac{\int_0^a x^3\sqrt{a^2-x^2}\,dx}{\int_0^a x^2\sqrt{a^2-x^2}\,dx}$$

Put $x = a \sin\theta$

$\therefore\ dx = a \cos\theta\, d\theta$

$$= \frac{\int_0^{\pi/2} a^3 \sin^3\theta\, a\cos\theta\, a\cos\theta\, d\theta}{\int_0^{\pi/2} a^2 \sin^2\theta\, a\cos\theta\, a\cos\theta\, d\theta} \qquad = \frac{a\int_0^{\pi/2} \sin^3\theta \cos^2\theta\, d\theta}{\int_0^{\pi/2} \sin^2\theta \cos^2\theta\, d\theta}$$

$$= \frac{\dfrac{a\Gamma(2)\Gamma\left(\frac{3}{2}\right)}{2\Gamma\left(\frac{7}{2}\right)}}{\dfrac{\Gamma\left(\frac{3}{2}\right)\Gamma\left(\frac{3}{2}\right)}{2\Gamma(3)}}$$

$$= \frac{\dfrac{a.\frac{1}{2}.\sqrt{\pi}}{2.\frac{5}{2}.\frac{3}{2}.\frac{1}{2}.\sqrt{\pi}}}{\dfrac{\frac{1}{2}\sqrt{\pi}\frac{1}{2}\sqrt{\pi}}{2.2}} \qquad = \frac{\frac{2a}{15}}{\frac{\pi}{16}} = \frac{32a}{15\pi}$$

If the C.P. coincides with the focus, then depth of C.P. below the free surface

$= ae$

$\therefore \quad \bar{x} = ae$

$\Rightarrow \quad 32a/15\pi = ae$

$\Rightarrow \quad e = 32/15\pi$

Example 8: Find the centre of pressure of a quadrant of an ellipse just immersed vertically in a homogeneous liquid with major axis in the surface.

Sol. Let OAB be a quadrant of an ellipse just immersed vertically in a homogeneous liquid with semi-major axis in the surface.

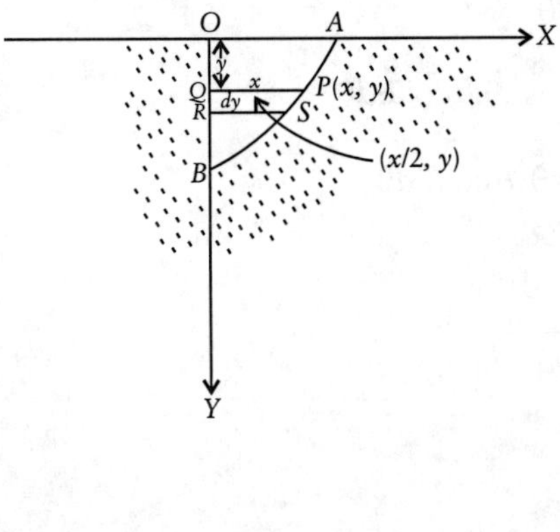

Take the centre O as origin, semi-major axis OA as X-axis and semi-minor axis OB as Y-axis. Let $OA = a$ and $OB = b$.

Then, the equation of the ellipse is

$$\frac{x^2}{a^2}+\frac{y^2}{b^2}=1$$

$$\Rightarrow \quad x=\frac{a}{b}\sqrt{b^2-y^2} \qquad ...(1)$$

Consider an elementary strip $PQRS$ of thickness dy and at a depth y below the free surface. Let $P \to (x, y)$. Then, the coordinates of the C.P. of the strip $PQRS$ will be $(x/2, y)$.

$$dS = \text{Area of the strip } PQRS$$
$$= xdy$$
$$= \frac{a}{b}\sqrt{b^2-y^2}\,dy \qquad | \text{ From (1)}$$

$$p = \text{Intensity of pressure at any point of the strip } PQRS$$
$$= g\rho y$$

Let $(\bar{x}, \bar{y})$ be the coordinates of the C.P. of the quadrant OAB referred to OX and OY. Then,

$$\bar{x}=\frac{\int xpds}{\int pds}=\frac{\int_0^b \frac{x}{2}g\rho y\frac{a}{b}\sqrt{b^2-y^2}\,dy}{\int_0^b \frac{x}{2}g\rho y\frac{a}{b}\sqrt{b^2-y^2}\,dy}=\frac{\frac{1}{2}\int_0^b xy\sqrt{b^2-y^2}\,dy}{\int_0^b y\sqrt{b^2-y^2}\,dy}$$

$$=\frac{\frac{1}{2}\int_0^b \frac{a}{b}\sqrt{a^2-y^2}\,y\sqrt{b^2-y^2}\,dy}{\int_0^b y\sqrt{a^2-y^2}\,dy} \qquad | \text{ From (1)}$$

$$=\frac{\frac{a}{2b}\int_0^b y(b^2-y^2)dy}{\int_0^b y\sqrt{b^2-y^2}\,dy}$$

Put $y = b\sin\theta$

$\therefore \quad dy = b\cos\theta\, d\theta$

$$=\frac{\frac{a}{2b}\int_0^{\pi/2} b\sin\theta b^? \cos^?\theta . b\cos\theta d\theta}{\int_0^{\pi/2} b\sin\theta b\cos\theta . b\cos\theta d\theta}$$

$$=\frac{\frac{a}{2}\int_0^{\pi/2}\sin\theta\cos^3\theta\, d\theta}{\int_0^{\pi/2}\sin\theta\cos^2\theta\, d\theta}$$

$$= \frac{\dfrac{a}{2}\dfrac{\Gamma(1)\Gamma(2)}{2\Gamma(3)}}{\dfrac{\Gamma(1)\Gamma\left(\frac{3}{2}\right)}{2\Gamma\left(\frac{5}{2}\right)}}$$

$$= \frac{\dfrac{a}{2}\dfrac{1}{2.2}}{\dfrac{\frac{1}{2}\sqrt{\pi}}{2.\frac{2}{3}.\frac{1}{2}\sqrt{\pi}}} \qquad = \frac{\frac{a}{8}}{\frac{1}{3}} = \frac{3a}{8}$$

and, $$\bar{y} = \frac{\int y p\, dS}{\int p\, dS}$$

$$= \frac{\int_0^b y g\rho y \frac{a}{b}\sqrt{b^2 - y^2}\; dy}{\int_0^b g\rho y \frac{a}{b}\sqrt{b^2 - y^2}\; dy} \qquad = \frac{\int_0^b y^2\sqrt{b^2 - y^2}\; dy}{\int_0^b y\sqrt{b^2 - y^2}\; dy}$$

Put $y = b\sin\theta$

$\therefore \quad dy = b\cos\theta\, d\theta$

$$= \frac{\int_0^{\pi/2} b^2\sin^2\theta\, b\cos\theta\, b\cos\theta\, d\theta}{\int_0^{\pi/2} b\sin\theta\, b\cos\theta\, b\cos\theta\, d\theta} \qquad = \frac{b\int_0^{\pi/2}\sin^2\theta\cos^2\theta\, d\theta}{b\int_0^{\pi/2}\sin\theta\cos^2\theta\, d\theta}$$

$$= \frac{b\dfrac{\Gamma\left(\frac{3}{2}\right)\Gamma\left(\frac{3}{2}\right)}{2\Gamma(3)}}{\dfrac{\Gamma(1)\Gamma\left(\frac{3}{2}\right)}{2\Gamma\left(\frac{5}{2}\right)}} = \frac{\dfrac{b.\frac{1}{2}\sqrt{\pi}.\frac{1}{2}.\sqrt{\pi}}{2.2}}{\dfrac{\frac{1}{2}\sqrt{\pi}}{2.\frac{3}{2}.\frac{1}{2}\sqrt{\pi}}} = \frac{\frac{b\pi}{16}}{\frac{1}{3}} = \frac{3b\pi}{16}$$

Hence, the required coordinates of the C.P. are ($3a/8$, $3b\pi/16$).

Example 9: A quadrant of a circle of radius a is immersed vertically with its bounding radius horizontal at a depth b. Find the centre of pressure.

Sol. Let OAB be a quadrant of a circle of radius a immersed vertically with its bounding radius OB horizontal and at a depth b.

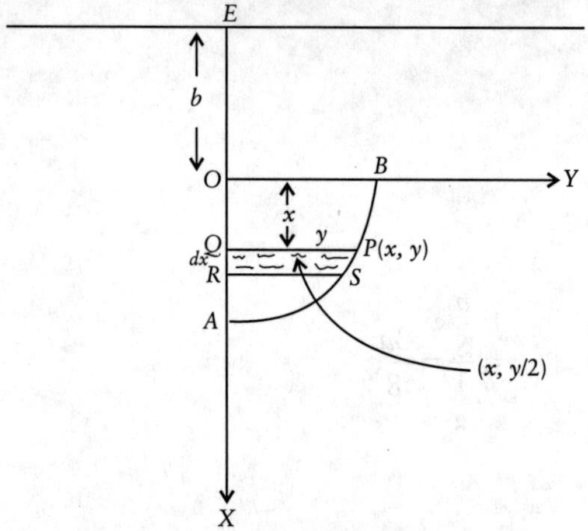

Take O as origin, OA as X-axis and OB as Y-axis.

Then, referred to these axes, the equation of the circle is $x^2 + y^2 = a^2$ ...(1)

Consider an elementary strip $PQRS$ of thickness dx and at a depth x below Y-axis. Let $P \to (x, y)$.

C.P. of the strip $PQRS$ will be $(x, y/2)$.

dS = elementary area of the strip $PQRS = y\,dx$

p = Intensity of pressure at any point of the strip $= g\rho\,(b + x)$

Let $(\bar{x}, \bar{y})$ be the coordinates of C.P. of the quadrant OAB. Then,

$$\bar{x} = \frac{\int xp\,dS}{\int p\,dS}$$

$$= \frac{\int_0^a x g\rho(b+x)y\,dx}{\int_0^a g\rho(b+x)y\,dx} = \frac{\int_0^a x(b+x)\sqrt{a^2-x^2}\,dx}{\int_0^a (b+x)\sqrt{a^2-x^2}\,dx}$$

$$= \frac{b\int_0^a x\sqrt{a^2-x^2}\,dx + \int_0^a x^2\sqrt{a^2-x^2}\,dx}{b\int_0^a \sqrt{a^2-x^2}\,dx + \int_0^a x\sqrt{a^2-x^2}\,dx}$$

Put $x = a\sin\theta$

$\therefore$ $dx = a\cos\theta\,d\theta$

$$= \frac{b\int_0^{\pi/2} a\sin\theta a\cos\theta a\cos\theta d\theta + \int_0^{\pi/2} a^2\sin^2\theta a\cos\theta\ a\cos\theta d\theta}{b\int_0^{\pi/2} a\cos\theta a\cos\theta d\theta + \int_0^{\pi/2} a\sin\theta a\cos\theta\ a\cos\theta d\theta}$$

$$= \frac{ab\int_0^{\pi/2} \sin\theta\cos^2\theta d\theta + a^2\int_0^{\pi/2} \sin^2\theta\cos^2\theta\,d\theta}{b\int_0^{\pi/2} \cos^2\theta\,d\theta + a\int \sin\theta\cos^2\theta\ d\theta}$$

$$= \frac{ab\dfrac{\Gamma(1)\Gamma\left(\frac{3}{2}\right)}{2\Gamma\left(\frac{5}{2}\right)} + a^2\dfrac{\Gamma\left(\frac{3}{2}\right)\Gamma\left(\frac{3}{2}\right)}{2\Gamma(3)}}{b.\dfrac{1}{2}.\dfrac{\pi}{2} + a.\dfrac{\Gamma(1)\Gamma\left(\frac{3}{2}\right)}{2\Gamma\left(\frac{5}{2}\right)}} \qquad = \frac{\dfrac{ab\frac{1}{2}\sqrt{\pi}}{2.\frac{3}{2}.\frac{1}{2}.\sqrt{\pi}} + \dfrac{a^2\frac{1}{2}\sqrt{\pi}\frac{1}{2}\sqrt{\pi}}{2.2}}{\dfrac{b\pi}{4} + \dfrac{a.\frac{1}{2}\sqrt{\pi}}{2.\frac{3}{2}.\frac{1}{2}\sqrt{\pi}}}$$

$$= \frac{\dfrac{ab}{3} + \dfrac{\pi a^2}{16}}{\dfrac{b\pi}{4} + \dfrac{a}{3}} \qquad = \frac{16ab + 3\pi a^2}{4(3b\pi + 4a)} = \frac{a(16b + 3\pi a)}{4(3b\pi + 4a)}$$

and, $\bar{y} = \dfrac{\int y p dS}{\int p dS}$

$$= \frac{\int_0^a \frac{y}{2}.g\rho(b+x).y\,dx}{\int_0^a g\rho(b+x)y\,dx} \quad = \frac{\int_0^a (by^2 + xy^2)dx}{\int_0^a (by + xy)dx} \quad = \frac{1}{2}\frac{\int_0^a \{b\,(a^2 - x^2) + x(a^2 - x^2)\}dx}{\int_0^a \left(b\sqrt{a^2 - x^2} + x\sqrt{a^2 - x^2}\right)dx}$$

Put $x = a \sin\theta$

$\therefore \quad dx = a\cos\theta\, d\theta$

$$= \frac{1}{2}\frac{\left[\int_0^{\pi/2} b\, a^2\cos^2\theta\, a\cos\theta\, d\theta + \int_0^{\pi/2} a\sin\theta\, a^2\cos^2\theta\, a\cos\theta\, d\theta\right]}{\int_0^{\pi/2} b\, a\cos\theta\, a\cos\theta\, d\theta + \int_0^{\pi/2} a\sin\theta\, a\cos\theta\, a\cos\theta\, d\theta}$$

$$= \frac{1}{2}\frac{\left[ba^3\int_0^{\pi/2}\cos^3\theta\, d\theta + a^4\int_0^{\pi/2}\sin\theta\cos^3\theta\, d\theta\right]}{ba^2\int_0^{\pi/2}\cos^2\theta\, d\theta + a^3\int_0^{\pi/2}\sin\theta\cos^2\theta\, d\theta}$$

$$= \frac{\dfrac{1}{2}\left[ba\dfrac{2}{3} + a^2\dfrac{\Gamma(1)\Gamma(2)}{2\Gamma(3)}\right]}{b\dfrac{1}{2}.\dfrac{\pi}{2} + a\dfrac{\Gamma(1)\Gamma\left(\frac{3}{2}\right)}{2\Gamma\left(\frac{5}{2}\right)}} \qquad = \frac{\dfrac{1}{2}\left[\dfrac{2ab}{3} + \dfrac{a^2}{4}\right]}{\dfrac{b\pi}{4} + \dfrac{a.\frac{1}{2}\sqrt{\pi}}{2.\frac{3}{2}.\frac{1}{2}\sqrt{\pi}}}$$

$$= \frac{\dfrac{ab}{3} + \dfrac{a^2}{8}}{\dfrac{b\pi}{4} + \dfrac{a}{3}} \qquad = \frac{a(8b + 3a)}{2(3b\pi + 4a)}$$

Hence, the coordinates of C.P. referred to OX and OY as axes are

$$\left[\frac{a(16b+3a\pi)}{4(3b\pi+4a)}, \frac{a(8b+3a)}{2(3b\pi+4a)}\right]$$

Example 10: An area bounded by curve $ay^2 = x^3$, the X-axis and the ordinate $x = a$, is immersed in water with the axis in the surface. Find the co-ordinates of the centre of pressure.

Sol. Let O be the origin, OX be the X-axis and OY be the Y-axis. The given curve $ay^2 = x^3$ is semi-cubical parabola. It is symmetrical about X-axis. It passes through the origin. The tangents at the origin are $y^2 = 0$ [obtained by equating the lowest degree term to zero], i.e., there are two real and coincident tangents at the origin. Hence, there is a cusp at the origin. [Note that two coincident tangents are $y = 0$ (X-axis) and $y = 0$ (X-axis)]. Also, the curve does not exist on the negating side of the origin.

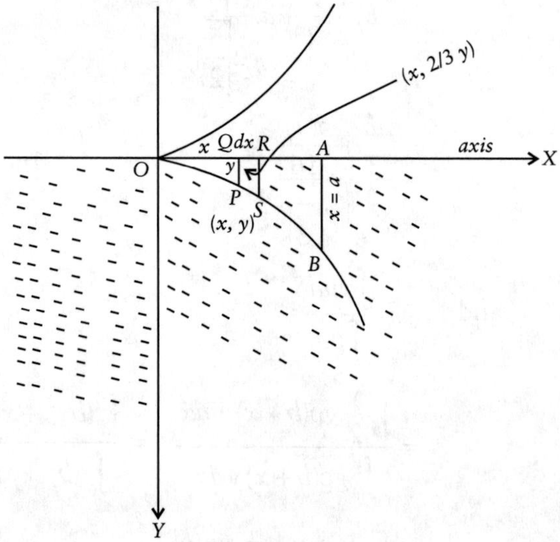

The area under consideration is bounded by the curve $ay^2 = x^3$, the X-axis and the coordinate AB $(x = a)$, therefore, the area whose coordinate of C.P. we are to find out is the area OAB as shown in the figure.

Consider an elementary strip $PQRS$ of thickness dx and at a distance x from the origin. Let $P \rightarrow (x, y)$.

dS = elementary area of the strip $PQRS = y\,dx$

p = Intensity of pressure at the C.G. of the strip $PQRS$

$= g\rho y/2$ | Refer chapter 5

The C.P. of the strip is $(x, 2/3y)$, it being rectangular in shape.

Let $(\bar{x}, \bar{y})$ be the coordinate of C.P. of the area OAB. Then,

$$\bar{x} = \frac{\int xpdS}{\int pdS} = \frac{\int_0^a x.g\rho\frac{y}{2}y\,dx}{\int_0^a g\rho\frac{y}{2}y\,dx}$$

$$= \frac{\int_0^a \frac{xy^2}{2}dx}{\int_0^a \frac{y^2}{2}dx} = \frac{\int_0^a xy^2\,dx}{\int_0^a y^2\,dx}$$

$$= \frac{\int_0^a x\frac{x^3}{a}dx}{\int_0^a \frac{x^3}{a}dx} \qquad \Big|\ \because ay^2 = x^3$$

$$= \frac{\int_0^a x^4\,dx}{\int_0^a x^3\,dx} = \frac{\left(\frac{x^5}{5}\right)_0^a}{\left(\frac{x^4}{4}\right)_0^a} = \frac{\frac{x^5}{4}}{\frac{x^4}{4}} = \frac{4a}{5}$$

and, $\bar{y} = \dfrac{\int yp\,dS}{\int p\,dS}$

$$= \frac{\int_0^a \frac{2}{3} yg\rho \frac{y}{2} y\,dx}{\int_0^a g\rho \frac{y}{2} y\,dx} = \frac{\frac{2}{3}\int_0^a y^3\,dx}{\int_0^a y^2\,dx} = \frac{\frac{2}{3}\int_0^a \frac{x^{9/2}}{a^{3/2}}\,dx}{\int_0^a \frac{x^3}{a}\,dx} = \frac{\frac{2}{3\sqrt{a}}\int_0^a x^{9/2}\,dx}{\int_0^a x^3\,dx}$$

$$= \frac{2}{3\sqrt{a}}\,\frac{\frac{2}{11}a^{11/2}}{\frac{a^4}{4}} = 16a/33$$

Hence, the coordinate of the C.P. of the given area are (4a/5, 16a/33)

Example 11: Find the coordinates of the centre of pressure of the area between the curve $\sqrt{\frac{x}{a}} + \sqrt{\frac{y}{b}} = 1$, and the axes, taking the axis to be rectangular and one of them in the surface.

Sol. The given curve is a parabola touching the axes at distances a and b from the origin. The shape of the curve is as shown in the figure.

The curve touches then X-axis at A $(a, 0)$ and the Y-axis at B $(0, b)$

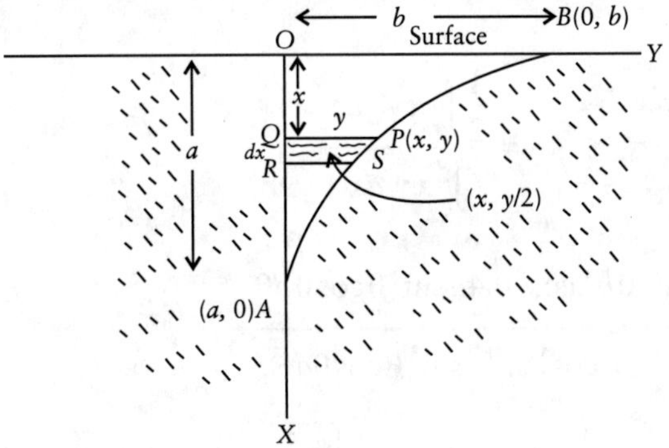

Consider an elementary strip $PQRS$ of thickness dx and at a depth x from O below the free surface. Let $P \rightarrow (x, y)$. Then, the coordinates of the C.P. of the strip $PQRS$ will be $(x, y/2)$.

dS = elementary area of the strip $PQRS$

$$= y\,dx = \frac{\sqrt{b}}{\sqrt{a}}\left(\sqrt{a} - \sqrt{x}\right) \qquad \left|\because \sqrt{\frac{x}{a}} + \sqrt{\frac{y}{b}} = 1\right.$$

p = Intensity of pressure at any point of the strip $PQRS$

$= g\rho x$

Let $(\bar{x}, \bar{y})$ be the coordinate of the C.P. of the area AOB. Then,

$$\bar{x} = \frac{\int xp\,dS}{\int p\,dS} = \frac{\int_0^a xg\rho xy\,dx}{\int_0^a g\rho xy\,dx} = \frac{\int_0^a x^2 y\,dx}{\int_0^a xy\,dx}$$

The parametric equations of the given curve are

$$x = a\sin^4\theta,\ y = b\cos^4\theta$$

$$\therefore \quad dx = 4a\sin^3\theta\cos\theta\,d\theta$$

$$= \frac{\int_0^{\pi/2} a^2\sin^8\theta\, b\cos^4\theta\ 4a\sin^3\theta\cos\theta\,d\theta}{\int_0^{\pi/2} a\sin^4\theta\, b\cos^4\theta\ 4a\sin^3\theta\cos\theta\,d\theta}$$

$$= \frac{\int_0^{\pi/2}\sin^{11}\theta\cos^5\theta\,d\theta}{\int_0^{\pi/2}\sin^7\theta\cos^5\theta\,d\theta}\,a = \frac{\dfrac{\Gamma(6)\,\Gamma(3)}{2\,\Gamma(9)}}{\dfrac{\Gamma(4)\,\Gamma(3)}{2\,\Gamma(7)}}\,a$$

$$= \frac{\dfrac{5.4.3.2.1.2.1}{2.8.7.6.5.4.3.2.1}}{\dfrac{3.2.1.2.1}{2.6.5.4.3.2.1}}\,a = \frac{a}{8.7.6}\cdot\frac{6.5.4}{1} = \frac{5a}{14}$$

and, $$\bar{y} = \frac{\int yp\,ds}{\int p\,ds}$$

$$= \frac{\int_0^a \frac{y}{2} g\rho xy\,dx}{\int_0^a g\rho xy\,dx} = \frac{\frac{1}{2}\int_0^a xy^2\,dx}{\int_0^a xy\,dx}$$

$$= \frac{\frac{1}{2}\int_0^{\pi/2} a\sin^4\theta\, b^2\cos^8\theta . 4a\sin^3\theta\cos\theta\,d\theta}{\int_0^{\pi/2} a\sin^4\theta\, b\cos^4\theta . 4a\sin^3\theta\cos\theta\,d\theta}$$

$$= \frac{\frac{b}{2}\int_0^{\pi/2}\sin^7\theta\cos^9\theta\,d\theta}{\int_0^{\pi/2}\sin^7\theta\cos^5\theta\,d\theta} = \frac{\dfrac{b}{2}\,\dfrac{\Gamma(4)\,\Gamma(5)}{2\,\Gamma(9)}}{\dfrac{\Gamma(4)\,\Gamma(3)}{2\,\Gamma(7)}}$$

$$= \frac{b}{2}\cdot\frac{\Gamma(5)}{\Gamma(9)}\cdot\frac{\Gamma(7)}{\Gamma(3)} = \frac{b}{2}\frac{4.3}{8.7} = \frac{3}{28}b$$

Hence, the coordinates of the C.P. are ($5a/14$, $3b/28$)

Example 12: A circular disc of radius a is completely immersed with its plane vertical in a homogeneous fluid. If h is the depth of the centre below the free surface of the liquid, prove that the distance between the centres of pressure of the two semi-circles into which disc is divided by horizontal diameter is $\dfrac{6\pi a(4h^2 - a^2)}{9\pi^2 h^2 - 16a^2}$

Sol. O is centre of the circular disc of radius a.

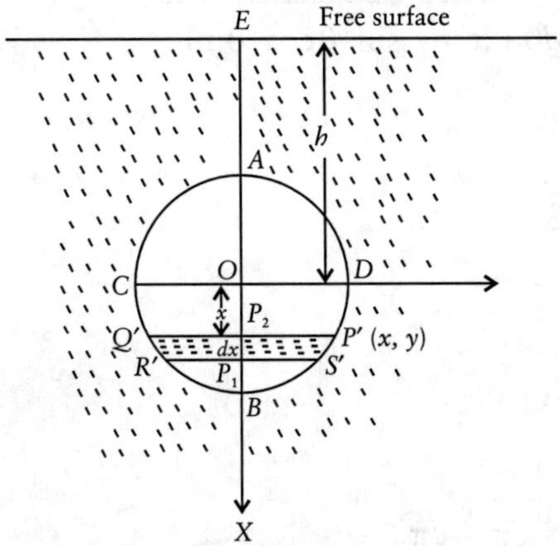

COD is the horizontal diameter of circular disc. The axes of reference are as shown in the figure.

Consider an elementary strip of thickness dx at a depth x below the centre O. Let $P' \to (x, y)$.

By symmetry, C.P. of semi-circle CAD will lie on vertical diameter AB. Also, C.P. of semi-circle CBD will lie on AB. Moreover, C.P. of the whole circle will lie on AB.

The equation of the circle is

$$x^2 + y^2 = a^2 \qquad ...(1)$$

dS = area of the elementary strip $P'Q'R'S' = 2y\,dx$

$= 2\sqrt{a^2 - x^2}\,dx$ | From (1)

p = Intensity of pressure at any point of the strip $P'Q'R'S'$

$= g\rho\,(h + x)$

Let $\bar{x}$ be the depth of semi-circle COD below O. Then,

$$\bar{x} = \frac{\int xp\,dS}{\int p\,dS} = \frac{\int_0^a xg\rho(h+x)2\sqrt{a^2 - x^2}\,dx}{\int_0^a g\rho(h+x)2\sqrt{a^2 - x^2}\,dx}$$

$$= \frac{h\int_0^a x\sqrt{a^2 - x^2}\,dx + \int_0^a x^2\sqrt{a^2 - x^2}\,dx}{h\int_0^a \sqrt{a^2 - x^2}\,dx + \int_0^a x\sqrt{a^2 - x^2}\,dx}$$

Put $x = a \sin \theta$

$\because dx = a \cos d\theta$

$$= \frac{h\int_0^{\pi/2} a\sin\theta\, a\cos\theta\, a\cos\theta\, d\theta + \int_0^{\pi/2} a^2 \sin^2\theta\, a\cos\theta\, a\cos\theta\, d\theta}{h\int_0^{\pi/2} a\sin\theta\, a\cos\theta\, d\theta + \int_0^{\pi/2} a\sin\theta\, a\cos\theta\, a\cos\theta\, d\theta}$$

$$= \frac{a^3 h\int_0^{\pi/2} \sin\theta\cos^2\theta\, d\theta + a^4\int_0^{\pi/2} \sin^2\theta\cos^2\theta\, d\theta}{a^2 h\int_0^{\pi/2} \cos^2\theta\, d\theta + a^3\int_0^{\pi/2} \sin^2\theta\cos^2\theta\, d\theta}$$

$$= \frac{ah\dfrac{\Gamma 1\,\Gamma\frac{3}{2}}{2\,\Gamma\frac{5}{2}} + a^2\dfrac{\Gamma\frac{3}{2}\,\Gamma\frac{3}{2}}{2\,\Gamma 3}}{h.\dfrac{1}{2}.\dfrac{\pi}{2} + \dfrac{a\Gamma 1\,\Gamma\frac{3}{2}}{2\,\Gamma\frac{5}{2}}}$$

$$= \frac{\dfrac{ah.\frac{1}{2}\sqrt{\pi}}{2.\frac{3}{2}.\frac{1}{2}.\sqrt{\pi}} + \dfrac{a^2\frac{1}{2}\sqrt{\pi}\frac{1}{2}\sqrt{\pi}}{2.2}}{\dfrac{h\pi}{4} + \dfrac{a.\frac{1}{2}\sqrt{\pi}}{2.\frac{3}{2}.\frac{1}{2}.\pi}}$$

$$= \frac{\dfrac{ah}{3} + \dfrac{a^2\pi}{16}}{\dfrac{h\pi}{4} + \dfrac{a}{3}} = \frac{a(16h + 3\pi a)}{4(3\pi h + 4a)}$$

Let P_1 be the C.P. of the semi-circle CAD. Then,

$$OP_1 = \frac{a(16h + 3\pi a)}{4(3\pi h + 4a)}$$

Let P be the C.P. of the whole circle whose centre is O. Then,

$$OP = a^2/4h$$

X = Fluid thrust on the whole circle

$= g\rho\pi a^2 h$ $\qquad\qquad | \; P = wS\,\bar{z}$

X_1 = Fluid thrust on the semi-circle CBD

$$= g\rho.\frac{\pi a^2}{2}\left(h + \frac{4a}{3\pi}\right)$$

$= 1/6\ (3\pi h + 4a)a^2 g\rho$

$\therefore$ Depth of the C.P. (P_2) of the semi-circle CAD below O

$$= \frac{X.OP - X_1.OP_1}{X - X_1}$$

$$= \frac{g\rho\pi a^2 h.\dfrac{a^2}{4h} - \dfrac{3\pi h + 4a}{6} a^2 g\rho \dfrac{a(16h + 3\pi a)}{4(3\pi h + 4a)}}{g\rho\pi a^2 h - \dfrac{(3\pi h + 4a)a^2 g\rho}{6}}$$

$$= \frac{\dfrac{1}{4}\pi a^2 - \dfrac{(16h + 3\pi a)a}{24}}{\pi h - \dfrac{1}{6}(3\pi h + 4a)} = \frac{a(3\pi a - 16h)}{4(3\pi h - 4a)}$$

$\therefore$ Required distance between the centres of pressure P_1 and P_2

$= OP_1 - OP_2$

$$= \frac{a(3\pi a + 16h)}{4(3\pi h + 4a)} - \frac{a(3\pi a - 16h)}{4(3\pi h - 4a)} = \frac{a}{4}\left[\frac{3\pi a + 16h}{3\pi h + 4a} - \frac{3\pi a - 16h}{3\pi h - 4a}\right]$$

$$= \frac{a}{4}\left[\frac{(3\pi a + 16h)(3\pi h - 4a) - (3\pi a - 16h)(3\pi h + 4a)}{(3\pi h + 4a)(3\pi h - 4a)}\right]$$

$$= \frac{a}{4}\left[\frac{9\pi^2 ah - 12\pi a^2 + 48\pi h^2 - 64ha - 9\pi^2 ah - 12\pi a^2 + 48\pi h^2 + 64ha}{9\pi^2 h^2 - 16a^2}\right]$$

$$= \frac{a(96\pi h^2 - 24\pi a^2)}{4(a\pi^2 h^2 - 16\pi a^2)} = \frac{a(24\pi h^2 - 6\pi a^2)}{9\pi^2 h^2 - 16a^2} = \frac{6\pi a(4h^2 - a^2)}{9\pi^2 h^2 - 16a^2}$$

EXERCISE 4

1. A rectangle is immersed vertically in a heavy homogeneous liquid with two of its sides horizontal and at depths α and β below the surface. Show that the depth of the centre of pressure below the free surface is $\dfrac{2}{3}\dfrac{\alpha^2 + \alpha\beta + \beta^2}{\alpha + \beta}$

   [**Hint:** Refer Ex. 1]

2. Show that the depth below the free surface of the centre of pressure of a parallelogram, two of whose sides are horizontal and at depths h and k below the free surface, is $\dfrac{2}{3}\dfrac{h^2 + hk + k^2}{h + k}$

   [**Hint:** Refer Ex. 1]

3. Prove that the depth of C.P. of a parallelogram whose two sides are horizontal and at depths h and k below the surface of a liquid whose density varies as the depth below the surface is $\frac{3}{4}\frac{h^3+h^2k+k^2h+k^3}{h^2+hk+k^2}$

   [**Hint:** Refer Ex. 1]

4. A rectangle is immersed vertically in a liquid whose density varies as the depth below the surface. If the two horizontal sides of the rectangle are at depths h and k below the surface, find the depth of the centre of pressure.

   Ans. $\frac{3}{4}\frac{(h+k)(h^2+k^2)}{h^2+hk+k^2}$

   [**Hint:** Refer Ex. 1]

5. Find the C.P. of a square lamina, held vertically in a fluid, whose density varies a the depth, with a vertex (angular point) in the surface, and the diagonal vertical.

   Ans. $AP = 9/14\ AC$

   Where A is in the surface, AC is the vertical diagonal and P is the C.P. of the square $ABCD$.

6. A semi-circular lamina of radius a is immersed in a liquid with the diameter in the surface. Find the depth of the centre of pressure below the free surface.

   Ans. $3/16\ \pi a$

   [**Hint:** Refer Ex. 2. Put $b = 0$]

7. A parabolic area is held vertical under a fluid of uniform density such that the latus rectum is in the surface. Find the centre of pressure.

   Ans. Depth of C.P. below the free surface = 4a/7 where $4a$ is the latus rectum of the parabola.

   [**Hint:** Refer Ex. 4.

   Here, $h = a$, since the distance between the vertex and the latus rectum is a for parabola $y^2 = 4ax$ and the latus rectum is in the surface.

   $\therefore$ Required depth = 4a/7]

8. Prove that the depth of C.P. of a parabolic lamina immersed vertically in a liquid with its vertex at a depth k below the water surface, and is bounded by an ordinate at a distance h from the vertex is $\frac{3}{7}\cdot\left(\frac{7k+5h}{5k+3h}\right)h$, when the vertex is above the ordinate.

   [**Hint:**

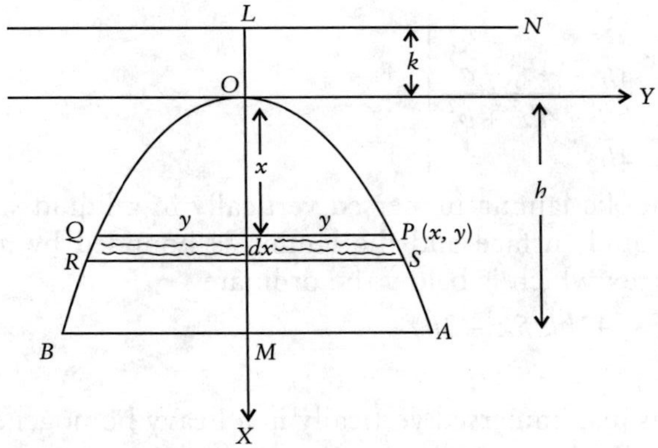

$y^2 = 4ax$

$p = g\rho\,(k + x)$

$dS = 2y\,dx = 2\sqrt{4ax}\,dx = 4\sqrt{ax}\,dx$

$\bar{y} = 0$

$$\bar{x} = \frac{\int xp\,dS}{\int p\,dS} = \frac{\int_0^h xg\rho(k+x)4\sqrt{ax}\,dx}{\int_0^h g\rho(k+x)4\sqrt{ax}\,dx} \qquad = \frac{3}{7}\cdot\left(\frac{7k+5h}{5k+3h}\right)h]$$

9. An ellipse is just immersed in water with its major axis vertical. Find the position of the centre of pressure.

   Ans. At a depth $a/4$ below the centre [At a depth $5a/4$ below the free surface]

   [**Hint:** Refer Ex. 5 Put $h = a$]

10. An ellipse is just immersed in water with its major axis vertical. Show that if the centre of pressure coincides with the focus, the eccentricity of the ellipse must be 1/4.

    [**Hint:** Refer Ex. 5 and Ex. 6

    Here, $a/4 = ae \quad \Rightarrow \quad e = 1/4$]

11. Find the depth of the centre of pressure of an elliptic lamina partly immersed in a liquid with its major axis in the surface.

    Ans. $3\pi b/16$

12. Show that the ratio of the depths of the centre of pressure (from the free surface) of an ellipse completely immersed in a liquid with its centre at a depth h below the free surface and with its minor axis horizontal in one case and vertical in another case is $4h^2 + a^2/4h^2 + b^2$

    [**Hint:** Refer Ex. 5

    Depth of C.P. below the free surface when the minor axis is horizontal and is at a depth h below the free surface $= h + a^2/4h$.

    Depth of C.P. below the free surface when the major axis is horizontal and is at a depth h below the free surface $= h + b^2/4h$.

$$\therefore \text{ Required ratio } = \frac{h+\dfrac{a^2}{4h}}{h+\dfrac{b^2}{4h}} = \frac{4h^2+a^2}{4h^2+b^2}\Bigg]$$

13. Find the C.P. of a parabolic lamina immersed vertically in a liquid with its vertex at a depth k below the liquid surface and the lamina is bounded by an ordinate at a distance h from the vertex which is below the ordinate

    Ans. $1/7 \,.\, (35k^2 + 15h^2 - 42hk/5k - 3h)$

    [**Hint:** Refer Q. 8]

14. A quadrant of a circle is just immersed vertically in a heavy homogeneous liquid with one edge in the surface. Find the centre of pressure.

    Ans. $\left(\frac{3\pi a}{16}, \frac{3}{8}a\right)$

    [**Hint:** Refer Ex. 9. Put $b = 0$]

15. A quadrant of a circle just immersed vertically with one edge in the surface in a liquid whose density varies as the depth. Find out the centre of pressure.

    Ans. $(32a/15\pi, 16a/15\pi)$

    [**Hint:** Refer Ex. 9]

16. An area bounded by curve $ay^2 = x^3$; an abscissa of length h and the ordinate at its extremity, is placed in water with the ordinate in the surface. Prove that the depth of the C.P. is $4/9h$.

    [**Hint:** Refer Ex. 10]

17. Find the coordinates of the centre of pressure of the area between the curve $\sqrt{x} + \sqrt{y} = \sqrt{a}$ and the axes, taking the axes to be rectangular and one of them in the surface.

    Ans. $(5a/14, 3a/14)$

    [**Hint:** Refer Ex. 11. Put $b = a$]

18. From a semi-circle whose diameter is in the surface of a liquid, a circle is cut out, whose diameter is the vertical radius of the semi-circle. Prove that the depth of the C.P. of the remainder is $9\pi a/8\ (16 - 3\pi)$.

12. C.P. of an Area Immersed in *n* Liquids of Different Densities and Which do not Mix

Let a given area, under consideration, be in contact with layers of n different liquids which do not mix. Let the densities of these n liquids in ascending order be $\rho_1, \rho_2, \rho_3, \ldots, \rho_n$ and let $A_1, A_2, A_3, \ldots, A_r$ be the parts of whole area in contact with these liquids respectively.

The whole pressure on the area will remain the same even if we consider that the liquid of density ρ_1 is in contact with the whole area $(A_1 + A_2 + A_3 + .. + A_n)$, the liquid of density $(\rho_2 - \rho_1)$ is in contact with the area $(A_2 + A_3 + \ldots + A_n)$, the liquid of density $(\rho_3 - \rho_2)$ $[= \rho_3 - \rho_1 - (\rho_2 - \rho_1)]$ is in contact with the area $(A_3 + A_4 + \ldots + A_n)$ and so on and lastly the liquid of density $(\rho_n - \rho_{n-1})$ is in contact with area A_n.

If $P_1, P_2, ..., P_n$ be the thrusts on the areas $(A_1 + A_2 + A_3 + ... + A_n)$, $(A_2 + A_3 + ... + A_n)$, $(A_3 + A_4 + ... + A_n)$, ..., A_n of the above liquids and $z_1, z_2, z_3, ..., z_n$ be the depths of their C.P.'s below the free surface respectively, then $\bar{z}$, the depth of C.P. of the whole area below the free surface, is given by

$$\bar{z} = \frac{P_1 z_1 + P_2 z_2 + P_3 z_3 + .. + P_n z_n}{P_1 + P_2 + P_3 + ... + P_n}$$

$$= \frac{\sum_{i=1}^{n} P_i z_i}{\sum_{i=1}^{n} P_i}$$

ILLUSTRATIVE EXAMPLES

Example 1: A rectangle is immersed in two liquids of density ρ and 2ρ which do not mix. The top of the rectangle is in the surface of the first liquid and the area immersed in each is the same. Prove that the centre of pressure divides the rectangle in the ratio 7 : 3.

Sol. Let $ABCD$ be a rectangle immersed in two liquids of density ρ and 2ρ such that the top of the rectangle AB is in the surface of the liquid of density ρ and the area immersed in each is the same.

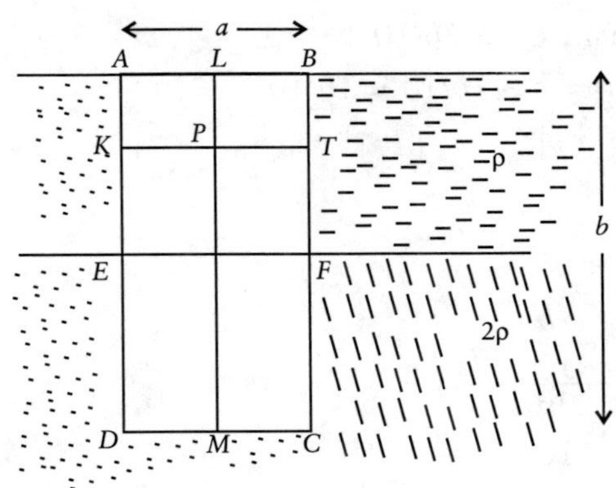

Let $AB = a$ and $BC = b$.

Then, area $ABFE$ = area $EFCD$ = $ab/2 = S$ (say)

Let us consider that the whole area $2S$ is in contact with the liquid of density ρ and the area S is in contact with the liquid of density $(2\rho - \rho) = \rho$.

P_1 = Thrust on rectangle $ABCD$ when it is in contact with liquid of density $\rho = g\rho \,.\, (2S) \,.\, b/2 = g\rho Sb$

z_1 = Depth of C.P. of rectangle $ABCD$ from the free surface when it is in contact with liquid of density $\rho = 2/3b$.

P_2 = Thrust on rectangle $EFCD$ when it is in contact with liquid of density $2\rho - \rho = \rho$

$= g\rho.\ S.\ b/4 = g\rho Sb/4$

| Note that the depth of C.G. has been measured from the common surface EF

z_2 = Depth of C.P. of rectangle $EFCD$ from the free surface when it is in contact with liquid of density $2\rho - \rho = \rho$

$$= \frac{b}{2} + \frac{2}{3}\left(\frac{b}{2}\right) = \frac{5b}{6}$$

Let $\bar{z}$ be the depth of the required C.P. below the free surface. Then,

$$\bar{z} = \frac{P_1 z_1 + P_2 z_2}{P_1 + P_2} \qquad | \text{ Refer § 7}$$

$$= \frac{g\rho Sb.\frac{2}{3} + \frac{g\rho Sb}{4}.\frac{5b}{6}}{g\rho Sb + \frac{g\rho Sb}{4}}$$

$$= \frac{\frac{2}{3}b + \frac{5b}{24}}{1 + \frac{1}{4}} = \frac{\frac{7b}{8}}{\frac{5}{4}} = \frac{7b}{10}$$

By symmetry, this C.P. will lie on LM where L and M are the mid-points of AB and DC respectively.

Let this C.P. be P. Then, $LP = 7b/10$

$$\therefore \quad PM = LM - LP = b - 7b/10 = 3b/10$$

$$\therefore \quad \frac{LP}{PM} = \frac{\frac{7b}{10}}{\frac{3b}{10}} = \frac{7}{3}$$

$$\therefore \quad \frac{\text{Area } ABTK}{\text{Area } KTCD} = \frac{a.\frac{7b}{10}}{a.\frac{3b}{10}} = \frac{7}{3}$$

Hence, the C.P. divides the rectangle in the ratio 7 : 3.

Example 2: Prove that the C.P. of a rhombus immersed in two liquids which do not mix with the vertex in the upper surface and a diagonal in the common surface, divides the other diagonal in the ratio 5 : 3 if the density of the lower liquid is thrice that of the upper liquid.

Sol. Let $ABCD$ be rhombus immersed in two liquids which do not mix with the vertex A in the upper surface and a diagonal BD in the common surface. Let the densities of the upper and lower liquids be ρ and 3ρ respectively.

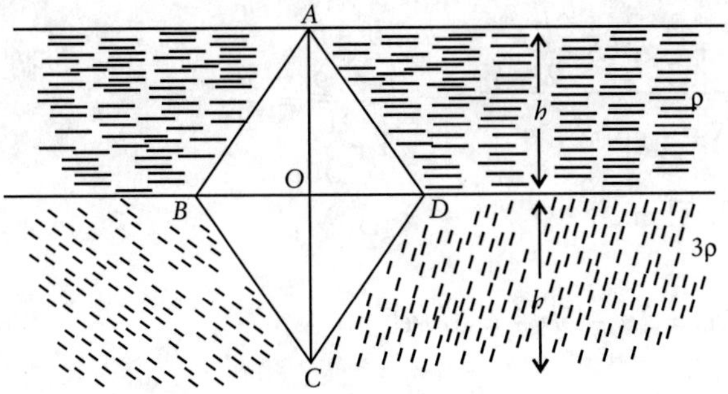

Let $AC = 2h$

Then, $AO = OC = h$

By symmetry, C.P. of rhombus $ABCD$ lies on AC and, depth of C.P. of the rhombus $ABCD$ = depth of C.P. of $\Delta\ ADC$

Let the area of $\Delta\ ADC = 2S$

Then, area of $\Delta\ AOD = S =$ area $\Delta\ COD$.

Let us consider that the whole area 2S (Δ ADC) is in contact with the liquid of density ρ and the area S ($\Delta\ COD$) is in contact will the liquid of density $(3\rho - \rho) = 2\rho$.

P_1 = Thrust on area ADC in contact with liquid of density ρ

$$= g\rho 2S \frac{0 + h + 2h}{3} \qquad | \; P = wS\,\bar{z}$$

$$= 2g\rho hS$$

z_1 = Depth of C.P. of the $\Delta\ ADC$ from the free surface when it is in contact with liquid of density ρ

$$= \frac{(0)^2 + h^2 + (2h)^2 + 0(h) + h(2h) + (2h)(0)}{2(0 + h + 2h)} \qquad | \text{ Refer § 10 case II for } \bar{z}$$

$$= 7h/6$$

P_2 = Thrust on area COD in contact with liquid of density 2ρ

$$= g2\rho.S.\frac{h}{3} \qquad | \; P = wS\,\bar{z}$$

$$= 2/3\ g\rho hS$$

z_2 = Depth of C.P. of $\Delta\ COD$ from the free surface when it is in contact with liquid of density 2ρ

$$= h + 1/2\ h = 3h/2$$

Let $\bar{z}$ be the depth of the required C.P. below the free surface. Then,

$$\bar{z} = \frac{P_1 z_1 + P_2 z_2}{P_1 + P_2} \qquad | \text{ Refer § 7}$$

$$= \frac{2ghS.\frac{7h}{6} + \frac{2}{3}g\rho hS.\frac{3h}{2}}{2g\rho hS + \frac{2}{3}g\rho hS} = \frac{\frac{7h}{6}+\frac{h}{2}}{1+\frac{1}{3}} = \frac{5h}{4}$$

Let this C.P. be P. Then,

$AP = 5h/4$

$\therefore \quad PC = AC - AP = 2h - 5h/4 = 3h/4$

$\therefore \quad AP : PC = 5 : 3$

Example 3: A fluid of depth $2a$ and uniform density ρ is superimposed on a liquid of density 2ρ and depth greater than a. A circular lamina of radius a is placed with its plane vertical and its centre in the surface common to the two liquids.

Prove that the depth of C.P. below the centre of lamina, neglecting the atmospheric pressure, is

$$9\pi a/16(3\pi + 1)$$

Sol. The figure shown below explains the problem.

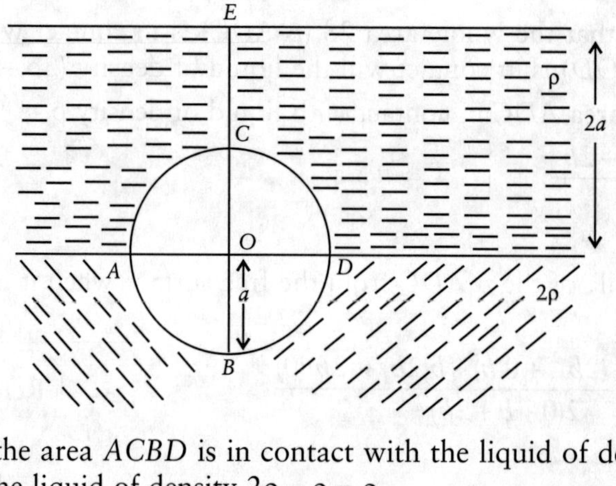

Let us consider that the area $ACBD$ is in contact with the liquid of density ρ and the area ADB is in contact with the liquid of density $2\rho - \rho = \rho$.

P_1 = Thrust on area $ACBD$ in contact with the liquid of density ρ.

$= g\rho \,.\, \pi a^2 \,.\, 2a$ | $P = wS\,\bar{z}$

$= 2g\rho\pi a^3$

z_1 = Depth of C.P. of the area $ACBD$ in contact with liquid of density ρ below O

$= \text{'}a^2/4h\text{'}$

$= a^2/4(2a) = a/8$

P_2 = Thrust on area ADB in contact with the liquid of density ρ.

$= g\rho.\pi a^2/2 \,.\, 4a/3\pi$ | $P = wS\,\bar{z}$

$= 2/3\; g\rho a^3$

z_2 = Depth of C.P. of the area ADB in contact with liquid of density ρ below O

$= 3/16\; \pi a$ | Refer Ex. 9 when $b = 0$

Let $\bar{z}$ be the depth of the C.P. of the lamina below the centre O. Then,

$$\bar{z} = \frac{P_1 z_1 + P_2 z_2}{P_1 + P_2} \quad \text{| Refer § 7}$$

$$= \frac{2g\rho\pi a^3 \frac{a}{8} + \frac{2}{3} g\rho a^3 . \frac{3\pi a}{16}}{2g\rho\pi a^3 + \frac{2}{3} g\rho a^3}$$

$$= \frac{\frac{\pi a}{4} + \frac{\pi a}{8}}{2\pi + \frac{2}{3}} = \frac{\frac{3\pi a}{8}}{2\frac{(3\pi + 1)}{3}} = \frac{9\pi a}{16(3\pi + 1)}$$

Example 4: A vessel contains three fluids of densities ρ, 2ρ and 3ρ. A triangle of area A is supported with one side in the surface of the fluid of density ρ and the opposite vertex in the fluid of density 3ρ. If $3h$, $2h$ and h are the depths of vertex below the upper surfaces of the three fluids, prove that, neglecting the atmospheric pressure, the thrust on each face of the triangle is 4/3 gρ Ah.

Also prove that the depth of the C.P. of the triangle from the free surface is $59/36h$.

Sol.

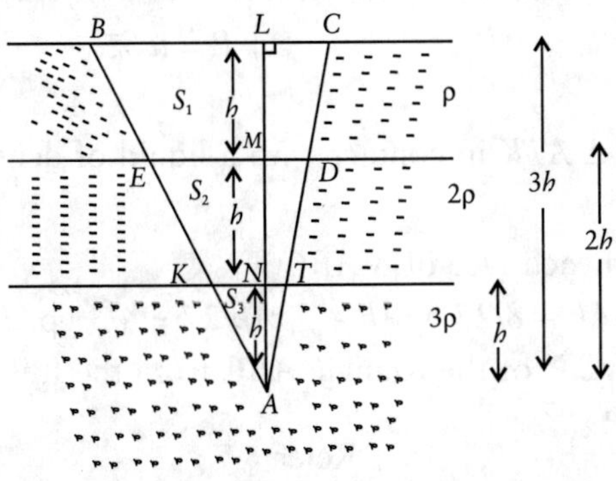

Let ABC be a triangle of area A supported with one side BC in the surface of liquid of density ρ and the opposite vertex A in the fluid of density 3ρ. Let the horizontal lines ED and KT divide the area of the triangle ABC in to three parts, each of which is in contact with one liquid; the part $BCDE$ of area S_1 (say) in contact with liquid of density ρ, the part $EDTK$ of area S_2 (say), in contact with liquid of density 2ρ and the part KTA of area S_3 (say) in contact with liquid of density 3ρ as shown in the figure.

Let us consider that the area $(S_1 + S_2 + S_3)$, i.e., area Δ ABC is in contact with a liquid of density ρ, the area $(S_2 + S_3)$, i.e., area Δ ADE is in contact with a liquid of density 2ρ – ρ = ρ and the area S_3, i.e., area Δ ATK is in contact with a liquid of density 3ρ – 2ρ = ρ.

Area of triangle ABC = A = $BC \times AL/2 = S_1 + S_2 + S_3$.

By a property of similarity of triangles,

$ED = 2/3\ BC$

$AM = 2/3\ AL$

$\therefore$ Area of $\Delta\, ADE = EX \times AM/2$

$= 2/3\ BC \times 2/3 AL/2 = 4/9\ BC \times AL/2 = 4/9\ A$

Similarly, area of $\Delta\, ATK = KT\ .\ AN/2$

$$= \frac{\frac{1}{3}BC.\frac{1}{3}AL}{2} \qquad | \ KT = 1/3\ BC,\ AN = 1/3\ AL$$

$$= \frac{1}{9}\frac{BC.AL}{2} = 1/9A$$

P_1 = Thrust on area $(S_1 + S_2 + S_3)$, i.e., $\Delta\, ABC$ in contact with a liquid of density ρ

$= g\rho\ .\ A\ .\ h$ $\qquad | \ P = wS\bar{z}$

z_1 = Depth of C.P. of $\Delta\, ABC$ in contact with a liquid of density ρ below BC

$= 3h/2$

P_2 = Thrust on area $(S_2 + S_3)$, i.e., $\Delta\, ADE$ in contact with a liquid of density ρ

$= g\rho\ .\ 4/9\ A\ .\ 2h/3 = 8/27\ g\rho\ Ah$ $\qquad | \ P = wS\bar{z}$

z_2 = Depth of C.P. of $\Delta\, ADE$ in contact with a liquid of density ρ below BC

$= h + 2h/2 = 2h$

P_3 = Thrust on area S_3, i.e., $\Delta\, ATK$ in contact with a liquid of density ρ

$= g\rho\ .\ A/9\ .\ h/3$ $\qquad | \ P = wS\bar{z}$

$= g\rho\ Ah/27$

z_3 = Depth of C.P. of $\Delta\, ATK$ in contact with a liquid of density ρ below BC

$= 2h + h/2 = 5h/2$

$\therefore$ Required thrust on each face of $\Delta\, ABC$

$= P_1 + P_2 + P_3 = g\rho Ah + 8/27\ g\rho Ah + g\rho Ah/27 = 4/3\ g\rho Ah$

Let $\bar{z}$ be the depth of C.P. of the triangle ABC from the free surface. Then,

$$\bar{z} = \frac{P_1z_1 + P_2z_2 + P_3z_3}{P_1 + P_2 + P_3} \qquad | \text{ Refer § 7}$$

$$= \frac{g\rho Ah.\frac{3h}{2} + \frac{8}{27}g\rho Ah.2h + \frac{1}{27}g\rho Ah.\frac{5h}{2}}{\frac{4}{3}g\rho Ah} = \frac{\frac{3h}{2} + \frac{16}{27}h + \frac{5h}{54}}{\frac{4}{3}} = \frac{\frac{59h}{27}}{\frac{4}{3}} = \frac{59}{36}h$$

Example 5: A rectangle is immersed in fluids of densities ρ, 2ρ, 3ρ, ..., $n\rho$; the top of the rectangle being in the surface of the first liquid and the area immersed in each liquid being the same. Show that the depth of the centre of pressure of the rectangle is

$$(3n + 1/2n + 1)\ .\ h/2,$$

where h is the depth of the lower side.

Sol. Let $ABCD$ be a rectangle immersed in n fluids of densities ρ, 2ρ, 3ρ, ..., $n\rho$; the top AB of the rectangle being in the free surface of the first liquid and the area immersed in each liquid being the same. Let $AB = a$. Clearly, depth of each liquid is h/n. Also area of each rectangle like ABB_1A_1, $A_1B_1B_2A_2$, $A_2B_2B_3A_3$, ..., $A_{n-1}B_{n-1}$ CD is the same say equal to S.

Let us consider that the rectangular area $ABCD$ is in contact with a liquid of density ρ, rectangular area A_1B_1 CD is in contact with a liquid of density $2\rho - \rho = \rho$, rectangular area A_2 B_2 CD is in contact with a liquid of density $3\rho - 2\rho = \rho$, rectangular area A_3B_3 CD is in contact with a liquid of density $4\rho - 3\rho = \rho$ and so on. Lastly, the rectangular area A_{n-1} B_{n-1} CD is in contact with a liquid of density $n\rho - (n-1)\rho = \rho$.

The thrusts on these rectangles are

$$g.\rho\, AB \,.\, h \,.\, h/2,\ g\rho A_1B_1(h - h/n) \,.\, 1/2\,(h - h/n),\ g\rho\, A_2B_2\,(h - 2h/n) \,.\, 1/2\,(h - 2h/n),\ ...,\ g\rho\, A_{n-1}\, B_{n-1}\, h/n \,.\, 1/2\,(h/n)$$

or $1/2\, g \,.\, \rho a h^2$, $1/2\, g\rho a\,(h - h/n)^2$, $1/2\, g\rho a\,(h - 2h/n)^2$, ..., $1/2\, g\rho a\,(h/n)^2$

Heights of C.P.'s of these areas above DC are $h/3$, $1/3\,(h - h/n)$, $1/3\,(h - 2h/n)$, ..., $h/3n$

Let $\bar{z}$ be the height of C.P. of whole rectangle $ABCD$ above DC. Then,

$$\bar{z} = \frac{\frac{1}{2}g\rho a h^2 \frac{h}{3} + \frac{1}{2}g\rho a\left(h - \frac{h}{n}\right)^2 \frac{1}{3}\left(h - \frac{h}{n}\right) + \frac{1}{2}g\rho a\left(h - \frac{2h}{n}\right)^2 \frac{1}{3}\left(h - \frac{2h}{n}\right) + ... + \frac{1}{2}g\rho a\left(\frac{h}{n}\right)^2 \frac{h}{3n}}{\frac{1}{2}g\rho a h^2 + \frac{1}{2}g\rho a\left(h - \frac{h}{n}\right)^2 + \frac{1}{2}g\rho a\left(h - \frac{2h}{n}\right)^2 + ... + \frac{1}{2}g\rho a\left(\frac{h}{n}\right)^2}$$

$$= \frac{h}{3}\left[\frac{1 + \left(1 - \frac{1}{n}\right)^3 + \left(1 - \frac{2}{n}\right)^3 + ... + \left(\frac{1}{n}\right)^3}{1 + \left(1 - \frac{1}{n}\right)^2 + \left(1 - \frac{2}{n}\right)^2 + ... + \left(\frac{1}{n}\right)^2}\right]$$

$$= \frac{h}{3n}\left[\frac{n^3 + (n-1)^3 + (n-2)^3 + ... + 1^3}{n^2 + (n-1)^2 + (n-2)^2 ..... + 1^2}\right]$$

$$= \frac{h}{3n}\frac{\Sigma n^3}{\Sigma n^2} = \frac{h}{3n}\frac{\left\{\frac{n(n+1)}{2}\right\}^2}{\frac{n(n+1)(2n+1)}{6}}$$

$$= \frac{h}{3n} \cdot \frac{n^2(n+1)^2}{4} \cdot \frac{6}{n(n+1)(2n+1)} = \frac{h}{2} \cdot \frac{n+1}{2n+1}$$

∴ Required depth of C.P. of the rectangle $ABCD$ below AB

$$= h - \frac{h}{2}\frac{n+1}{2n+1} = h\left\{1 - \frac{n+1}{2(2n+1)}\right\}$$

$$= \frac{h}{2(2n+1)}\{2(2n+1) - (n+1)\} = \frac{3n+1}{2n+1} \cdot \frac{h}{2}$$

EXERCISE 5

1. Prove that the centre of pressure of a rhombus immersed in two liquids which do not mix, with a vertex in the upper surface and a diagonal in the common surface, divides the other diagonal in the ratio of 17 : 11 if the density of the lower liquid is twice that of the upper liquid.
   [**Hint:** Refer Ex. 2]
2. A rectangle is immersed in three liquids of densities ρ, 2ρ and 3ρ, the top of the rectangle being in the surface of the first liquid and the area immersed in each liquid being the same. Show that the depth of the centre of pressure of the rectangle is $5h/7$, where h is the depth of the lower side.
   [Hint: Refer Ex. 5, Put $n = 3$]

HARDER ILLUSTRATIVE EXAMPLES

Example 1: The angular points of a triangle immersed in a liquid whose density varies as the depth are at distances, α, β, γ respectively below the surface. Show that the centre of pressure is at a depth

$$\frac{3}{5}\frac{(\alpha+\beta+\gamma)(\alpha^2+\beta^2+\gamma^2)+\alpha\beta\gamma}{\alpha^2+\beta^2+\gamma^2+\alpha\beta+\beta\gamma+\gamma\beta}$$

Let ABC be a triangle immersed in a liquid whose density varies as the depth. Let the depths of the vertices A, B, C of the $\Delta\, ABC$ below the free surface be α, β, γ respectively.

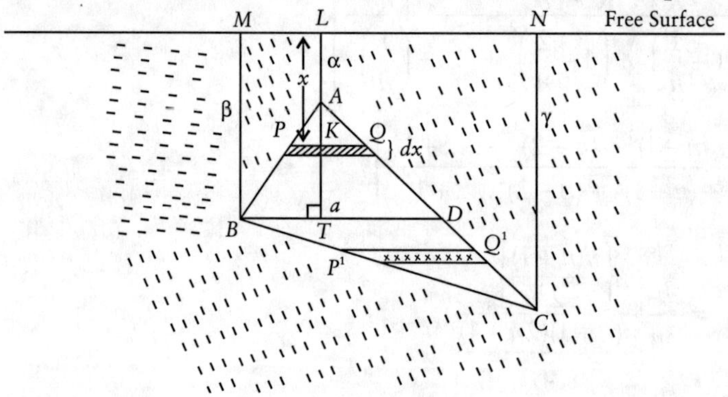

Draw a horizontal line BD. Then Δ ABC is divided into two triangles $\Delta\, ABD$ and $\Delta\, BCD$. Let the depth of the strip PQ of breadth dx below the free surface be x. Let $BD = a$.

$\therefore \quad \Delta\, APQ \sim \Delta\, ABD$

$\therefore \quad \dfrac{PQ}{BD} = \dfrac{AK}{AT}$ | by property of similar triangles

$\Rightarrow \quad \dfrac{PQ}{a} = \dfrac{x-\alpha}{\beta-\alpha}$

$\Rightarrow \quad PQ = \dfrac{a(x-\alpha)}{\beta-\alpha}$ ...(1)

$\therefore$ Whole pressure on $\Delta\, ABD$

$$(P_1) = \int_\alpha^\beta PQ.dx\ g(\lambda x)x \qquad \mid P = wS\bar{z}$$

$$= \int_\alpha^\beta \frac{a(x-\alpha)}{\beta-\alpha} g\lambda x^2\ dx \qquad = \frac{ga\lambda}{\beta-\alpha}\int_\alpha^\beta x^2(x-\alpha)\,dx$$

$$= \frac{ga\lambda}{\beta-\alpha}\left(\frac{x^4}{4} - \alpha\frac{x^3}{3}\right)_\alpha^\beta \qquad = \frac{ga\lambda}{\beta-\alpha}\left\{\frac{\beta^4-\alpha^4}{4} - \frac{\alpha(\beta^3-\alpha^3)}{3}\right\}$$

$$= \frac{ga\lambda}{12(\beta-\alpha)}\left[3(\beta^4-\alpha^4) - 4\alpha(\beta^3-\alpha^3)\right]$$

$$= \frac{ga\lambda}{12(\beta-\alpha)}(3\beta^4 - 4\alpha\beta^3 + \alpha^4) \qquad \text{...(2)}$$

Depth of C.P. of $\Delta\, ABD$ below the free surface (z_1)

$$= \frac{\int xpdS}{\int pdS} \qquad = \frac{\int_\alpha^\beta xg(\lambda x)x\,PQ\,dx}{\int_\alpha^\beta g(\lambda x)x\,PQ\,dx}$$

$$= \frac{\int_\alpha^\beta x^3(x-\alpha)dx}{\int_\alpha^\beta x^2(x-\alpha)dx} \qquad \mid \text{From (1)}$$

$$= \frac{\left(\frac{x^5}{5}\right)_\alpha^\beta - \alpha\left(\frac{x^4}{4}\right)_\alpha^\beta}{\left(\frac{x^4}{4}\right)_\alpha^\beta - \alpha\left(\frac{x^3}{3}\right)_\alpha^\beta} \qquad = \frac{\frac{\beta^5-\alpha^5}{5} - \alpha\frac{\beta^4-\alpha^4}{4}}{\frac{\beta^4-\alpha^4}{4} - \alpha\frac{\beta^3-\alpha^3}{3}}$$

$$= \frac{3}{5}\frac{4(\beta^5-\alpha^5) - 5\alpha(\beta^4-\alpha^4)}{3(\beta^4-\alpha^4) - 4\alpha(\beta^3-\alpha^3)}$$

$$= \frac{3}{5}\frac{4\beta^5 - 5\alpha\beta^4 + \alpha^5}{3\beta^4 - 4\beta^3 + \alpha^4} \qquad \text{...(3)}$$

From (2) and (3),

$$P_1 z_1 = \frac{ga\lambda(3\beta^4 - 4\alpha\beta^3 + \alpha^4)}{12(\beta-\alpha)}\frac{3}{5}\frac{4\beta^5 - 5\alpha\beta^4 + \alpha^5}{3\beta^4 - 4\alpha\beta^3 + \alpha^4}$$

$$= \frac{ga\lambda}{12(\beta-\alpha)}\frac{3}{5}(4\beta^5 - 5\alpha\beta^4 + \alpha^5) \qquad \text{...(4)}$$

Take a horizontal strip $P'Q'$ in $\Delta\, BCD$, whose breadth is dx and where depth from A is x. Then,

$$P'Q' = \frac{a(\gamma - x)}{\gamma - \beta} \qquad \text{...(5)}$$

$\therefore$ Whole pressure on $\Delta\, BCD$ (P_2)

$$= \int P'Q'\, dx\, g(\lambda x)x \qquad \mid P = wS\bar{z}$$

$$= \int_\beta^\gamma \frac{a(\gamma - x)}{\gamma - \beta} g\lambda x^2\, dx \qquad = \frac{ga\lambda}{\gamma - \beta}\int_\beta^\gamma (\gamma - x)x^2\, dx$$

$$= \frac{ga\lambda}{\gamma - \beta}\left(\gamma\frac{x^3}{3} - \frac{x^4}{4}\right)_\beta^\gamma = \frac{ga\lambda}{\gamma - \beta}\left[\frac{\gamma^4 - \gamma\beta^3}{3} - \frac{\gamma^4 - \beta^4}{4}\right]$$

$$= \frac{ga\lambda}{12(\gamma - \beta)}[4\gamma^4 - 4\gamma\beta^3 - 3\gamma^4 + 3\beta^4]$$

$$= \frac{ga\lambda}{12(\gamma - \beta)}[3\gamma^4 - 4\gamma\beta^3 + \gamma^4] \qquad \text{...(6)}$$

Depth of C.P. of $\Delta\, BCD$ below the free surface (z_2)

$$= \frac{\int xp\,dS}{\int p\,dS} \qquad = \frac{\int_\beta^\gamma x.g(\lambda x)xP'Q'\, dx}{\int_\beta^\gamma g(\lambda x)xP'Q'\, dx}$$

$$= \frac{\int_\beta^\gamma x^3(\gamma - x)dx}{\int_\beta^\gamma x^2(\gamma - x)dx} \qquad \mid \text{From (5)}$$

$$= \frac{\gamma\left(\frac{x^4}{4}\right)_\beta^\gamma - \left(\frac{x^5}{5}\right)_\beta^\gamma}{\gamma\left(\frac{x^3}{3}\right)_\beta^\gamma - \left(\frac{x^4}{4}\right)_\beta^\gamma} = \frac{\frac{\gamma}{4}(\gamma^4 - \beta^4) - \left(\frac{\gamma^5 - \beta^5}{5}\right)}{\gamma\left(\frac{\gamma^3 - \beta^3}{3}\right) - \left(\frac{\gamma^4 - \beta^4}{4}\right)}$$

$$= \frac{3}{5}\frac{5\gamma(\gamma^4 - \beta^4) - 4(\gamma^5 - \beta^5)}{4\gamma(\gamma^3 - \beta^3) - 3(\gamma^4 - \beta^4)} \qquad = \frac{3}{5}\frac{5\gamma^5 - 5\beta^4\gamma - 4\gamma^5 + 4\beta^5}{4\gamma^4 - 4\gamma\beta^3 - 3\gamma^4 + 3\beta^4}$$

$$= \frac{3}{5}\frac{4\beta^5 - 5\beta^4\gamma + \gamma^5}{3\beta^4 - 4\gamma\beta^3 + \gamma^4} \qquad \text{...(7)}$$

From (6) and (7)

$$P_2 z_2 = \frac{ga\lambda}{12(\gamma - \beta)}(3\beta^4 - 4\gamma\beta^3 + \gamma^4).\frac{3}{5}\frac{4\beta^5 - 5\beta^4\gamma + \gamma^5}{3\beta^4 - 4\gamma\beta^3 + \gamma^4}$$

$$= \frac{ga\lambda}{12(\gamma - \beta)}.\frac{3}{5}(4\beta^5 - 5\beta^4\gamma + \gamma^5) \qquad \text{...(8)}$$

$\therefore \quad \bar{z} = \dfrac{P_1 z_1 + P_2 z_2}{P_1 + P_2}$ where $\bar{z}$ is the depth of the C.P. of $\Delta\, ABC$ below the free surface

$$\frac{\frac{ga\lambda}{12(\beta-\alpha)}\frac{3}{5}(4\beta^5-5\alpha\beta^4+\alpha^5)+\frac{ga\lambda}{12(\gamma-\beta)}\frac{3}{5}(4\beta^5-5\beta^4\gamma+\gamma^5)}{\frac{ga\lambda}{12(\beta-\alpha)}(3\beta^4-4\alpha\beta^3+\alpha^4)+\frac{ga\lambda}{12(\gamma-\beta)}(3\beta^4-4\gamma\beta^3+\gamma^4)}$$

$$=\frac{3}{5}\cdot\frac{(4\beta^5-5\alpha\beta^4+\alpha^5)(\gamma-\beta)+(4\beta^5-5\gamma\beta^4+\gamma^5)(\beta-\alpha)}{(\gamma-\beta)(3\beta^4-4\alpha\beta^3+\alpha^4)+(\beta-\alpha)(3\beta^4-4\gamma\beta^3+\gamma^5)}$$

$$=\frac{3}{5}\cdot\frac{(\beta-\gamma)(\gamma-\alpha)(\alpha-\beta)\{(\alpha+\beta+\gamma)(\alpha^2+\beta^2+\gamma^2)+\alpha\beta\gamma\}}{(\alpha-\beta)(\beta-\gamma)(\gamma-\alpha)(\alpha^2+\beta^2+\gamma^2+\alpha\beta+\beta\gamma+\gamma\alpha)}$$

$$=\frac{3}{5}\cdot\frac{(\alpha+\beta+\gamma)(\alpha^2+\beta^2+\gamma^2)+\alpha\beta\gamma}{\alpha^2+\beta^2+\gamma^2+\alpha\beta+\beta\gamma+\gamma\alpha}$$

Example 2: A system of coaxial circles is immersed in water with the line of centres at a given depth. Prove that the centres of pressure of these circular areas which are completely immersed, lie on a parabola.

Sol. Take the line of centres as X-axis and the radical line as the Y-axis. Then, the equation of any circle of the system of coaxial circles can be written as

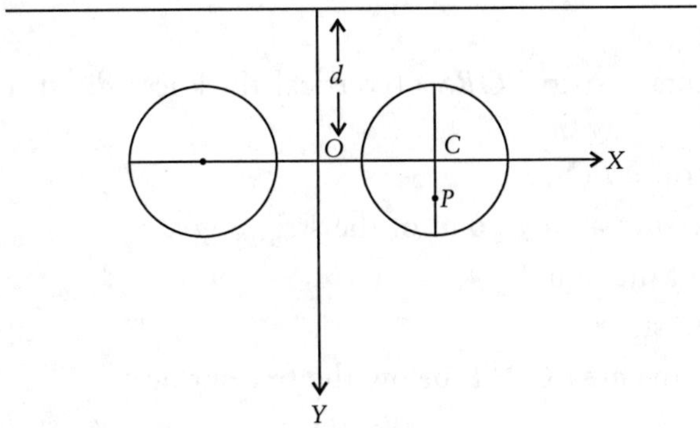

$$x^2 + y^2 - 2gx + c = 0 \qquad \text{...(1)}$$

where g is a parameter, i.e., it varies from circle to circle and c is a constant.

The centre of circle (1) is $(g, 0)$. Let r be the radius of circle (1). Then,

$$r^2 = g^2 - c \qquad \text{...(2)}$$

Let d be the depth of the line of centres below the free surface. Let the C.P. of this circle be $P(x, y)$. Then,

$x = OC = g$

and, $y = CP$ | $\because$ C.P. lies below C

$= a^2/4h$ $= r^2/4d$

$= g^2 - c/4d$ | From (2)

$= x^2 - c/4d$ | $\because g = x$

$$\Rightarrow 4dy = x^2 - c \qquad \text{...(3)}$$

which is the equation of locus of P. Clearly (3) represents a parabola.

Example 3: The asymptote of a hyperbola lies in the surface of a fluid. Find the depth of the centre of pressure of the area included between the immersed asymptote, the curve, and two given horizontal lines in the plane of the hyperbola.

Sol. The asymptote OA is in the free surface. The asymptote OB is in the liquid. Let CD and EF be two horizontal lines at depths a and b respectively. We are required to find out the C.P. of the area $CDFE$.

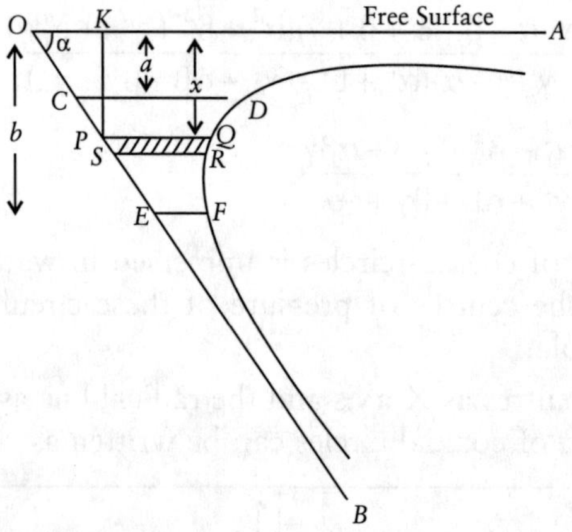

Consider an elementary strip $PQRS$ of vertical thickness dx at a depth x below the free surface. Then,

dS = Area of the strip = $PQ \,.\, dx$

p = Intensity of pressure at any point of the strip = $g\rho x$

where ρ is the density of the liquid.

$= g\rho PK = g\rho \,.\, OP \sin\alpha$

$\therefore$ Depth of C.P. of the area $CDFE$ below the free surface

$$= \frac{\int xpdS}{\int pdS} = \frac{\int_a^b xg\rho OP\sin\alpha\, PQ\,dx}{\int_a^b g\rho OP\sin\alpha\, PQ\,dx}$$

$$= \frac{\int_a^b xc\,dx}{\int_a^b c\,dx} \qquad \Big|\ \because OP \,.\, PQ \sin\alpha = \text{constant} = c \text{ (say) for a hyperbola}$$

$$= \frac{\left(\dfrac{x^2}{2}\right)_a^b}{(x)_a^b} = \frac{\left(\dfrac{b^2-a^2}{2}\right)}{b-a} = \frac{b+a}{2}$$

which is the arithmetic mean of the depths of the given horizontal lines.

Example 4: A semi-circular lamina is completely immersed in water with its plane vertical, so that the extremity A of its bounding diameter is in the surface and the diameter makes with the surface an angle α. Prove that, if E be the centre of pressure and ϕ the angle between AE and the diameter, then

$$\tan\phi = \frac{3\pi + 16\tan\alpha}{16 + 15\pi\tan\alpha}$$

Sol. Let ABC be a semi-circular lamina completely immersed in water with its plane vertical, so that the extremity A of its bounding diameter AC is in the surface and the diameter makes with the surface an angle α.

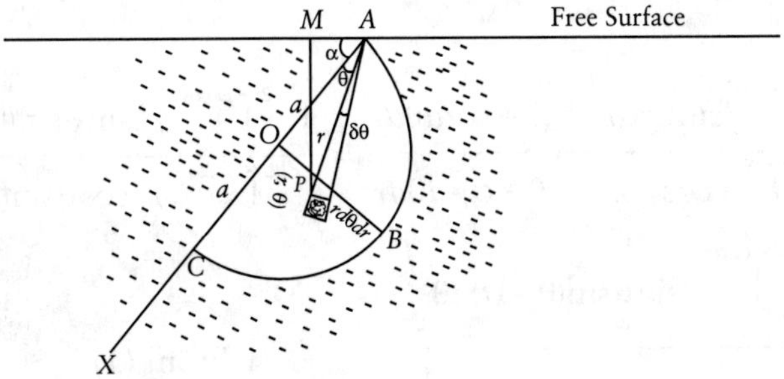

Let a be the radius of the semi-circle.

Take A as pole and AC as initial line.

Then, referred to these, the polar equation of the semi-circle is

$$r = 2a\cos\theta \qquad \text{...(1)}$$

Take any point ρ (r, θ) in the lamina.

Surrounding the point P, consider an elementary area $r\,d\theta\,dr$. From P, draw perpendicular PM to the free surface.

Depth of P below the free surface

$= MP = AP\sin(\theta + \alpha)$

$= r\sin(\theta + \alpha)$

$\therefore$ p = Intensity of pressure at P

$= g\rho r\sin(\theta + \alpha)$, where ρ is the density of water.

$dS = rd\theta dr$

Let $(\overline{x}, \overline{y})$ be the coordinates of E, the C.P. of the lamina. Then,

$$\overline{x} = \frac{\int xpdS}{\int pdS}$$

$$= \frac{\int_{\theta=0}^{\pi/2}\int_{r=0}^{2a\cos\theta} r\cos\theta g\rho r\sin(\theta+\alpha)rd\theta dr}{\int_{\theta=0}^{\pi/2}\int_{r=0}^{2a\cos\theta} g\rho r\sin(\theta+\alpha)rd\theta dr} \qquad \text{...(2)}$$

and, $\bar{y} = \dfrac{\int ypdS}{\int pdS}$

$$= \frac{\int_{\theta=0}^{\pi/2}\int_{r=0}^{2a\cos\theta} r\sin\theta g\rho r\sin(\theta+\alpha) rd\theta dr}{\int_{\theta=0}^{\pi/2}\int_{r=0}^{2a\cos\theta} g\rho r\sin(\theta+\alpha) rd\theta dr} \quad ...(3)$$

If ϕ is the angle between AE and the diameter AC, then,

$$\tan\phi = \frac{\bar{y}}{\bar{x}}$$

$$= \frac{\int_{0}^{\pi/2}\int_{0}^{2a\cos\theta} r\sin\theta g\rho r\sin(\theta+\alpha) rd\theta dr}{\int_{0}^{\pi/2}\int_{0}^{2a\cos\theta} r\cos g\rho r\sin(\theta+\alpha) rd\theta dr} = \frac{\int_{0}^{\pi/2}\int_{0}^{2a\cos\theta} r^3\sin\theta\sin(\theta+\alpha)d\theta dr}{\int_{0}^{\pi/2}\int_{0}^{2a\cos\theta} r^3\cos\theta\sin(\theta+\alpha)d\theta dr}$$

$$= \frac{\int_{0}^{\pi/2}\left(\frac{r^4}{4}\right)_0^{2a\cos\theta}\sin\theta\sin(\theta+\alpha)d\theta}{\int_{0}^{\pi/2}\left(\frac{r^4}{4}\right)_0^{2a\cos\theta}\cos\theta\sin(\theta+\alpha)d\theta} \quad | \text{ From (1)}$$

$$= \frac{\int_{0}^{\pi/2}\cos^4\theta\sin\theta\sin(\theta+\alpha)d\theta}{\int_{0}^{\pi/2}\cos^5\theta\sin(\theta+\alpha)d\theta} = \frac{\int_{0}^{\pi/2}\cos^4\theta\sin\theta(\sin\theta\cos\alpha+\cos\theta\sin\alpha)d\theta}{\int_{0}^{\pi/2}\cos^5\theta(\sin\theta\cos\theta+\cos\theta\sin\alpha)d\theta}$$

$$= \frac{\cos\alpha\int_{0}^{\pi/2}\sin^2\theta\cos^4\theta d\theta+\sin\alpha\int_{0}^{\pi/2}\sin\theta\cos^5\theta d\theta}{\cos\alpha\int_{0}^{\pi/2}\sin\theta\cos^5\theta d\theta+\sin\alpha\int_{0}^{\pi/2}\cos^6\theta d\theta}$$

$$= \frac{\cos\alpha\dfrac{\Gamma\frac{3}{2}\Gamma\frac{5}{2}}{2\,\Gamma 4}+\sin\alpha\dfrac{\Gamma 1\,\Gamma 3}{2\,\Gamma 4}}{\cos\alpha\dfrac{\Gamma 1\,\Gamma 3}{2\,\Gamma 4}+\sin\alpha\left(\dfrac{5}{6}\cdot\dfrac{3}{4}\cdot\dfrac{1}{2}\cdot\dfrac{\pi}{2}\right)} = \frac{\cos\alpha\dfrac{\frac{1}{2}\sqrt{\pi}.\frac{3}{2}.\frac{1}{2}.\sqrt{\pi}}{2.3.2.1}+\sin\alpha.\dfrac{1.2}{2.3.2.1}}{\cos\alpha\dfrac{1.2}{2.3.2.1}+\dfrac{5}{32}\pi\sin\alpha}$$

$$= \frac{\cos\alpha\dfrac{\pi}{32}+\dfrac{\sin\alpha}{6}}{\dfrac{\cos\alpha}{6}+\dfrac{5\pi\sin\alpha}{32}} = \frac{\dfrac{3\pi\cos\alpha+16\sin\alpha}{96}}{\dfrac{16\cos\alpha+15\pi\sin\alpha}{96}}$$

$$= \frac{3\pi\cos\alpha+16\sin\alpha}{16+15\pi\sin\alpha} = \frac{3\pi+16\tan\alpha}{16+15\pi\tan\alpha}$$

Example 5: Show that if a lamina totally immersed in a liquid is a quadrant of a circle of radius a, of which the centre is in the surface, the locus of the C.P. lies in a straight line.

$x + y = 3/16a\ (\pi + 2)$.

Sol. Let OAB be a lamina in the form of a quadrant of a circle of radius a, which is totally immersed in a liquid, with centre O in the free surface. Let OA make an angle α with the free surface. Let the radius of the circle be a.

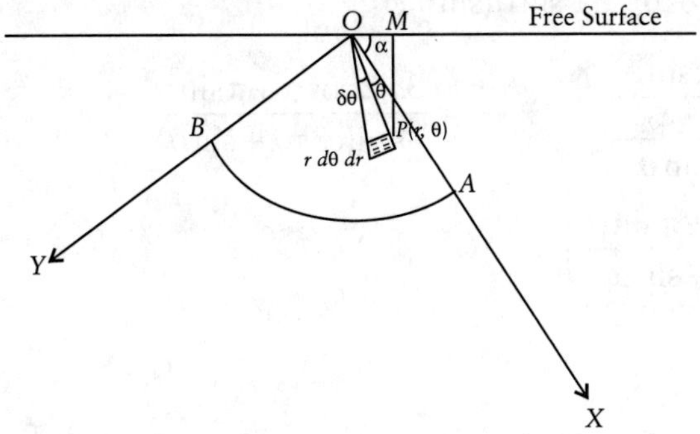

Take O as pole and OA as initial line. Then, referred to these, the polar equation of the circle is

$$r = a \qquad \text{...(1)}$$

Take any point $P(r, \theta)$ in the lamina.

Surrounding the point P, consider an elementary area $r\, d\theta\, dr$. From P, draw perpendicular to the free surface.

Then, depth of P below the free surface

$= PM$

$= OP \sin(\theta + \alpha)$

$= r \sin(\theta + \alpha)$

$\therefore$ p = Intensity of pressure at $P = g\alpha r \sin(\theta + \alpha)$ where ρ is the density of the liquid.

$dS = r d\theta dr$

Let $(\bar{x}, \bar{y})$ be the coordinates of the C.P. of the quadrant OAB. Then,

$$\bar{x} = \frac{\int xp dS}{\int p dS}$$

$$= \frac{\int_{\theta=0}^{\pi/2}\int_{r=0}^{a} r\sin\theta\, g\rho r\sin(\theta+\alpha) r\, d\theta\, dr}{\int_{\theta=0}^{\pi/2}\int_{r=0}^{a} g\rho r\sin(\theta+\alpha) r\, d\theta\, dr} = \frac{\int_0^{\pi/2}\int_0^a r^3\cos\theta\sin(\theta+\alpha)\, d\theta\, dr}{\int_0^{\pi/2}\int_0^a r^2\sin(\theta+\alpha)\, d\theta\, dr}$$

$$= \frac{\int_0^{\pi/2}\left(\frac{r^4}{4}\right)_0^a \cos\theta\sin(\theta+\alpha)\, d\theta}{\int_0^{\pi/2}\left(\frac{r^3}{3}\right)_0^a \sin(\theta+\alpha)\, d\theta} = \frac{\frac{3a}{4}\int_0^{\pi/2}\cos\theta(\sin\theta\cos\alpha+\cos\theta\sin\alpha)\, d\theta}{\int_0^{\pi/2}(\sin\theta\cos\alpha+\cos\theta\sin\alpha)\, d\theta}$$

$$= \frac{3a}{4}\left[\frac{\cos\alpha\int_0^{\pi/2}\sin\theta\cos\theta d\theta + \sin\alpha\int_0^{\pi/2}\cos^2\theta\, d\theta}{\cos\alpha\int_0^{\pi/2}\sin\theta d\theta + \sin\int_0^{\pi/2}\cos\theta\, d\theta}\right]$$

$$= \frac{3a}{4} \frac{\cos\alpha.\left(\frac{\sin^2\theta}{2}\right)_0^{\pi/2} + \sin\alpha.\frac{1}{2}.\frac{\pi}{2}}{\cos\alpha.(-\cos\theta)_0^{\pi/2} + \sin\alpha(\sin\theta)_0^{\pi/2}}$$

$$= \frac{3a}{4} \frac{\frac{\cos\alpha}{2} + \frac{\pi\sin\alpha}{4}}{\cos\alpha + \sin\alpha} \qquad = \frac{3a}{4} \frac{2\cos\alpha + \pi\sin\alpha}{4(\cos\alpha + \sin\alpha)}$$

$$= \frac{3a}{16}\left(\frac{2\cos\alpha + \pi\sin\alpha}{\cos\alpha + \sin\alpha}\right) \qquad ...(2)$$

and, $\bar{y} = \dfrac{\int y p dS}{\int p dS}$

$$= \frac{\int_{\theta=0}^{\pi/2}\int_{r=0}^{a} r\sin\theta\, g\rho r\sin(\theta+\alpha) r\, d\theta\, dr}{\int_{\theta=0}^{\pi/2}\int_{r=0}^{a} g\rho r\sin(\theta+\alpha) r\, d\theta\, dr} = \frac{\int_0^{\pi/2}\int_0^a r^3 \sin\theta\sin\theta\sin(\theta+\alpha) d\theta\, dr}{\int_0^{\pi/2}\int_0^a r^2 \sin(\theta+\alpha) d\theta\, dr}$$

$$= \frac{\int_0^{\pi/2}\left(\frac{r^4}{4}\right)_0^a \sin\theta\sin(\theta+\alpha) d\theta}{\int_0^{\pi/2}\left(\frac{r^3}{3}\right)_0^a \sin(\theta+\alpha) d\theta} = \frac{\frac{3a}{4}\int_0^{\pi/2} \sin\theta(\sin\theta\cos\alpha + \cos\theta\sin\alpha) d\theta}{\int_0^{\pi/2} (\sin\theta\cos\alpha + \cos\theta\sin\alpha) d\theta}$$

$$= \frac{3a}{4}\left[\frac{\cos\alpha\int_0^{\pi/2} \sin^2 d\theta + \sin\theta\int_0^{\pi/2} \sin\theta\cos\theta d\theta}{\cos\alpha\int_0^{\pi/2}\sin\theta d\theta + \sin\alpha\int_0^{\pi/2}\cos\theta d\theta}\right]$$

$$= \frac{3a}{4}\left[\frac{\cos\alpha.\frac{1}{2}.\frac{\pi}{2} + \sin\alpha\left(\frac{\sin^2\theta}{2}\right)_0^{\pi/2}}{\cos\alpha(-\cos\theta)_0^{\pi/2} + \sin\theta(\sin\theta)_0^{\pi/2}}\right]$$

$$= \frac{3a}{4} \frac{\frac{\pi}{4}\cos\alpha + \frac{\sin\alpha}{2}}{\cos\alpha + \sin\alpha}$$

$$= \frac{3a}{16}\left(\frac{\pi\cos\alpha + 2\sin\alpha}{\cos\alpha + \sin\alpha}\right) \qquad ...(3)$$

$$\therefore \quad \bar{x} + \bar{y} = \frac{3a}{16}\left[\frac{2\cos\alpha + \pi\sin\alpha}{\cos\alpha + \sin\alpha} + \frac{2\sin\alpha + \pi\cos\alpha}{\cos\alpha + \sin\alpha}\right]$$

$$= \frac{3a}{16} \frac{2(\cos\alpha + \sin\alpha) + \pi(\sin\alpha + \cos\alpha)}{\cos\alpha + \sin\alpha} = \frac{3a}{16}(\pi + 2)$$

EXERCISE 6

1. Using the concept of super-incumbent liquid as explained in §4, prove that the centre of pressure of a rectangular plane area lies on the middle line of the rectangle at a distance down equal to two-thirds of it.
2. A cube, with edges of length $2a$, is immersed in a liquid and has one edge in the surface and two faces through that edge equally inclined to the horizontal. Find the centres of pressure of all the faces.

   Ans. Vertical face at depth $7a/2\sqrt{2}$; upper slant faces 1/3 a from centres of faces lower face 1/9 a from centre.
3. Given that the centre of pressure of a circular disc of radius r with one point in the surface, is at a distance p from the centre. Show that for a disc of radius p wholly immersed with its centre at a distance h from the surface, the distance between the centre of the circle and the centre of pressure is

   $$pR^2/hr$$
4. A circular sector, centre O and radius a, symmetrical about the radius OP, is completely immersed with P in the surface and OP vertical. Determine the depth of its centre of pressure.

   Ans. $\dfrac{a(15\alpha + 3\sin\alpha\cos\alpha - 16\sin\alpha)}{4(3\alpha - 2\sin\alpha)}$, where 2α is the angle of the sector.

   [**Hint:**

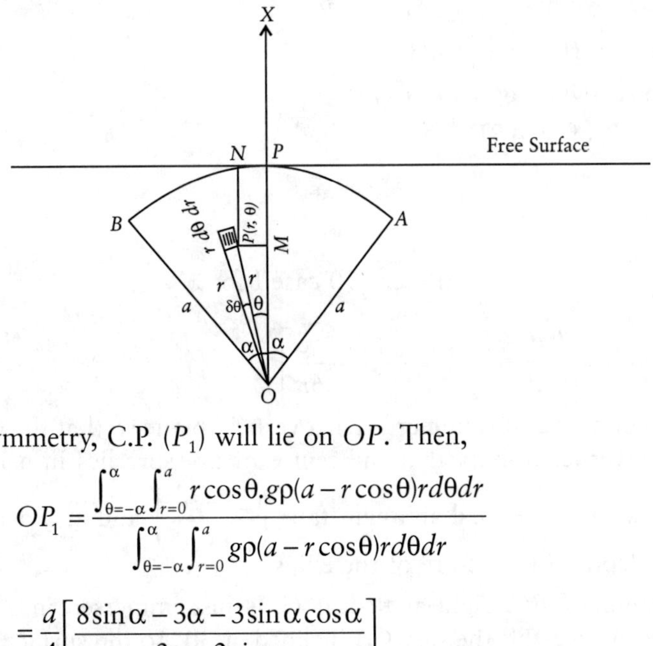

By symmetry, C.P. (P_1) will lie on OP. Then,

$$OP_1 = \frac{\int_{\theta=-\alpha}^{\alpha}\int_{r=0}^{a} r\cos\theta . g\rho(a - r\cos\theta) r d\theta dr}{\int_{\theta=-\alpha}^{\alpha}\int_{r=0}^{a} g\rho(a - r\cos\theta) r d\theta dr}$$

$$= \frac{a}{4}\left[\frac{8\sin\alpha - 3\alpha - 3\sin\alpha\cos\alpha}{3\alpha - 2\sin\alpha}\right]$$

$$\therefore PP_1 = OP - OP_1$$

$$= a - \frac{a}{4}\left[\frac{8\sin\alpha - 3\alpha - 3\sin\alpha\cos\alpha}{3\alpha - 2\sin\alpha}\right]$$

$$= \frac{[15\alpha - 3\sin\alpha\cos\alpha - 16\sin\alpha]a}{4(3\alpha - 2\sin\alpha)}\Bigg]$$

5. A plane quadrilateral $ABCD$ in entirely immersed in water with the side AB in the surface. If the depths of C and D below the surface are γ and δ respectively, and that of the C.G. is h, prove that the depth of the centre of pressure is

$$\frac{\gamma+\delta}{2} - \frac{\gamma\delta}{6h}$$

[Hint:

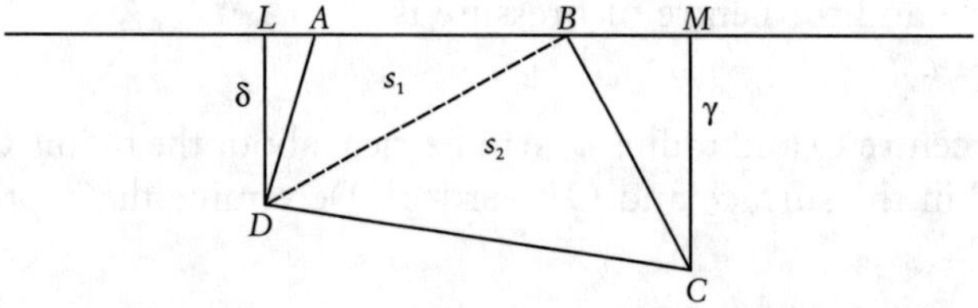

$$h = \frac{s_1 . \frac{1}{3}\delta + s_2 . \left(\frac{\gamma+\delta}{3}\right)}{s_1 + s_2} \qquad \Rightarrow \frac{s_1}{s_2} = \frac{\gamma + \delta - 3h}{3h - \delta} \qquad ...(1)$$

P_1 = Thrust on $\Delta\, ABD = g\rho s_1 \,.\, \delta/3$

P_2 = Thrust on $\Delta\, BDC = g\rho s_2 \,.\, (\gamma + \delta/3)$

p_1 = Depth of C.P. of $\Delta\, ABD$

$= \delta/2$

p_2 = Depth of C.P. of $\Delta\, BDC$

$$= \frac{\delta^2 + \gamma^2 + \gamma\delta}{2(\gamma+\delta)} \qquad \text{| Refer §10 case I for } \bar{z}$$

$$\therefore \text{Required depth } = \frac{P_1 p_1 + P_2 p_2}{P_1 + P_2} = \frac{\gamma+\delta}{2} - \frac{\gamma\delta}{6h}\Bigg]$$

6. An ellipse of semi-axes a, b is immersed in a vertical plane so that the axes make equal angles with the vertical. Show that the centre of pressure lies in a line through the centre making with the vertical an angle $\tan^{-1}\left(\frac{a^2 - b^2}{a^2 + b^2}\right)$ and at a depth $h + \frac{a^2 + b^2}{8h}$, where h is the depth of the centre of the ellipse.

7. A triangular lamina ABC, right-angled at C, is just immersed in a liquid with the vertex C in the surface and the side CA inclined at 30° to the surface. If the centre of pressure is vertically below C, prove that the angle B is $\frac{1}{2}\tan^{-1}(2\sqrt{3})$.

Chapter 7

Thrust on Curved Surfaces

1. Thrust on a Curved Surface

Let a curved surface S be in contact with a fluid at rest under gravity. Take a point P on the surface and consider a very small elementary area surrounding this point P. Since the fluid is perfect, therefore, the thrust of the fluid on this elementary area will act along to the normal to it. The same will hold good for other elementary areas too.

Thus, these fluid thrusts form a system of forces which are neither in the same direction, nor generally in the same plane. Therefore, such a system can be reduced, in general, to a single force together with a single couple, and not necessarily to a single force. But this totality of pressure is infact equivalent to certain forces in the horizontal and vertical directions and can be obtained in the following manner.

2. Resultant Vertical Thrust

Let the fluid thrust on each element by first resolved into vertical and horizontal components. Evidently all the vertical components will form a system of parallel forces which can be compounded into a single resultant vertical thrust, say Z. This resultant Z is called the *resultant vertical thrust* on the curved surface.

3. Resultant Horizontal Thrust

Let us now consider the horizontal components of the fluid thrusts on various elements. These are not necessarily parallel to each other [Note that horizontal lines can be drawn in infinite directions]. Therefore, we conveniently choose two horizontal directions OX and OY perpendicular to each other and resolve the horizontal components of the fluid thrusts on every element along these directions OX and OY. Then, the components resolved parallel to OX will form a system of parallel forces which can be compounded into a single resultant force, say X, acting parallel to OX. Similarly, the components resolved parallel to OY will also form a system of parallel forces which can be compounded into a single resultant force, say Y, acting parallel to OY. These resultant forces X and Y which can be thus determined in magnitude and direction are called the resultant horizontal thrusts parallel to OX and OY respectively.

Thus, the three resultants X, Y and Z, taken together, give the resultant thrusts of the fluid thrusts on the given curved surface. In general, these resultant thrusts act along non-coplanar lines.

Note 1: If in any particular case, these three resultant fluid thrusts, X, Y and Z are concurrent, i.e., pass through the same point, then they can be compounded into a single resultant force R of magnitude $\sqrt{X^2 + Y^2 + Z^2}$, which is then called the *resultant fluid thrust* on the curved surface.

Note 2: If only two resultant thrusts X and Y are concurrent, then they can be reduced to a single resultant force H of magnitude $\sqrt{X^2 + Y^2}$, which is then called the *resultant horizontal thrust* on the curved surface.

4. Method of Finding Resultant Vertical Thrust

Let a curved surface $ABCD$ be immersed in a heavy homogeneous fluid. From each and every point of the boundary of the surface $ABCD$, draw vertical straight lines to meet the effective surface (or the free surface of the fluid if the atmospheric pressure is not taken into account) in a closed curve $A'B'C'D'$.

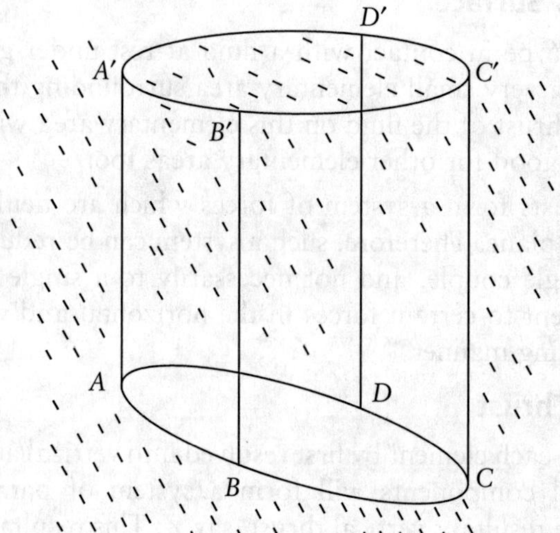

These vertical lines enclose a fluid cylinder which rests on the given curved surface. The fluid is this cylinder upto the effective surface is called the *super-incumbent fluid*. The super-incumbent fluid is in equilibrium under the action of the following vertical forces:

(i) the weight of the super-incumbent fluid, acting vertically downwards through its centre of gravity.

(ii) the vertical component of the reaction of the surface $ABCD$ upon it acting upwards. But, for equilibrium, the reaction of the surface is equal and opposite to the thrust of the fluid upon it. Therefore, the vertical component of the resultant thrust of the fluid upon the surface is equal to weight of the super-incumbent fluid.

Hence, the resultant vertical thrust on any curved surface immersed in a fluid is equal to the weight of the super-incumbent fluid and acts through the centre of gravity of this super-incumbent fluid.

Particular Cases

Case I: When the fluid presses upwards: If the fluid lies below the given surface at every point of it, then the fluid presses the surface upwards. In this case, the components of the thrusts over various elements are upwards and therefore the resultant vertical thrust is also upwards.

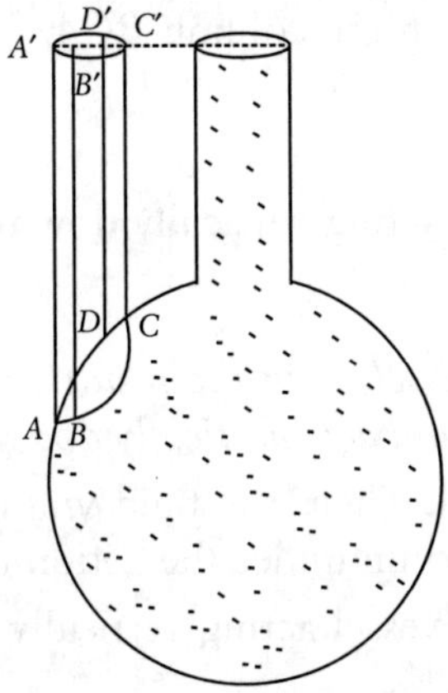

Let $ABCD$ be such a surface as shown in the figure. We draw vertical straight lines through each and every point of the boundary of the surface $ABCD$ to meet the horizontal plane through the surface of zero pressure in the closed curve $A'B'C'D'$.

The cylinder enclosed above contains no fluid. Since the pressure at each point of the surface $ABCD$ is proportional to the depth of the point below the surface of zero pressure of the liquid, therefore the vertical component of the upward thrust of the fluid on the surface $ABCD$ is equal to the weight of the super-incumbent fluid which would fill the cylinder $ABCD$, $A'B'C'D'$.

Hence, the resultant vertical thrust on the given portion of the surface is equal to the weight of the fluid that could lie upon it upto the level of surface of the liquid and acts vertically upwards through the centre of gravity of the liquid.

Case II: When the fluid presses partly upwards and partly downwards: Let ABC be the surface such that area BC is pressed downwards and area AB is pressed upwards.

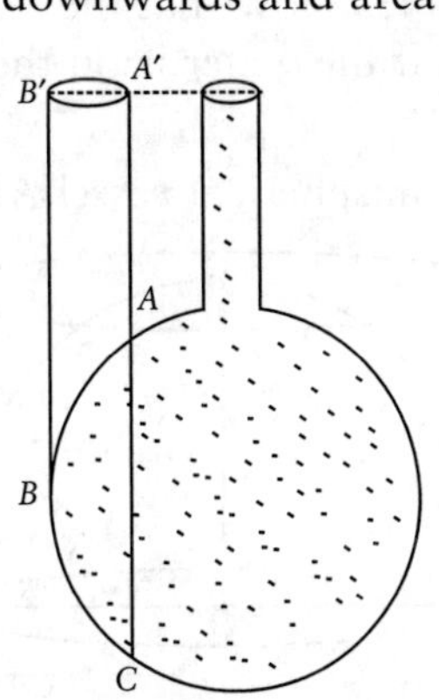

By case I, the resultant vertical thrust on the area AB is upwards and equal to the weight of the fluid that would fill the column $ABB'A'$ vertically above it. Again, the resultant vertical thrust on the area BC is downwards and is equal to the weight of the fluid that would fill the column $BCA\ A'B'$ vertically above it.

Hence, the vertical thrust acting downwards on the surface ABC

= vertical thrust on BC – vertical thrust on AB

= weight of the fluid contained in the column $BCAA'B'$ – weight of the fluid contained in the column $ABB'A'$

= weight of the fluid in CBA.

= weight of the contained fluid acting vertically downwards

5. Fluid Contained in a Vessel

If the fluid be contained in a vessel, then the resultant fluid thrust on the vessel is equal to the weight of the fluid in the vessel and acts vertically downwards through the C.G. of this fluid.

Consider a vessel full of a fluid. Then, the fluid will exert pressure on its sides. The fluid contained in the vessel is in equilibrium under the action of two forces acting on it:

(i) Weight of the fluid in the vessel acting vertically downwards through the C.G. of the fluid.

and (ii) The reaction of the vessel on the fluid.

Hence, for equilibrium, these two must be equal and opposite. | by statics.

But by Newton's Third Law, the reaction of the vessel upon the fluid is equal and opposite to the thrust of the vessel upon the fluid. Hence, the fluid thrust on the sides of the vessel is equal to the weight of the fluid contained in it and acts vertically downwards through the C.G. of the fluid.

Note: The direction of the resultant fluid thrust on the sides of vessel is the same as the direction of the weight of the fluid contained in the vessel, i.e., it is vertically downwards.

ILLUSTRATIVE EXAMPLES

Example 1: A hemispherical vessel is filled with water and inverted and placed with its plane base in contact with a horizontal table. Show that the resultant vertical thrust on its surface is one-third of the thrust on the table. If the water is on the point of escaping, find the weight of the vessel.

Sol. Let r be the radius of the hemispherical vessel ABC.

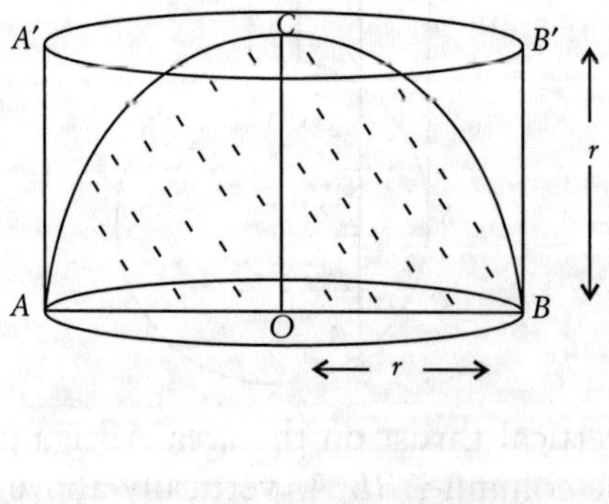

Thrust on the table = weight of the super-incumbent liquid

= weight of the water in the cylinder $ABB'A'$ raised on the base upto the height of the free surface

$= \pi r^2 . r. g\rho$

where ρ is the density of water

$= \pi g \rho r^3$ ...(1)

Resultant vertical thrust on the curved surface of the hemisphere

= weight of the super-incumbent liquid

= weight of the water in the cylinder $ABB'A'$ raised on the curved surface upto the height of the free surface

= weight of water in the cylinder $ABB'A'$ – weight of water contained in the hemisphere ACB

$= \pi r^2 . r. g\rho - 2/3\ \pi r^3 g\rho$

$= 1/3\ \pi r^3 g\rho$

= 1/3 thrust on the table.

If the water is on the point of escaping, then

weight of the vessel = upward vertical thrust on the vessel

$= 1/3\ \pi r^3 g\rho$

$= 1/2\ (2/3\ \pi r^3 . g\rho)$

= 1/2 weight of the contained water

Example 2: A conical wine glass is filled with water and placed in an inverted position upon a table. Show that the resultant vertical thrust of the water on the glass is two-thirds that on the table.

Sol. Let r and h be respectively the radius and height of the conical glass. Then, resultant vertical thrust on the table

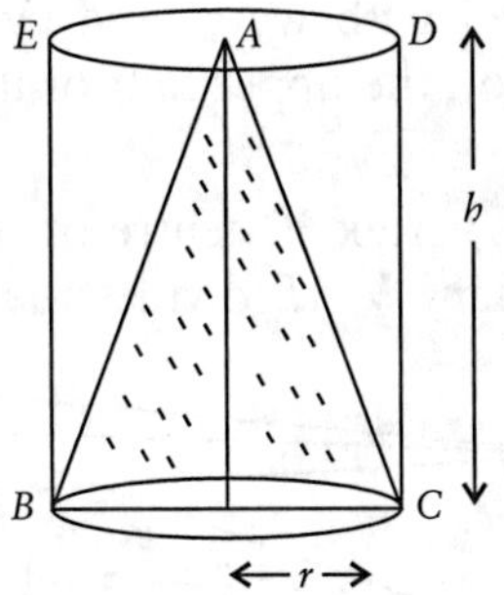

= weight of the super-incumbent liquid

= (volume of the cylinder $BCDE$) $g\rho$

$= \pi r^3 h g\rho$ ...(1)

$= 1/2\ (2/3\ \pi r^3 . g\rho)$

where ρ is the density of water

Resultant vertical thrust of water on the curved surface of glass

= weight of the super-incumbent water

= weight of the water that fills the space between the curve ABC and the circumscribing cylinder $BCDE$.

= $(\pi r^3 h - 1/3 \pi r^2 h)\, g\rho$

= $2/3\ \pi r^2 h g\rho$

= 2/3 (the resultant vertical thrust on the table)

Example 3: Find the resultant vertical fluid thrust on the lower half of the curved surface of a cylindrical pipe of length h and radius r when the pipe is full.

Sol. Let $ABCD$ be a horizontal plane through the axis of the cylinder which divides the cylinder into two halves. Then, downward vertical thrust on the lower half of the cylinder.

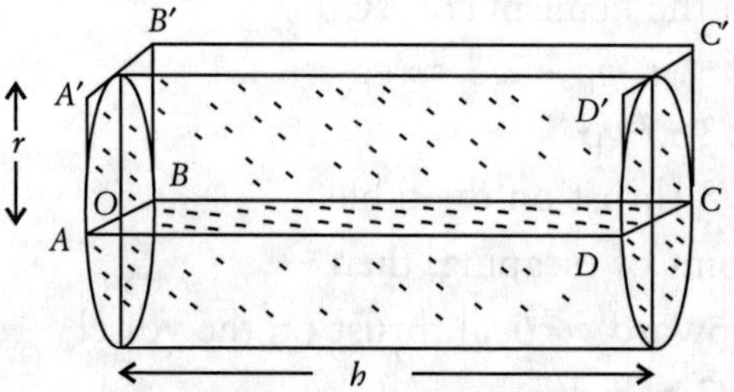

= weight of the super-incumbent fluid which will fill the lower half of the cylinder and the prism whose base is $ABCD$ and whose height is $AA' = r$.

= [volume of the lower half of the cylinder + volume of the prism whose base is $ABCD$ and height is r] $g\rho$, where ρ is the density of the fluid.

= $[1/2\pi\, r^2 h$ + Area of the rectangle $ABCD \times AA'$] $g\rho$

= $[1/2\pi r^2 h + 2rh.r]\, g\rho$

= $(1/2\pi + 2)\, r^2 h g\rho$

Example 4: A hollow cone filled with water and closed, is held with its axis horizontal. Find the resultant vertical pressure on the upper half of its curved surface. Find it on the lower half as well.

Sol. Let (V, AB) be a cone with vertex V, centre of base O, radius a and height h. Its axis VO is horizontal. The horizontal plane $V AB$ divides the cone into two halves.

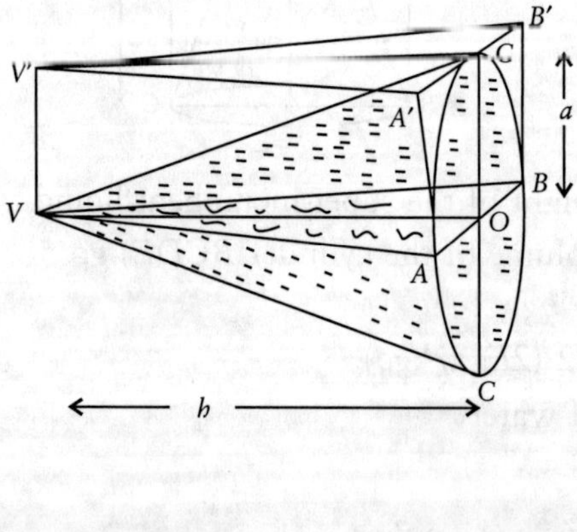

Resultant vertical thrust on the upper half (V, $C'BA$) of the curved surface of the cone.

= weight of the super-incumbent water

= weight of water contained in the prism (V, AOB, V', $A'C'B'$) – weight of water contained in the upper half cone (V, $C'BA$)

= Area of Δ $VAB \times OC' \times g\rho - 1/2\ (1/3\ \pi\ r^2\ h)\ g\rho$, where ρ is the density of water.

= $1/2.2r\ .\ h.rg\rho - 1/6\ \pi r^2 hg\rho$

= $r^2hg\rho - (1 - \pi/6)$

Resultant vertical thrust on the lower half (V, CAB) of the cone

= weight of the super-incumbent water

= weight of water contained in the prism (V, AOB, V', $A'C'B'$) + weight of water contained in the lower half (V, CBA) of the cone

= Area of Δ $VAB \times OC' \times g\rho + 1/2\ (1/3\ \pi r^2 h)\ g\rho$

= $1/2.2r\ .\ h.\ rg\rho + 1/6\ \pi r^2 hg\rho$

= $r^2hg\rho\ (1 + \pi/6)$

Example 5: A bucket in the form of a frustum of a cone, the radii of its top and bottom being a and b ($b < a$), is of height h and is full of water. Find the resultant vertical thrust on its curved surface.

Sol. Vertical thrust on the curved surface of the bucket

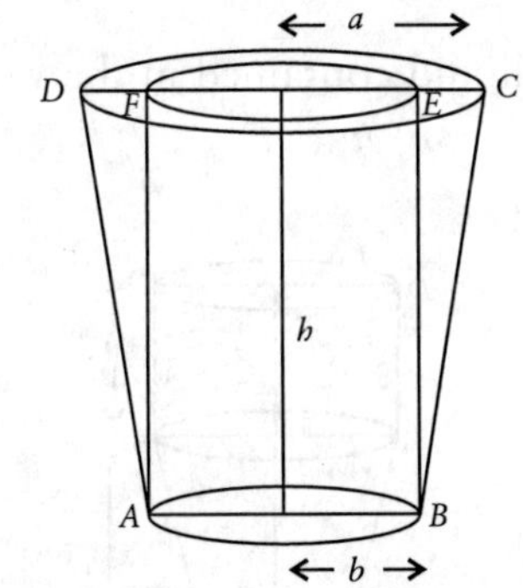

= weight of the super-incumbent liquid

= weight of the liquid that will fill the space between the frustum $ABCD$ and the cylinder $ABEF$

= (Volume of the frustum $ABCD$) $g\rho$ – (volume of the cylinder $ABEF$) $g\rho$, where ρ is the density of the liquid

= $1/3\pi h\ (a^2 + b^2 + ab)\ g\rho - \pi b^2 hg\rho$

= $1/3\pi h\ (a^2 + ab - 2b^2)\ g\rho$

Example 6: A hollow closed vessel in the shape of a cylinder surmounted by a cone is filled with liquid. If the axis of the cone be three times as long as that of the cylinder, prove that the resultant thrust on the surface of the cone will be the same in the two positions in which the vessel can be placed with its axis vertical.

Case I: When the vessel in erect

Resultant vertical thrust

$=$ weight of the super-incumbent liquid

$=$ (volume $ABB'A'$) w – (volume of cone OAB) w

$= (\pi a^2 3hw - 1/3\ \pi a^2 \,.\, 3hw)$

$= 2\pi a^2 hw$ ...(1)

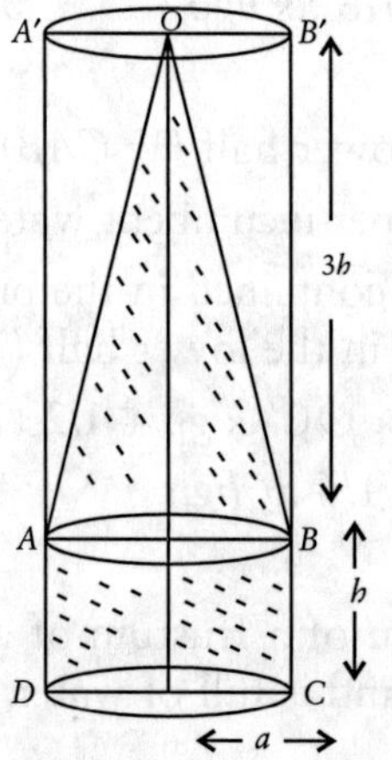

Case II: When the vessel in inverted

Resultant vertical thrust

$=$ weight of the liquid contained in the vessel

$= (\pi a^2 h + 1/3\ \pi a^2 \,.\, 3h)\ w$

$= 2\pi a^2 hw$ ...(2)

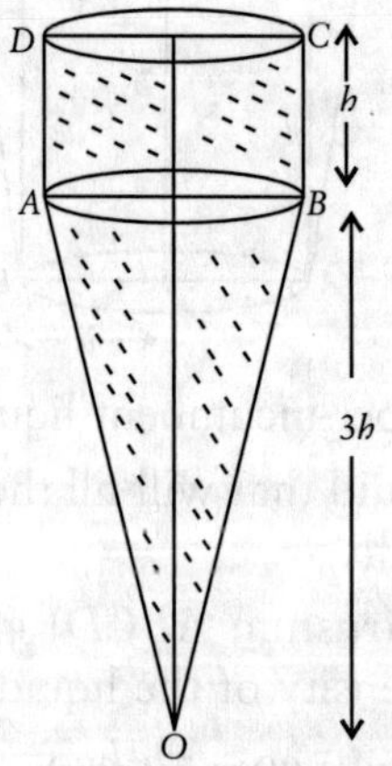

Hence, in both the cases the resultant vertical thrust is the same.

Example 7: A hollow cone is placed with its vertex upwards on a horizontal table and a liquid is poured in through a small hole in the vertex. If the cone begins to rise when the weight of the liquid poured in is equal to its own weight, prove that its weight to the weight of the liquid required to fill the cone is $9 - 3\sqrt{3}\ : 4$.

Sol. Let VAB be the cone of height h and semi-vertical angle α. Let the cone begin to rise, when the level of the liquid in the cone is CD such that CD is at a depth x below V.

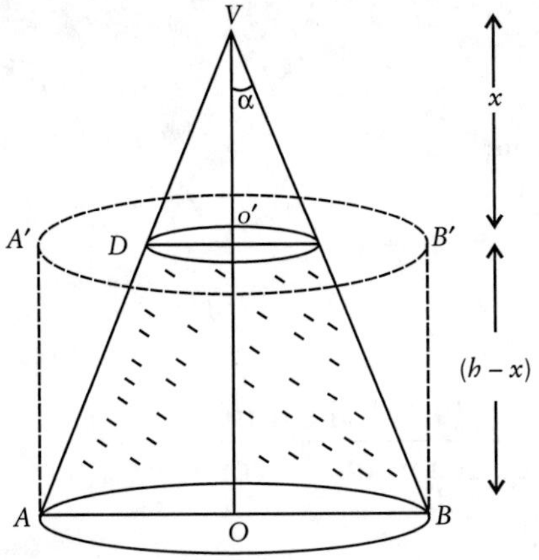

weight of the cone

$$= (1/3\pi\ OB^2\ OV - 1/3\ \pi\ O'C^2 .\ O'V)\ w$$

where w is weight per unit volume of the liquid

$$= (1/3\pi h^2 \tan^2 \alpha .\ h - 1/3\ \pi x^2 \tan^2 \alpha .\ x)\ w$$

$$= 1/3\pi \tan^2 \alpha w\ (h^3 - x^3) \qquad ...(1)$$

Since the cone begins to rise when the height of the liquid in the cone is $(h - x)$, therefore, the weight of the cone must be equal to the resultant vertical thrust on the curved surface. So, weight of the cone

= weight of the super-incumbent liquid

– weight of the liquid that will fill the space between the cylinder $ABB'A'$ and the frustum $ABCD$

$$= \pi\ OB^2 .OO' .\ w - 1/3\ \pi \tan^2 \alpha\ w\ (h^3 - x^3) \qquad | \text{ From (1)}$$

$$= \pi\ (h \tan \alpha)^2\ (h - x)\ w - 1/3\ \pi \tan^2 \alpha .\ w .\ (h^3 - x^3) \qquad ...(2)$$

Since the cone begins to rise when the height of liquid in the cone is $h - x$, therefore, the weight of the cone must be equal to the resultant vertical thrust on the curved surface.

$$\therefore 1/3\ \pi \tan^2 \alpha w\ (h^3 - x^3) = \pi\ h^2 \tan^2 \alpha\ (h - x)\ w - 1/3\ \pi \tan^2 \alpha .\ w\ (h^3 - x^3) \quad | \text{ From (1) and (2)}$$

$\Rightarrow (h^3 - x^3)/3 = h^2 (h - x) - (h^3 - x^3)/3 \quad \Rightarrow \quad 2(h^3 - x^3)/3 = h^2 (h - x)$

$\Rightarrow 2h^3 - 2x^3 = 3h^3 - 3h^2x \quad \Rightarrow \quad h^3 - 3h^2x + 2x^3 = 0$

$\Rightarrow h^3 - h^2x - 2h^2x + 2x^3 = 0 \quad \Rightarrow \quad h^2 (h - x) - 2x (h^2 - x^2) = 0$

$\Rightarrow (h - x) \{h^2 - 2x (h + x)\} = 0 \quad \Rightarrow \quad (h - x) (h^2 - 2xh - 2x^2) = 0$

$\Rightarrow h - x = 0$ or $h^2 - 2xh - 2x^2 = 0$

But $h - x = 0$ is inadmissible

$\therefore \quad h^2 - 2xh - 2x^2 = 0$

$\Rightarrow \quad 2h^2 + 2hx - h^2 = 0$

$$\therefore \quad x = \frac{-2h \pm \sqrt{4h^2 + 8h^2}}{4}$$

$$= \frac{-2h \pm 2\sqrt{3}h}{4}$$

$\because$ –ve sign is inadmissible

$$\therefore \quad x = \frac{-2h + 2\sqrt{3}h}{4}$$

$$= \frac{h}{2}(\sqrt{3} - 1) \qquad ...(3)$$

$$\therefore \quad \frac{\text{weight of liquid required to fill the frustum}}{\text{weight of liquid required to fill the cone}}$$

$$= \frac{\frac{1}{3}\pi(h^3 - x^3)\tan^2 \alpha w}{\frac{1}{3}\pi h^3 \tan^2 \alpha w}$$

$$= \frac{h^3 - x^3}{h^3} \qquad = 1 - \left(\frac{x}{h}\right)^3$$

$$= 1 - \left(\frac{\sqrt{3} - 1}{2}\right)^3 \qquad | \text{ From (3)}$$

$$= 1 - \frac{1}{8}\{3\sqrt{3} - 1 - 3\sqrt{3}(\sqrt{3} - 1)\}$$

$$= 1 - \frac{1}{8}\{6\sqrt{3} - 10\}$$

$$= 1 - \frac{3\sqrt{3} - 5}{4} \qquad = \frac{9 - 3\sqrt{3}}{4}$$

Example 8: A vessel in the shape of a hollow hemisphere surmounted by a cone is held with the axis vertical and vertex uppermost. If it be filled with a liquid so as to submerge half the axis of the cone in the liquid, and the height of the cone be double the radius of its base, show that the resultant downward thrust of the liquid on the vessel is 15/8 times the weight of the liquid that the hemisphere can hold.

Sol. Let r be the radius of the base of the hemisphere (or cone). Then, the height of the cone is $2r$. So, $FK = 1/2\ FO = r$

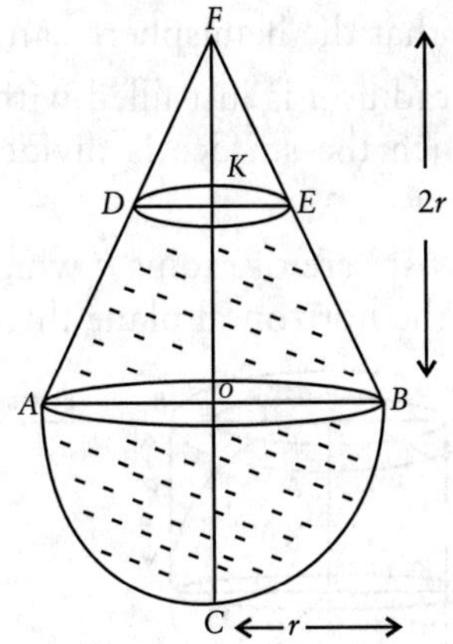

$\because \ \Delta FKE \sim \Delta FOB$

$\therefore \ \dfrac{KE}{OB} = \dfrac{FK}{FO}$ | By property of similar triangles

$\Rightarrow \ \dfrac{KE}{OB} = \dfrac{r}{2r} = \dfrac{1}{2} \Rightarrow \quad KE = \dfrac{1}{2} OB = \dfrac{1}{2} r$

Volume of the frustum *ABED*

$$= \frac{1}{3}\pi h(r_1^2 + r_2^2 + r_1 r_2) = \frac{1}{3}\pi r\left\{r^2 + \left(\frac{1}{2}r\right)^2 + r\left(\frac{1}{2}r\right)\right\}$$

$$= \frac{\pi r}{3}\frac{7r^2}{4}$$

$$= \frac{7\pi r^3}{12} \qquad \text{...(1)}$$

Volume of the hemisphere *ACB*

$$= 2/3\ \pi r^3 \qquad \text{...(2)}$$

$\therefore$ weight of the liquid contained in the vessel

$$= \left(\frac{7\pi r^3}{12} + \frac{2\pi r^3}{3}\right) g\rho, \text{ where } \rho \text{ is the density of the liquid}$$

$$= 5/3\ \pi r^3 g\rho \qquad \text{...(3)}$$

Since the liquid in the vessel presses partly upwards and partly downwards, therefore the resultant vertical thrust acts downwards and its magnitude will be equal to the weight of the liquid contained in the vessel.

$\therefore$ Resultant downward thrust of the liquid of the vessel

$$= \frac{5}{4}\pi r^3 g\rho \qquad = \frac{15}{8}\left(\frac{2}{3}\pi r^3 g\rho\right)$$

= 15/8. Weight of the liquid that the hemisphere can hold.

Example 9: A hollow sphere of radius a is just filled with water. Find the resultant vertical thrust on the two portions into which the surface is divided by horizontal plane at depth c below the centre.

Sol. O is the centre of the hollow sphere of radius a which is just filled with water. E is the highest point of the sphere. Clearly, the horizontal plane through E is the free surface of water.

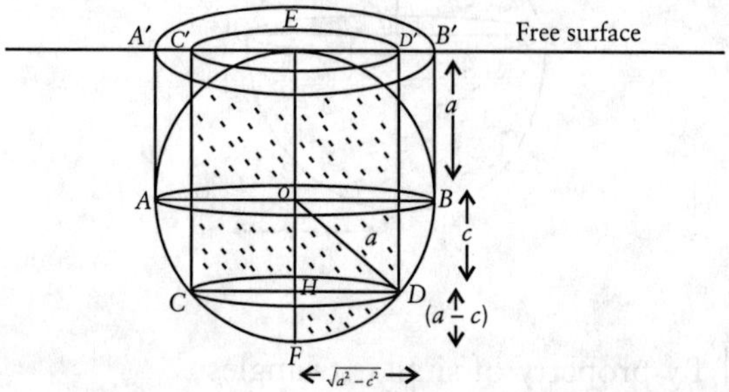

Let CD be the horizontal plane at a depth c below the centre which divides the surface of the sphere into two portions. AB is the horizontal plane passing through O.

$$OD = a$$

$$OH = c$$

$$\therefore \quad HD = \sqrt{a^2 - c^2}$$

Also, $EH = EO + OH = a + c$

The downward vertical thrust on the lower portion CFD

= weight of the super-incumbent water $CFDD'C'$

= weight of the cylinder $CDD'C'$ + weight of spherical segment CFD

$= [\pi (a^2 - c^2)(a + c) + \pi (a - c)(a - (a - c)/3)]\, g\rho$,

where ρ is the density of water

$= \pi/3\, (a - c)\{3(a + c)^2 + (a - c)(2a + c)\}\, g\rho$

$= \pi/3\, (a - c)\{3a^2 + 3c^2 + 6ac + 2a^2 + ac - 2ac - c^2\}\, g\rho$

$= \pi/3\, (a - c)(5a^2 + 5ac + 2c^2)\, g\rho$ ...(1)

The upper portion $CAEBD$ consists of two parts, the upper hemisphere AEB and the zone $ACDB$. The vertical upward thrust on the hemisphere AEB

= weight of the super-incumbent water $A'AEB'B'$ – weight of water in hemisphere AEB

$= \pi a^2 \cdot ag\rho - 2/3\, \pi a^3 g\rho$

$= 1/3\, \pi a^3 g\rho$ ...(2)

The vertical downward thrust on the zone $ACDB$

= weight of the super-incumbent water $A'ACC'$, $D'DBB'$

= weight of water in cylinder $AA'B'B$ + weight of water in zone $ACDB$ – weight of water in cylinder $CC'D'D$

$= \pi a^2 . a . g\rho + 1/3\ \pi c\ (3a^2 - c^2)\ g\rho - (a^2 - c^2)\ (a + c)\ g\rho$

$= \pi/3\ \{3a^3 + 3a^2c - c^3 - 3(a^2 - c^2)\ (a + c)\}\ g\rho$

$= \pi/3\ (3a^3 + 3a^2c - c^3 - 3a^3 - 3a^2c + 3c^2a + 3)\ g\rho$

$= \pi/3\ (3ac^2 + 2c^3)\ g\rho$ ...(3)

From (2) and (3), the resultant vertical thrust on the upper portion *CAEBD*

$= 1/3\ \pi a^3 g\rho \sim \pi/3\ (3ac^2 + 2c^3)\ g\rho$

$= 1/3\ \pi g\rho\ \{a^3 \sim (3ac^2 + 2c^3)\}$ ...(4)

which would be upwards or downwards according as (2) is greater or less than (3), depending upon the values of a and c.

Note: From (1) and (4), it is evident that the resultant vertical thrust on the whole surface of the sphere is downwards and is equal to $4/3\ \pi a^3 g\rho$ which is equal to the weight of the contained water in the hollow sphere.

Example 10: A double funnel is formed by joining two equal hollow cones at their vertices and stands on a horizontal plane with the common axis vertical. Liquid is poured into the cone until its surface bisects the axis of the upper cone. If the liquid be on the point of escaping between the lower cone and the table, prove that the weight of either cone is to that of the liquid it can held is 27 : 16.

Sol. Let the base radius of each cone be r and the height of each cone be h. Let EF be the free surface of the liquid that bisects the axis of the upper cone.

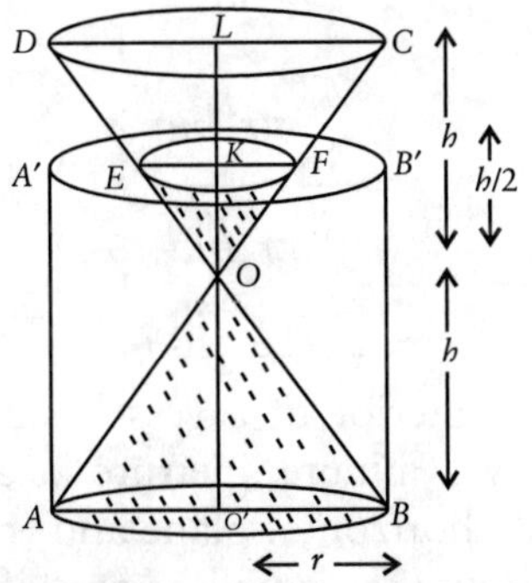

$\therefore$ Height of the cone $OEF = OK = h/2$

and, radius of the base of the cone $OEF = KF = r/2$

Height of the liquid $= O'K = O'O + OK$

$= h + h/2 = 3h/2$

The resultant vertical thrust on the curved surface of the lower cone, acting vertically upwards

= Weight of the super-incumbent liquid

= Weight of the liquid in the cylinder $ABB'A'$ – Weight of liquid in the cone OAB

$= \pi r^2\ (h + h/2)\ w - 1/3\ \pi r^2 hw,$

where w is the weight per unit volume of the liquid

$= 3/2\pi r^2 hw - 1/3\pi r^2 hw$

$= 7/6\pi r^2 hw$ ...(1)

The resultant vertical thrust on the curved surface of the upper cone, acting vertically downwards where of the super-incumbent liquid

= weight of the liquid contained in the cone OEF it self

= $1/3\ \pi\ KF^2\ OK\ w = 1/3\ \pi\ (r/2)^2\ h/2\ w$

$$= \frac{\pi r^2 hw}{24} \qquad \text{...(2)}$$

From (1) and (2), the resultant vertical upward thrust on the curved surface of the funnel

= Resultant vertical thrust on the lower cone – Resultant vertical thrust on the upper cone

= $7/6\ \pi r^2 hw - 1/24\ \pi r^2 hw$

$$= \frac{27}{24}\pi r^2 hw \qquad \text{...(3)}$$

Since the upper cone is filled to a height $h/2$, therefore the liquid will be at the point of escaping between the lower cone and the table. Hence the resultant vertical thrust acting upwards must be equal to the weight of the funnel. So, if W is the weight of each cone, then

$2W = 27/24\ \pi r^2 hw$

$\Rightarrow\ W = 27/48\ \pi r^2 hw$ ...(1)

$$\therefore \frac{\text{weight of each cone}}{\text{weight of the liquid it can hold}} = \frac{W}{\frac{1}{3}\pi r^2 hw}$$

$$= \frac{\frac{27}{48}\pi r^2 hw}{\frac{1}{3}\pi r^2 hw} \qquad \text{| From (4)}$$

$$= 27/16$$

Example 11: The shape of the interior of a vessel is a double cone, the ends being open and the two portions connected by a minute aperture at a common vertex. It is placed with one circular rim fitting close upon a horizontal plane and is filled with water. Find the resultant vertical thrust on the vessel and show that it will be zero if the length of the axis of the upper portion be twice that of the lower.

Sol. Refer Ex. 10.

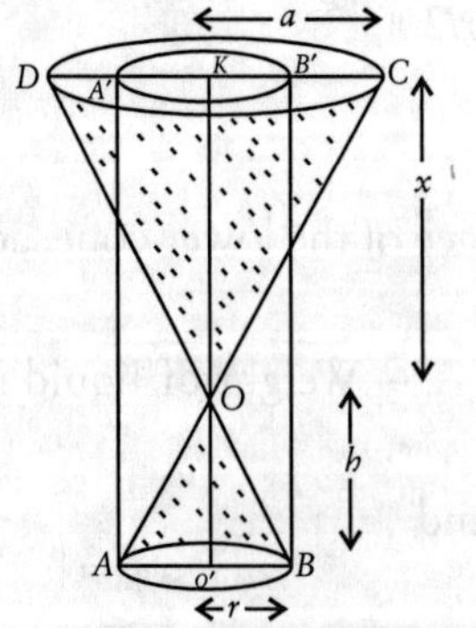

Let a and r be the radii of the bases of the upper and lower cones respectively. Let their heights be x and h respectively.

Then, $h/x = r/a$ | $\because \Delta\, OAO' \sim \Delta\, OCK$ (by property of similar triangles)

vertical upward thrust on the curved surface of the lower cone

$$= \pi r^2 (h + x)\, w - 1/3\, \pi r^2 h w \qquad \text{...(2)}$$

vertical downward thrust on the curved surface of the upper cone

= weight of the liquid contained in the upper cone

$$= 1/3\, \pi a^2 x w$$

$$= \frac{1}{3}\left(\frac{rx}{h}\right)^2 xw$$

$$= \frac{1}{3}\frac{\pi r^2 x^3 w}{h^2} \qquad \text{...(3)}$$

$\therefore$ Resultant vertical thrust on the vessel

$$= \left\{\pi r^2 (h + x) w - \frac{\pi r^2 h w}{3}\right\} - \frac{\pi r^2 x^3 w}{3h^2}$$

$$= \frac{\pi r^2 w}{3}\left(3h + 3x - h - \frac{x^3}{h^2}\right)$$

$$= \frac{\pi r^2 w}{3}\left(2h + 3x - \frac{x^3}{h^2}\right)$$

If it is zero, then $2h + 3x - x^3/h^2 = 0 \Rightarrow \quad 2h^3 + 3xh^2 - x^3 = 0$

$\Rightarrow \ (x + h)^2 (2x - h) = 0\ h^2 = 0 \quad \Rightarrow \quad 2x - h = 0$ | $\because x + h = 0$ is inadmissible

Hence, the height of the upper cone is twice that of the lower cone.

EXERCISE 1

1. A hollow right circular cone of height h is just full of liquid of density ρ and rests with its base of radius a in a horizontal position, the vertex being uppermost. Find the resultant vertical thrust on the curved surface of the cone.

   Ans. $2/3\, \pi a^2 h g \rho$

   [**Hint:** Refer Ex. 2]

2. A conical vessel, filled with water, stands with its plane base on a horizontal table. Prove that the thrust of the liquid on the base of the cone is three times the weight of the contained water. How do you explain the result?

   Ans. Thrust of water on the base > weight of the water contained in the cone

3. A right circular cylinder is just immersed in water with its axis horizontal. Compare the vertical thrusts on the two parts of the curved surface into which it is divided by the horizontal plane through the axis.

Or

If a hollow cylinder be filled with water and be closed at both ends and held with its axis horizontal, find the vertical thrust on the lower half of the curved surface and compare the vertical thrust into which it is divided by the horizontal plane through the axis.

Ans. $4 + \pi : 4 - \pi$

[**Hint:** Refer Ex. 3

Thrust on the lower half = $(1/2\ \pi + 2)\ r^2 hg\rho = (4 + \pi)\ r^2 hg\rho/2$

Thrust on the upper half = $(2 - \pi/2)\ r^2 hg\rho = (4 - \pi)\ r^2 hg\rho/2$

$\therefore$ Required ratio = $(4 + \pi) : (4 - \pi)$]

4. A solid right cone, height h and radius a (of the base), is completely immersed in a liquid with its axis horizontal and at depth b.

Compare the vertical thrusts on the two halves into which the curved surface is divided by a horizontal plane through the axis.

Ans. $(6b - \pi a) : (6b + \pi a)$

[**Hint:** Refer Ex. 4

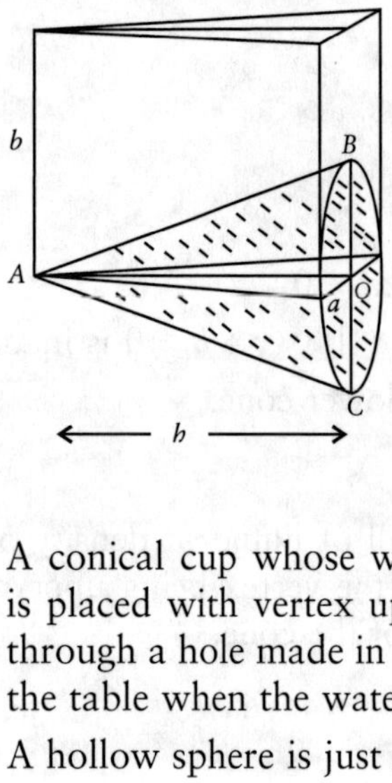

Thrust on lower half (P_1)

$$= \left(\frac{2a.h}{2} b + \frac{1}{2}.\frac{1}{3}\pi a^2 h\right) g\rho$$

$$= (6b + \pi a)\frac{ahg\rho}{6}$$

Thrust on lower half (P_2)

$$= (6b - \pi a)\frac{ahg\rho}{6}$$

$$\because P_2 : P_1 = (6b - \pi a) : (6b + \pi a)]$$

5. A conical cup whose weight is 5/8 of the weight of water which would just fill it, is placed with vertex upwards on a smooth table and water is gradually poured in through a hole made in the top. Show that the cup will be on the point of rising from the table when the water reaches half the height of the cup.

6. A hollow sphere is just filled with water. Prove that the resultant vertical pressure on the upper and lower halves of the internal surface are 1/4W and 5/4W respectively, where W is the weight of the water.

[**Hint:** Refer Ex. 9

vertical pressure on the upper half

$= 1/3\ \pi a^3 g\rho$

$$= \frac{1}{4}\left(\frac{4}{3}\pi a^3 g\rho\right) \qquad \Big|\ \text{From (4)}$$

$= 1/4\ W$

vertical pressure on the lower half

$= 5/3\ \pi a^3 g\rho$ | From (1)

$= \frac{5}{4}\left(\frac{4}{3}\pi a^3 g\rho\right)$

$= 5/4\ W$]

7. A vessel is in the shape of the greater segment of a sphere of radius a cut off by a plane at a distance c from the centre, rests with its base on a table. If the segment is filled with water, show that the resultant vertical thrust on the curved surface will be an upward force if $c < a/2$.

   [**Hint:** Refer Ex. 9.

   The resultant vertical thrust on the curved surface will be an upward force if (2) > (3), i.e.,

   $$1/3\ \pi a^3 g\rho > \pi/3\ (3ac^2 + 2c^3)\ g\rho$$
   $$\Rightarrow \quad a^3 > 3ac^2 + 2c^3 \Rightarrow \quad 2c^3 + 3ac^2 - a^3 < 0$$
   $$\Rightarrow \quad (a + c)^2 (2c - a) < 0$$
   $$\Rightarrow \quad 2c - a < 0 \qquad | \because a + c < 0 \text{ is inadmissible}$$
   $$\Rightarrow \quad c < a/2$$]

8. A double funnel, formed of two equal cones with a common axis communicating at their common vertex is placed on a horizontal table with the axis vertical and is filled with water. Prove that the resultant vertical thrust on the curved surface of the lower cone is $2\frac{1}{2}$ times the weight of the water.

   [**Hint:** Refer Ex. 10

   Resultant vertical thrust on the curved surface of the lower cone

   $= \pi r^3\ (h + h)\ w - 1/3\ \pi r^2 hw$ | Here, $OK = OL = h$; (*i.e.*, K coincides with L)

   $= 5/3\ \pi r^2 hw = 5/2\ (2/3\ \pi r^2 hw)$

   $= 5/3\ (1/3\ \pi r^2 hw + 1/3\ \pi r^2 hw)$

   $= 5/3$ weight of the water]

9. A pyramid with a square base and with sides which are equilateral triangles, is placed on a horizontal plane and filled with a liquid through an aperture in the vertex. Find the pressure on one of the sides.

   If the pyramid has no base, find its least weight consistent with it not being raised.

   Ans. $= \frac{1}{12}\sqrt{6}\, a^3 g\rho,\ \frac{1}{3}\sqrt{2}\ a^3 g\rho$

   where a is the length of each sides of the base and ρ is the density of the liquid.

6. Resultant Horizontal Thrust

The resultant horizontal thrust on a curved surface is defined as the sum of the components of thrusts on various elements of the curved surface in a particular horizontal direction.

Note: The resultant horizontal thrust on a curved surface may be different in different horizontal directions.

7. Method of Finding Resultant Horizontal Thrust

Let $ABCD$ be the given curved surface on which we are required to find out the resultant horizontal thrust in a given horizontal direction.

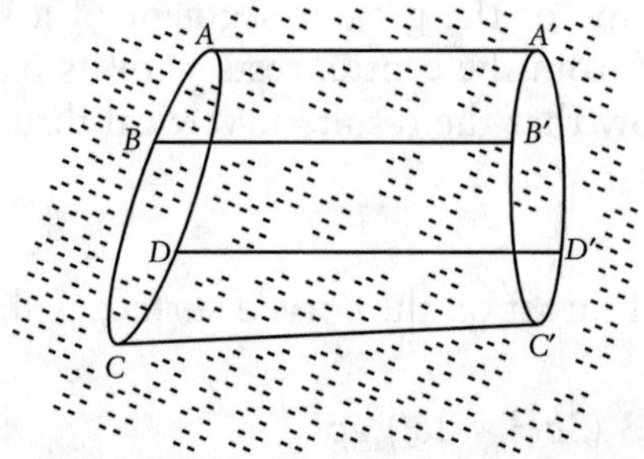

Let the curved surface $ABCD$ be immersed in a homogeneous fluid. From each and every point on the boundary of the surface, draw horizontal lines in the given (pre-assigned) direction to meet a vertical plane perpendicular to the given direction in a closed curve $A'B'C'D'$. These horizontal lines, which are parallel, will form a cylinder. Let us take a vertical cross-section $A'B'C'D'$ of this cylinder.

Now, we shall consider the equilibrium of the mass of the fluid in this cylinder enclosed by these parallel horizontal lines which generate the cylindrical surface, the given curved surface $ABCD$ and its projection $A'B'C'D'$ on a vertical plane. The fluid in the cylinder is in equilibrium under the action of the following horizontal forces along the given direction.

(i) The horizontal component of the thrust on the given curved surface $ABCD$ in the assigned direction.

(ii) The resultant thrust (or the whole pressure) on the plane end $A'B'C'D'$ acting through its centre of pressure parallel to the assigned direction.

For equilibrium, these two fluid thrusts must be equal in magnitude and act along the same line but in opposite directions.

Hence, *the resultant horizontal thrust on a given surface in contact with a fluid in any pre-assigned given direction is equal in magnitude and line of action to the whole pressure on the projection of the surface upon a vertical plane perpendicular to the given direction, provided any line drawn in this direction does not intersect the surface in more than one point.*

Note 1: The direction of the resultant horizontal thrust passes through the centre of pressure of area $A'B'C'D'$ from the liquid towards (to the side of) the surface $ABCD$ under consideration.

Note 2: In proving the above result, we have assumed that each of the horizontal lines AA', BB', CC', DD', etc. drawn in the given direction intersects the given curved surface $ABCD$ only once, i.e., in one point only. However, if these lines cut the given surface in more than one point, then we divide the given surface into two or more parts so that the horizontal lines cut each such part only once. Then the above result is applied to find the horizontal component in the given direction on each part separately and the resultant of these is obtained by using the

Laws of Statics. This finally gives the desired resultant horizontal thrust on the curved surface, under consideration.

Note 3: If a continuous surface can be divided into two parts only, the projections of which on the plane perpendicular to the given direction are identical then there will be no horizontal thrust on the surface in that direction because then the horizontal thrusts on each of the two parts shall be equal in magnitude, but opposite in directions and therefore they cancel each other. Equivalently, if a surface is symmetrical about a vertical plane perpendicular to the pre-assigned given direction, then the horizontal thrust alone that assigned direction is always zero.

8. Resultant Thrust on a Curved Surface

The resultant thrust on a curved surface is found out by using the resultant vertical thrust (Z) and the resultant horizontal thrusts (X, Y) in two horizontal directions, perpendicular to each other.

If the surface is symmetrical, these three thrusts are concurrent and therefore may be compounded into a single force. This single force is called the *Resultant Thrust* and its magnitude is $\sqrt{X^2+Y^2+Z^2}$ inclined at an angle $\tan^{-1}\dfrac{Z}{\sqrt{X^2+Y^2}}$ to the horizontal. Also, it will make angles $\cos^{-1}\dfrac{X}{R}, \cos^{-1}\dfrac{Y}{R}, \cos^{-1}\dfrac{Z}{R}$ with the direction of X, Y, Z axes respectively.

If, on the contrary, the surface is not symmetrical, then the resultant thrust consists of a force and a couple.

9. Principle of Archimedes—Resultant Thrust on a Solid Immersed in a Fluid

The principle of Archimedes may be stated as follow:

When a solid is wholly or partly immersed in a fluid at rest, the resultant thrust of the fluid on the solid is equal and opposite to the weight of the fluid displaced by the solid and acts vertically upwards through the centre of gravity of the fluid displaced.

Proof: Let a solid B be immersed in a fluid. We know that the thrust of fluid on a solid depends only upon the shape of the solid immersed and not on the substance (matter) of which it is made of. Therefore, the thrust on the solid B will remain unchanged, if instead of B, another solid of exactly the same shape be placed in that very position in the fluid, the fluid thrust on this replaced solid will be the same.

Now we imagine that the solid B is removed and the space which it occupies is filled up by the same fluid which may be called the displaced fluid. This fluid occupying the space of the solid is subjected to the same resultant thrust due to the surrounding fluid as the original solid. This fluid, which is supposed to occupy the position of the solid, will be in equilibrium under the action of the following forces:

(i) Its own which acting vertically downwards through its centre of gravity.

(ii) Thrust upon its bounding surface due to surrounding fluid.

For equilibrium, the thrust due to surrounding fluid must be equal to the weight of the displaced fluid.

Hence, the resultant thrust, to which the solid is subjected, must be equal to the weight of the displaced fluid. This thrust would act through the centre of gravity of the fluid displaced.

Note: Principle of Archimedes also holds good when the fluid into which the solid is immersed in not homogeneous. If the fluid consists of a number of layers of immiscible fluids of different densities, then the displaced fluid should be taken to be of the same density at any point as that of the surrounding fluid at the same horizontal level.

10. Force of Buoyancy

The resultant fluid thrust on a body wholly or partially immersed in a fluid is called the *Force of Buoyancy*. The other way round, the force of buoyancy is equal to the weight of the fluid displaced by the body and acts vertically upwards. That is way it is also called the upthrust.

11. Centre of Buoyancy

The centre of gravity of the fluid displaced is called the centre of buoyancy.

ILLUSTRATIVE EXAMPLES

Example 1: A solid right circular cone is placed with its axis horizontal and at a depth h below the surface of water. Find the horizontal thrust on half of the cone imagined to be cut off by a vertical plane through the axis.

Sol. Let r be the radius of the base and h' be the height of the cone. Let X and Y be the horizontal thrusts on the curved surface of half of the cone imagined to be cut off by a vertical plane through the axis along two mutually perpendicular directions VO and OC. Then,

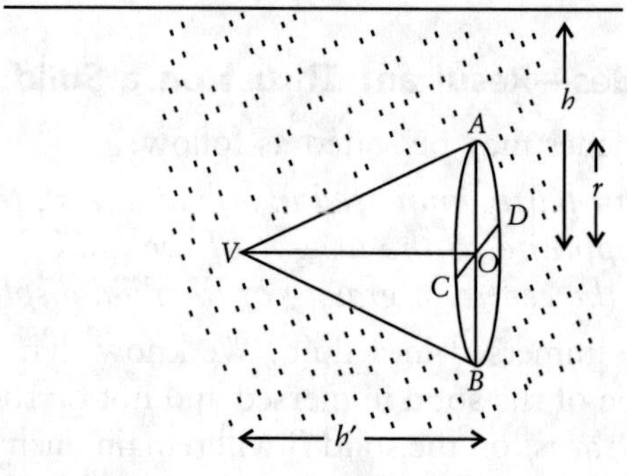

X = Horizontal thrust along VO = Thrust on the semi-circle ACB

$= w\,(1/2\ \pi r^2)\,h$

$= 1/2\ w\pi r^2 h$ ...(1)

Y = Horizontal thrust along OC

= Thrust on $\Delta\, AVB$

$$= w\left(\frac{2r.h^1}{2}\right)\left\{\frac{(h-r)+(h+r)+h}{3}\right\}$$

$= w\,r\,h^1 h$ ...(2)

$\therefore$ Resultant horizontal thrust

$$= \sqrt{X^2+Y^2} = \sqrt{(\frac{1}{2}w\pi r^2 h)^2 + (wrh'h)^2} = wrh\sqrt{\frac{1}{4}\pi^2 r^2 + h'^2}$$

Example 2: A right circular cone is just immersed in a liquid with its axis horizontal. Find the resultant thrust on the curved surface of half of the cone cut off by a vertical plane through the axis.

Sol. Let r be the radius of the base and h be the height of the cone. Let X and Y be the horizontal thrusts on the curved surface of half of the cone cut off by a vertical plane through the axis. Then,

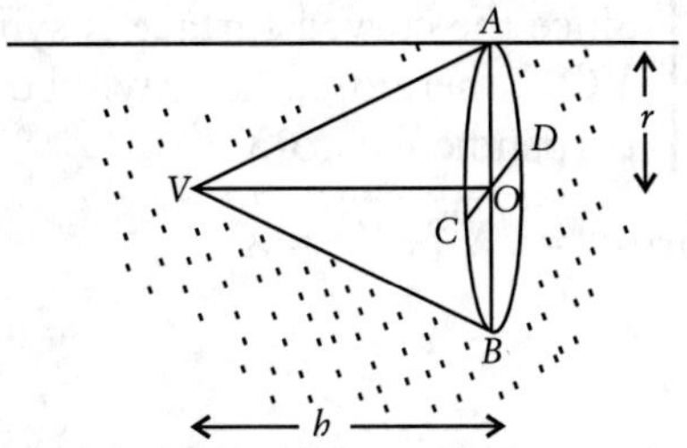

X = Horizontal thrust along VO

= Thrust on the semi-circle ACB

$= w\,(1/2\ \pi r^2)\ h$

$= w\ \pi r^3/2$ ...(1)

Y = Horizontal thrust along OC

= Thrust on $\Delta\ AVB$

$= w2r\ .\ h/2\ .\ r$

$= wr^2\ h$ ...(2)

$\therefore$ Resultant horizontal thrust

$$= \sqrt{X^2 + Y^2} \quad = \sqrt{\left(\frac{w\pi r^3}{2}\right)^2 + (wr^2 h)^2} \quad = wr^2\sqrt{\frac{\pi^2 r^2}{4} + h^2}$$

Example 3: A right circular cone is just immersed in a liquid with its axis horizontal. Find the resultant horizontal thrust on the curved surface of the lower half of the cone cut off by a horizontal plane through the axis.

Sol. Let r be the radius of the base and h be the height of the cone. Let X and Y be the horizontal thrusts on the curved surface of the lower half of the cone cut off by a horizontal plane through the axis. Then,

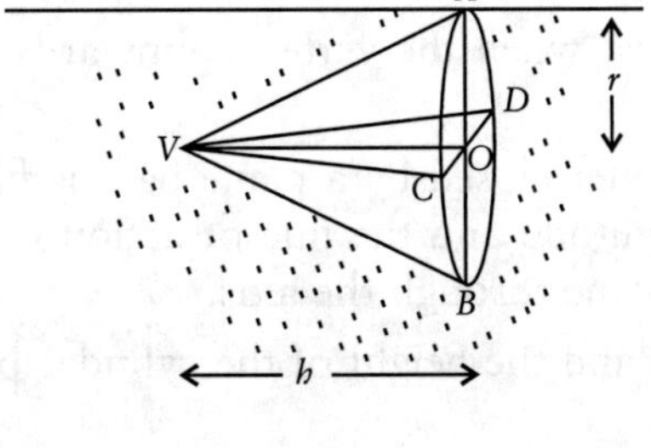

X = Horizontal thrust along VO

= whole pressure on the semi-circle CBD

$= w\,(1/2\,\pi\, r^2)\,(r + 4r/3\pi)$

$= w\pi r^3/2\,(1 + 4/3\pi)$ ...(1)

Y = Thrust along CD

$= 0$ | Since the curved surface is symmetrical about the plane VOC, therefore, there will be no horizontal component perpendicular to X. ...(2)

$\therefore$ Resultant horizontal thrust $= \sqrt{X^2 + Y^2} = X$

$$= \frac{1}{2} w\pi r^3 \left(1 + \frac{4}{3\pi}\right)$$

Example 4: A right circular cone filled with liquid is held with its axis vertical. Prove that the horizontal thrust on half the curved surface cut off by a plane through the axis, when the vertex is upwards, is twice that when the vertex is downwards.

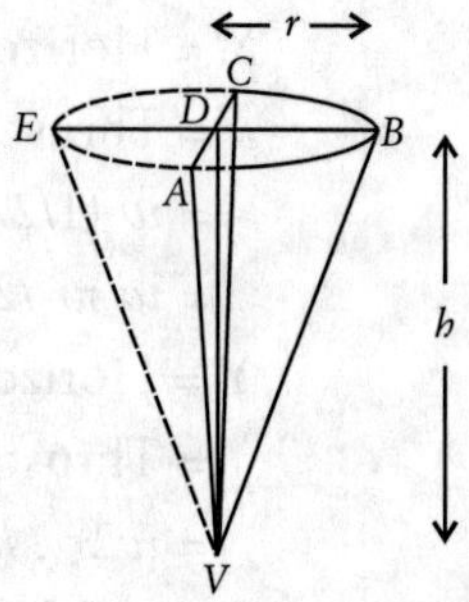

Sol. Let r be the radius of the base and h be the height of the cone.

Case I: When vertex is downwards

X = Horizontal thrust along BD

= Thrust on $\Delta\, OAC$

$= w 2r \,.\, h/2 \,.\, h/3$

$= 1/3\, wrh^2$ ...(1)

$Y = 0$ | By symmetry there is no component in the direction AC ...(2)

Case II: When vertex is upwards

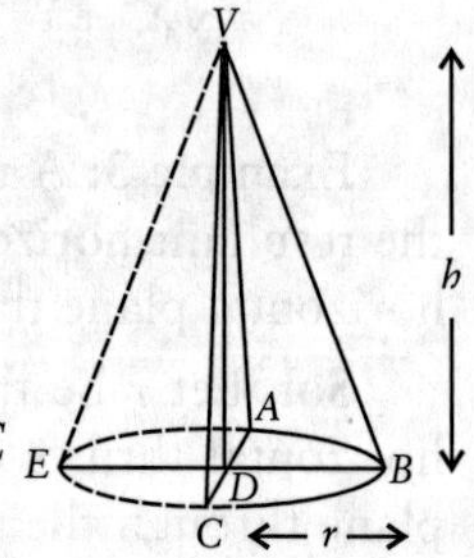

X' = Horizontal thrust along BD

= Thrust on $\Delta\, OAC$

$= w\,(2r \,.\, h/2)\, 2h/3$

$= 2/3\, wrh^2$ ...(3)

$Y' = 0$ | By symmetry there is no component in the direction AC ...(4)

From (1) and (3), we get $X' = 2X$

Hence, the horizontal thrust when the vertex is upwards is twice that when the vertex is downwards.

Example 5: A hollow cylinder closed by a plane base is filled with liquid and is held with its axis vertical. Find the magnitude and the line of action of the resultant thrust on half the cylinder cut off by a vertical plane through the axis.

Sol. Let the radius of base and the height of the cylinder be r and h respectively.

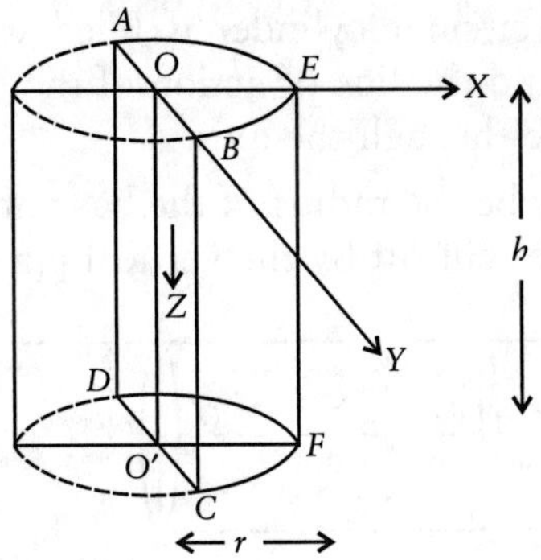

Let $ABCD$ be the vertical plane which divides the cylinder into two halves. We consider the right half.

Let X and Y be the horizontal thrusts on the curved surface of this half portion along two mutually perpendicular directions OE and OB respectively.

Then,

X = Horizontal thrust along OE

= Thrust on rectangle $ABCD$

$= w\,(2r\,.\,h)\,h/2$

$= wrh^2$ ...(1)

and, Y = Horizontal thrust along OB

$= 0$ | since the curved surface is symmetrical about the plane $OE\,FO'$ ...(2)

Let Z be the resultant vertical thrust on the curved surface. Then,

Z = weight of the super-incumbent liquid which would fill half cylinder

$= w\ 1/2\ \pi r^2 h$ ...(3)

The lines of action of X and Z lie in the plane $OGFO'$ and are intersecting.

∴ Resultant thrust

$$= \sqrt{X^2 + Y^2 + Z^2} \qquad = \sqrt{w^2 r^2 h^4 + 0 + \frac{w^2 \pi^2 r^4 h^2}{4}}$$

$$= wrh\sqrt{h^2 + \frac{\pi^2 r^2}{4}}$$

$$= \frac{wrh}{2}\sqrt{4h^2 + \pi^2 r^2} \qquad \text{...(4)}$$

Let θ be the angle which the resultant thrust makes with the horizontal. Then,

$$\tan\theta = \frac{Z}{X} \qquad = \frac{\dfrac{w\pi r^2 h}{2}}{wrh^2}$$

$$= \frac{\pi r}{2h} \qquad \Rightarrow \qquad \theta = \tan^{-1}\left(\frac{\pi r}{2h}\right)$$

Example 6: A hollow right circular cylinder is filled with liquid and held with its axis horizontal. Find the magnitude and the line of action of the resultant thrust on half the curved surface cut off by a vertical plane through the axis.

Sol. Let h be the height and r be the radius of the base of the cylinder. Let $AEDCFB$ be the curved surface of the half cylinder cut off by the vertical plane $ABCD$.

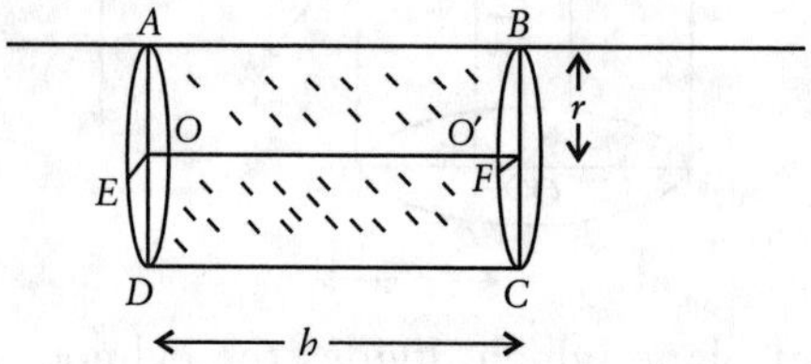

Let X and Y be the components of the horizontal thrust on the given curved surface $AEDCFB$ in the two mutually perpendicular directions OE and OO' respective. Then

X = Horizontal thrust in the direction of OE

= Whole pressure on the rectangle $ABCD$ which is the projection of the curved surface of the half cylinder on a vertical plane perpendicular to OE.

$= (2r \cdot h)\, rg\rho$, where ρ is the density of the liquid

$= 2r^2hg\rho$ ...(1)

and, Y = Horizontal thrust in the direction of OO'

= Whole pressure on the projection of the curved surface of the half cylinder on a vertical plane perpendicular to OO'

$= 0$ | by symmetry ...(2)

∴ Resultant horizontal thrust (H) on half the curved surface

$= \sqrt{X^2 + Y^2}$

$= 2r^2hg\rho$ ...(3)

Let Z be the resultant vertical thrust on this curved surface. Then,

Z = weight of the super-incumbent liquid

= weight of the liquid contained in the half cylinder

$= 1/2\ \pi r^2 hg\rho$ ...(4)

The lines of action of H and Z are coplanar and as these are not parallel, they meet in a point.

$$\therefore \quad \text{Resultant thrust (R)} = \sqrt{H^2 + Z^2}$$

$$= \sqrt{4r^4h^2g^2\rho^2 + \frac{\pi^2 r^4 h^2 g^2 \rho^2}{4}}$$

$$= \frac{1}{2} r^2 hg\rho\sqrt{16 + \pi^2}$$

If the resultant thrust makes on angle ϕ with the horizontal, then,

$$\tan\phi = \frac{Z}{H} = \frac{\frac{1}{2}\pi r^2 h g\rho}{2r^2 h g\rho} = \frac{\pi}{4} \Rightarrow \phi = \tan^{-1}\left(\frac{\pi}{4}\right)$$

Example 7: A cylindrical pipe of circular cross-section is half full of water. If the pipe be imagined to be divided into two halves by a vertical plane along the middle, show that the water will tend to push them asunder horizontal with a force (W/π), where W is the weight of the water contained. Show that the resultant thrust on either half of the pipe makes with the vertical an angle $\cot^{-1}\left(\frac{\pi}{2}\right)$.

or

A horizontal trough is semi-circular in section and is filled with water whose weight is W. If the trough be imagined to be divided into two halves along the middle, show that the water will tend to push them asunder horizontally with a force W/π.

Show also that the resultant thrust of the water on either half of the trough makes with the vertical an angle θ so that

$$\cot\theta = \pi/2.$$

Sol. Let the length of the pipe be h and the radius of its base be r. W is the weight of the water contained in half of the pipe.

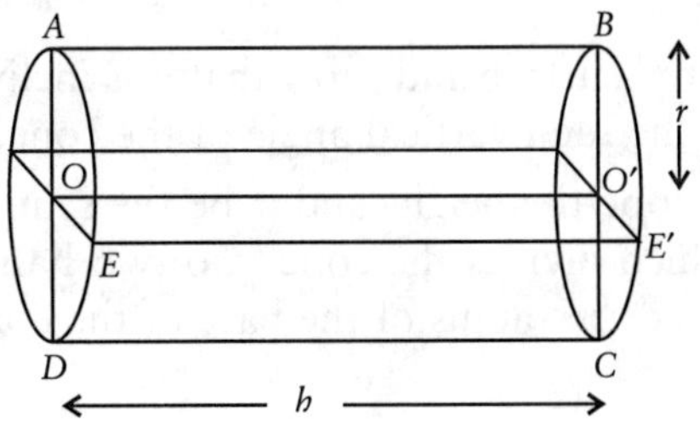

$\therefore \quad W = 1/2\ \pi r^2 h g\rho$

where ρ is the density of water. ...(1)

Let X and Y be the horizontal thrusts on the curved surface of half the cylinder in the directions OE and OO'. Then,

X = Horizontal thrust on the curved surface of half the cylinder in the direction of OE

= Thrust on the projection of this curved surface on a plane perpendicular to OE

= Thrust on rectangle $ODCO'$

= $hr\ 1/2\ rg\rho$, where ρ is the density of water

= $1/2\ r^2 h g\rho$

= W/π | From (1) ...(2)

and, Y = Horizontal thrust on the curved surface of half the cylinder in the direction of OO'

= 0 | by symmetry ...(3)

Hence, the resultant horizontal thrust (H) on the curved surface of half the cylinder.

$= \sqrt{X^2 + Y^2}$

$= W/\pi$ | From (2) and (3) ...(4)

Resultant vertical thrust on the curved surface (DOE, $CO'E'$) of half the pipe (Z)

= weight of the super-incumbent water

= weight of the water contained

$= 1/2\ (1/2\ \pi r^2 hg\rho)$

$= 1/2\ W$ | From (1) ...(5)

Let θ be the angle which the resultant thrust on the curved surface under consideration makes with the vertical. Then,

$$\cot\theta = \frac{Z}{X} = \frac{\frac{1}{2}W}{\frac{W}{\pi}} \quad \text{| From (5) and (2)}$$

$= \pi/2$

$$\Rightarrow \quad \theta = \cot^{-1}\left(\frac{\pi}{2}\right)$$

Example 8: A solid right circular cone is divided into two parts by a plane through its axis and one of these portions is just immersed, vertex downwards and axis vertical in water. Find the resultant thrust on its curved surface and show that it is inclined at an angle $\tan^{-1}\left(\frac{\pi}{2}\tan\alpha\right)$ to the horizontal, where α is the semi-vertical angle of the cone.

Sol. Let V be the vertex, h be the height and α be the semi-vertical angle of the cone. Let VAC be the vertical plane which divides the cone into two halves. Let $VCBA$ be one of them immersed in the liquid. Let r be the radius of the base of the cone.

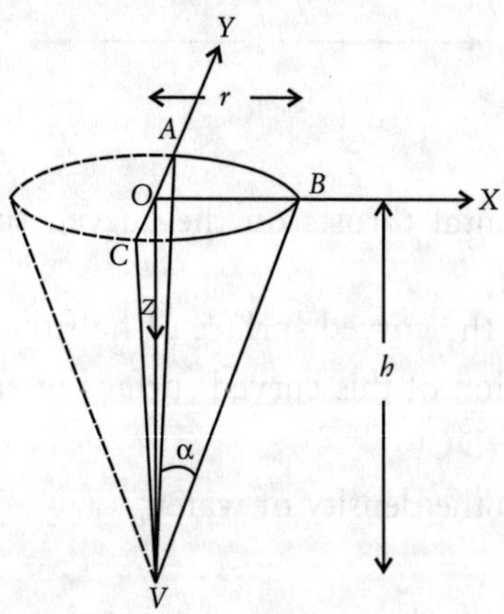

Let X and Y be the horizontal thrusts on the curved surface of this half in two mutually perpendicular directions OB and OA respectively. Then,

X = Horizontal thrust in the direction of OB

= Thrust on the projection of the curved surface under consideration on a vertical plane perpendicular to OB

= *Thrust on* Δ *VAC* = $2r \cdot h/2$ 1/3 $hg\rho$, where ρ is the density of water

= 1/3 $r\, h^2 g\rho$ ...(1)

and, Y = Horizontal thrust in the direction of OA

= 0 | since the curved surface is symmetrical about the plane VOB perpendicular to OB ...(2)

Resultant vertical thrust on the curved surface under consideration (Z)

= weight of the super-incumbent water = weight of water that will fill half the cone

1/2 (1/3 $\pi r^2 h$) $g\rho$

1/6 $\pi\, r^2 h g\rho$ ...(3)

Since the lines of action of H and Z lie in one plane and are not parallel to each other, therefore, they will intersect each other essentially.

$\therefore$ Resultant thrust $= \sqrt{X^2 + Y^2 + Z^2} = \sqrt{X^2 + Z^2}$ | From (2)

$$= \sqrt{\left(\frac{1}{3} r h^2 g\rho\right)^2 + \left(\frac{1}{6}\pi r^2 h g\rho\right)^2} = \frac{1}{3} g\rho r h \sqrt{h^2 + \frac{1}{4}\pi^2 r^2}$$

$$= \frac{1}{3} g\rho (h\tan\alpha) h \sqrt{h^2 + \frac{1}{4}\pi^2 h^2 \tan^2\alpha} \qquad | \because \tan\alpha = r/h$$

$$= \frac{1}{3} g\rho h^3 \tan\alpha \sqrt{1 + \frac{\pi^2 \tan^2\alpha}{4}} = \frac{1}{6} g\rho h^3 \tan\alpha \sqrt{4 + \pi^2 \tan^2\alpha}$$

$$= \frac{1}{6} g\rho h^3 \tan^2\alpha \sqrt{4\cot^2\alpha + \pi^2}$$

$$= \frac{1}{6} g\rho r^3 \cot^3\alpha \tan^2\alpha \sqrt{4\cot^2\alpha + \pi^2} \qquad | \because h = r\cot\alpha$$

$$= \frac{1}{6} g\rho r^3 \cot^3\alpha \sqrt{4\cot^2\alpha + \pi^2}$$

Let θ be the inclination of the resultant thrust to the horizontal. Then,

$$\tan\theta = \frac{Z}{X} = \frac{\frac{1}{6}\pi r^2 h g\rho}{\frac{1}{3} r h^2 g\rho} \qquad | \text{ From (3) and (1)}$$

$$= \left(\frac{\pi}{2}\frac{r}{h}\right) = \left(\frac{\pi}{2}\tan\alpha\right) \qquad | \because \tan\alpha = r/h$$

$$\Rightarrow \theta = \tan^{-1}\left(\frac{\pi}{2}\tan\alpha\right)$$

Example 9: A closed cylindrical vessel with hemispherical ends is filled with water, and placed with its axis horizontal. Find the resultant thrust on each of the ends and determine its line of action.

Sol. Let r be the radius of the base of the cylinder.

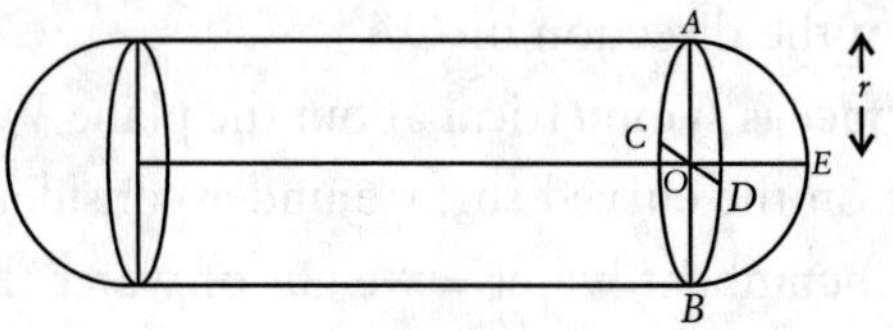

Let X and Y be the horizontal thrusts on each end in two mutually perpendicular directions OE and OD respectively. Then,

X = Horizontal thrust on the curved surface of the hemispherical end in the direction of OE

= Thrust on the circle $ACBD$, which is the projection of the surface under consideration on a vertical plane perpendicular to OE

$= \pi r^2 . r . g\rho$, where ρ is the density of water

$= \pi r^3 g\rho$ ...(1)

and, Y = Horizontal thrust on the curved surface under consideration in the direction of OD

$= 0$ | due to symmetry about the plane BEA ...(2)

Z = Resultant vertical thrust on the above curved surface

= weight of the super-incumbent water

$= 2/3\ \pi r^3 g\rho$ ...(3)

Let R be the resultant thrust on the above curved surface. Then,

$$R = \sqrt{X^2 + Y^2 + Z^2}$$

$$= \sqrt{(\pi r^3 g\rho)^2 + (0)^2 + \left(\frac{2}{3}\pi r^3 g\rho\right)^2}$$

$$= \frac{\sqrt{13}}{3}\pi r^3 g\pi \quad ...(4)$$

If θ is the angle which the resultant thrust makes with the horizontal, then,

$$\tan\theta = \frac{Z}{H} = \frac{Z}{X} \qquad \text{|Refer Ex. 8, } H = \sqrt{X^2 + Y^2} = X \ |\because Y = 0$$

$$= \frac{\frac{2}{3}\pi r^3 g\pi}{\pi r^3 g\pi} \qquad \text{| From (3) and (1)}$$

$= 2/3 \qquad \Rightarrow \qquad \theta = \tan^{-1}(2/3)$

Example 10: A solid hemisphere, of radius a, is placed with its centre at a depth h below the surface of water, and has its plane base in a vertical plane. What in the horizontal thrust on its curved surface? Find also the resultant thrust on it.

Sol.

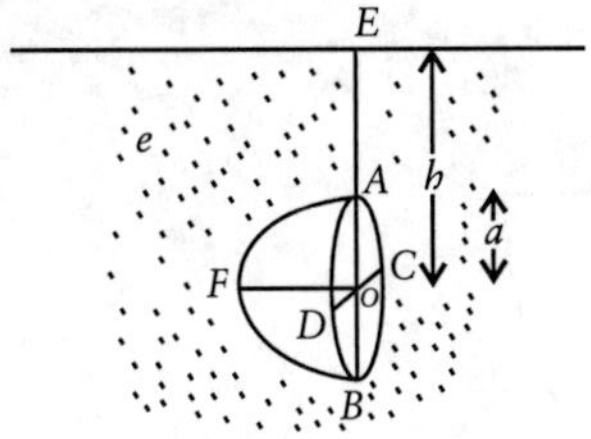

Let X and Y be the components of the horizontal thrust on the curved surface in two mutually perpendicular directions FO and DO respectively. Then,

X = horizontal thrust in the direction of FO

= Thrust on the projection of the curved surface perpendicular to FO

= Thrust on the circle $DBCA$

$= \pi a^2 . h . g\rho$, where ρ is the density of water ...(1)

and, Y = Horizontal thrust in the direction of $DO = 0$ ...(2)

| since the surface is symmetrical about the plane ABF

$\therefore$ Required resultant horizontal thrust (H) on the curved surface

$= \sqrt{X^2 + Y^2} = X$ | From (2)

$= \pi a^2 h g\rho$ ...(3)

Also, Z = Resultant vertical thrust on the curved surface

= weight of the super-incumbent water which would fill the hemisphere

$= 2/3\ \pi a^3 g\rho$ ...(4)

[**Aliter** for Z : Z = weight of water displaced by the hemisphere | By principle of Archimedes

$= 2/3\ \pi a^3 g\rho$]

Hence, the required resultant thrust (R) on the curved surface is given by

$R = \sqrt{H^2 + Z^2}$

$= \sqrt{(\pi a^2 h g\rho)^2 + (2/3\pi a^3 g\rho)^2}$ | From (3) and (4)

$= \dfrac{\pi a^2}{3}(9h^2 + 4a^2) g\rho$ ...(5)

Let this resultant thrust make an angle θ with the horizontal. Then,

$\tan\theta = Z/H = Z/X = 2/3\ \pi a^3 g\rho/\pi a^2 h g\rho = 2a/3h \quad \Rightarrow \theta = \tan^{-1}(2a/3h)$

Example 11: A hemispherical bowl is filled with water. Find the horizontal fluid pressure on one half of the surface divided by a vertical diameteral plane, and show that it is $1/\pi$ of the magnitude of the resultant fluid thrust on the whole surface.

Sol. CED is the vertical diameteral plane dividing the hemisphere into two halves.

Let a be the radius of the base of the hemisphere.

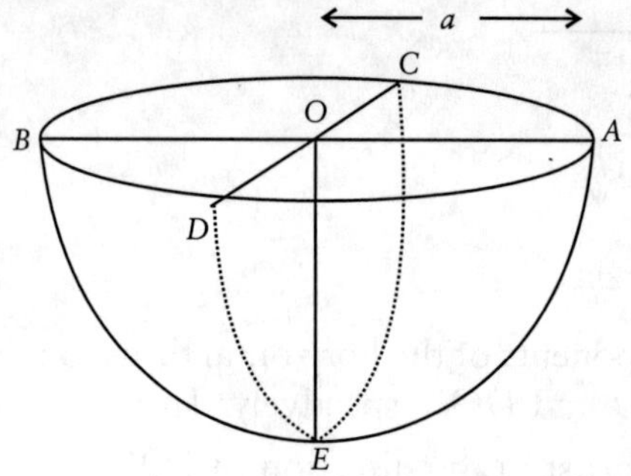

Let X and Y be the components of the horizontal thrust on half of the curved surface in two mutually perpendicular directions OB and OD respectively. Then,

X = Horizontal thrust in the direction of OB

= Thrust on the projection of half of the curved surface under consideration perpendicular to OB = Thrust on semi-circle CED

= $1/2\ \pi a^2 \cdot 4a/3\pi \cdot g\rho$, where ρ is the density of water = $2a^3\ g\rho/3$ ...(1)

| $P = wS\bar{z}$; Depth of C.G. of the semi-circle from the bounding diameter = $4a/3\pi$

and, Y = Horizontal thrust in the direction of OD = 0 ...(2)

| since the surface is symmetrical about the plane BOE which is perpendicular to OD

$\therefore$ Resultant horizontal thrust on one half of the curved surface of the hemisphere

$= \sqrt{X^2 + Y^2} = X = 2/3a^3\ g\rho = 1/2\ (2/3\ \pi a^3\ g\rho) = 1/\pi$ (weight of the water contained)

= $1/\pi$ (magnitude of the resultant fluid thrust on the whole surface).

Example 12: A hemispherical bowl with its lowest point downwards and the plane base horizontal is filled with water. The water is poured into a cylindrical tube, the radius of whose base is equal to that of the hemisphere. Prove that the horizontal thrust on half of their curved surfaces in which they may be divided by vertical planes through their axes, are in the ratio of 3 : 2.

Sol.

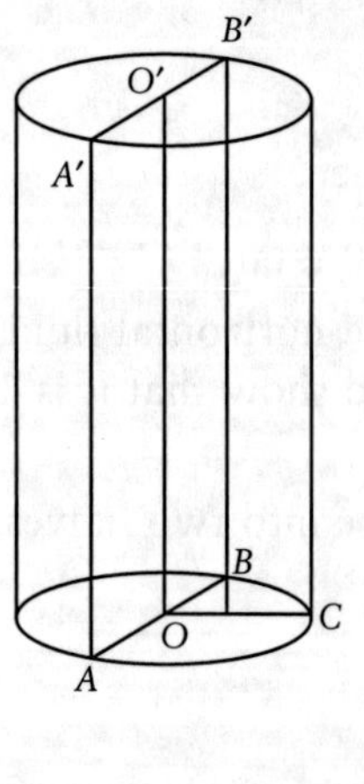

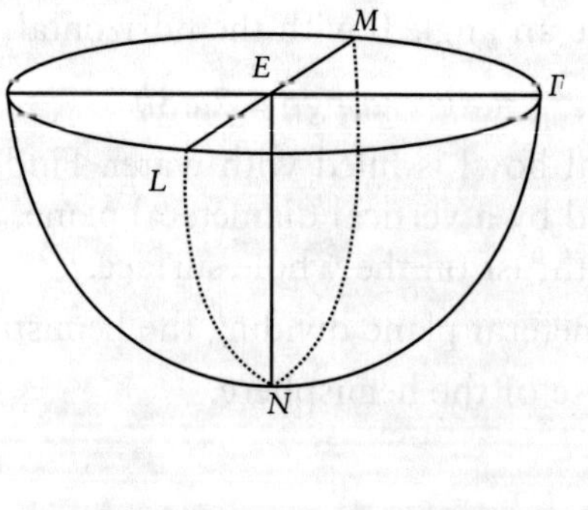

Let the radius of the base of cylinder be r.

Then, the radius of the base of hemisphere = r

Let the water stand to a height h in the cylinder. Then, volume of water in the cylinder = $\pi r^2 h$

Since the volume of water in the cylinder is equal to the volume of water in the hemisphere, therefore,

$$\pi r^2 h = 2/3 \pi r^3$$

$$\Rightarrow \quad h = 2/3 r \qquad ...(1)$$

Let X and Y be the components of the horizontal thrust on half of the curved surface of the hemisphere in two mutually perpendicular directions EF and EM respectively. Then,

X = Horizontal thrust in the direction of EF

= Thrust on the projection of the curved surface on a vertical plane perpendicular to EF

= Thrust on semi-circle LNM

= $1/2\ \pi a^2 \cdot 4r/3\pi \cdot g\rho$, where ρ is the density of water ...(2)

and, Y = Horizontal thrust in the direction of $EM = 0$ | by symmetry ...(3)

$\therefore$ Resultant horizontal thrust (H) on half of the curved surface of the hemisphere

$= \sqrt{X^2 + Y^2} = X$ | From (3) $= 2r^3/3 g\rho$ ...(4)

Again, let X' and Y' be the components of the horizontal thrust on half of the curved surface of the cylinder in two mutually perpendicular directions OC and OA respectively. Then,

X' = Horizontal thrust in the direction of OC

= Thrust on the projection of the curved surface under consideration on a vertical plane perpendicular to OC

= Thrust on rectangle $ABB'A'$

$= 2r \cdot h \cdot h/2 = rh^2 g\rho = r\,(2r/3)^2\, g\rho$ | From (1)

$= 4/9\ r^3 g\rho$ ...(5)

and, Y' = Horizontal thrust in the direction of $OA = 0$ | by symmetry ...(6)

$\therefore$ Resultant horizontal thrust (H$'$) on the curved surface under consideration of the cylinder

$= \sqrt{X'^2 + Y'^2}$

$= X'$ | From (6)

$= 4/9\ r^3 g\rho$ ...(7)

Hence, the required ratio = H : H$'$

$= 2r^3/3\ g\rho : 4/9\ r^3 g\rho = 3 : 2$

Example 13: The stem of a funnel has a radius a and its mouth has a radius b, the height of the stem is h and that of the slanting part is ℓ. The mouth is placed in a horizontal plane (being greased so as to remain water-tight) and the funnel is filled with water to the top of the stem. Prove that the upward pressure on the funnel is equal to

$$\pi w\,(b - a)\,\{h(a + b) + 1/3\ \ell\,(a + 2b)\},$$

where w is the weight of unit volume of water.

Sol.

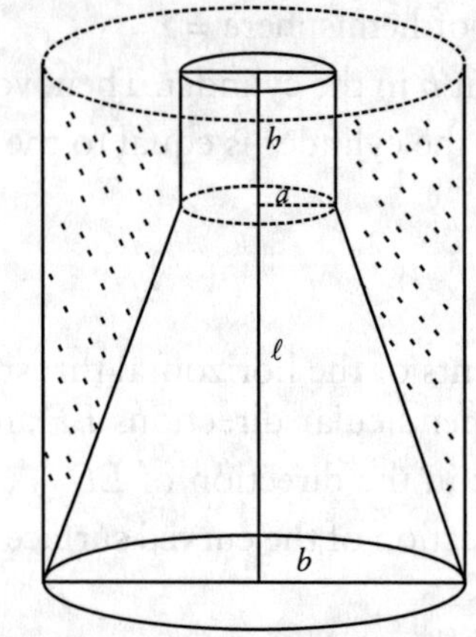

Upward pressure = weight of the super-incumbent water

= weight of the water contained in the shaded portion

= weight of the water contained in the cylinder of radium b and height $(h + \ell)$ – weight of the water contained in the frustum of the cone – weight of the water contained in the cylinder of radius a and height h

$= \pi b^2 (h + \ell) w - 1/3 \pi \ell (a^2 + ab + b^2) w - \pi a^2 hw = \pi w (b^2 - a^2) h + 1/3 \pi \ell (2b^2 - a^2 - ab) w$

$= \pi w (b - a) (b + a) h + 1/3 \pi \ell (b - a) (a + 2b) w = \pi w (b - a) \{h (a + b) + 1/3 \ell (a + 2b)\}$

Example 14: The end of a horizontal pipe is closed by a sphere of the same radius a as the internal section of the pipe. The sphere is hinged at its highest point. If the pipe is just full of liquid of density ρ, prove that the moment about the hinge of the liquid pressure on the sphere is $g\rho\pi a^4$

Sol.

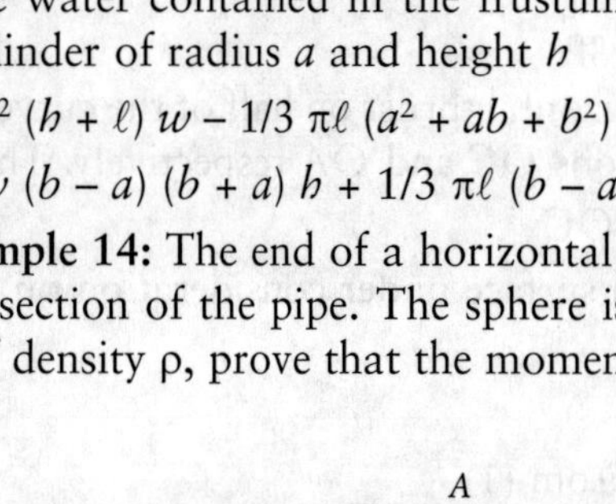

Let us suppose that the liquid is on the left side of the sphere. Then, only the curved surface of the hemisphere ACB will be in contact with the liquid. Hence, the fluid thrust will act only this half surfacc.

Let X and Y be the components of the horizontal thrust on this curved surface in two mutually perpendicular directions CO and EO respectively. Then,

X = Horizontal thrust on the surface under consideration in the direction of CO

= whole pressure on the projection of the above surface on a vertical plane perpendicular to CO

= whole pressure on the circle $AEBD = \pi a^2 \cdot a \cdot g\rho = \pi a^3 g\rho$

which acts at P, the centre of pressure of the circle $AEBD$, such that $AP = a + a^2/4a = 5a/4$

and, Y = Horizontal thrust on the above surface in the direction of EO

$= 0$ | by symmetry, since the surface under consideration is symmetrical about the plane $AOBC$

Let Z be the resultant vertical upward thrust on the above curved surface. Then,

Z = weight of the liquid displaced | by Principle of Archimedes

$= 2/3\ \pi a^3 g\rho$

which acts vertically upwards through G, the C.G. of the hemisphere (or the C.G. of the liquid displaced) such that $OG = 3/8a$.

Hence, the required moment of the liquid pressure about the hinge at A

$= X \,.\, AP - Z \,.\, OG = \pi a^3 g\rho \,.\, 5a/4 - 2/3\ \pi a^3 g\rho \,.\, 3a/8 = 5/4\ \pi a^4 g\rho - \pi a^4 g\rho/4 = \pi a^4 g\rho$

Example 15: A right cone whose base is an ellipse of semi-axes a, b is divided into two equal parts by a plane through its axis and the major axis of the base. One part is then removed and just immersed in a liquid with its vertex upwards and the axis of the cone vertical. Find the horizontal thrust on one half of its curved surface made by plane perpendicular to the vertical face of the solid.

Sol.

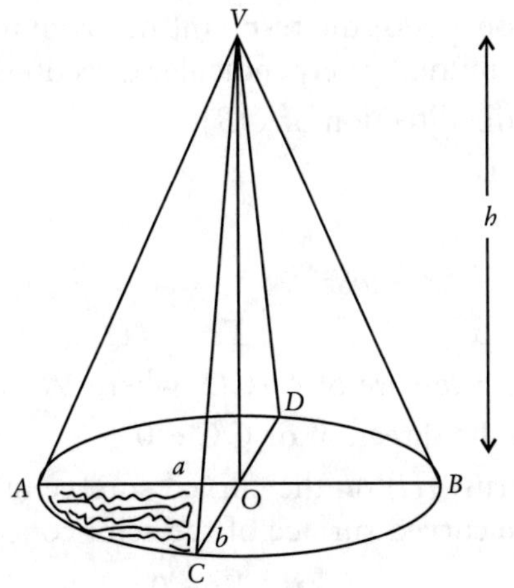

Let the height of the cone be h.

X = Horizontal thrust along AO = Thrust on $\Delta\ OVC$

$= bh/2 \,.\, 2h/3\ w = bh^2/3\ w$ ...(1)

Y = Horizontal thrust along OC = Thrust on $\Delta\ AVO$

$= ah/2 \,.\, 2h/3\ w = ah^2/3\ w$ ...(2)

$\therefore$ Resultant horizontal thrust (H)

$$= \sqrt{X^2 + Y^2} = \sqrt{\left(\frac{bh^2}{3}w\right)^2 + \left(\frac{ah^2}{3}w\right)^2} = \frac{1}{3}h^2 w\sqrt{a^2 + b^2}$$

Example 16: A thin hollow vessel in the shape of a right cone with a circular base is cut in two by a plane through the axis, and the two parts are hinged together at the vertex and the edges greased so as to be water-tight. The vessel is then hung up by the hinge and filled up with water through a small aperture near the hinge. Show that the water will not flow out if the vertical angle of the cone exceeds 120°.

Sol. Let h be the height and α be the semi-vertical angle of the cone.

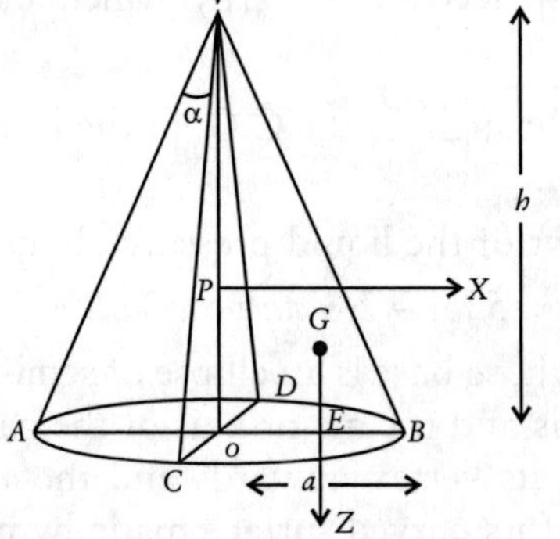

Let a be the radius of the base of the cone. Then, $\tan \alpha = a/h$ ...(1)

The vertical plane VCD divides the cone into two halves.

Let X and Y be the components of the resultant horizontal thrust on the curved surface of the semi-cone (V, BCD) in two mutually perpendicular directions OB and OC respectively. Then,

X = Horizontal thrust in the direction of OB

= whole pressure on $\Delta\ VCD$

$= g\rho \dfrac{2a.h}{2} \dfrac{2}{3} h$ | '$P = wS\bar{z}$'

$= 2/3\ g\rho a h^2 = 2/3\ g\rho h^3 \tan \alpha$ | From (1) ...(2)

acting through P, the centre of pressure of ΔVCD, where $VP = 3h/4$

and, Y = Horizontal thrust in the direction of $OC = 0$ | by symmetry ...(3)

∴ Resultant horizontal thrust (H) on the curved surface under consideration. Let Z be the resultant vertical thrust on the curved surface of the semi-cone (V, BCD). Then,

$= \sqrt{X^2 + Y^2} = X$ | $\because Y = 0$

$= 2/3\ g\rho h^3 \tan \alpha$ ...(4)

Z – Weight of the super-incumbent water = weight of water contained in this semi-cone

$= 1/2\ (1/3\ \pi a^2 h)\ g\rho$

$= 1/6\ \pi h^3 \tan^2\alpha g\rho$ | From (1) ...(5)

acting vertically downwards through C.G. G of the water contained, where the horizontal distance of G from the axis of the cone = OE

$= a/\pi = h \tan \alpha/\pi$ | From (1)

Now, from statics, we know that the water will not flow out, if

the moment of Z about $V >$ moment of H (or X) about V

$\Rightarrow 1/6\ \pi h^3 \tan^2 \alpha g\rho \,.\, h \tan a/\pi > 2/3\ g\rho h^3 \tan \alpha,\ 3h/4$

$\Rightarrow 1/3 \tan^2 \alpha > 1$ $\qquad \Rightarrow \tan^2 \alpha > 3$

$\Rightarrow \tan \alpha > \sqrt{3}$ $\qquad \Rightarrow \tan \alpha > \tan 60°$

$\Rightarrow \alpha > 60°$ $\qquad \Rightarrow 2\alpha > 120°$, i.e., the vertical angle of the cone exceeds 120°

Example 17: A solid sphere of density ρ is placed at the bottom of a vessel which is horizontal, and a liquid of density σ ($< \rho$) is poured in so as just to cover up the sphere. The sphere is then cut along the vertical diametral plane. Prove that the two parts will not separate if $\rho < 4\sigma$

Sol.

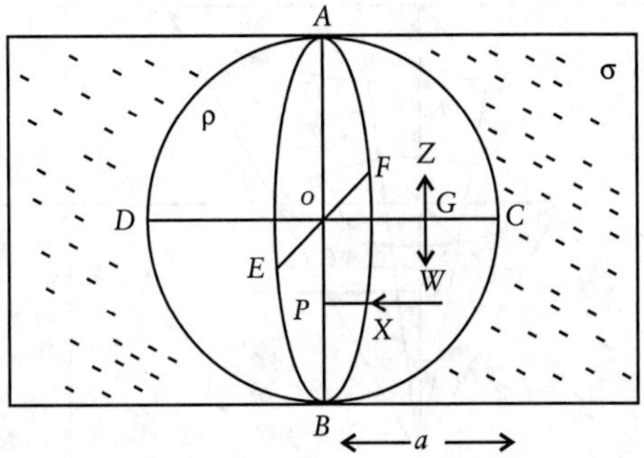

Let the centre of the sphere be O. Let a be the radius of the sphere.

A is the highest point and B is the lowest point of the sphere.

Let $AEBF$ be the vertical diametral plane dividing the sphere into two halves. The forces acting on the right hemisphere are:

(i) X = Horizontal thrust in the direction of CO = whole pressure on the circle $AEBF$

$= g\sigma\pi a^2 \,.\, a$ $\qquad | \text{'}P = wS\bar{z}\text{'}$

$= g\sigma\pi a^3$, acting through P, the C.P. of the circle, where $AP = a + a/4 = 5a/4$ so that $BP = AB - AP = 2a - 5a/4 = 3a/4$

(ii) Y = Horizontal thrust in the directum of EO (a direction perpendicular to CO)

$= 0$ $\qquad$ | by symmetry

(iii) Z = Resultant vertical thrust on the surface ACB

= weight of the liquid displaced $\qquad$ | by Principle of Archimedes

$= 2/3\ \pi a^3 g\sigma$, acting vertically upwards through G, the C.G. of the hemisphere ACB, where $OG = 3a/8$

(iv) W = weight of the sold hemisphere ACB

$= 2/3\ \pi a^3 g\rho$, acting through its centre of gravity G, where $OG = 3a/8$

(v) The reaction of the vessel in which the sphere has been placed

If the two parts separate, then they will take turn about the lowest point B. Hence they will not separate from each other, if the sum of the moments of X and Z about A is greater than the moment of W about A.

$\Rightarrow\ g\sigma\rho a^3 \,.\, 3a/4 + 2/3\ \pi a^3 g\sigma \,.\, 3a/8 > 2/3\ \rho a^3 g\rho \,.\, 3a/8$

$\Rightarrow\ 3\sigma/4 + \sigma/4 > \rho/4$ $\qquad \Rightarrow \sigma > \rho/4$ $\qquad \Rightarrow 4\sigma > \rho$ $\qquad \Rightarrow \rho < 4\sigma$

Example 18: A solid circular cone of uniform material and height h and of vertical angle 2α, is made of uniform material and floats in water with its axis vertical and vertex downwards and a length h' of the axis is immersed. The cone is bisected by a vertical plane through the axis and the two parts are hinged together at the vertex. Show that the two parts will remain in contact if $h' > h \sin^2 \alpha$ or $\tan^2 \alpha < h'/(h - h')$.

Sol.

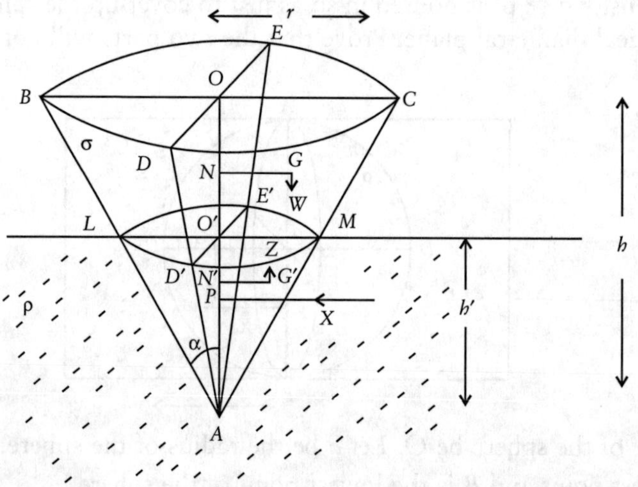

Let A be the vertex, O be the centre of base and r be the radius of the base of the solid cone. Let $AO = h$,

$AO' = h'$ and $O'M = r'$. Then, $\tan \alpha = r/h = r'/h'$ ...(1)

$ADEC$ is the half cone out off by the vertical plane ADE.

Let ρ be the density of water.

Since the cone is floating in water, therefore, its weight = weight of the water displaced (W'). Let W be the weight of half of the cone. Then, $W = W'$

$\Rightarrow$ $W = 1/2\ (1/3\ \pi r^2 h' g\rho)$

$= 1/6\ \pi h'^3 \tan^2 \alpha\ g\rho$ | From (1) ...(2)

The forces acting on the right half are

(i) X = Horizontal thrust in the direction of CO

= whole pressure on $\Delta\ AD'E' = g\rho\ 2r' \cdot h'/2 \cdot h'/3$

$= g\rho/3\ r'h'^2 = g\rho/3\ h'^3 \tan^2 \alpha$ | From (1)

acting through P. The C.P. of $\Delta\ AD'E'$, where $O'P = h'/2$

(ii) Y = Horizontal thrust in the direction of $OD = 0$ | by symmetry

$\therefore$ Resultant horizontal thrust $(H) = \sqrt{X^2 + Y^2} = X = \dfrac{g\rho}{3} h'^3 \tan^2 \alpha$

(iii) Z = Resultant vertical thrust on the surface under consideration

= weight of water displaced

$= 1/6\ \pi\ h'^3 \tan^2 \alpha\ g\rho$, acting upwards through the C.G. G' of the water displaced, where the horizontal distance of G' from the axis of cone $AO = G'N' = r'/\pi = h' \tan \alpha/\pi$ | From (1)

(iv) W = weight of the right half of the cone $ADCE$ = $1/6\ \pi h'^3 \tan^2 \alpha g\rho$ | From (2)
acting vertically downwards through G, the C.G. of the semi-cone (A, DCE), where the horizontal distance of G from the axis of the cone $AO = GN = r/\pi = h \tan \alpha/\pi$ | From (1)

The two parts will remain in contact (i.e. will not separate) if the sum of the moments of H and Z whose tendency is to keep the two parts together about A is greater than the moment of W whose tendency is to separate the two parts about A, i.e., if

$$\frac{g\rho}{3}h'^3 \tan^2 \alpha \frac{h'}{2} + \frac{1}{6}\pi h'^3 \tan^2 \alpha g\rho . \frac{h' \tan \alpha}{\pi} > \frac{1}{6}\pi h'^3 \tan^2 \alpha g\rho \frac{h \tan \alpha}{\pi}$$

$\Rightarrow h' (1 + \tan^2 \alpha) > h \tan^2 \alpha \qquad \Rightarrow h' > h \tan^2\alpha/1 + \tan^2 \alpha$

$\Rightarrow h' > h \tan^2 \alpha \cos^2 \alpha \Rightarrow h' > h \sin^2 \alpha \Rightarrow \tan^2 \alpha < h'/h - h'$

Example 19: A hollow right circular cone, filled with water is held with its axis vertical and vertex downward. Find the resultant pressure on the portion of its surface contained between two vertical planes through its axis and show that if the inclination of these planes to each other be 2β and the vertical angle of the cone 2α, the direction of the resultant pressure makes with the vertical an angle equal to $\tan^{-1}\left(\frac{\sin\beta}{\beta\tan\alpha}\right)$

Sol.

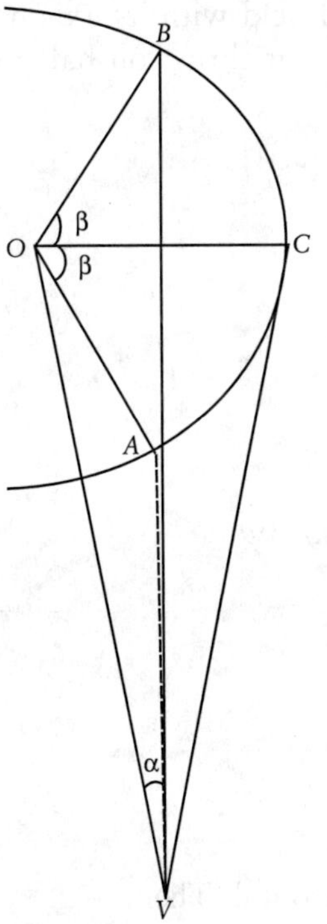

Let the radius of the base and the height of the cone be r and h respectively. Let (V, $OACB$) be the portion of the cone between two vertical planes VOA and VOB.

$\angle AOB = 2\beta$ | Given

Let C be the middle point of the arc AB. Then,

$\angle AOC = \angle BOC = \theta$, $\angle OVA = \alpha = \tan^{-1} r/h$

Area of the sector $OA\ CB = 1/2\ r^2\theta = 1/2\ r^2\ 2\beta = r^2\beta$

Let X and Y be the components of the resultant horizontal thrust on the curved surface of the portion of the cone under consideration in two mutually perpendicular directions OC and a line perpendicular to OC. Then,

X = Horizontal thrust on the curved surface of the portion under consideration in the direction OC

= Thrust on the projection of this curved surface on a vertical plane perpendicular to OC

= Thrust on $\Delta\ VAC = g\rho\ (1/2 . h . 2r \sin \beta)\ h/3$ | '$P = wS\bar{z}$'

$= 1/3\ h^2\ r \sin \beta\ g\rho$

Y = Horizontal thrust in a direction perpendicular to $OC = 0$ | by symmetry

Z = Resultant downward vertical thrust on the curved surface of the portion under consideration

= weight of water contained = $g\rho$. (1/3 area of the sector $OACB$) $h = g\rho . 1/3\ r^2\ \beta h$

$\therefore$ The required angle $\phi = \tan^{-1}(X/Z) = \tan^{-1}\left(\dfrac{\frac{1}{3}h^2 r\sin\beta g\rho}{g\rho\frac{1}{3}r^2\beta h}\right)$

$$= \tan^{-1}\left(\frac{h\sin\beta}{r\beta}\right)$$

$$= \tan^{-1}\left(\frac{h\sin\beta}{h\tan\alpha.\beta}\right) \qquad | \because r = h\tan\alpha$$

$$= \tan^{-1}\left(\frac{\sin\beta}{\beta\tan\alpha}\right)$$

EXERCISE 2

1. An unsymmetrical cask with plane horizontal ends of equal areas is just filled with water. Prove that the thrust on its curved surface is the same in magnitude whichever end is upper most.
2. If a hollow right circular cone is filled with liquid and held with its axis horizontal, find the magnitude and the line of action of the resultant thrust on half the curved surface cut off by a vertical plane through the axis.

   Ans. Magnitude $= wr^2\sqrt{\dfrac{\pi^2 r^2}{4} + h^2 + \dfrac{\pi^2 h^2}{36}}$;

   line of action is given by $\tan\theta = \dfrac{\pi h}{3\sqrt{\pi^2 r^2 + 4h^2}}$

   [**Hint:** Refer Ex. 2

   Resultant horizontal thrust $(H) = wr^2\sqrt{\dfrac{\pi^2 r^2}{4} + h^2}$

   Resultant vertical thrust $(Z) = 1/2\,.\,(1/3\ \pi r^2 hw) = 1/6\ \pi r^2 hw$

   $\therefore$ Resultant thrust $= \sqrt{H^2 + Z^2}$

   $$= \sqrt{w^2 r^4\left(\frac{\pi^2 r^2}{4} + h^2\right) + \frac{\pi^2 r^4 h^2 w^2}{36}}$$

   $$= wr^2\sqrt{\frac{\pi^2 r^2}{4} + h^2 + \frac{\pi^2 h^2}{36}}$$

   Let this resultant thrust make an angle θ with the horizontal. Then,

   $$\tan\theta = \frac{Z}{H} = \frac{\frac{1}{6}\pi r^2 hw}{wr^2\sqrt{\frac{\pi^2 r^2}{4} + h^2}} = \frac{\pi h}{3\sqrt{\pi^2 r^2 + 4h^2}}\ ]$$

3. A hollow cylinder closed by a plane base is filled with liquid and held with its axis vertical. Find the resultant horizontal thrust on half the cylinder cut off by a vertical plane through the axis.

   Ans. $ah^2g\rho$, where h is the height and a is the radius of the base of the cylinder.

   [**Hint:** Refer Ex. 5]

4. A solid right circular cylinder is divided in two equal parts by a plane through the axis. If it is held just immersed in a liquid with the axis horizontal and the plane section vertical, what will be the resultant horizontal thrust on the curved surface?

   Ans. $2r^2hg\rho$, where ρ is the density of the liquid, h is the height and r is the radius of the base of the cylinder.

   [**Hint:** Refer Ex. 6]

5. A hollow right circular cone filled with liquid is held with its axis vertical and vertex downwards. Prove that the magnitude of the resultant fluid thrust on half the surface of the cone cut off by a vertical plane through the axis is $\frac{1}{6}wrh\sqrt{\pi^2r^2+4h^2}$ and that it is inclined at an angle $\tan^{-1}(\pi r/2h)$ to the horizontal.

   [**Hint:** Refer Ex. 8. Use $\tan\alpha = r/h$]

6. A right circular cylinder is just immersed in water with its axis horizontal. Compare the horizontal fluid thrusts on the four parts of the curved surface made by the horizontal and vertical planes through the axis.

   Ans. Thrust on a lower part = 3 thrust on an upper part

7. A sphere of radius r is filled with a liquid of density ρ. Determine the resultant horizontal thrust upon half the curved surface of the sphere divided by a vertical plane through its centre.

   Ans. $\pi r^3 g\pi$

   [**Hint:** Refer Ex. 9]

8. A spherical shell formed of two halves in contact along a vertical plane is filled with water. Show that the resultant pressure an either half of the shell is $\frac{\sqrt{13}}{4}$ of the total weight of the liquid.

   [**Hint:** Refer Ex. 9]

   $$R = \frac{\sqrt{13}}{3}\pi r^3 g\rho$$

   $$= \frac{\sqrt{13}}{4}(4/3\ \pi r^3 g\rho)$$

   $$= \frac{\sqrt{13}}{4}[\text{total weight of the liquid}]$$

9. A hemisphere of radius a is immersed in a liquid of density σ. The plane of the base is vertical and its centre at a depth $a\sqrt{5}$, below the surface. Show that the resultant force

on the curved surface is 7/3 $\pi\sigma g a^3$ and that its direction makes with the horizontal an angle θ where $\tan\theta = \dfrac{2}{\sqrt{45}}$.

[**Hint:** Refer Ex. 10

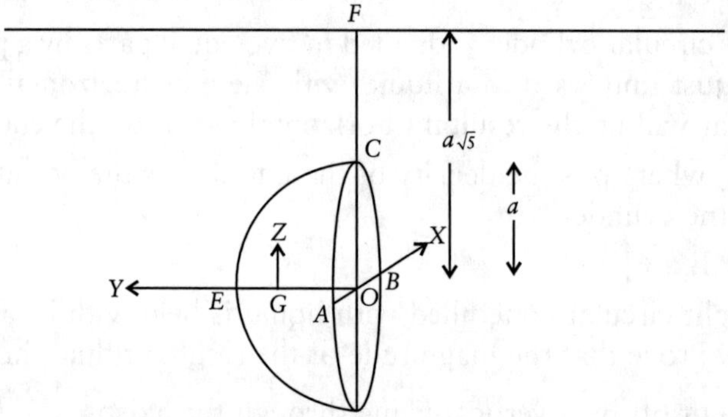

$X = 0$ | By symmetry about the plane *CDE*

$Y = \pi a^2 . a\sqrt{5}\, g\sigma = \pi a^3 g\sigma\sqrt{5}$

$\therefore\ H = \sqrt{X^2 + Y^2} = Y = \pi a^3 g\sigma\sqrt{5}$

$Z = \dfrac{2}{3}\pi a^3 g\sigma,\ R = \sqrt{X^2 + Y^2 + Z^2}$ | ∵ Lines of action of *Y* and *Z* lie in the plane *CDE* and are not parallel. Hence, they will be intersecting.

$= \sqrt{X^2 + Z^2} = \dfrac{7}{3}\pi a^3 g\sigma,\ \tan\theta = \dfrac{Z}{H} = \dfrac{Z}{Y} = \dfrac{2}{\sqrt{45}}$]

10. A solid hemisphere of radius *a* is placed with its plane base in a vertical plane. What is the horizontal thrust on the curved surface?

    Find also the resultant thrust on it. Assume that the centre of the hemisphere is at a depth *h* below the surface of water.

    Ans. $H = \pi a^2 h g\rho$; $R = \pi a^2 g\rho\sqrt{h^2 + \dfrac{4a^2}{9}}$; $\theta = \tan^{-1}(2a/3h)$

    [**Hint:** Refer Ex. 10.

    $X = \pi a^2 . h . g\rho$ ...(1)

    $Y = 0$

    $\therefore\ H = \sqrt{X^2 + Y^2} = X = \pi a^2 h g\rho$

    $Z = \dfrac{2}{3}\pi\, a^3 g\rho$

    $R = \sqrt{H^2 + Z^2} = \sqrt{X^2 + Z^2}$

$$= \sqrt{(\pi a^2 hg\rho)^2 + \left(\frac{2}{3}\pi a^3 g\rho\right)^2}$$

$$= \pi a^2 g\rho\sqrt{h^2 + \frac{4a^2}{9}}\ ;\ \tan\theta = \frac{Z}{H} = \frac{Z}{X} = \frac{\frac{2}{3}\pi a^3 g\rho}{\pi a^2 hg\rho} = \frac{2a}{3h}\,]$$

11. A right hollow cylinder with a circular base is filled with liquid and closed at the top. Prove that the ratio of the height of the cylinder to the radius of the base is 2 : 1 if the resultant horizontal thrust on half the curved surface cut off by a vertical plane through the axis, when the axis is horizontal is equal to that when the axis is vertical.

[Hint: Case I. Refer Ex. 6

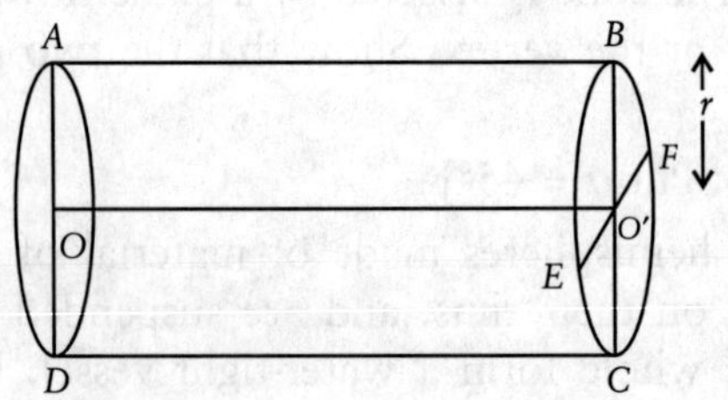

X = Horizontal thrust along $O'E$

= Thrust on rectangle $ABCD$

$= 2r \,.\, h \,.\, r \,.\, w = 2r^2hw$, where w is the weight per unit volume of the liquid.

Y = Horizontal thrust along OO'

$= 0$ | by symmetry

$$\therefore H = \sqrt{X^2 + Y^2} = X = 2r^2hw \qquad ...(1)$$

Case II. Refer Ex. 12

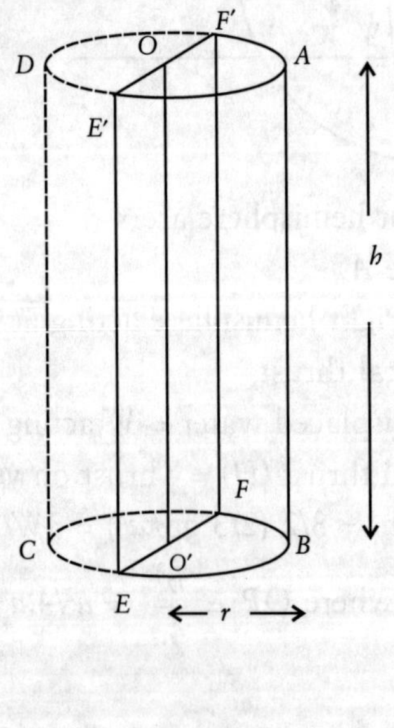

X' = Horizontal thrust along OA = Thrust on rectangle $EFF'E'$

= $2r \cdot h \cdot h/2w$, where w is the weight per unit volume of the liquid.

= rh^2w

Y' = Horizontal thrust along $E'F'$

= 0 | by symmetry

$\therefore H' = \sqrt{X'^2 + Y'^2} = X' = rh^2w$ ...(2)

According to the question, $H = H'$

$\Rightarrow 2r^2hw = rh^2w \quad \Rightarrow 2r = h \quad \Rightarrow h/r = 2/1]$

12. A solid right circular cone of uniform material and height h and of vertical angle 90°, floats in water with its axis vertical and vertex downwards and a length h' of its axis immersed. The cone is bisected by a plane through the axis and the two parts are hinged together at the vertex. Show that the two parts will remain in contact if $h' > h/2$.

[**Hint:** Refer Ex. 18. Put $\alpha = 45°$]

13. Two closely fitting hemispheres made of material of uniform thickness are hinged together at a point on their rims, and are suspended from the hinge. The rims are greased so that the whole form a water-tight vessel. It is being filled with a liquid through a small hole near the hinge. Prove that the contact will not give away if the weight of the shell exceeds three times that of the water it contains.

[**Hint:**

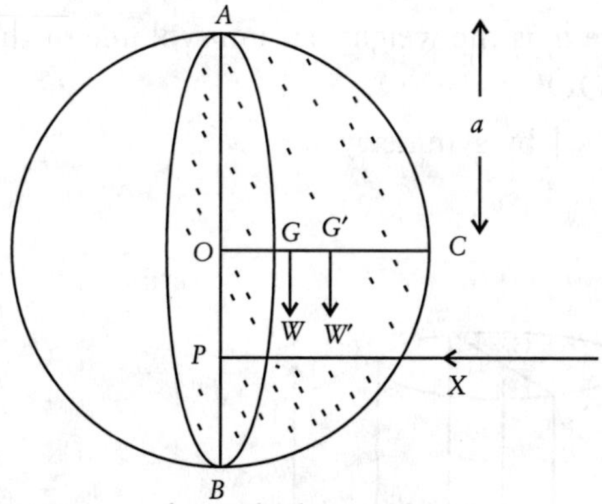

Force acting on the right hemisphere are:

(i) Reaction at the hinge A

(ii) W' = weight of the right hemisphere acting at G where $OG = a/2$

(iii) Z = Resultant vertical thrust

= Weight of the displaced water = W acting at G' where $OG' = 3a/8$

(iv) Resultant horizontal thrust (H) = Thrust on vertical circle

= $g\rho\pi a^2 \cdot a = g\rho\pi a^3 = 3/2\ (2/3\ g\rho\pi a^3) = 3W/2$

acting through P where $OP = \dfrac{\text{'}h^2\text{'}}{4a} = a^2/4.a = a/4$

The contact will not give away if the sum of the moments of (ii) and (iii) about $A \geq$ moment of (iv) about A

$\Rightarrow$ $W' \,.\, a/2 + W.\, 3a/8 \geq 3W/2\,(a + a/4)$

$\Rightarrow$ $W' \geq 3W$

$\Rightarrow$ weight of the shell $\geq$ 3 weight of water contained.]

14. A weightless sphere is divided by a vertical plane into two halves, which are hinged together at their lowest point and it is just filled with water. Show that the tension of the string, which ties together the highest point of the two halves is 3/8 of the weight of water that the sphere would contain.

[**Hint:**

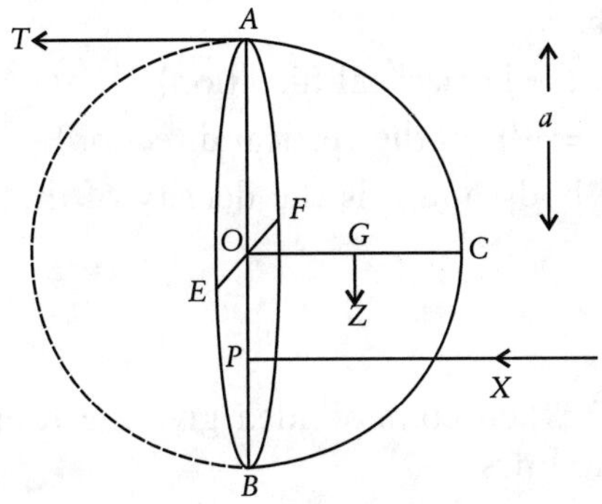

The forces acting on right hemisphere ACB

(1) X = Horizontal thrust along $OC = \pi a^2 \,.\, a \,.\, g\rho = \pi a^3 g\rho$, acting through P, the C.P. of the circle $AEBF$, where OP = '$h^2/4a$' $= a^2/4a = a/4$

$\therefore\ AP = 5a/4 \ \therefore\ BP = 2a - 5a/4 = 3a/4$

(2) Y = Horizontal thrust along $OE = 0$ | by symmetry

(3) $Z = 2/3\ \pi a^3 g\rho$ acting at G where $OG = 3a/8$

(4) Action of the hinge at B

(5) Tension T of the string at A

Taking moment about B, we get

$T \,.\, 2a - X.\, 3/4\, a - Z \,.\, 3a/8 = 0$

$\Rightarrow$ $T = 1/2\ \pi a^3 g\rho = 3/8\ (4/3\ \pi a^3 g\rho) = 3/8$ weight of water contained.]

12. Thrust on a Curved Surface Bounded by a Plane Curve

Let a solid body immersed completely in a fluid be composed partly of a curved surface and partly of known plane area. Then, the resultant fluid thrust on the whole body (including curved surface and plane area) will be equal to the weight of the fluid displaced [Refer Principle of Archimedes] and act vertically upwards through the C.G. of the displaced fluid.

Let X and Y be the components of the resultant thrust R on the curved surface in the horizontal and vertical directions respectively. Also, let X' and Y' be the components of the resultant thrust R' (whole pressure) on the plane surface in the horizontal and vertical directions respectively.

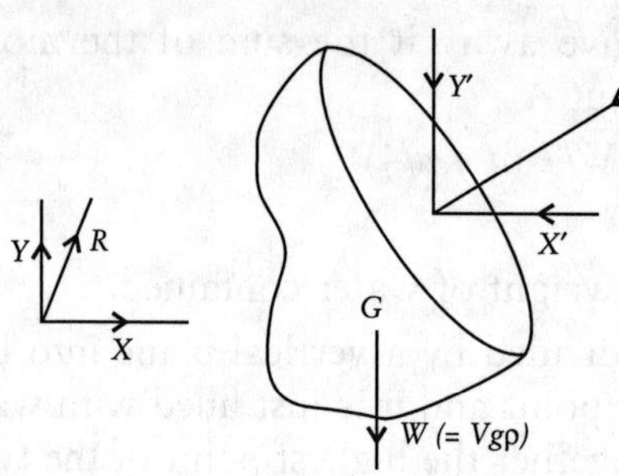

Since the resultant thrust on the solid body as a whole is a vertical force $Vg\rho$ acting through the C.G. of the solid and obviously there is no horizontal component of the resultant thrust on the solid body, therefore, we have

$$X - X' = 0 \text{ (in the horizontal direction)}$$

and, $$Y - Y' = Vg\rho \ (= W) \text{ (in the vertical direction)}$$

where V is the volume of the body and ρ is the density of the fluid so that $Vg\rho$ (= W) is the weight of the fluid displaced

$$\Rightarrow \quad X = X' \qquad \text{...(1)}$$

and, $$Y = Y' + Vg\rho \qquad \text{...(2)}$$

The equations (1) and (2) when compounded give the resultant thrust R'' on the curved surface of immersed body as follows:

$$R'' = \sqrt{X^2 + Y^2} \qquad \text{...(3)}$$

If ϕ is the angle which this resultant thrust R'' makes with the horizontal, then,

$$\tan \phi = Y/X$$

$$\Rightarrow \quad \phi = \tan^{-1} (Y/X) \qquad \text{...(4)}$$

ILLUSTRATIVE EXAMPLES

Example 1: A solid cone, whose vertical angle is 2α, is immersed in a liquid with its vertex in the surface and axis vertical. Prove that if P be the whole pressure on the curved surface and base and P^1 the resultant pressure, then

$$P/P^1 = (2 + 3 \sin \alpha)/\sin \alpha$$

Sol.

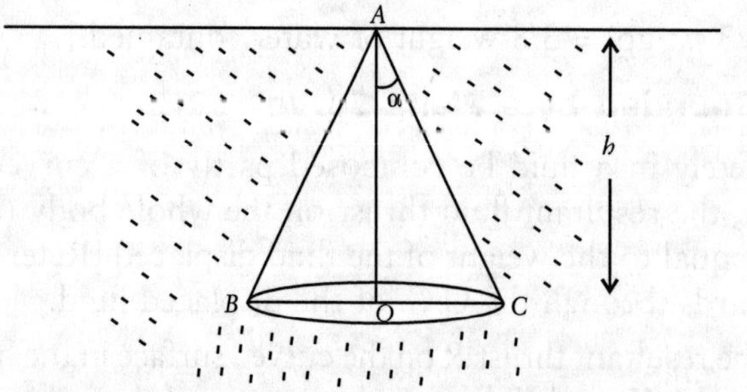

Let h be the height of the cone. Let ρ be the density of the liquid.

Whole pressure on the base of the cone

$= g\rho \cdot \pi (h \tan \alpha)^2 \cdot h$ | '$P = wS\bar{z}$'

$= \pi h^3 \tan^2 \alpha g\rho$ ...(1)

Whole pressure on the curved surface of the cone

$= g\rho (\pi rl)\, 2/3\, h$ | '$P = wS\bar{z}$'

$= g\rho (\pi h \tan \alpha h \sec \alpha)\, 2/3\, h$

$= 2/3\, \pi h^3 \tan \alpha \sec \alpha g\rho$ ...(2)

$\therefore P$ = whole pressure on the curved surface and base

$= 2/3\, \pi h^3 \tan \alpha \sec \alpha g\rho + \pi h^3 \tan^2 \alpha g\rho$

$= 1/3\, \pi h^3 \tan \alpha\, (2 \sec \alpha + 3 \tan \alpha)\, g\rho$ ...(3)

Again, P' = Resultant pressure on the surface of the cone

= weight of the liquid displaced

$= 1/3\, \pi h^3 \tan^2 \alpha g\rho$ ...(4)

$$\therefore \quad \frac{P}{P'} = \frac{\frac{1}{3}\pi h^3 \tan\alpha(2\sec\alpha + 3\tan\alpha)g\rho}{\frac{1}{3}\pi h^3 \tan^2\alpha g\rho} = \frac{2\sec\alpha + 3\tan\alpha}{\tan\alpha} = \frac{2 + 3\sin\alpha}{\sin\alpha}$$

Example 2: A solid hemisphere is immersed in a liquid with the highest point of its plane base in the surface of liquid and the base is inclined at an angle $\tan^{-1} 2$ to the horizontal. Show that the resultant thrust on the curved surface is equal to twice the weight of the displaced liquid.

Sol.

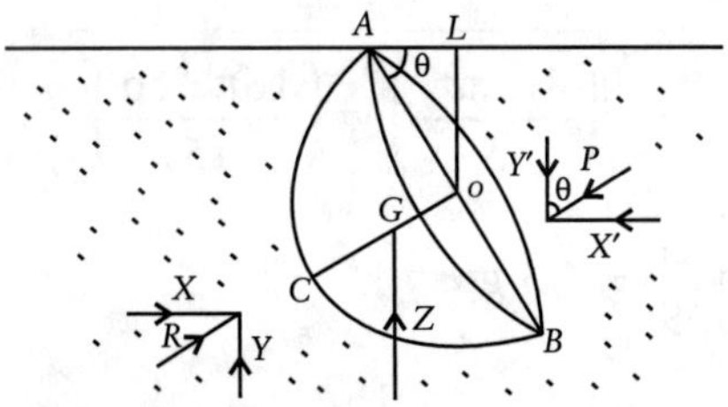

Let a be the radius of the hemisphere. Let ρ be the density of the liquid. Let the base AB be inclined at an angle θ to the horizontal. Then,

$\tan\theta = 2$ | Given ...(1)

From (2),

$$\left.\begin{aligned} \sin\theta &= \frac{2}{\sqrt{5}} \\ \text{and,} \quad \cos\theta &= \frac{1}{\sqrt{5}} \end{aligned}\right\} \quad ...(2)$$

Let P be the resultant thrust on the plane base of the hemisphere. Then,

$P = g\rho \cdot \pi a^2 \cdot OL = g\rho\pi a^2\, a \sin\theta = g\rho\pi a^3 \sin\theta$

$$= \frac{2g\rho\pi a^3}{\sqrt{5}} \qquad \text{| From (2)} \qquad ...(3)$$

Let X' and Y' be the components of P in the horizontal and vertical directions. Then,

$$\left.\begin{aligned} X' &= P\sin\theta \\ Y' &= P\cos\theta \end{aligned}\right\} \qquad ...(4)$$

Let Z be the resultant thrust on the whole surface of the hemisphere. Then,

$$Z = \text{Weight of the displaced liquid} \qquad \text{| By Principle of Archimedes}$$

$$= 2/3\ \pi a^3 g\rho \qquad ...(5)$$

Let X and Y be the horizontal and vertical components of the resultant thrust on the curved surface of the hemisphere.

Since the hemisphere is in equilibrium, therefore, resolving the forces acting on the hemisphere horizontally and vertically, we get

$$X = X' = P\sin\theta$$

$$Z = Y - P\cos\theta$$

$$\therefore \qquad X = \frac{2g\rho\pi a^3}{\sqrt{5}} \cdot \frac{2}{\sqrt{5}} = \frac{4g\rho\pi a^3}{5}$$

and, $\qquad Y = Z + P\cos\theta$

$$= \frac{2}{3}\pi a^3 g\rho + \frac{2g\rho\pi a^3}{\sqrt{5}} \cdot \frac{1}{\sqrt{5}} = \left(\frac{2}{3} + \frac{2}{5}\right)\pi a^3 g\rho = 16/15\ \pi a^3 g\rho$$

Let R be the resultant thrust on the curved surface of the hemisphere. Then,

$$R = \sqrt{X^2 + Y^2} = \sqrt{\left(\frac{4g\rho\pi a^3}{5}\right)^2 + \left(\frac{16\pi a^3 g\rho}{15}\right)^2}$$

$$= \frac{4}{5} g\rho\pi a^3 = \sqrt{1 + \frac{16}{9}} = 4/3\ g\rho\pi a^3$$

$$= 2\left(\frac{2}{3}\pi a^3 g\rho\right)$$

= Twice the weight of liquid displaced.

Again, let this resultant thrust makes an angle ϕ with the horizontal. Then,

$$\tan\phi = Y/X$$

$$= \frac{16\pi a^3 g\rho/15}{4g\rho\pi a^3/5} = \frac{4}{3}$$

Example 3: A hemisphere, radius a, is entirely submersed in a liquid of density ρ, so that its diametral plane makes an angle θ with the horizontal and has its centre at a depth h. Prove that the resultant force on the curved surface is

$$\pi a^2 g\rho\ (4/9\ a^2 + h^2 \pm 4/3\ ah\cos\theta)^{1/2}$$

or

A hemispherical surface of radius a is immersed in liquid of density ρ with its centre at a depth h and its base inclined at an angle θ to the horizontal. Find the resultant thrust on the curved surface and its direction.

Sol.

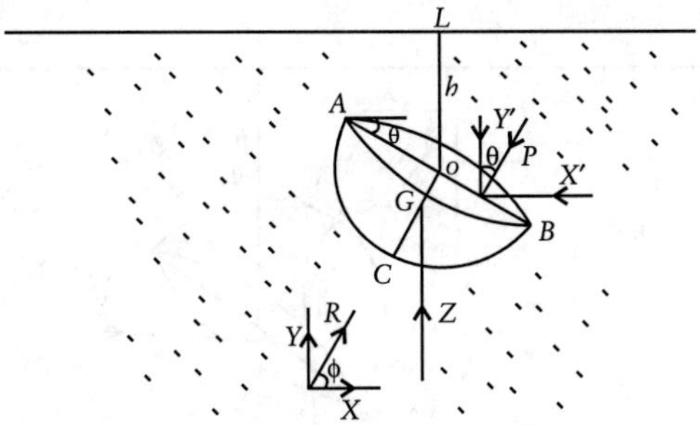

Case I (Base upwards): Let P be the whole pressure on the plane base of the hemisphere acting at the centre of pressure of the base. Then,

$$P = g\rho\pi a^2 h \qquad | \text{ `}P = wS\bar{z}\text{'}$$
$$= \pi a^2 hg\rho \qquad ...(1)$$

Let X' and Y' be the horizontal and vertical components of P. Then,

$$X' = P\sin\theta = \pi a^2 hg\rho \sin\theta \qquad ...(2)$$
$$Y' = P\cos\theta = \pi a^2 hg\rho\cos\theta \qquad ...(3)$$

Let Z be the resultant thrust on the whole hemisphere. Then,

$$Z = \text{weight of the liquid displace} \qquad | \text{ By Principle of Archimedes}$$
$$= 2/3\ \pi a^3 g\rho, \text{ acting vertically upwards} \qquad ...(4)$$
through G, the C.G. of the liquid displaced

Let X and Y be the components of the thrust R on the curved surface of the hemisphere in horizontal and vertical directions respectively. Then, resolving horizontally and vertically all the forces acting on the hemisphere as a whole, we get

$$X = X' = P\sin\theta = \pi a^2 h g \rho \sin\theta \qquad ...(5)$$
$$Y - Y' = Z$$
$$\Rightarrow \quad Y = Y' + Z$$
$$= \pi a^2 hg\rho\cos\theta + 2/3\ \pi a^3 hg\rho \qquad ...(6)$$

$\therefore \quad R$ = Resultant thrust on the curved surface

$$= \sqrt{X^2 + Y^2} = \sqrt{(\pi a^2 hg\rho\sin\theta) + (\pi a^2 hg\rho\cos\theta + 2/3\pi a^3 g\rho)^2}$$

$$= \sqrt{\pi^2 a^4 h^2 g^2 \rho^2 + 4/3\pi^2 a^5 hg^2\rho^2\cos\theta + 4/9\pi^2 a^6 g^2 \rho^2}$$

$$= \pi a^2 g\rho\sqrt{h^2 + 4/9a^2 + 4/3ah\cos\theta} \qquad ...(7)$$

Let ϕ be the angle which the resultant thrust makes with the horizontal. Then, $\tan\phi = Y/X$

$$= \frac{\pi a^2 hg\rho\cos\theta + 2/3\pi a^3 g\rho}{\pi a^2 hg\rho\sin\theta} = \frac{2a + 3h\cos\theta}{3h\sin\theta}$$

Case II (Base downwards)

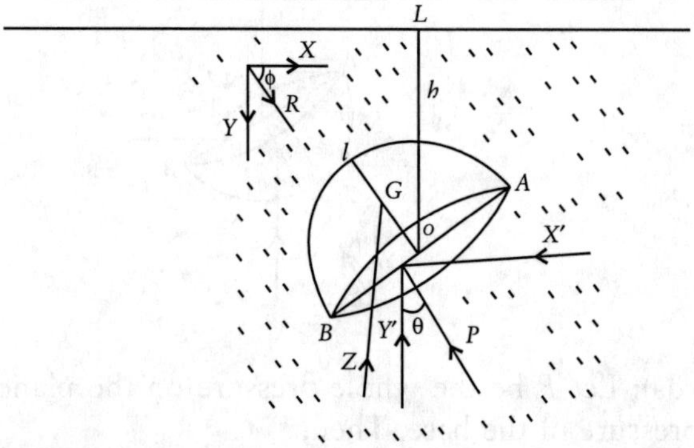

$$P = g\rho\pi a^2 h = \pi a^2 hg\rho \quad \text{...(1)}$$
$$X' = P\sin\theta = \pi a^2 hg\rho\sin\theta \quad \text{...(2)}$$
$$Y' = P\cos\theta = \pi a^2 hg\rho\cos\theta \quad \text{...(3)}$$
$$Z = 2/3\ \pi a^3 hg\rho \quad \text{...(4)}$$

Resolving all the force acting on the hemisphere horizontally and vertically, we get

$$X = X' = P\sin\theta = \pi a^2 hg\rho\sin\theta \quad \text{...(5)}$$
$$Y' - Y = Z \qquad \Rightarrow Y = Y' - Z$$
$$= \pi a^2 hg\rho\cos\theta - 2/3\ \pi a^2 g\rho \quad \text{...(6)}$$

$\therefore$ R = Resultant thrust on the curved surface

$$= \sqrt{X^2 + Y^2} = \sqrt{(\pi a^2 hg\rho\sin\theta) + (\pi a^2 hg\rho\cos\theta - 2/3\pi a^3 g\rho)^2}$$
$$= \pi a^2 g\rho\sqrt{h^2 + 4/9a^2 - 4/3ah\cos\theta} \quad \text{...(7)}$$

Let ϕ be the angle which the resultant thrust makes with the horizontal. Then,

$$\tan\phi = Y/X$$
$$= \frac{\pi a^2 hg\rho\cos\theta - 2/3\pi a^3 g\rho}{\pi a^2 hg\rho\sin\theta} = \frac{3h\cos\theta - 2a}{3h\sin\theta}$$

Hence, the required resultant thrust on the curved surface is

$$\pi a^2 g\rho\sqrt{\frac{4}{9}a^2 + h^2 \pm \frac{4}{3}ah\cos\theta}$$

Example 4: A hollow weightless hemisphere with a plane base is filled with water and hung up by means of a string. Prove that the whole pressure on the curved surface and on the base are in the ratio of 19 : 8.

Also, find the resultant thrust on the curved surface and prove that the resultant thrust on its curved surface makes on angle $\tan^{-1}\dfrac{109}{96}$ with the horizontal.

Sol.

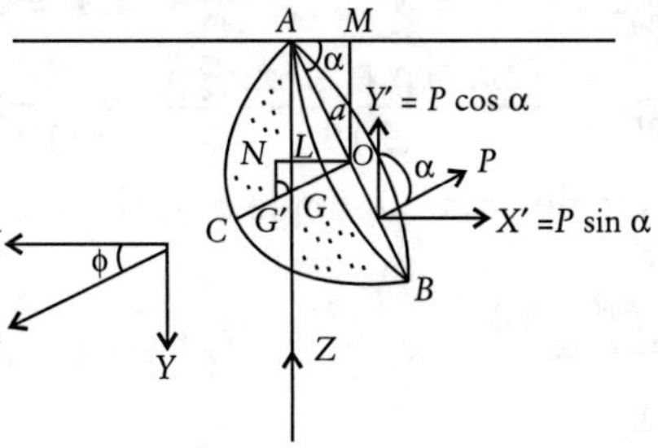

Let O be the centre and a be the radius of the hemisphere. A is the point of suspension. Let G be the C.G. of the liquid contained in the hemisphere. Then, by statics, for equilibrium AG is vertical where $OG = 3a/8$

Let the diameter AB be inclined at an angle α to the horizontal. Then,

$$\tan(90° - \alpha) = \frac{OG}{OA} \qquad \Rightarrow \cot\alpha = \frac{3a/8}{a}$$

$$\Rightarrow \quad \cot\alpha = 3/8 \qquad \Rightarrow \tan\alpha = 8/3$$

$$\therefore \quad \sin\alpha = \frac{8}{\sqrt{73}}$$

and, $\cos\alpha = \dfrac{3}{\sqrt{73}}$

Depth of C.G. (O) of the plane circular base below the horizontal line through A

$$= OM = OA \sin\alpha = a \sin\alpha = \frac{8a}{\sqrt{73}}$$

Let P be the whole pressure on the plane circular base acting at the C.P. of the base. Then,

P = 'wSz'

$$= g\rho\pi a^2 \frac{8a}{\sqrt{73}} = \frac{8\pi a^3}{\sqrt{73}} g\rho$$

Let X' and Y' be the horizontal and vertical components of P. Then,

$$X' = P\sin\alpha = \frac{8\pi a^3}{\sqrt{73}} g\rho \sin\alpha = \frac{8\pi a^3}{\sqrt{73}} g\rho . \frac{8}{\sqrt{73}} = \frac{64}{73}\pi a^3 g\rho$$

$$Y' = P\cos\alpha = \frac{8\pi a^3}{\sqrt{73}} g\rho \cos\alpha = \frac{8\pi a^3}{\sqrt{73}} g\rho . \frac{8}{\sqrt{73}} = \frac{24}{73}\pi a^3 g\rho$$

Let Z be the resultant thrust on the whole hemisphere. Then,

Z = Weight of the water displaced | By Principle Archimedes

= $2/3\ \pi a^3 g\rho$ acting vertically upwards through G, the C.G. of the water displaced.

Let X and Y be the components of the thrust R on the curved surface of the hemisphere in horizontal and vertical directions respectively. Then, resolving horizontally and vertically all the forces acting on the hemisphere as a whole, we get

$X = X' = 64/73\ \pi\ a^3 g\pi;\ Y - Y' = z \qquad \Rightarrow Y = Y' + z$

$$= \frac{24}{73}\pi a^3 g\rho + \frac{2}{3}\pi a^3 g\rho = \frac{72\pi a^2 g\rho + 146\pi a^3 g\rho}{219} = \frac{218}{219}\pi a^3 g\rho$$

If the resultant thrust R is inclined at an angle ϕ to the horizontal, then

$$\tan\phi = \frac{Y}{X} = \frac{\frac{218}{219}\pi a^3 g\rho}{\frac{64}{73}\pi a^3 g\rho} = \frac{109}{96} \qquad \Rightarrow \phi = \tan^{-1}\left(\frac{109}{96}\right)$$

Also, $R = \sqrt{X^2 + Y^2} = \sqrt{(P\sin\alpha)^2 + (P\cos\alpha + Z)^2} = \sqrt{P^2 + Z^2 + 2PZ\cos\alpha}$

$$= \sqrt{\left(\frac{8}{\sqrt{73}}\pi a^3 g\rho\right)^2 + \left(\frac{2}{3}\pi a^3 g\rho\right)^2 + 2\frac{8}{\sqrt{73}}\pi a^3 g\rho.\frac{2}{3}\pi a^3 g\rho.\frac{3}{\sqrt{73}}}$$

$$= \pi a^3 g\rho\sqrt{\frac{64}{73} + \frac{4}{9} + \frac{32}{73}} = \pi a^3 g\rho\sqrt{\frac{576 + 292 + 288}{9 \times 73}} = \pi a^3 g\rho\sqrt{\frac{1156}{9 \times 73}}$$

$$= 2\pi a^3 g\rho\sqrt{\frac{289}{9 \times 73}} = 2\pi a^3 g\rho.\frac{17}{3\sqrt{73}} = \frac{34}{3\sqrt{73}}\pi a^3 g\rho$$

Again, let G' be the C.G. of the curved surface of the hemisphere. Then, G' lies on the symmetrical radius OC such that $OG' = a/2$. Its depth from the horizontal line through A

$$= AL + NG' = MO + OG' \cos\alpha = \frac{89}{\sqrt{73}} + \frac{a}{2}.\frac{3}{\sqrt{73}} = \frac{19a}{\sqrt{73}}$$

$$\therefore \text{ Required ratio } = \frac{\frac{19a}{\sqrt{73}}}{\frac{8a}{\sqrt{73}}} = \frac{19}{8} = 19:8$$

Example 5: A cylindrical vessel full of water is held with its axis inclined at an angle 45° to the vertical. Find the magnitudes of the pressures on the ends and show that the resultant pressure in the curved surface is equal to the difference between the pressures on the ends.

Sol.

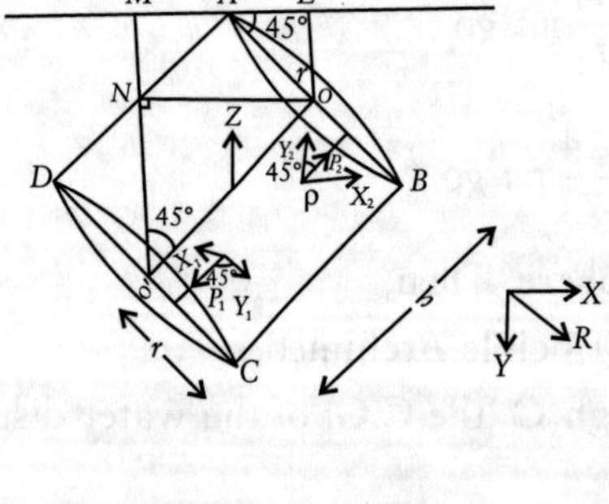

Let r and h be the radius of the base and the height of the cylinder respectively. Let P_1 and P_2 be the pressures on the lower and upper ends respectively.

Then,

$$P_1 = \text{‘}w\,S\,\bar{z}\text{’}$$
$$= g\rho\pi r^2\, O'M, \text{ where } \rho \text{ is the density of water}$$
$$= g\rho\pi r^2\,(O'N + NM) = g\rho\pi r^2\,(O'N + OL) = g\rho\pi r^2\,(h \cos 45° + r \sin 45°)$$
$$= \frac{\pi r^2 g\rho(h+r)}{\sqrt{2}}$$
$$P_2 = \text{‘}wS\bar{z}\text{’}$$
$$= g\rho\pi r^2\, OL = g\rho\pi r^2\, r \sin 45° = \frac{\pi r^3 g\rho}{\sqrt{2}}$$

∴ Difference between the pressures at the ends

$$= P_1 - P_2 = \frac{\pi r^2 g\rho(h+r)}{\sqrt{2}} - \frac{\pi r^3 g\rho}{\sqrt{2}} = \frac{\pi r^2 g\rho h}{\sqrt{2}}$$

Let Z be the resultant thrust on the whole surface of the cylinder. Then,

$$Z = \text{weight of water in the cylinder} \qquad | \text{ By Principle of Archimedes}$$
$$= \pi r^2 h g\rho$$

Let X and Y be the horizontal and vertical components of the resultant thrust on the curved surface of the cylinder.

Since the water in the cylinder is in equilibrium, therefore the weight of the water in the cylinder must balance the pressures of water on both ends and the curved surface.

∴ Resulting horizontal and vertically all the forces acting on the whole cylinder, we get

$$X + P_2 \cos 45° - P_1 \cos 45° = 0$$

and, $Y + P_1 \cos 45° - P_2 \cos 45° = Z$

$$\therefore \quad X = P_1 \cos 45° - P_2 \cos 45° = \frac{P_1 - P_2}{\sqrt{2}} = \frac{\pi r^2 g\rho h}{2}$$

and, $Y = Z + P_2 \cos 45° - P_1 \cos 45°$

$$= Z + \frac{P_2 - P_1}{\sqrt{2}} = Z - \frac{P_1 - P_2}{\sqrt{2}} = \pi r^2 h g\rho - \frac{\pi r^2 g\rho h}{2} = \frac{\pi r^2 g\rho h}{2}$$

Let R be the resultant pressure on the curved surface of the cylinder. Then,

$$R = \sqrt{X^2 + Y^2} = \frac{\pi r^2 g\rho h}{2}\sqrt{2} = \frac{\pi r^2 g\rho h}{\sqrt{2}} = P_1 - P_2$$

Example 6: A solid cone is just immersed with a generating line in the surface. If θ be the inclination to the vertical of the resultant thrust on the curved surface, prove that

$$(1 - 3 \sin^2 \alpha) \tan\theta = 3 \sin \alpha \cos \alpha, \; 2\alpha \text{ being the vertical angle of the cone.}$$

Sol.

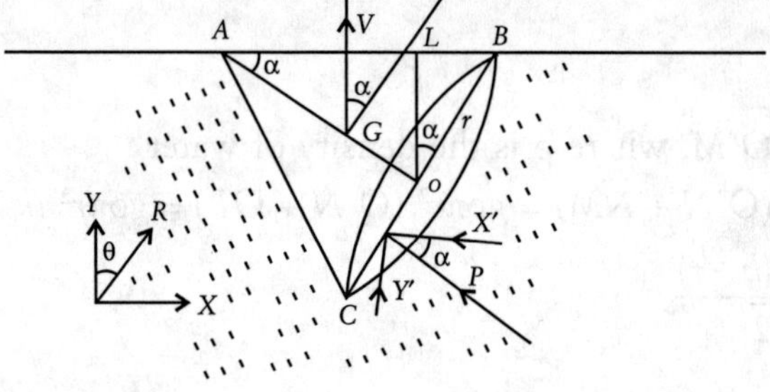

Let AB be a generating line of the cone ABC in the free surface. Let r be the radius of the base of the cone. Then, height of the cone = $r \cot \alpha$

By the principle of Archimedes, the resultant thrust on the whole surface of the cone is equal to the weight of the liquid displaced.

$\therefore \quad V = 1/3\ \pi r^2\ (r \cot \alpha)\ g\rho$, where ρ is the density of the liquid.

$\Rightarrow \quad V = \pi r^3/3 \cot \alpha g\rho$, acting vertically upwards through G, the C.G. of the liquid displaced

Let P be the whole pressure on the plane base of the cone.

Then, $\quad P = \text{'}wS\bar{z}\text{'} = g\rho\ .\ \pi r^2\ .\ OL = g\rho\ .\ \pi r^2\ .\ r \cos \alpha = \pi r^3 \cos \alpha g\rho$,

acting at right angles to the plane end through the C.P. of the plane end.

Let X' and Y' be the horizontal and vertical components respectively of the whole pressure P. Then,

$$X' = P \cos \alpha = \pi r^3 \cos^2 \alpha g\rho \quad ; \quad Y' = P \sin \alpha = \pi r^3 \cos \alpha \sin \alpha g\rho$$

Let X and Y be the horizontal and vertical components respectively of the resultant thrust R on the curved surface. Then resolving the forces horizontally and vertically, we get

$$X = X' = \pi r^3 \cos^2 \alpha g\rho$$

and, $\quad Y + Y' = V \quad \Rightarrow Y = V - Y' = \pi r^3/3 \cot \alpha g\rho - \pi r^3/3 \cos \alpha \sin \alpha g\rho$

$$= \frac{\pi r^3 \cot\alpha}{3}(1 - 3\sin^2 \alpha)g\rho$$

Since the resultant thrust on the curved surface makes an angle θ with the vertical, therefore,

$$\tan\theta = \frac{X}{Y} = \frac{\pi r^3 \cos^2 \alpha g\rho}{\frac{\pi r^3 \cot\alpha}{3}(1 - 3\sin^2 \alpha)g\rho} = \frac{3\sin\alpha\cos\alpha}{1 - 3\sin^2 \alpha}$$

$$\Rightarrow \quad (1 - 3 \sin^2 \alpha) \tan\theta = 3 \sin \alpha \cos \alpha$$

Example 7: A hollow cone without weight, closed and filled with some liquid, is suspended from liquid, is suspended from a point in the rim of its base. If ϕ be the angle which the direction of the resultant pressure makes with the vertical, then show that $\cot\phi = \dfrac{28\cot\alpha + \cot^3 \alpha}{48}$, α being the semi-vertical angle of the cone.

Sol.

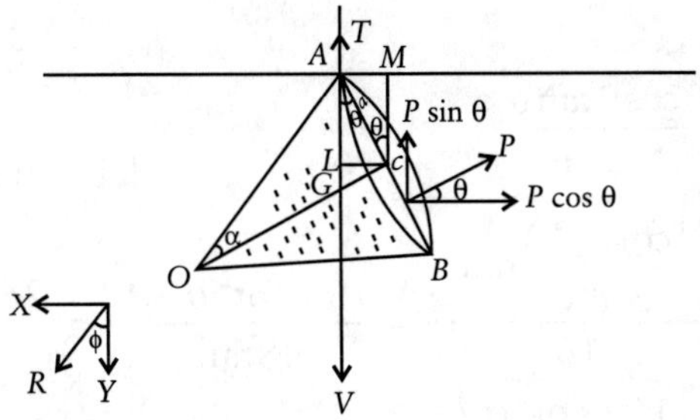

Let a be the radius of the base of cone and A be the point of suspension of the cone OAB. Then, in the equilibrium position, A and G, the C.G. of the cone filled with liquid, must be in the same vertical line. Let h be the height of the cone. Let θ be the angle which the base of the cone makes with the vertical.

Then, $$\tan\theta = \frac{GC}{AC} = \frac{\frac{h}{4}}{a} = \frac{h}{4a} = \frac{a\cot\alpha}{4a} \quad | \because h = a\cot\alpha$$

$$= \frac{\cot\alpha}{4} \qquad \text{...(1)}$$

$$\therefore \quad \sin\theta = \frac{\cot\alpha}{\sqrt{16+\cot^2\alpha}} \qquad \text{...(2)}$$

and, $$\cos\theta = \frac{4}{\sqrt{16+\cot^2\alpha}} \qquad \text{...(3)}$$

Let P be the thrust on the base of the cone. Then,

P = 'wSz' = $g\rho\pi a^2\, a\cos\theta$, where ρ is the density of the liquid

$= g\rho\pi a^3 \cos\theta$

$= g\rho\pi h^3 \tan^3\alpha \cos\theta$ ...(4)

Let V be the resultant vertical thrust on the cone as a whole. Then,

V = weight of the liquid contained | By Principle of Archimedes

$= 1/3\ \pi a^2 hg\rho$

$= 1/3\ \pi h^3 \tan^2\alpha g\rho$, acting vertically downwards through G. ...(5)

Let X and Y be the horizontal and vertical components of the resultant thrust on the curved surface of the cone. Then, resolving the forces acting on the cone in the horizontal and vertical directions, we have,

$X = P\cos\theta$

$= g\rho\pi h^3 \tan^3\alpha \cos^2\theta$ | From (4) ...(6)

$V = Y - P\sin\theta \quad \Rightarrow Y = V + P\sin\theta$

$= 1/3\ \pi h^3 \tan^2\alpha g\rho + g\rho\pi h^3 \tan^3\alpha \cos\theta \sin\theta$ | From (5) and (4)

$= 1/3\ \pi h^3 \tan^2\alpha g\rho\ (1 + 3\sin\theta\cos\theta\tan\alpha)$ ...(7)

If ϕ is the angle which the direction of the resultant pressure makes with the vertical, then

$$\cot\phi = Y/X$$

$$= \frac{1+3\sin\theta\cos\theta\tan\alpha}{3\tan\alpha\cos^2\theta} \quad | \text{ From (6) and (7)}$$

$$= \frac{1+3\left(\frac{4\cot\alpha}{16+\cot^2\alpha}\right)\tan\alpha}{3\tan\alpha\left(\frac{16}{16+\cot^2\alpha}\right)} = \frac{16+\cot^2\alpha+12}{48\tan\alpha} = \frac{28\cot\alpha+\cot^3\alpha}{48}$$

Example 8: A cone floats with its axis horizontal in a liquid of density double its own. Find the pressure on its base and prove that if θ be the inclination to the vertical of the resultant thrust on the curved surface, and α the semi-vertical angle of the cone, then $\tan\theta = 4/\pi \tan\alpha$.

Sol. Let h be the height of the cone. Since the density of the liquid is double that of cone, the cone will float with half of its part inside the fluid. Hence, the horizontal plane OCD passing through the axis OO′ of the cone will be in the surface of the liquid.

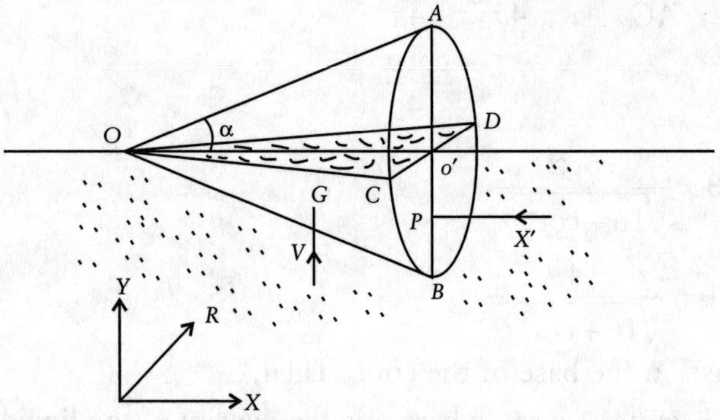

By principle of Archimedes, the resultant thrust (V) on the curved surface of the semi-cone immersed in the liquid will be equal to the weight of the liquid displaced acting vertically upwards through G, the C.G. of the liquid displaced.

$$\therefore\ V = 1/2\,(1/3\ \pi h^3 \tan^2\alpha g\rho) \quad ...(1)$$

Let X and Y be the horizontal and vertical components of the resultant thrust on the curved surface of semi-cone. Then, resolving horizontally and vertically, we get

$$X = X' = \text{Horizontal thrust on the semi-circle } BCD$$

$$= \text{‘}w S \bar{z}\text{’} = g\rho.\frac{\pi r^2}{2}.\frac{4r}{3\pi} = \frac{2}{3}g\rho r^3 = \frac{2}{3}g\rho h^3\tan^3\alpha \quad ...(2)$$

$$Y = V \quad | \because \text{ There is no vertical thrust on the semi-circle } BCD$$

$$= 1/6\ \pi h^3 \tan^2\alpha g\rho \quad | \text{ From (1)} \quad ...(3)$$

Let R be the resultant thrust on the curved surface of the semi-cone making an angle θ with the vertical. Then,

$$R = \sqrt{X^2+Y^2} = \sqrt{\left(\frac{2}{3}g\rho h^3\tan^3\alpha\right)+\left(\frac{1}{6}\pi h^3\tan^2\alpha g\rho\right)^2}$$

$$= \frac{1}{6} h^3 \tan^2 \alpha g\rho \sqrt{16 \tan^2 \alpha + \pi^2}$$

and, $$\tan\theta = \frac{X}{Y} = \frac{\frac{2}{3} g\rho h^3 \tan^3 \alpha}{\frac{1}{6} \pi h^3 \tan^2 \alpha g\rho} = \frac{4}{\pi} \tan\alpha$$

Example 9: A solid octant of a sphere is immersed with one plane face in the surface. Prove that the resultant pressure on the curved surface is $(1 + 8/\pi^2)^{1/2}$ times the weight of the water displaced by the octant.

Sol.

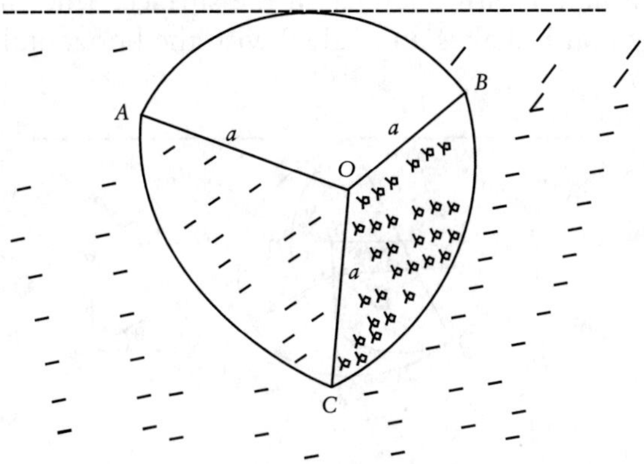

Let a plane face *OAB* of the octant be in the surface of water. Let *X* and *Y* be the components of the horizontal thrust on the curved surface of the octant in the directions *OA* and *OB* respectively. Then,

X = Horizontal thrust along *OA*

= whole pressure on the projection of the curved surface upon the plane at right angles to *OA*

= whole pressure on area *OBC*

$= g\rho . \frac{4a}{3\pi} \frac{\pi a^2}{4}$ | '$P = wS\bar{z}$'

$= 1/3\ a^3 g\rho$

Y = Horizontal thrust along *OB*

= whole pressure on the projection of the curved surface upon the plane at right angles to *OB*

= whole pressure on area *OAC*

$= g\rho . \frac{4a}{3\pi} \frac{1}{4} \pi a^2$ | '$P = wS\bar{z}$'

$= 1/3\ a^3 g\rho$

Let *Z* be the vertical thrust of the fluid on the curved surface. Then,

Z = weight of the liquid displaced by the octant | By Principle of Archimedes

$= 1/8\ (4/3\ \pi a^3)\ g\rho = 1/6\ \pi a^3 g\rho$

$\therefore$ Resultant pressure on the curved surface

$$= \sqrt{X^2 + Y^2 + Z^2} = \sqrt{\left(\frac{1}{3}a^3 g\rho\right)^2 + \left(\frac{1}{3}a^3 g\rho\right)^2 + \left(\frac{1}{6}\pi a^3 g\rho\right)^2}$$

$$= \frac{1}{3}a^3 g\rho\sqrt{1+1+\frac{\pi^2}{4}} = \frac{1}{3}a^3 g\rho\frac{1}{2}\sqrt{8+\pi^2} = \frac{1}{6}\pi a^3 g\rho\sqrt{1+\frac{8}{\pi^2}}$$

$$= \left(1+\frac{8}{\pi^2}\right)^{1/2} \text{ times the weight of the water displaced by the octant}$$

Example 10: Portion of a sphere cut-off by two planes through its centre inclined at an angle $\pi/4$ is just immersed in a liquid with one face in the surface. Find the resultant thrust on the curved surface and show that it makes an angle θ with the horizontal so that $\tan\theta = \pi/2 - 1$.

Sol.

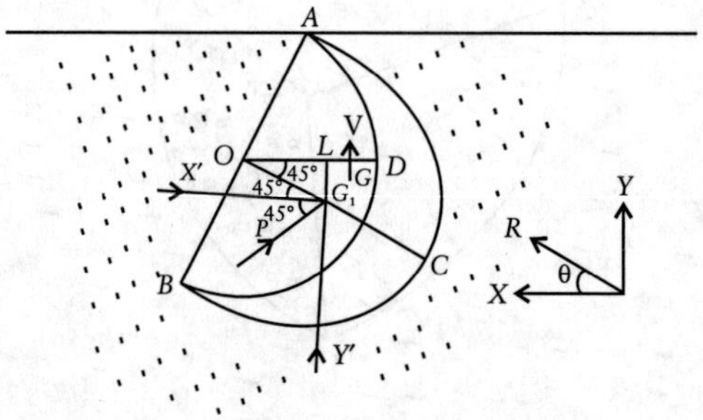

Let ABD and ABC be the two plane faces of the solid. Let the plane face ABD be in the surface of the liquid. The lower plane face ABC is inclined at an angle $\pi/4$ with the horizontal. Let a be the radius of the sphere. Let P be the whole pressure on the plane face ABC. Then,

P = '$wS\bar{z}$'

$= g\rho\,(1/2\pi a^2)\, G_1L = g\rho\,(1/2\pi a^2)\,(OG_1 \sin \pi/4)$

$= g\rho\,(1/2\pi a^2)\,(4a/3\pi\; 1/\sqrt{2}) = 1/3\; a^2\, \sqrt{2}\; g\rho$ acting normally to the plane force ABC.

Let X' and Y' be the horizontal and vertical components respectively of the whole pressure P. Then,

$$X' = P\cos 45° = \frac{P}{\sqrt{2}} = \frac{a^3}{3}g\rho$$

$$Y' = P\sin 45° = \frac{P}{\sqrt{2}} = \frac{a^3}{3}g\rho$$

Let V be the resultant thrust on the body as a whole.

Then, V = weight of the liquid displaced | By Principle of Archimedes

$= 1/8$ (volume of the sphere) $g\rho$

$= 1/8\; 4/3\; \pi a^3 g\rho$

$= 1/6\; \pi a^3 g\rho$, acting vertically upwards through G, the C.G. of the liquid displaced

Let X and Y be the horizontal and vertical components respectively of the resultant thrust on the curved surface of the body. Then, resolving the forces acting on the body in the horizontal and vertical directions, we get

$$X = X' = a^3 g\rho/3$$

and, $\quad Y + Y' = V \qquad \Rightarrow Y = V - Y' = 1/6\ \pi a^3 g\rho - a^3 g\rho/3$

Let R be the resultant thrust on the curved surface. Then,

$$R = \sqrt{X^2 + Y^2} = \sqrt{\left(\frac{a^3 g\rho}{3}\right)^2 + \left(\frac{\pi a^3 g\rho}{6} - \frac{a^3 g\rho}{3}\right)^2}$$

$$= \frac{a^3 g\rho}{6}\sqrt{4 + (\pi - 2)^2} = \frac{a^3 g\rho}{6}\sqrt{\pi^2 - 4\pi + 8}$$

Let R make an angle θ with the horizontal.

Then,

$$\tan\theta = \frac{Y}{X} = \frac{\frac{1}{6}\pi a^3 g\rho - \frac{a^3 g\rho}{3}}{\frac{a^3 g\rho}{3}} = \frac{\frac{\pi}{6} - \frac{1}{3}}{\frac{1}{3}} = \frac{\pi}{2} - 1$$

EXERCISE 3

1. A solid hemisphere is immersed in a liquid with the highest point of its plane base in the surface and the base is inclined at a given angle α to the horizontal. Show that the resultant thrust on the curved surface will be equal to twice the weight of the liquid displaced, if $\tan\alpha = 2$.

   [**Hint:** Refer Ex. 2

   $R^2 = (P \sin\alpha)^2 + (Z + P\cos\alpha)^2 = P^2 \sin^2\alpha + Z^2 + P^2\cos^2\alpha + 2ZP\cos\alpha$

   $= P^2 + Z^2 + 2ZP\cos\alpha = (g\rho\pi a^3 \sin\alpha)^2 + (2/3\ \pi a^2 g\rho)^2 + 2.2/3\ \pi a^3 g\rho . g\rho\pi a^3 \sin\alpha . \cos\alpha$

   $= (\pi a^3 g\rho)^2\ [\sin^2\alpha + 4/9 + 4/3 \sin\alpha\cos\alpha]$

   $\Rightarrow$ R $= \pi a^3 g\rho\ (\sin^2\alpha + 4/9 + 4/3\sin\alpha\cos\alpha)^{1/2}$

   If $R = 2$ weight of the liquid displaced, then

   $\pi a^3 g\rho\ (\sin^2\alpha + 4/9 + 4/3\sin\alpha\cos\alpha)^{1/2} = 2\ (2/3\ \pi a^3 g\rho)$

   $\Rightarrow \sin^2\alpha + 4/9 + 4/3\sin\alpha\cos\alpha = (4/3)^2 = 16/9 \qquad \Rightarrow 9\sin^2\alpha + 4 + 12\sin\alpha\cos\alpha = 16$

   $\Rightarrow 9\sin^2\alpha + 12\sin\alpha\cos\alpha - 12 = 0 \qquad \Rightarrow 3\sin^2\alpha + 4\sin\alpha\cos\alpha - 4 = 0$

   $\Rightarrow 3\sin^2\alpha + 4\sin\alpha\cos\alpha - 4(\sin^2\alpha + \cos^2\alpha) = 0 \Rightarrow 4\sin\alpha\cos\alpha - \sin^2\alpha - 4\cos^2\alpha = 0$

   $\Rightarrow \sin^2\alpha - 4\sin\alpha\cos\alpha + 4\cos^2\alpha = 0 \qquad \Rightarrow \tan^2\alpha - 4\tan\alpha + 4 = 0$

   $\Rightarrow (\tan\alpha - 2)^2 = 0 \qquad \Rightarrow \tan\alpha = 2$]

2. Find the direction and magnitude of the resultant thrust on the curved surface of a hemisphere of radius 3 cm, placed with its base vertical and centre at a depth of 6 cm below the surface of a liquid of which one cubic cm weighs w grams.

   Ans. $\tan^{-1} 1/3$ with the horizontal, 178.88 w grams.

3. A spherical shell is filled with liquid. Find the magnitude and line of action of the resultant pressure on each of the hemispheres divided by a diametral plane making an angle θ to the horizon.

Ans. $R = \frac{1}{3}\pi a^3 g\rho\sqrt{13+12\cos\theta},\ \phi = \tan^{-1}\left(\frac{2+3\cos\theta}{3\sin\theta}\right)$ (Base upwards)

$R = \frac{1}{3}\pi a^3 g\rho\sqrt{13-12\cos\theta},\ \phi = \tan^{-1}\left(\frac{3\cos\theta-2}{3\sin\theta}\right)$ (Base downwards)

[**Hint:** Refer Ex. 3. Put $h = a$]

4. A cylinder closed at both ends is entirely immersed in water with its axis inclined at a given angle θ to the horizontal. If its height be h and the radius of its base be a, prove that the resultant thrust on its curved surface is $\cos\theta$ times the weight of the liquid displaced inclined at are angle θ to the vertical.

5. A closed cylinder whose base diameter is equal to its length is full of water and hangs freely from a point in its upper rim. If the weight of the cylinder is neglected, prove that the vertical and horizontal components of the resultant thrust on the curved surface are each half the weight of the water.

6. A right cone of semi-vertical angle 30° has its upper most generator in the surface of a liquid. Show that the resultant thrust on its curved surface passes through the centre of the base.

7. A hollow weightless cone of vertical angle 120° is filled with water and suspended freely from a point in the rim of its base. Find the ratio of the thrust on the base to the weight of water, that the cone would contain. If it is held with its lowest generator horizontal, prove that the resultant pressure on the curved surface is 7/2 times the weight of the water contained.

Ans. $\tan^{-1}(72/7)$ [**Hint:** Refer Q. 13]

8. A solid right circular cone of vertical angle 2α is just immersed in water so that one generator is in the surface of the liquid. Prove that the resultant pressure on the curved surface of the cone is to weight of the fluid displaced by the cone, as $\sqrt{1+3\sin^2\alpha} : 1$ and that it is inclined to the axis of the cone at an angle $\cot^{-1}(2\tan\alpha)$.

[**Hint:** Refer Ex. 6

$X = X' = P\cos\alpha \qquad Y = V - P\sin\alpha$

$$V = \frac{\pi r^3 \cot\alpha\, g\rho}{3}$$

$P = \pi r^3 \cos\alpha g\rho = (\pi r^3 \cot\alpha g\rho)\sin\alpha = 3V\sin\alpha$

$$\therefore \quad R = \sqrt{X^2+Y^2} = \sqrt{(P\cos\alpha)^2+(V-P\sin\alpha)^2} = \sqrt{P^2+V^2-2VP\sin\alpha}$$

$$= \sqrt{9V^2\sin^2\alpha+V^2-6V^2\sin\alpha} = V\sqrt{9\sin^2\alpha+1-6\sin^2\alpha} = V\sqrt{3\sin^2\alpha+1}$$

$$\therefore \quad \frac{R}{V} = \frac{\sqrt{1+3\sin^2\alpha}}{1} = \sqrt{1+3\sin^2\alpha} : 1$$

Resolving the forces acting upon the curved surface along and perpendicular to the axis of the cone, we get

$X'' = P - V \sin\alpha = \pi r^3 \cos\alpha g\rho - \pi (r^3/3) \cot\alpha g\rho \sin\alpha = 2/3\, \pi r^3 \cos\alpha g\rho$

$$Y'' = V\cos\alpha = \frac{\pi r^3}{3}\cot\alpha\, g\rho\cos\alpha = \frac{\pi r^3}{3}\frac{\cos^2\alpha}{\sin\alpha}g\rho$$

∴ Required angle = $\cot^{-1}(X''/Y'')$

$$= \cot^{-1}\left(\frac{\frac{2}{3}\pi r^3\cos\alpha g\rho}{\frac{\pi r^3}{3}\frac{\cos^2\alpha}{\sin\alpha}g\rho}\right) = \cot^{-1}(2\tan\alpha)$$

Aliter [For angle]: Let the resultant thrust make an angle ϕ with the horizontal. Then,

$\tan\phi = \cot\theta \qquad | \because \theta + \phi = \pi/2$

$$= \frac{1-3\sin^2\alpha}{3\sin\alpha\cos\alpha} = \frac{\sin^2\alpha - 3\tan^2\alpha}{3\tan\alpha} = \frac{1-2\tan^2\alpha}{3\tan\alpha}$$

∴ Angle between the resultant thrust R and the axis of the cone

$= \phi + \alpha = \tan^{-1}\{\tan(\phi + \alpha)\}$

$$= \tan^{-1}\left(\frac{\tan\phi + \tan\alpha}{1-\tan\phi\tan\alpha}\right) = \tan^{-1}\left(\frac{\frac{1-2\tan^2\alpha}{3\tan\alpha}+\tan\alpha}{1-\frac{1-2\tan^2\alpha}{3\tan\alpha}.\tan\alpha}\right) = \tan^{-1}\left(\frac{1+\tan^2\alpha}{2\tan\alpha + 2\tan^3\alpha}\right)$$

$$= \tan^{-1}\left(\frac{1}{2\tan\alpha}\right) = \tan^{-1}\left(\frac{\cot\alpha}{2}\right) = \cot^{-1}\left(\frac{2}{\cot\alpha}\right)$$

$= \cot^{-1}(2\tan\alpha)$]

9. A solid cone is just immersed in water with a generating line in the surface. Prove that the inclination to the vertical of the resultant thrust on the curved surface is $\tan^{-1}\left(\frac{3\tan\alpha}{1-2\tan^2\alpha}\right)$, where 2α is the vertical angle of the cone.

   [**Hint:** Refer Aliter Q. 8]

10. A solid right circular cone of vertical angle 60° is just immersed in water with a generating line in the surface. Find the angle which the resultant thrust on the curved surface makes with the vertical.

    Ans. $\tan^{-1}(3/\sqrt{3})$

    [**Hint:** Refer Aliter Q. 8.

    $2\alpha = 60° \qquad \Rightarrow \alpha = 30°$

$$\tan\phi = \frac{1-2\tan^2\alpha}{3\tan\alpha} = \frac{1-2\tan^2 30°}{3\tan 30°} = \frac{1-2\left(\frac{1}{\sqrt{3}}\right)^2}{3\left(\frac{1}{\sqrt{3}}\right)} = \frac{1-\frac{2}{3}}{\sqrt{3}} = \frac{1}{3\sqrt{3}}$$

Required angle = $\pi/2 - \phi$

$= \tan^{-1} \{\tan (\pi/2 - \phi)\} = \tan^{-1} \{\tan (\cot \phi) = \tan^{-1} (3/\sqrt{3})]$

11. A right cone of vertical angle 2α is filled with water. It is then closed and laid with a generating line in contact with a table. Prove that the resultant vertical and horizontal thrusts upon the curved surface are respectively.

$W(1 + 3\sin^2\alpha)$ and $3W \sin\alpha \cos\alpha$, where W is the weight of the contained liquid.

[**Hint:** Refer Ex. 7; Put $\theta = \alpha$

$X = g\rho\pi h^3 \tan^3\alpha \cos^2\alpha$

$= 3(1/3\, \pi h^3 \tan^3 \alpha g\rho) \tan\alpha \cos^2\alpha$

$= 3W \sin\alpha \cos\alpha$ | $\because W = 1/3\, \pi h^3 \tan^2 \alpha g\rho$

$Y = 1/3\, \pi h^3 \tan^2 \alpha g\rho\, (1 + 3 \sin\alpha \cos\alpha \tan\alpha)$

$= W(1 + 3\sin^2\alpha)$]

12. A cone, whose vertical angle is 2α, has its lowest generator horizontal and is filled with liquid. Prove that the resultant pressure on the curved surface is where $W\sqrt{1+15\sin^2\alpha}$, where W is the weight of the liquid.

[**Hint:** Refer Q. 11

$R^2 = X^2 + Y^2$

$= 9W^2 \sin^2\alpha \cos^2\alpha + W^2 (1 + 6\sin^2\alpha + 9\sin^4\alpha)$

$= W^2 [1 + 9\sin^2\alpha (\sin^2\alpha + \cos^2\alpha) + 6\sin^2\alpha]$

$= W^2 (1 + 15\sin^2\alpha)$ $\Rightarrow R = W\sqrt{1+15\sin^2\alpha}$]

13. A hollow weightless cone of vertical angle 2α, is filled with a liquid and suspended freely from a point on the rim of its base. Show that the thrust at the base is to the weight of fluid the cone would contain in the ratio.

$12\sin^2\alpha : \cos^2\alpha\sqrt{1+15\sin^2\alpha}$]

[**Hint:** Refer Ex. 7

$P = g\rho\pi h^3 \tan^3\alpha \cos\theta$

$W = 1/3\, \pi h^3 \tan^2 \alpha g\rho$

$\therefore$ $P/W = 3\tan\alpha\cos\theta$

$$= 3\tan\alpha \frac{4}{\sqrt{16+\cot^2\alpha}} = \frac{12\tan^2\alpha}{\sqrt{1+15\sin^2\alpha}} = \frac{12\sin^2\alpha}{\cos^2\alpha\sqrt{1+15\sin^2\alpha}}]$$

14. A right cone of semi-vertical angle α is just immersed with a slant side of length l in the free surface. Show that the resultant thrust on the curved surface will cut the slant side at a distance.

$3l/4\ (1 - 3\sin^2 \alpha)$ from the vertex, and find the magnitude of the thrust.

Ans. $1/3\ \pi l^3 \sin^2 \alpha \cos \alpha\ \sqrt{1+3\sin^2 \alpha w}$

15. A solid is formed by the revolution of a semi-circle of radius a about the bounding diameter through an angle α, and the solid is immersed with one plane face in the surface of a liquid. Prove that the magnitude of the resultant thrust on the curved surface of the solid is

$$\frac{2a^3 g\rho}{3}\{(\alpha - \sin\alpha\cos\alpha)^2 + \sin^4 \alpha\}^{1/2}$$

Chapter 8

Equilibrium of Floating Bodies

1. Conditions of Equilibrium of a Body Freely Floating in a Liquid

Let us consider the equilibrium of a body floating freely wholly or partly immersed in a heavy homogeneous liquid which is at rest under gravity. The only forces acting on the body are as follows:

(1) The weight w (say) of the body acting vertically downwards through G, the C.G. of the body.

(2) The resultant fluid thrust (i.e., the force of buoyancy) w' (say) which is equal to the weight of the liquid displaced, acting vertically upwards through H, the centre of buoyancy, which is actually the C.G. of the liquid displaced.

We know that two forces acting on a body keep it in equilibrium if and only if they are equal in magnitude but opposite in direction and have the same line of action.

Hence, in our case, the necessary and sufficient conditions of equilibrium for a body floating freely in heavy homogeneous liquid are as follows:

(1) The weight of the body must be equal to the weight of the liquid displaced by it.

(2) The centre of gravity G of the body and the centre of buoyancy H must be in the same vertical line.

2. Volume Immersed

When a solid of volume v and mean density ρ is floating in a liquid of density σ, then the volume of the solid immersed is $v\rho/\sigma$.

Let v' be the volume of the solid immersed in the liquid. Then,

weight of the liquid displaced = $v'\sigma g$ ...(1)

Also, weight of the solid = $v\rho g$ ...(2)

For equilibrium,

weight of the liquid displaced = weight of the solid | Refer condition (1) of §1

$\Rightarrow \quad v'\sigma g = v\rho g \qquad \Rightarrow v' = v\rho/\sigma$ ...(3)

Note 1: Since v' cannot be greater than v, therefore, from (3) we conclude that σ cannot be less than ρ, i.e., a solid cannot float in liquid of density less than its own. Therefore, if the density of the liquid is less than the density of the solid, then the solid will sink in the liquid.

For example, a piece of iron dropped in water.

Note 2: If the density of the solid is less than the density of the liquid, then the solid is partially immersed. If the density of the solid is equal to the density of the liquid, then it is wholly immersed.

Note 3: By density of a solid, we mean density of the solid. Note that mean density of solid = mass of the solid/volume of the solid.

Note 4: Refer § 2.

weight of the solid (w) = $v\rho g \qquad \Rightarrow \rho = w/vg$

Hence, for a given weight w of a solid, ρ will depend upon the volume v. This density ρ is called the mean density and will differ from actual density of the solid. In finding out the mean density of a solid, we have to take into account the hollow space inside. We know that a ship made of iron floats in water whereas the density of iron is greater than that of water. The reason is that the mean density of the ship is less than that of water. Similarly a hollow iron ball may be made to float in water.

Example 1: A mean of weight 80 kg floats in water with 400 cubic cm of his body above the surface. What is his volume?

Sol. Let v' cubic cm be the volume of the man immersed in water. Then,

weight of water displaced by the man = weight of the man

$\Rightarrow \quad v'.i.g = 80$ kg wt | $\because$ Density of water in C.G.S. system is 1 gm/cubic cm

$= 80 \times 1000$ gm wt $= 80 \times 1000 \times g$ gm

$\Rightarrow \quad v' = 8000$ cubic cm

Again, volume of the body of the man above the surface of water = 400 cubic cm

$\therefore$ Total volume of the man = 80000 + 400 = 80400 cubic cm

Example 2: A cylinder floats vertically with 8 cms of its length above the fluid. Find the whole length of the cylinder, if the sp.gr. of the fluid is three times that of the cylinder.

Sol. Let the sp.gr. of the cylinder be ρ.

Then, the sp.gr. of the fluid = 3ρ

Let A be the area of cross-section of the cylinder. Let h cms be the whole length of the cylinder.

We know that

weight of the body = weight of the fluid displaced

$\Rightarrow$ $Ahg\rho = A\,(h - 8)\,g3\rho$ $\Rightarrow h = 3h - 24$ $\Rightarrow h = 12$ cms.

Hence, the whole length of the cylinder is 12 cms.

Example 3: A piece of wood weighing 24 grammes floats in water with 2/3rds of its volume immersed. Find the density and the volume of the wood.

Sol. Let v be the volume and ρ be the density of the piece of wood. Then,

volume of the piece of wood immersed in water = $2v/3$

Weight of the piece of wood = $v\rho g$ = 24 grammes wt.

Also, weight of water displace = $2v/3 \,.\, 1.\, g.$ | Density of water = 1 gm/c.c.

For floatation in equilibrium, weight of the piece of wood = weight of water displaced

$\Rightarrow$ $v\rho g = 2/3 \text{ v } 1 \,.\, g$

$\Rightarrow$ $\rho = 2/3$

Also, $v\rho g$ = 24 grammes wt. $\Rightarrow v\rho = 24$ $\Rightarrow v \,.\, 2/3 = 24$ $\Rightarrow v = 36$ c.c.

Hence, the density of the wood is 2/3 gm/cc. and the volume is 36 c.c.

Example 4: What fraction of a piece of cork will be outside water in which it floats, the specific gravity of cork being 0.24?

Sol. Let v the volume of the cork and v' be the volume of the water displaced by it.

Let the density of water be ρ. Then, the density of the cork = $0.24\ \rho$.

For floatation in equilibrium,

weight of the cork = weight of water displaced by it

$\Rightarrow$ $v\,(0.24\ \rho)\ g = v'.1.\ g\rho$ $\Rightarrow v' = 24v/100 = 6v/25$

$\therefore$ Volume of the cork outside the water

$= v - 6v/25 = 19v/25$

$\therefore$ Required fraction = Volume of the cork outside the water / v

$$= \frac{\frac{19v}{25}}{v} = \frac{19}{25} = 0.76$$

Example 5: A steamer in going from salt water into fresh water was observed to sink 2 inches, but after burning 50 tons of coal to rise 1 inch. Supposing the densities of salt and fresh water to be as 65 : 64, find the displacement (weight) of the steamer in metric tons.

(1 metric ton = 2240 lbs.)

Sol. Let us assume that the mean area of cross-section of the steamer near the water level is A. Let w be the displacement (weight) of the steamer in metric tons. Let the steamer sink a inches in the salt water. Let the densities of salt and fresh water be 65 ρ and 64 ρ respectively. Then, is salt water,

weight of the steamer = weight of salt water displaced

$\Rightarrow$ $w = A\,(a{+}2/12)\ 65\ \rho g$ ...(1)

In fresh water, the steamer sinks 2 inches more, i.e., the steamer sinks $(a + 2)$ inches in fresh water. So, in fresh water,

weight of the steamer = weight of fresh water displaced

$\Rightarrow \quad w = A\ (a + 2/12)\ 64\ \rho g$ ...(2)

From (1) and (2), we get

$65a = 64a + 128$

$\Rightarrow \quad a = 128$ ...(3)

It is given that after burning 50 tons of coal, the steamer rises 1 inch, i.e., when the weight of the steamer is $(w - 50 \times 2240)$, the steamer sinks $(a + 2 - 1)$, i.e., $(a + 1)$ inches in fresh water

$$\therefore \quad w - 50 \times 2240 = A\frac{a+1}{12}64\rho g \qquad | \text{ By the same principle}$$

$$= A\frac{a+2}{12}64\rho g - \frac{A}{12}64\rho g = w - \frac{A}{12}64\rho g \qquad | \text{ From (2)}$$

$$\Rightarrow \quad 50 \times 2240 = \frac{A}{12}64\rho g \Rightarrow \frac{A\rho g}{12} = \frac{50 \times 2240}{64} = 1750 \qquad \text{...(4)}$$

$\therefore$ From (1),

$$w = (A\rho g/12)\ a.\ 65 = 1750 \times 128 \times 65 \qquad | \text{ From (4) and (3)}$$

$$= \frac{1750 \times 128 \times 65}{2240} \text{ metric tons} = 6500 \text{ metric tons}$$

Hence, the displacement of the steamer is 6500 metric tons.

Example 6: A cubical box of 1 foot external dimension is made of a material of thickness one inch and floats in water immersed to a depth of $3\frac{3}{4}$ inches. Not taking air into consideration, determine the weight of the box and the amount of water that must be poured in so that water inside and outside may stand at the same level. (Given 12 inches = 1 foot, 1 cubic ft. of water weighs 62.5 pounds)

Sol.

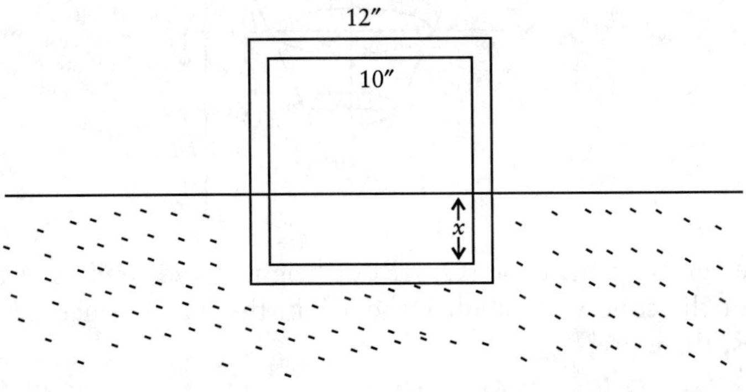

Depth immersed in water = $3\frac{3}{4}$ inches = 15/4 × 12 foot = 5/16 foot

∴ Volume of water displaced = 1 × 1 × 5/16 = 5/16 cu. ft.

∴ Weight of water displaced = 5/16 $g\rho$, where ρ is the density of water ...(1)

= 15/16 × 62.5 lbs. | ∵ 1 cubic ft. of water weighs 62.5 lbs.

= 5/16 × 125/2 = 65/32 = 19.5 lbs. (nearly)

Now let the water be poured in the box to a depth of x ft. Then, in order that the water inside and outside may stand at the same level, the box will be immersed to a depth (x + 1/12) ft. in the water as the thickness of the box is 1″ (= 1/12 ft).

Also, the inner dimensions of the box are 10″ × 10″.

∴ The volume of water inside the box = 10/12 × 10/12 × x = 25/36 x cu. ft.

The volume of water displaced by the box = 1 × 1 × (x + 1/12) cu. ft.

From (1),

Weight of the box = 5/16 $g\rho$

For floatation in equilibrium, we have,

weight of the box + weight of water inside the box

= weight of water displaced by the box

$$\Rightarrow \quad \frac{5}{16}g\rho + \frac{25}{36}xg\rho = \left(x + \frac{1}{12}\right)g\rho$$

$$\Rightarrow \quad \left(1 - \frac{25}{36}\right)x = \frac{5}{16} - \frac{1}{12} \quad \Rightarrow \frac{11x}{36} = \frac{44}{192} \quad \Rightarrow x = \frac{3}{4} \text{ ft.} = 9''$$

∴ Amount of water inside the box = 10″ × 10″ × 9″ = 900 cu. inches

Example 7: A frustum of a right circular cone, cut off by a plane bisecting the axis perpendicularly, floats with its smaller end in water and its axis just half immersed. Prove that the sp.gr. of the cone is 19/56.

Sol.

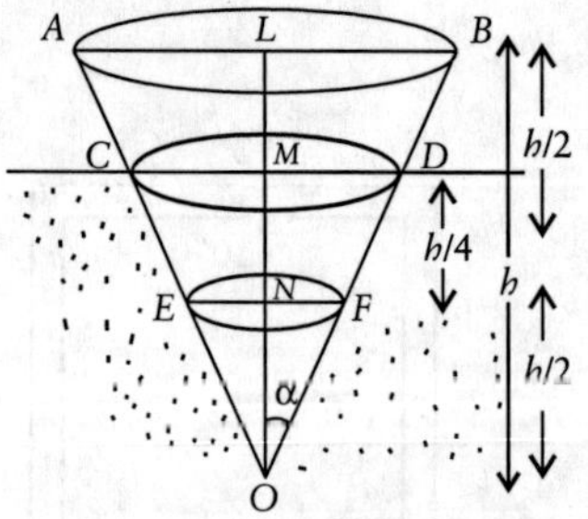

Let O be the vertex of the cone (O, AB) of height h and semi-vertical angle α. Let AE FB be the frustum of the cone with smaller end EF in the water. Since EF is the bisecting plane bisecting the axis, therefore,

$ON = NL = 1/2\, h = 1/2\, OL$.

Since the frustum is just half immersed, therefore

$NM = ML = 1/2\ NL = 1/4\ h$

Radius of the lower end of the frustum = $NF = h/2 \tan \alpha$

Radius of the upper end of the frustum = $LB = h \tan \alpha$

Radius of the circular section in which the water level cuts the frustum

$= MD = OM \tan \alpha = (ON + NM) \tan \alpha = (h/2 + h/4) \tan \alpha = 3h/4 \tan \alpha$

Volume of the frustum of the cone

$= 1/3\ \pi\ .\ h/2\ \{(h \tan \alpha)^2 + (h/2 \tan \alpha)^2 + (h \tan \alpha)\ (h/2 \tan \alpha)\}$

$= \pi h/6\ .\ 7/4 h^2 \tan^2 \alpha = 7\alpha\ h^3/24 \tan^2 \alpha$

$\therefore$ Weight of the frustum of the cone

$= 7\pi h^3 \tan^2\alpha/24\ sg\rho$ | 'W = vsw'

where s is the sp.gr. of the cone and ρ is the density of water.

Volume of water displaced

$$= \frac{1}{3} . \pi . \frac{h}{4}\left\{\left(\frac{h}{2}\tan\alpha\right)^2 + \left(\frac{3h}{4}\tan\alpha\right)^2 + \left(\frac{h}{2}\tan\alpha\right)\left(\frac{3h}{2}\tan\alpha\right)\right\}$$

$$= \frac{\pi h}{12} . \frac{19}{16} h^2 \tan^2 \alpha = \frac{19\pi}{192} h^3 \tan^2 \alpha$$

$\therefore$ Weight of water displaced $= \dfrac{19\pi}{192} h^3 \tan^2 \alpha g\rho$ | 'W = $vg\rho$'

For equilibrium,

weight of the frustum of the cone = weight of water displaced

$$\Rightarrow \frac{7}{24}\pi h^3 \tan^2 \alpha sg\rho = \frac{19\pi}{192} h^3 \tan^2 \alpha g\rho \qquad \Rightarrow s = 19/56$$

Example 8: A thin rod of weigh w is loaded at one end with a weight P of insignificant volume. If the rod floats in an inclined position with $1/n$ of its length out of the water, prove that

$$(n - 1)\ P = W$$

Sol.

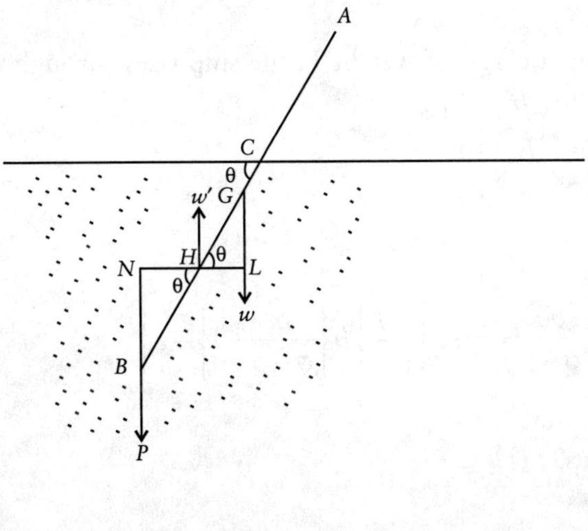

Let AB (= $2a$) be a thin rod of weight w loaded at one end B with a weight P with its length AC outside the water. Let θ be the inclination of the rod with the horizontal.

$$CA = 1/n\ AB = 2a/r \qquad \therefore BC = AB - CA = 2a - 2a/n = 2a\ (n{-}1/n)$$

Let G and H be the C.G. of the rod AB and the centre of buoyancy. Let w' be the force of buoyancy. The portion BC of the rod AB is in water.

$$BH = 1/2\ BC = a\ (n - 1/n),\ HG = a - a\ (n - 1/n) = a/n$$

The only forces acting on the rod arc P and w vertically downwards and w' vertically upwards.

Taking moments about H, we get

$$w\ .\ HL = P\ .\ HN$$

$$\Rightarrow \quad w\ .\ HG \cos\theta = P\ .\ BH \cos\theta \qquad \Rightarrow w\ .\ HG = P\ .\ BH$$

$$\Rightarrow \quad w\ .\ a/n = P\ .\ a\ (n - 1/n) \qquad \Rightarrow w = (n - 1)\ P$$

Example 9: A ship sailing out of sea into a river sinks through *a* distance *b* feet. On unloading a cargo of weight *P*, the ship rises through *c* feet. Show that the weight of the ship after unloading is

$$\left\{\frac{b\sigma}{c(\sigma-\rho)}-1\right\}P,$$

where σ and ρ are sp.gr. of sea water and river water respectively. The cross-section of the ship near the level of water way be assumed to be uniform.

Sol. Let the area of cross-section of the ship near the level of water be s. Let w be the weight of the ship before unloading. Let h be the height to which the ship is already sinking in sea-water.

In sea-water, by Principle of Archimedes, weight of ship = weight of sea-water displaced

$$\Rightarrow \quad w = (Ah)\ \sigma g \qquad \text{...(1)}$$

In sea-water, by Principle of Archimedes, weight of ship = weight of river water displaced

$$\Rightarrow \quad w = A\ (h + b)\ \rho g \qquad | \text{ the ship sinks through } b \text{ feet more} \qquad \text{...(2)}$$

From (1) and (2), we feet

$$Ah\sigma g = A\ (h + b)\ \rho g$$

$$\Rightarrow \quad h\sigma = (h + b)\ \rho$$

$$\Rightarrow \quad h\ (\sigma - \rho) = b\rho$$

$$\Rightarrow \quad h = b\rho/\sigma - \rho \qquad \text{...(3)}$$

Since refer unloading a cargo of weight P, the ship rises through c feet, therefore,

$$w - P = A\ (h + b - c)\ \rho g \qquad \text{...(4)}$$

Subtracting (4) from (2), we get

$$P = Ac\rho g$$

$$\Rightarrow \quad A\rho g = P/c \qquad \text{...(5)}$$

(3), (4) and (5) give,

$$w - P = \frac{P}{c}\left[\frac{b\rho}{\sigma-\rho}+b-c\right] = \frac{P}{c}\left[b\left\{\frac{\rho}{\sigma-\rho}+1\right\}-c\right]$$

$$= \frac{P}{c}\left[\frac{b\sigma}{\sigma - \rho} - c\right] = \left\{\frac{b\sigma}{c(\sigma - \rho)} - 1\right\}\rho$$

Example 10: A body is composed of a hemisphere of radius r, with a cone of height $2r$ on the same plane base. In a liquid of density ρ_1, it floats with the whole body just immersed and a liquid of density ρ_2, with the hemisphere just immersed. Prove that

$$\rho_2 = \rho_1$$

Sol. Volume of cone = $1/3\ \pi r^2.2r = 2/3\ \pi r^3$

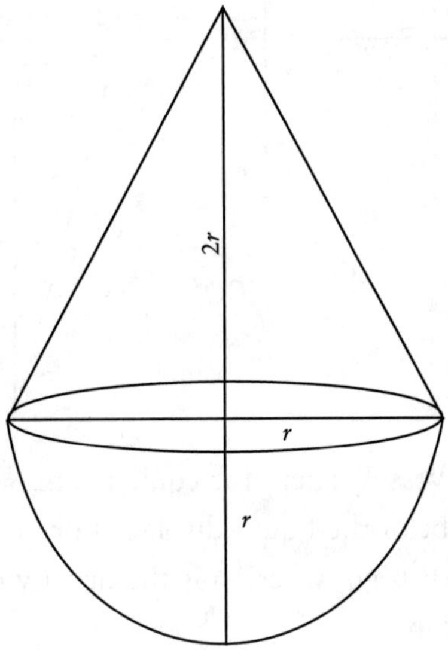

Volume of hemisphere = $2/3\ \pi r^3$

Let ρ be the density of the material of the cone and the hemisphere.

In a liquid of density ρ_1,

weight of the whole body = weight of the liquid displaced by the whole body

$\Rightarrow\ (2/3\ \pi r^3 + 2/3\ \pi r^3)\ g\rho = (2/3\ \pi r^3 + 2/3\ \pi r^3)\ \rho_1 g$

$\Rightarrow \rho = \rho_1$ ...(1)

In a liquid of density ρ_2,

weight of the whole body = weight of the liquid displaced by the hemisphere

$$\Rightarrow \left(\frac{2}{3}\pi r^3 + \frac{2}{3}\pi r^3\right) g\rho = \left(\frac{2}{3}\pi r^3\right)\rho_2 g$$

$\Rightarrow\ 2\rho = \rho_2$ ...(2)

$\Rightarrow\ 2\rho_1 = \rho_2$ | From (1) $\Rightarrow \rho_2 = 2\rho_2$

Example 11: A hollow conical vessel floats in water with its vertex downwards and certain depth of its axis immersed. When water is poured in to it upto the level originally immersed, it sinks till its mouth is on a level with the surface of water. What portion of the axis was originally immersed?

Sol. Let V be the vertex and α the semi-vertical angle of the hollow conical vessel (V, LM). Let is height by h. Let it be immersed to a depth h' in water. Then, $VO = h$ and $VC = h'$ in the original position.

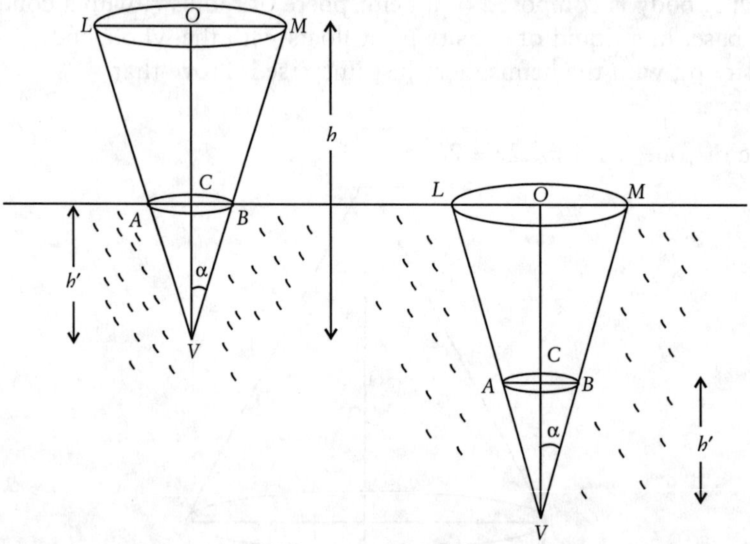

Let w be the weight of the vessel. Then, for equilibrium,

weight of the vessel = weight of the liquid displaced originally

$\Rightarrow$ $w = 1/3\ \pi\ (h' \tan \alpha)^2\ h' g\rho$, where ρ is the density of the liquid

$\Rightarrow$ $w = 1/3\ \pi h'^3 \tan^2 \alpha g\rho$ ...(1)

Let the water be poured into the vessel upto a height h'.

Then, in this case for equilibrium,

weight of the vessel with water = weight of the water displaced in this case

$\Rightarrow$ weight of the vessel + weight of water contained in it = weight of water displaced

$\Rightarrow$ $1/3\ \pi h'^3 \tan^2 \alpha g\rho + 1/3\ \pi h'^3 \tan^2 \alpha g\rho = 1/3\ \pi h'^3 \tan^2 \alpha g\rho$

$\Rightarrow$ $2h'^3 = h^3$ $\quad\Rightarrow h' = \dfrac{h}{\sqrt[3]{2}}$

Example 12: A solid homogeneous cone of sp.gr. σ and height h floats in a given liquid of sp.gr. ρ. Find the position of equilibrium when

(i) the vertex is down and the base up

(ii) the base is down and the vertex up

Sol. Let V be the vertex and σ be the semi-vertical angle of the cone. Let h be the height of the cone.

Case I: The vertex is down and the base up

Let h' be the length of the axis immersed in the liquid. Then, for equilibrium,

weight of the cone = weight of the liquid displaced by it

$\Rightarrow$ $1/3\ \pi\ (h \tan \alpha)^2\ hg\sigma = 1/3\ \pi\ (h \tan \alpha)^2\ h' g\rho$

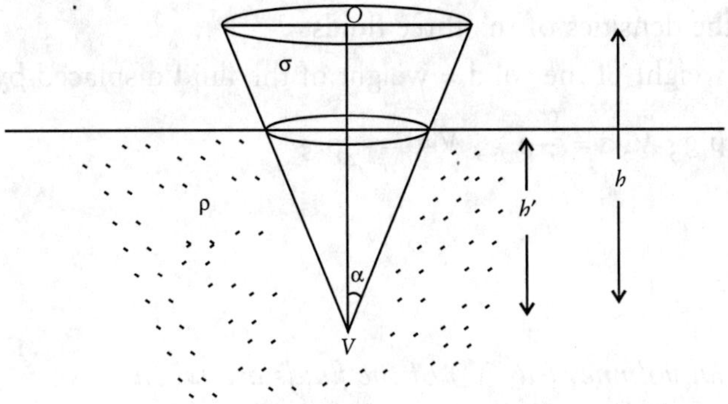

where σ is the density of the cone and ρ is the density of the liquid.

$\Rightarrow \quad h^3\,\sigma = h'^3\rho \qquad \Rightarrow h^3 = h\,(\sigma/\rho)^{1/3}$

Case II: The base is down and the vertex up

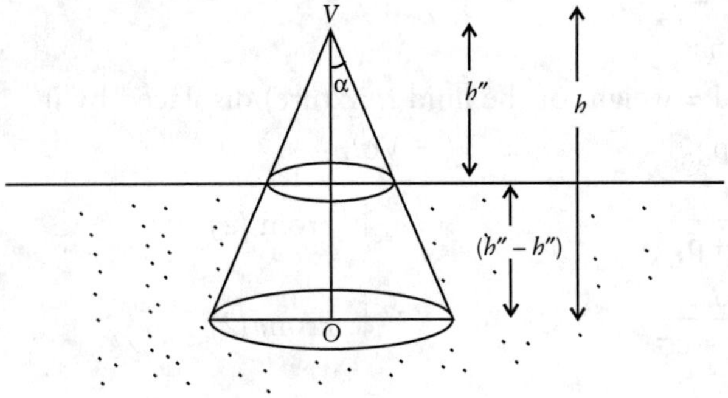

Let h'' be the length of axis out side the liquid.

Then, the length of the axis immersed in the liquid.

$= h - h'$

Then, for equilibrium, weight of the cone = weight of the liquid displaced by it

$$\Rightarrow \frac{1}{3}\pi(h\tan\alpha)^2 hg\sigma = \left[\frac{1}{3}\pi(h\tan\alpha)^2 h - \frac{1}{3}\pi(h''\tan\alpha)^2 h''\right]g\rho$$

$$\Rightarrow h^3\sigma = (h^3 - h''^3)\rho \quad \Rightarrow h''^3\rho = h^3(\rho-\sigma) \quad \Rightarrow h''^3 = \frac{\rho-\sigma}{\rho}h^3 \quad \Rightarrow h'' = \left(\frac{\rho-\sigma}{\rho}\right)^{1/3} h$$

$\therefore$ The length of the axis immersed in the liquid

$$= h - h'' = h - \left(\frac{\rho-\sigma}{\rho}\right)^{1/3} h = h\left\{1 - \left(\frac{\rho-\sigma}{\rho}\right)^{1/3}\right\}$$

Example 13: A solid displaces $1/a$, $1/b$, $1/c$ of its volume respectively when it floats in three of different fluids. Find the volume it displaces when it floats in a mixture formed of

(i) equal volumes of the fluids

(ii) equal weights of the fluids

Sol. Let V be the volume of the solid and σ its density.

Let ρ_1, ρ_2, ρ_3 be the densities of the three fluids.

For equilibrium, weight of the solid = weight of the fluid displaced by is

$$\Rightarrow \quad Vg\sigma = \frac{V}{a}\rho_1 g,\ Vg\sigma = \frac{V}{b}\rho_2 g,\ Vg\sigma = \frac{V}{c}\rho_3 g$$

$$\Rightarrow \quad \left.\begin{array}{l}\rho_1 = a\sigma \\ \rho_2 = b\sigma \\ \rho_3 = c\sigma\end{array}\right\} \qquad \text{...(1)}$$

Case I: *When equal volumes (say V') of the fluids are taken*

Let ρ be the density of the mixture of these fluids and V' be the volume displaced by the solid in this mixture. Then,

$$\rho = \frac{V'\rho_1 + V'\rho_2 + V'\rho_3}{V' + V' + V'} = \frac{\rho_1 + \rho_2 + \rho_3}{3} \qquad \text{...(2)}$$

Also, we known that

weight of the solid = weight of the fluid (mixture) displaced by it

$$\Rightarrow \quad Vg\sigma = V'\rho g \qquad \Rightarrow V' = V\sigma/\rho$$

$$= \frac{3V\sigma}{\rho_1 + \rho_2 + \rho_3} \qquad | \text{ From (2)}$$

$$= \frac{3V\sigma}{a\sigma + b\sigma + c\sigma} \qquad | \text{ From (2)}$$

$$= \frac{3V}{a + b + c}$$

Case II: *When equal weights (say w') of the fluids are taken*

Let ρ' be the density of the mixture of these fluids and V'' be the volume displaced by the solid in this mixture. Then,

$$\rho' = \frac{w' + w' + w'}{\frac{w'}{\rho_1} + \frac{w'}{\rho_2} + \frac{w'}{\rho_3}} = \frac{3}{\frac{1}{\rho_1} + \frac{1}{\rho_2} + \frac{1}{\rho_3}}$$

$$= \frac{3}{\frac{1}{a\sigma} + \frac{1}{b\sigma} + \frac{1}{c\sigma}} \qquad | \text{ From (2)}$$

$$= \frac{3\sigma}{\frac{1}{a} + \frac{1}{b} + \frac{1}{c}} \qquad \text{...(3)}$$

Also, we know that

weight of the solid = weight of the fluid (mixture) displaced by it

$$\Rightarrow \quad Vg\sigma = V''\rho' g$$

$$\Rightarrow \quad V'' = \frac{V\sigma}{\rho'} \Rightarrow V'' = \frac{V\sigma}{\dfrac{3\sigma}{\dfrac{1}{a}+\dfrac{1}{b}+\dfrac{1}{c}}} \qquad | \text{ From (3)}$$

$$= \frac{V}{3}\left(\frac{1}{a}+\frac{1}{b}+\frac{1}{c}\right) \qquad \Rightarrow \qquad V'' = \frac{(bc+ca+ab)V}{3abc}$$

Example 14: A hollow cylinder of external and internal radii a and b is open at the top and thickness of the base is c. The cylinder floats in water with its axis vertical and a depth h immersed. If the cylinder develop a small leak, show that it will never sink if its height is greater than $\dfrac{a^2h - b^2c}{a^2 - b^2}$.

Sol.

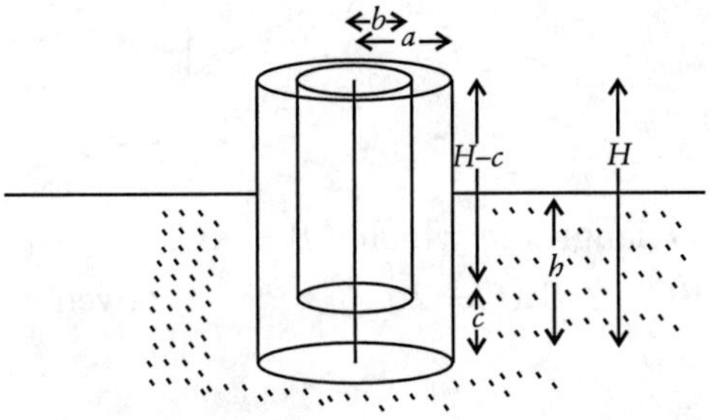

Let H be the height of the cylinder. Let ρ be the density of its material. Then, volume of the hollow cylinder

= volume of the external cylinder of base radius a and height H – volume of the internal cylinder of base radius b and height $(H - c)$

$$= \pi a^2 H - \pi b^2 (H - c)$$

$$= \pi (a^2 - b^2) H + \pi b^2 c \qquad \text{...(1)}$$

When that is no leak, then for equilibrium,

weight of the cylinder = weight of the water displaced by it with a depth h immersed

$$\Rightarrow \quad [\pi (a^2 - b^2) H + \pi b^2 c]\, g\rho = (\pi a^2 h) \,.\, g.1$$

$$\Rightarrow \quad \rho = \frac{a^2 h}{(a^2 - b^2)H + b^2 c} \qquad \text{...(2)}$$

Now if the cylinder develops a small leak, then it will never sink if

the weight of the cylinder < weight of water displaced by the whole cylinder

$$\Rightarrow \quad [\pi (a^2 - b^2) H + \pi b^2 c]\, g\rho = [\pi (a^2 - b^2) \mathrm{H} + \pi b^2 c].\, g.1$$

$$\Rightarrow \quad \rho < 1$$

$$\Rightarrow \quad \frac{a^2 h}{(a^2 - b^2)H + b^2 c} < 1 \qquad | \text{ From (2)}$$

$\Rightarrow \quad a^2h < (a^2 - b^2)\,\mathrm{H} + b^2c \qquad \Rightarrow a^2h - b^2c < (a^2 - b^2)\,\mathrm{H}$

$\Rightarrow \quad \dfrac{a^2h - b^2c}{a^2 - b^2} < H \qquad \Rightarrow H > \dfrac{a^2h - b^2c}{a^2 - b^2}$

Example 15: A uniform prism, whose cross-section is an isosceles triangle of vertical angle 2α, floats freely in a liquid with its base just immersed, one edge being in the surface. Show that the ratio of its density to that of the liquid is 2 sin²α : 1.

Sol.

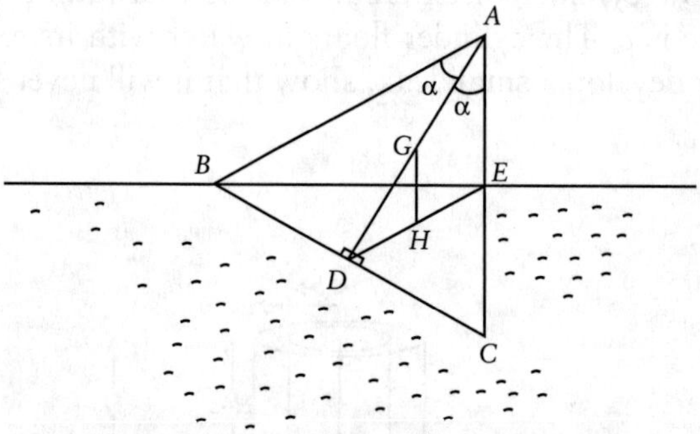

Let ABC be an isosceles triangles in which $AB = AC$.

AD is a median of ΔABC . $\angle BAC = 2\alpha$ | Given

$\therefore \angle BAD = \angle CAD = \alpha$

By Geometry, $AD \perp AC$

$\therefore \angle ADC = \angle ADB = 90°$ $\qquad \therefore \angle ACD = 90° - \alpha$

Let G be the C.G. of ΔABC and H be the centre of buoyancy (C.G. of ΔBCE). Then,

$DG = 1/3\ AD \qquad DH = 1/3\ DE$

For equilibrium,

DH must be vertical

$\therefore$ In ΔDGH and ΔDAE,

$DG/DA = DH/DE\ (=1/3)$

$\therefore GH$ is parallel to AE

But GH is vertical

$\therefore AE$ is vertical $\qquad \therefore AC$ is vertical

Let ρ and σ be the density of the prism and the liquid respectively. Let h be the length of the prism.

For equilibrium,

weight of the prism = weight of the liquid displaced by the prism

$\Rightarrow \quad (\Delta ABC) \times hg\rho = (\Delta ABC) \times hg\sigma$

$\Rightarrow \quad \rho/\sigma = \Delta BCE/\Delta ABC$

$= 1/2\ .\ BE\ .\ EC/1/2\ .\ BE.\ AC = EC/AC$ ...(1)

In $\Delta\, BCE$, cot C = EC/BE

$$\Rightarrow \quad \cot(90° - \alpha) = \frac{EC}{BE} \quad \Rightarrow \tan\alpha = \frac{EC}{BE} \quad \Rightarrow EC = BE \tan \alpha \qquad ...(2)$$

In $\Delta\, ABE$,

$$\sin A = BE/AB$$

$$\Rightarrow \quad \sin 2\alpha = \frac{BE}{AB} \quad \Rightarrow BE = AB \sin 2\alpha \qquad ...(3)$$

From (2) and (3),

$$EC = AB \sin 2\alpha \tan \alpha \qquad ...(4)$$

From (1) and (4),

$$\frac{\rho}{\sigma} = \frac{AB \sin 2\alpha \tan\alpha}{AC} = \sin 2\alpha \tan \alpha \qquad | \because AB = AC$$

$$= 2 \sin \alpha \cos \alpha \,.\, \sin \alpha/\cos \alpha = 2\sin^2 \alpha/1$$

$$\Rightarrow \quad \rho : \sigma = 2 \sin^2 \alpha : 1$$

Example 16: A prism, whose cross-section is a triangle ABC, is made of uniform material, and floats freely in water with vertex C in the surface. Prove that is sp.gr. is either $\dfrac{\sin A \cos B}{\sin C}$ or $\dfrac{\sin B \cos A}{\sin C}$.

Sol.

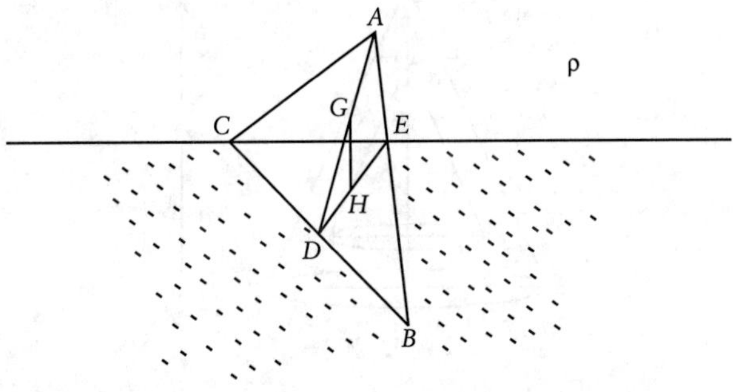

Refer Ex. 16.

Let ρ be the density of the prism. Let h be the length of the prism.

As proved in Ex. 15, GH must be vertical.

For equilibrium,

weight of the prism = weight of water displaced by the prism

$$\Rightarrow \quad \Delta\, ABC \times h \times g\rho = \Delta\, BCE \times h \times g \,.\, 1$$

$$\Rightarrow \quad \rho = \frac{\Delta BCE}{\Delta ABC} = \frac{\frac{1}{2} BE.CE}{\frac{1}{2} AB.CE} = \frac{BE}{AB} \qquad ...(1)$$

In $\Delta\ ABC$, by sine formula

$$\frac{AB}{\sin C} = \frac{BC}{\sin A} \qquad \Rightarrow AB = BC\frac{\sin C}{\sin A} \qquad ...(2)$$

From (1) and (2),

$$\rho = \frac{BE.\sin A}{BC.\sin C} = \frac{\sin A \cos B}{\sin C} \qquad \Big| \because \text{In } \Delta\ ABC,\ \cos B = BE/BC$$

or

If the vertex A is inside the water and the vertex B is outside the water, then density of the prism $\rho = \dfrac{\sin A \cos B}{\sin C}$

Example 17: A hollow closed cone of semi-vertical angle $\sin^{-1} 1/3$ of metal whose specific weight is w' is made of such uniform thickness that it will float in all positions wholly submerged in a liquid of specific gravity w. Show that the thickness must be

$1/4\ h\ \{1-(1 - w/w')^{1/3}\}$, where h is the external height of the cone.

Sol. Let (V, AB) be the external cone of height h. Let $(V', A'B')$ be the internal cone. Let d be the thickness of the metal and α be the semi-vertical angle of the cone. Then,

$$\sin \alpha = 1/3 \qquad | \text{ Given}$$

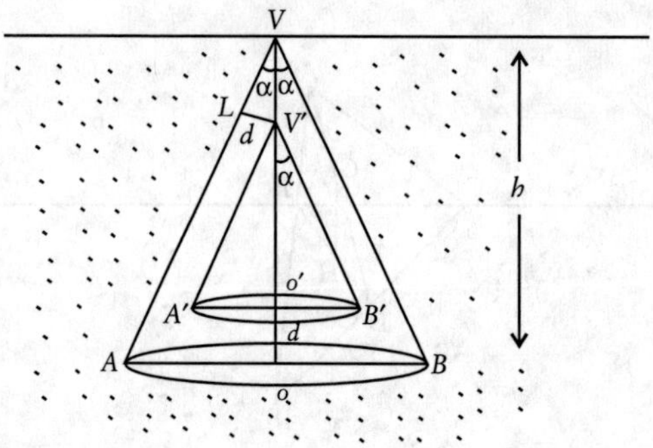

Let O be the centre of base of the external cone.

Then, $VO = h.$

$V'L = d$

In $\Delta\ VLV'$,

$\sin \alpha = d/VV'$

$\Rightarrow 1/3 = d/VV' \qquad |$ From (1)

$\Rightarrow VV' = 3d \qquad ...(2)$

$\therefore$ Height of the internal cone

$= OO' = VO - VV' - OO' = h - 3d - d = h - 4d \qquad ...(3)$

$\therefore$ Volume of the metal

= Volume of the external cone – Volume of the internal cone

$= 1/3\pi\ (h \tan \alpha)^2\ h - 1/3\pi\ \{(h - 4d) \tan \alpha\}^2\ (h - 4d)$

$= \pi/3\ [h^3 - (h - 4d)^3] \tan^2 \alpha$ ...(4)

When the cone is wholly immersed in the liquid, then, for equilibrium,

weight of the cone = weight of the liquid displaced by the cone

$\Rightarrow \pi/3\ [h^3 - (h - 4d)^3 \tan^2 \alpha\ w' = 1/3\ h^3 \tan^2 \alpha\ w$

$\Rightarrow h^3 - (h - 4d)^3 = h^3\ w/\ w'$ $\Rightarrow h^3\ (1 - w/w') = (h - 4d)^3$

$\Rightarrow h\ (1 - w/\ w')^{1/3} = h - 4d$

$$\Rightarrow 4d = h\left[1-\left(1-\frac{w}{w'}\right)^{1/3}\right] \qquad \Rightarrow d = \frac{h}{4}\left[1-\left(1-\frac{w}{w'}\right)^{1/3}\right]$$

Example 18: A sin hollow cone with a base floats completely immersed in water wherever it is placed. Show that the vertical angle is 2 sin (1/3).

Sol. Let r be the radius of the base, h be the vertical height and l be the slant height of the cone.

Then,

curved surface area of the cone = πrl, area of the base = πr^2

C.G. of the base of the cone = O (the centre of the base)

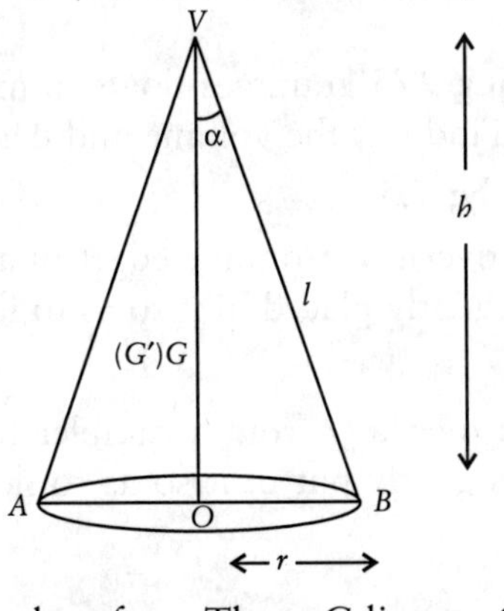

Let G be the C.G. of the curved surface. Then, G lies on OV such that $OG = h/3$

Hence, the distance of the C.G. of hollow cone together with base from the centre of the base (O)

$$= \frac{\pi rl.\frac{1}{3}h + \pi r^2.0}{\pi rl - \pi r^2} = \frac{lh}{3(l+r)} \qquad ...(1)$$

When the cone is completely submerged in water, the displaced water assumes the shape of a solid cone.

Hence, the C.G. (G') of the liquid displaced is at a distance $h/4$ from the centre of the base of the cone.

If the cone floats completely immersed in water wherever it is placed, then the C.G. of the cone together with base must coincide with the C.G. (G') of the liquid displaced.

So, $\dfrac{lh}{3(l+r)} = \dfrac{h}{4} \quad \Rightarrow 4l = 3(l+r) \quad \Rightarrow l = 3r \quad \Rightarrow \dfrac{r}{l} = \dfrac{1}{3}$

$\Rightarrow \sin\alpha = \dfrac{1}{3} \quad \Rightarrow \alpha = \sin^{-1}\left(\dfrac{1}{3}\right)$

$\therefore$ Vertical angle of the cone $= 2\alpha = 2\sin^{-1}\left(\dfrac{1}{3}\right)$.

EXERCISE 1

1. What fraction of a solid (sp.gr. 0.7) must be outside the water in which it floats?

   Ans. 3/10

   [**Hint:** Refer Ex. 4]
2. What fraction of the volume of the body will remain immersed if the density of the liquid is three times the mean density of the body?

   Ans. $\sigma - \rho/\sigma$
3. The specific gravity of sea-water is 1.028 and of ill is 0.918. What fraction of the volume of an iceberg floats out of the water?

   Ans. 55/514
4. A piece of iron weighing 275 grammes floats in mercury of density 13.6 with 5/9 of its volume immersed. Find out the volume and density of the iron.

   Ans. 36.4 cu. cm. nearly; 7.5
5. How much water will overflow from the edges of a cup just full of water when a cork 2 cu. cm. in volume is gently placed in it so as to float? The sp.gr. of the cork is 0.24.

   Ans. 0.48 cu. cm.
6. How much water will overflow from a tumbler full of water when a piece of ice 1 cubic cm. in volume is gently put in it so as to float? (specific gravity of ice may be taken to 0.9).

   Ans. 0.9 cu. cm.
7. A specific shell of copper (sp.gr. 0.8) floats in water with half of its surface immersed. Find its internal radius if the outer one be 10 cms.

   Ans. $10\ (15/16)^{1/3}$ cm.
8. A uniform cylinder, when floating with its axis vertical in distilled water, sinks to a depth of 3.2 cm, and when floating in alcohol sinks to a depth of 4 cm. Find the specific gravity of alcohol.

   Ans. 0.8
9. The specific gravity of ice is 0.92, that of the sea-water is 1.025. What depth of water will be required to float a cubical iceberg whose side is 100 feet?

   Ans. 89.756 feet

10. A wooden solid cube of side one metre floats in water with 3/4ths of its volume immersed. Calculate the depth to which it will sink in a liquid of sp.gr. 0.8.
Ans. 93.75 cm.

11. A man whose weight is equal to 6.9 kg and whose sp.gr. is 1.1 can just float in water with his head above the surface by the aid of a piece of cork which is wholly immersed. Having given that the volume of his head is 1/16th of his whole volume and that the specific gravity of the cork is 0.24, find the volume of cork, the weight of 1 c.c. of water being 1 gm.
Ans. $1343\frac{113}{209}$ c.c.

12. A man whose weight is equal to 160 lbs. and whose sp.gr. is 1.1 can just float in water with his head above the surface by the aid of a piece of cork which is wholly immersed. Having given that the volume of his head is one sixteenth of his whole volume and that the sp.gr. of the cork is 0.24, find the volume of the cork, the weight of a cubic foot of water being 62.5 lbs.
And. 104/209 cubic feet.

13. A ship sailing from the sea into a river sinks a inches and on discharging x tons of her cargo rises to b inches. If the sea water be one fortieth heavier than river water, prove that the weight of the ship is 41 ax/b tons. Assume the sides of the ship to be vertical at water level. (Given 12 inches = 1 foot, 2240 lbs. = 1 ton)

14. A cubical box one metre external dimension is made of a material of thickness 2 cm and floats is water immersed to a depth of 3 $\frac{3}{4}$ cm. Not taking air into consideration, determine the weight of the box and the amount of water that must be poured in so that the water inside and outside may stand at the same level.
Ans. 37.5 kg, 1417500/19 cm³.

15. A rod, of small section and of density ρ, has a small portion of metal of weight $1/n$th that of the rod attached to one extremity, with this extremity immersed. Prove that the rod will float at any inclination in a liquid of density σ if $(n + 1)^2 \rho = n^2\sigma$

[**Hint:** Refer Ex. 8.

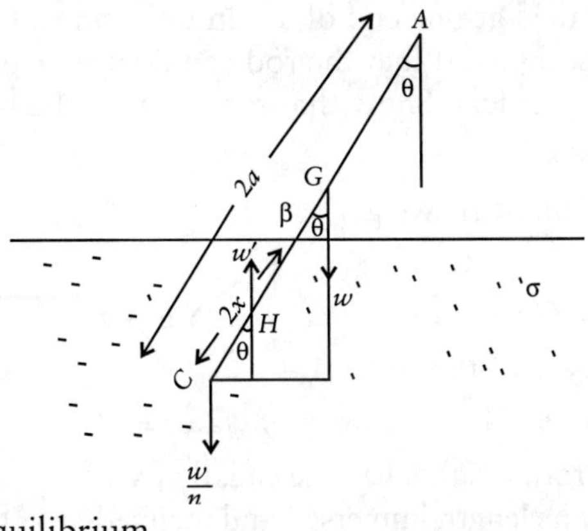

For floatation in equilibrium,
$w' = w + w/n$

$\Rightarrow$ $2x \,.\, S.\,\sigma g = 2a \,.\, S.\,\rho g + 2a \,.\, S.\,\rho g/n$ where S is the area of cross-section of the rod.

$\Rightarrow$ $x = a\rho/\sigma\ (n+1/n)$ ...(1)

Taking moments about C, we get,

$2x\ S\ \sigma gx \sin\theta = 2a\ S\ \rho\ g\ a \sin\theta$

$\Rightarrow$ $x^2\,\sigma = a^2\rho$ ...(2)

From (1) and (2),

$a^2\rho^2/\sigma^2\ (n+1/n)^2\ \sigma = a^2\rho/\sigma$

$\Rightarrow$ $(n+1/n)^2\ \rho = n^2\sigma$]

16. A thin uniform rod of weight W has a particle of weight w attached to one end. It is floating in an inclined position in water, with this end immersed. Prove that the length of the rod above water is w/w+W times its whole length and that the specific gravity of the rod is W^2/w+W$)^2$.

[**Hint:** Refer Ex. 8

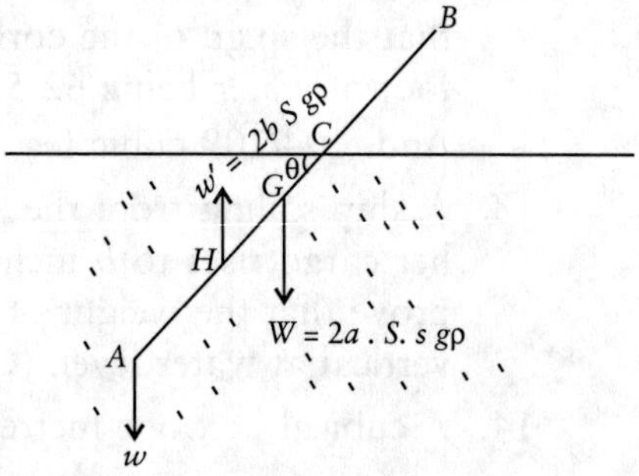

$AC = 2b$ | $\therefore AH = HC = b$

$GH = a - b$

S = Area of cross-section of the rod

Taking moments about H, we get,

$w \,.\, AH \cos\theta = W.\ HG \cos\theta$

$\Rightarrow$ $w\ AH = W.\ HG \Rightarrow w\ b = W.\ (a-b)$

$\Rightarrow$ $b = aW/w+W$ ...(1)

$BC = 2a - 2b = 2a - 2aW/w+W = 2aw/w+W = w/w+W$ (whole length of rod)

For floatation in equilibrium,

$W' = w + W$

$\Rightarrow$ $W'/w = w + W/W \Rightarrow 2bSg\rho/2aSsg\rho = w + W/W$

$\Rightarrow$ $b/as = w + W/W \Rightarrow s = b/a\ W/W + w = (W/W+w)^2$ | From (1)]

17. A small hole is drilled at one end of a thin uniform rod and it is filled with some such denser metal. It is observed that the rod can float in water half immersed and inclined at an angle to the vertical. Show that the sp.gr. of the rod is 1/4.

[**Hint:** Refer Ex. 8

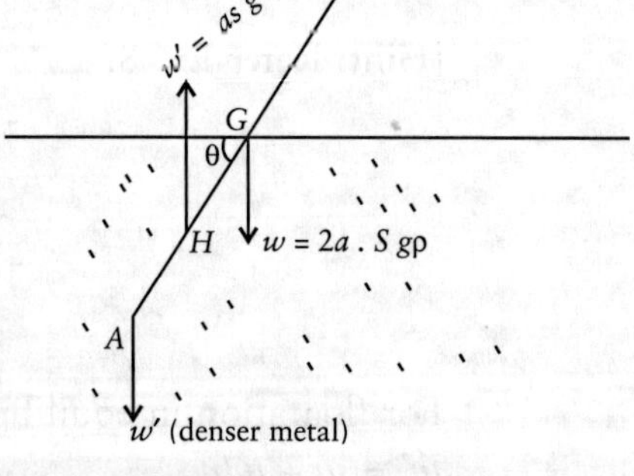

Taking moments about A, we get

$w' \,.\, AH \cos\theta = w.\ AG \cos\theta$

$\Rightarrow$ $w' \,.\, AH = w.\ AG \Rightarrow a\ S\ g \,.\, 1/2\ a = 2a\ S\ g\ \rho\ a$

$\Rightarrow$ $\rho = 1/4 \Rightarrow$ density of the rod = 1/4

$\Rightarrow$ sp.gr. of the rod = 1/4 | $\because$ density of water = 1]

18. A thin cylindrical rod, weighted at one end, floats in water with half of its length immersed and inclined at any angle to the horizon. Prove that the weight which is added is equal to the weight of the rod.

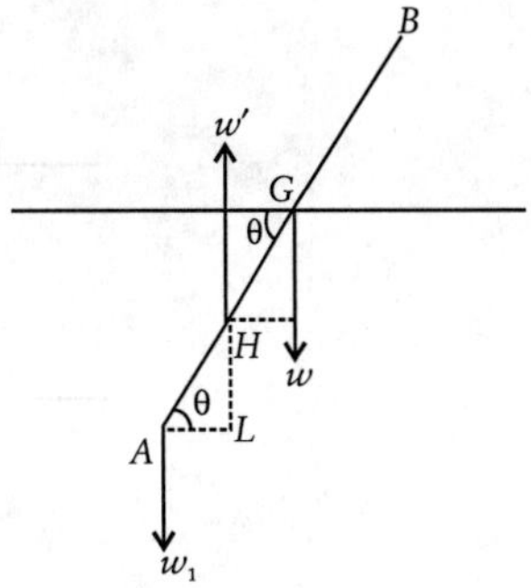

[**Hint:** Refer Ex. 8

$AB = 2a$, $AG = GB = a$, $AH = HG = a/2$

Taking moments about H, we get

$w \,.\, a/2 \cos\theta = w_1 \, a/2 \cos\theta \Rightarrow w = w_1$]

19. Show that a homogeneous body, in the shape of a right circular cone can float in a liquid of twice its own density with its axis horizontal.

[**Hint:** Let ρ and 2ρ be the densities of the solid cone and the liquid in which it floats.

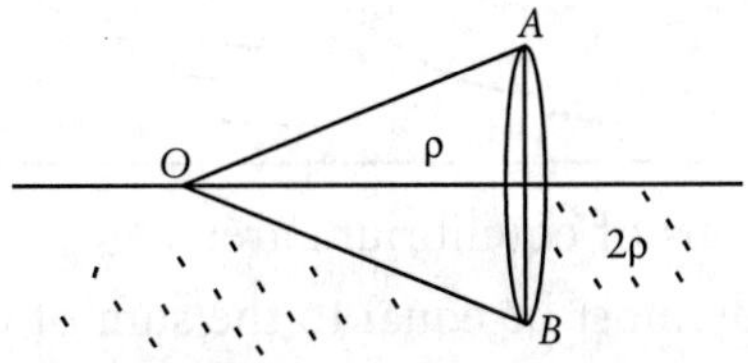

Let v and v^1 be the volumes of the cone and part of the cone immersed in water. Then, weight of the cone = weight of liquid displaced | For equilibrium

$v\, g\rho = v^1\, g \,.\, 2\rho$

$\Rightarrow v^1 = v/2$ which is true as the axis is horizontal.

Hence the cone can float in a liquid of twice its own density when its axis is horizontal]

20. A triangular lamina ABC of density ρ floats in a liquid of density σ with its plane vertical, the angle B being in the surface of the liquid, and the angle A not immersed. Show that $\rho/\sigma = \sin A \cos C/\sin B = a^2 + b^2 - c^2/2b^2$

[**Hint:** Refer Ex. 15, 16]

21. A triangular lamina ABC of specific gravity s, floats with its plane vertical in water, A being outside the liquid and BC not horizontal. The angle A is a right angle and $AB = AC$. If θ be the inclination of AB to the vertical, prove that

$$\sin 2\theta = \frac{2-2s}{2s-1} \quad \text{given } s > 3/4$$

3. Bodies Floating in more than one Liquids

A body floats partly immersed in one liquid and partly in another, to find the conditions of equilibrium.

Let a body be floating in two liquids, partly immersed in one liquid and partly in another. Let w be the weight of the body acting vertically downwards through its C.G., say G. Let w_1 and w_2 be the weights of the two displaced liquids. Then, w_1 and w_2 are also the forces of buoyancy acting vertically upwards at H_1 and H_2 (say) respectively.

Now, there are only three forces acting on the body, all of them vertical.

For equilibrium,

$w = w_1 + w_2$ | By statics

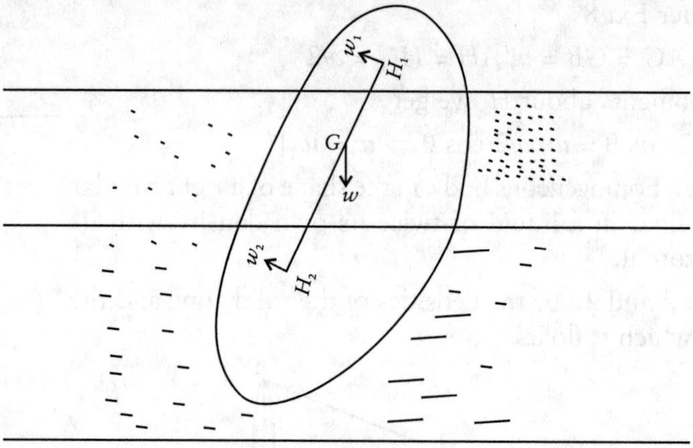

Hence, the required conditions of equilibrium are:

(1) The weight of the body must be equal to the sum of the weights of the two displaced liquids.

(2) The C.G. of the body must lie on the line joining the centres of gravity of the two displaced liquids.

Note 1: The points H_1 and H_2 are two centres of buoyancy. They are also the C.G.'s of the two displaced liquids respectively.

Note 2: The above theorem also holds good when a body is floating partly immersed in a liquid and partly in air.

Note 3: The above theorem is also applicable in case there are more than two liquids.

ILLUSTRATIVE EXAMPLES

Example 1: A mass composed partly of gold (sp.gr. 19.25) and partly of silver (sp.gr. 10.5) floats with 15/16th of its volume immersed in mercury (sp.gr. 13.6) and the remainder in water. Compare the weights of gold and silver in the mass.

Sol. Let the volumes of gold and silver in the alloy be v_1 and v_2 respectively. Let ρ be the density of water.

$\therefore$ Total volume = $v_1 + v_2$

$\therefore$ Weight of the mass = $v_1\, g\rho\, (19.25) + v_2\, g\rho\, (10.5)$

Volume of mercury displaced = $15/16\, (v_1 + v_2)$

Volume of water displaced = $(v_1 + v_2) - 15/16\, (v_1 + v_2) = v_1 + v_2/16$

$\therefore$ Weight of mercury displaced = $15/16\, (v_1 + v_2)\, 13.6\, g\rho$

Weight of water displaced = $1/16\, (v_1 + v_2)\, 1\, g\rho$

For equilibrium,

weight of the mass = weight of mercury displaced + weight of water displaced

$\Rightarrow v_1\, g\rho\, (19.25) + v_2\, g\rho\, (10.5) = 15/16\, (v_1 + v_2)\, 13.6\, g\rho + 1/16\, (v_1 + v_2)\, .\, 1.\, g\rho$

$\Rightarrow 19.25\, v_1 + 10.5\, v_2 = 205/16\, (v_1 + v_2)$

$\Rightarrow v_1\ (1925/100 - 205/16) = v_2\ (205/16 - 105/10)$

$\Rightarrow v_1\ (77/4 - 205/16) = v_2\ (205/16 - 21/2)$

$\Rightarrow 103/16\ v_2 = 37/16\ v_2$

$\Rightarrow v_1/v_2 = 37/103$ ...(1)

$\therefore$ Weight of gold/Weight of silver $= v_1\ g\rho\ (19.5)/v_2\ g\rho\ (10.5)$

$= 37 \times 19.25/103 \times 10.5$

$= 37 \times 1925/103 \times 1050$

$= 37 \times 77/103 \times 42 = 37 \times 11/103 \times 6$

$= 407/618$

Example 2: A cylinder of specific gravity σ floats with its axis partly in a liquid of sp.gr. σ_1 and partly in another of sp.gr. σ_2. Show that the common surface divides the axis in the ratio $\sigma - \sigma_2 : \sigma_1 - \sigma$.

Sol. Let the heights of the cylinder in contact with the two liquids of sp.gr. σ_1 and σ_2 be h_1 and h_2 respectively. Let ρ be the density of the standard substance. Let r be radius of the base of the cylinder of sp.gr. σ.

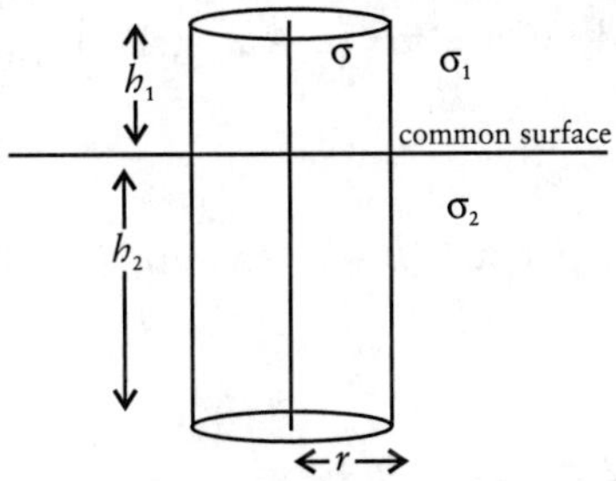

Weight of the cylinder $= \pi r^2\ (h_1 + h_2)\ \sigma g\rho$ ...(1)

Sum of the weights of the two liquids displaced

$= \pi r^2\ h_1\ \sigma_1\ g\rho + \pi r^2\ h_2 \cdot \sigma_2\ g\rho$

$= \pi r^2\ g\rho\ (h_1\ \sigma_1 + h_2\ \sigma_2)$ ...(2)

For equilibrium,

weight of the cylinder = Sum of the weight of the two liquids displaced by the cylinder

$\Rightarrow \pi r^2\ (h_1 + h_2)\sigma g\rho = \pi r^2 g\rho\ (h_1\sigma_1 + h_2\sigma_2) \Rightarrow (h_1 + h_2)\ \sigma = h_1\sigma_2 + h_2\sigma_2$

$\Rightarrow h_1\sigma - h_1\sigma_1 = h_2\sigma_2 - h_2\sigma \Rightarrow h_1\ (\sigma - \sigma_1) = h_2\ (\sigma_2 - \sigma)$

$\Rightarrow h_1/h_2 = \sigma_2 - \sigma/\sigma - \sigma_1 \Rightarrow h_1/h_2 = \sigma - \sigma_2/\sigma_1 - \sigma$

Example 3: A cylinder of wood (sp.gr. 3/4) of length h, floats with its axis vertical in water and oil (sp.gr. 1/2), the length of the solid in contact with oil being a (< 1/2 h). Find how much of the wood is above the liquids. Also find to what additional depth must oil be added so as just to cover the cylinder?

Sol.

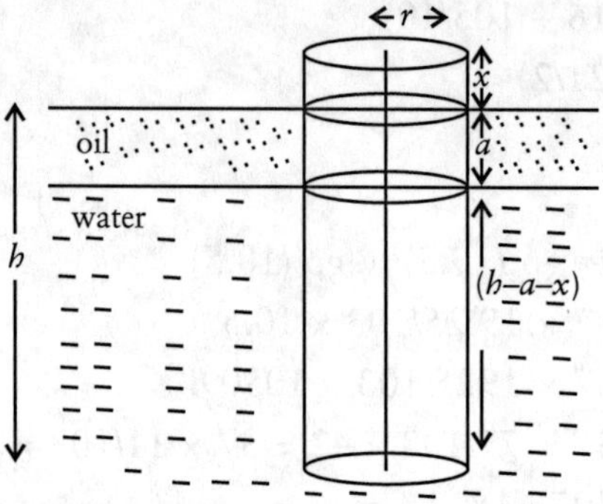

First Part: Let x be the length of the cylinder of wood above the two liquids.

Height of cylinder $= h$ | Given

$\therefore$ Length of the cylinder of wood in contact with the water $= h - a - x$

Let r be the radius of the base of the cylinder. Let ρ be the density of the standard substance.

Weight of the cylinder of wood $(W) = \pi r^2 h\ 3/4\ g\rho$ | '$W = vsw$'

Weight of the oil displaced $(W_1) = \pi r^2 a\ 1/2\ g\rho$ | '$W = vsw$'

Weight water displaced $(W_2) = \pi r^2\ (h - a - x)\ .\ 1.\ g\rho$ | '$W = vsw$'

For equilibrium,

$$W = W_1 + W_2$$

$$\Rightarrow \pi r^2 h\ 3/4\ \rho g = \pi r^2 a\ 1/2\ g\rho + \pi r^2\ (h - a - x)\ g\rho \quad \text{... (1)}$$

$$\Rightarrow 3h/4 = a/2 + h - a - x$$

$$\Rightarrow x = h/4 - a/2$$

Here, x is positive as $a < h/2 \Rightarrow a/2 < h/4$

Second Part: Let y be the additional depth of the oil which is necessary to just cover the cylinder.

$\therefore$ New length of the cylinder in contact with the oil $= a + y$

$\therefore$ New length in contact with the water $= h - a - y$

$\therefore$ In equilibrium, we have from (1),

$$\pi r^2\ h\ 3/4\ \rho g = \pi r^2\ (a + y)\ 1/2\ g\rho + \pi r^2\ (h - a - y)\ g\rho$$

$$\Rightarrow 3/4h = a + y/2 + (h - a - y) \Rightarrow 3h = 2a + 2y + 4h - 4a - 4y$$

$$\Rightarrow 2y = h - 2a \Rightarrow y = h/2 - a$$

Example 4: A cylinder floats completely with its axis vertical immersed in two liquids, the densities of the upper and lower liquids being ρ and 2ρ and the density of the cylinder is $5/4\ \rho$. Find what portion of the cylinder is in the lower liquid?

Sol.

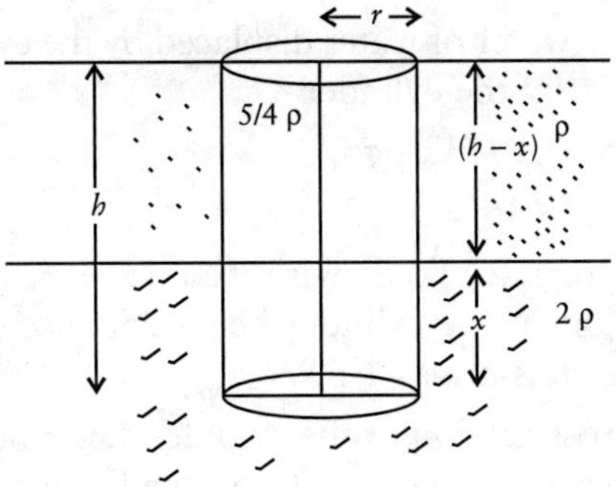

Let h be the height of the cone. Let r be the base radius of the cone.

Let x be the length of the cylinder in contact with the lower liquid.

Then, length of the cylinder in contact with the upper liquid = $h - x$

For equilibrium,

weight of the cylinder = weight of liquid of density ρ displaced by the cylinder + weight of liquid of density 2ρ displaced by the cylinder

$\Rightarrow \pi r^2 hg \,.\, 5/4\rho = \pi r^2 x \,.\, g \,.\, 2\rho + \pi r^2 (h - x)\, g\rho$ | '$W = vg\rho$'

$\Rightarrow 5h/4 = 2x + h - x \qquad \Rightarrow \qquad h/4 = x$

Hence, one fourth of the length (height) of the cylinder is immersed in the lower liquid.

Example 5: A right circular cylinder of sp.gr. σ floats in water with its axis vertical, one third being above the water. If ρ be the sp.gr. of air, prove that $3\sigma = 2 + \rho$

Sol. Let A be the area of cross-section of the cylinder. Let he be the height of the cylinder. Let ρ_1 be the density of standard substance (water).

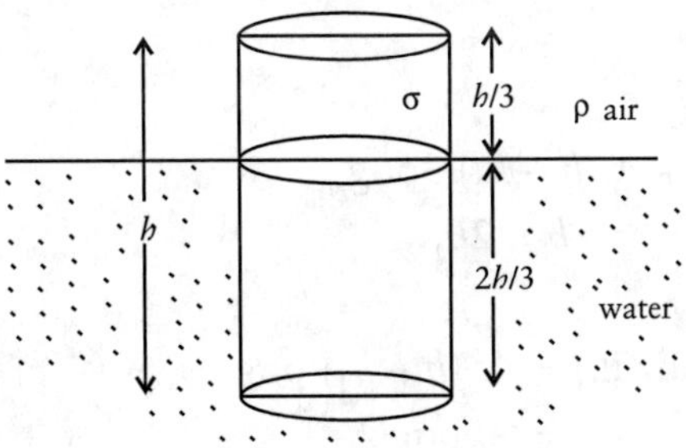

Length of the cylinder above water = $h/3$

Length of the cylinder in water = $h - h/3 = 2/3\ h$

Weight of the cylinder = $Ah \,.\, \sigma g\rho_1$ | '$W = vsw$'

Weight of water displaced by the cylinder

$= A \,.\, 2h/3 \,.1.\, g\rho_1$ | sp.gr. of water = 1

Weight of air displaced by the cylinder

$= A \,.\, h/3\ \rho g\rho_1$

For equilibrium,

weight of the cylinder = water of water displaced by the cylinder + weight of air displaced by the cylinder

$\Rightarrow$ $Ah\sigma g\rho_1 = A2h/3 \,.\, g\rho_1 + A\, h/3 \,.\, \rho g\rho_1$

$\Rightarrow$ $\sigma = 2/3 + \rho/3$ $\qquad \Rightarrow 3\sigma = \rho + 2$

Example 6: A cylinder of sp.gr. 2ρ floats with its axis vertical between two liquids of sp.gr. ρ and 3ρ respectively, the height of the cylinder being equal to the depth of the upper liquid. Prove that the pressures on its ends are as 1 : 5.

Sol. Let A be the area of cross-section of the cylinder. Let h be the height of the cylinder. Let the cylinder be immersed to depths h_1 and $(h - h_1)$ in the liquids of sp.gr. ρ and 3ρ respectively. Let σ be the density of the standard substance (water).

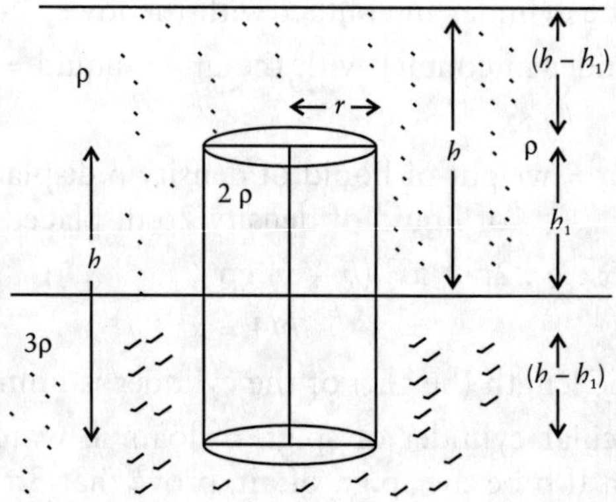

Weight of the cylinder (W) = $Ah2\rho\sigma g$ | '$W = vgw$'

Weight of the liquid of sp.gr. ρ displaced by the cylinder (W_1) = $A\,(h - h_1)\,3\rho\sigma g$

For equilibrium,

$W = W_1 + W_2$

$\Rightarrow$ $Ah2\rho\sigma g = Ah_1\rho\sigma g + A\,(h - h_1)\,3\rho\sigma g$

$\Rightarrow$ $2h = h_1 + 3\,(h - h_1)$ $\Rightarrow h = 2h_1$

$\Rightarrow$ $h_1 = h/2$ ...(1)

Pressure on the upper end (P_1) = $A.\,(h - h_1)\,\rho\sigma g$ | '$P = wS\bar{z}$'

= $Ah\rho\sigma g/2$ | From (1) ...(2)

Pressure on the lower end (P_1) = $A\,[h\rho\sigma g + (h - h_1)\,3\rho\sigma g]$

= $A\,[h\rho\sigma g + 3h/2\,\rho\sigma g]$ | From (1)

= $5/2\,Ah\rho\sigma g$

$\therefore$ $P_1/P_2 = 1/5$ | From (2) and (3)

Example 7: Two liquids which do not mix are placed in the same vessel, the density of the lower liquid is ρ and that of the upper is $n\rho$. A cylinder floats in these with its axis vertical and is completely submerged, its density being $n\rho$. Prove that the condition that it may be half in the upper and half in the lower liquid is $n = 1/2\,(m + 1)$.

Sol. Let r be the radius of the base of the cylinder and h be the height of the cylinder.

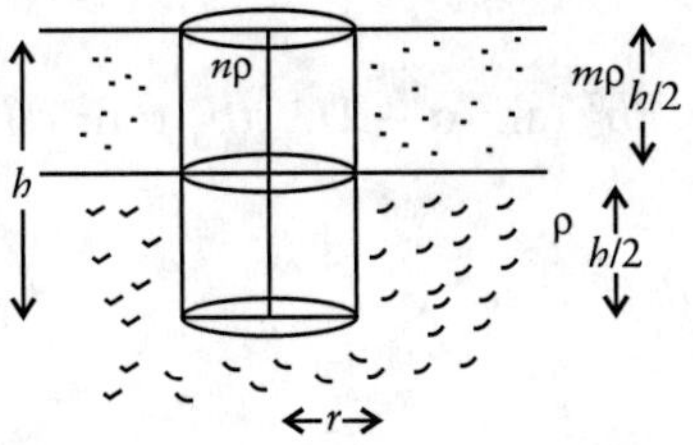

Weight of the cylinder $(W) = \pi r^2 hn\rho g$ ...(1)

Weight of the liquid of density $m\rho$ displaced by cylinder $(W_1) = \pi r^2\, h/2\, m\rho g$ ...(2)

Weight of the liquid of density ρ displaced by cylinder $(W_2) = \pi r^2\, h/2\, \rho g$ ...(3)

For equilibrium,

$$W = W_1 + W_2$$

$$\Rightarrow \pi r^2 hn\rho g = \pi r^2 hm\rho g/2 + \pi r^2 h\rho g/2 \qquad \Rightarrow n = m + 1/2$$

Example 8: A right circular cylinder of density ρ, floats just immersed with its vertex downwards in a vessel containing two liquids of densities σ_1 and σ_2 respectively. Show that the plane of separation of the two liquids cuts off

$$\left(\frac{\rho-\sigma_2}{\sigma_1-\sigma_2}\right)^{1/3} \text{ of its length.}$$

Sol. Let $V\,AB$ be the given cone. Let its height be h and semi-vertical angle be α. The base AB of the cone is in the surface and the vertex is downwards. CD is the surface of separation of the two liquids of densities σ_1 and σ_2. O is the centre of the base AB. O[1] is the centre of the surface of separation.

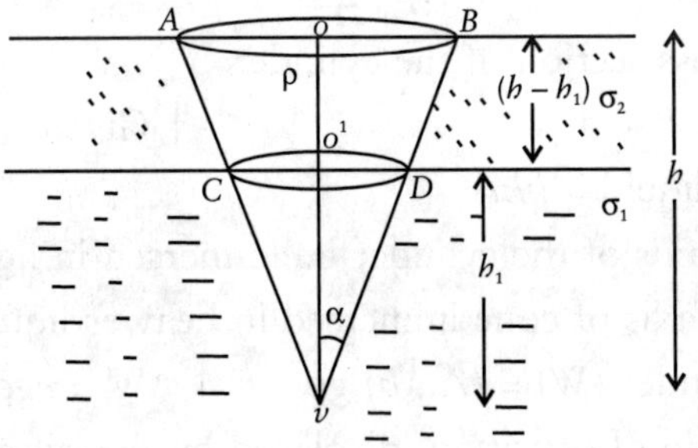

Let $V\,o' = h_1$

Then, $oo' = h - h_1$

Weight of the cone $(W) = 1/3\, \pi h^3 \tan^2 \alpha g\rho$

Weight of the liquid of density σ_2 displaced by the cone (W_1)

$$= (1/3\, \pi h^3 \tan^2 \alpha + 1/3\, \pi\, h_1^3 \tan^2 \alpha)\, g\sigma_2$$

Weight of the liquid of density σ_1 displaced by the cone $(W_2) = 1/3\, \pi h_1^3 \tan^2 \alpha g\sigma_1$

For equilibrium,

$$W = W_1 + W_2$$

$$\Rightarrow \; 1/3\ \pi h^3 \tan^2 \alpha g\rho = (1/3\ \pi h^3 \tan^2 \alpha - 1/3\ \pi h_1^3 \tan^2 \alpha)\, g\sigma_2 + 1/3\ \pi h_1^3 \tan^2 \alpha g\sigma_1$$

$$\Rightarrow \; h^3\rho = (h^3 - h_1^3)\, \sigma_2 + h_1^3\sigma_1$$

$$\Rightarrow \; h^3\,(\rho - \sigma_2) = h_1^3\,(\sigma_1 - \sigma_2)$$

$$\Rightarrow \; h_1 = h\left\{\frac{\rho - \sigma_2}{\sigma_1 - \sigma_2}\right\}^{1/3}$$

Example 9: A vertical cylinder of density ρ, floats in two liquids, the density of the upper liquid being ρ_1 and that of the lower ρ_2. If the length h of the cylinder be n times the depth of the upper liquid, prove that the depth of immersion of the upper face of the cylinder is

$$\frac{h}{n} - h\left(\frac{\rho_2 - \sigma}{\rho_2 - \rho_1}\right), \text{ provided}$$

$$\rho_2 - \rho_2 - \rho_1/n < \rho < \rho_2 \text{, where } h \text{ is the height of the cylinder.}$$

Sol.

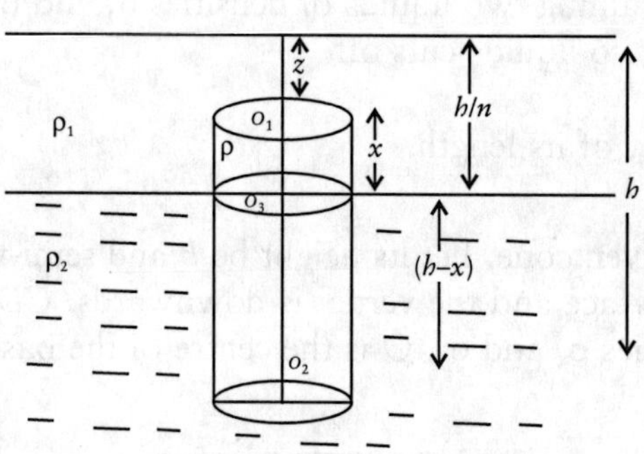

Let A be the area of cross-section of the cylinder.

Height of the cylinder $= h$ | Given

$\therefore$ Depth of the lighter liquid $= h/n$

Let the length x of the axis of the cylinder be immersed in lighter liquid of density ρ_1.

Then, the length of the axis of cone immersed in heavier liquid of density $\rho_2 = h - x$

Weight of the cylinder $(W) = (A\,.\,h)\, g\rho$ | '$W = vg\rho$' ...(1)

Weight of the liquid of density ρ_1 displaced by the cylinder (W_1)

$= (A\,x)\, g\rho_1$ | '$W = vg\rho$' ...(2)

Weight of the liquid of density ρ_2 displaced by the cylinder (W_2)

$= A\,(h - x)\, g\rho_2$ | '$W = vg\rho$' ...(3)

For equilibrium,

weight of the cylinder = water of the liquid of density ρ_1 displaced by the cylinder + weight of the liquid of density ρ_2 displaced by the cylinder

$\Rightarrow\ W = W_1 + W_2$

$\Rightarrow\ Ahg\rho = Axg\rho_1 + A\,(h - x)\,g\rho_2$

$\Rightarrow\ h\rho = x\rho_1 + (h - x)\,\rho_2$

$\Rightarrow\ h\,(\rho - \rho_2) = x(\rho_1 - \rho_2)$

$$\Rightarrow\ x = \frac{h(\rho - \rho_2)}{\rho_1 - \rho_2} = \frac{h(\rho_2 - \rho)}{\rho_2 - \rho_1}$$

$\therefore$ Required depth of O_1, the centre of the upper end

$$= z = \frac{h}{n} - x = \frac{h}{n} - \frac{h(\rho_2 - \rho)}{\rho_2 - \rho_1}$$

Given that $\rho < \rho_2$. In case otherwise, the cylinder will wholly sink in the liquid of density ρ_2. As the result, no part of it will be in the lighter liquid of density ρ_1. Again, z is essentially positive.

$\therefore\ h/n > h\rho_2 - \rho/\rho_2 - \rho_1 \Rightarrow \rho_1 - \rho_1 > n\,(\rho_1 - \rho) \Rightarrow n\rho > n\rho_2 - \rho_2 + \rho_1 \Rightarrow \rho > \rho_2 - (\rho_2 - \rho_1/n)$

Hence the validity of the given relations.

Example 10: A solid sphere of radius a is just immersed in three liquids whose densities are 1: 3: 5; the two surfaces of separation of the liquids are at depths $2a/3$ and $5a/3$ from the top. Prove that the density of the sphere to that of the topmost liquid are in the ratio of 71 : 27.

Sol.

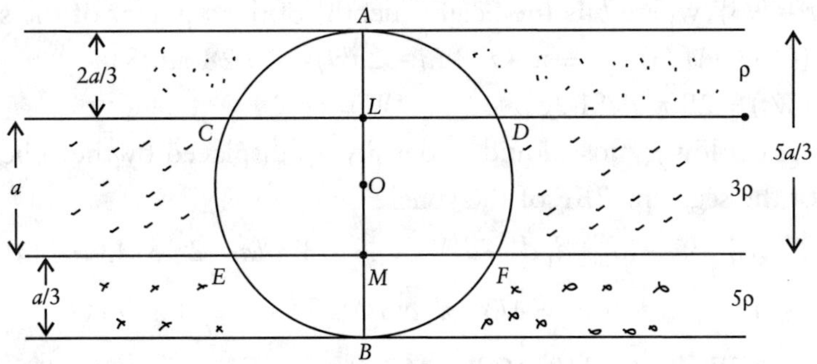

Let the density of the sphere be σ. Let the densities of the liquids from the top be ρ, 3ρ and 5ρ. Let CD and EF the surfaces of separation of the three liquids meeting the vertical diameter AB at L and M respectively. Then,

$AL = 2a/3,\ OL = OA - AL = a - 2a/3 = a/3$

$AM = 5a/3$

$\therefore\ LM = AM - AL = 5a/3 - 2a/3 = a$

$MB = AB - AM = 2a - 5a/3 = a/3$

$OM = OB - MB = a - a/3 = 2a/3$

Volume of the sphere $(V) = 4/3\ \pi a^3$

Weight of the sphere $(W) = 4/3\ \pi a^3 g\sigma$ | '$W = vg\rho$' ... (1)

Volume v_1 of the upper most liquid of density ρ displaced by the sphere

= volume of the segment CAD of the sphere

= '$1/3\ \pi\ (d_1 - d_2)\ \{3r^2 - (d_1^2 + d_1 d_2 + d_2^2)\}$' $= 1/3\ \pi\ (a - a/3)\ \{3a^2 - (a^2 + a\ .\ a/3 + a^2/9)\}$

$= 1/3\ \pi\ 2a/3\ (14a^2/9)$ $= 28\ \pi a^3/81$

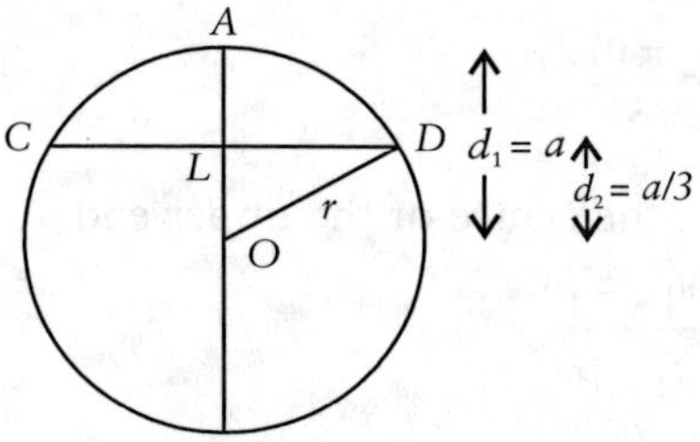

Aliter:

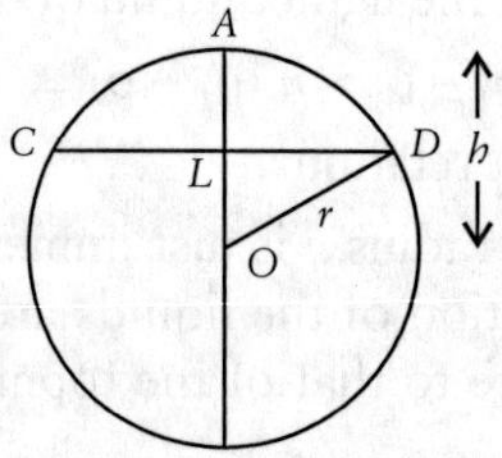

v_1 = '$\pi h^2\ (r - h/3)$' where h is the height (depth) of the segment of the sphere

$= \pi AL^2\ (OA - AL/3)$ $= \pi\ 4a^2/9\ (a - 2a/9)$ $= 28\ \pi a^3/81$

Its weight $(W_1) = 28\ \pi\ a3/81\ .\ g\ \rho$ | '$W = vg\rho$' ... (2)

Volume v_2 of the lower most liquid of density 5ρ displaced by the sphere

= volume of the segment EBF of the sphere

= '$1/3\ \pi\ (d_1 - d_2)\ \{3r^2 - (d_1^2 + d_1 d_2 + d_2^2)\}$' $= 1/3\ \pi\ (a - 2a/3)\ \{3a^2 - (a^2 + a\ .\ 2a/3 + 4a^2/9)\}$

$= 1/3\ \pi\ a/3\ (3a^2 - 19a^2/9)$ $= 1/3\ .\ \pi.\ a/3\ .\ 8a^2/9$ $= 8\ \pi a^3/81$

v_2 = '$\pi h^2\ (r - h/3)$' $= \pi\ MB^2\ (OB - MB/3) = \pi\ .\ a^2/9\ (a - a/9) = 8\ \pi a^3/81$

Its weight $(W_2) = 8\ \pi a^3/81\ g\ .\ 5\rho$ | '$W = vg\rho$' ... (3)

Volume v_3 of the middle liquid of density 3ρ displaced by the sphere

= volume of the zone $CDFE$ of the sphere

$= v - (v_1 + v_2)$

$= 4/3\ \pi a^3 - (28\ \pi a^3/81 + 8\ \pi a^3/81)$ $= 4/3\ \pi a^3 - 36\ \pi a^3/81$ $= 4/3\ \pi a^3 - 4/9\ \pi a^3$

$= 8/9\ \pi a^3$

Its weight $(W_3) = 8/9\ \pi a^3 g 3\rho$ | '$W = vg\rho$' ... (4)

For equilibrium,

weight of the sphere = sum of the weights of the three liquids displaced by it

$\Rightarrow W = W_1 + W_2 + W_3$

$\Rightarrow$ $4/3\ \pi a^3 g\sigma = 28\ \pi a^3/81\ g\rho + 8\ \pi a^3/81\ g\ 5\rho + 8/9\ \pi a^3 g\ 3\rho$

$\Rightarrow$ $108\ \sigma = 28\ \rho + 40\ \rho + 216\ \rho$ $\quad\Rightarrow 108\ \sigma = 284\ \rho$ $\quad\Rightarrow \sigma/\rho = 284/108 = 71/27$

Example 11: A solid sphere floats just immersed in heterogeneous liquid composed of three liquids which do not mix and whose densities are as 1 : 2 : 3. If the thickness of the two upper liquids be each one-third of the diameter of the sphere, show that the density of the liquid in the middle is equal to the density of the sphere.

Sol.

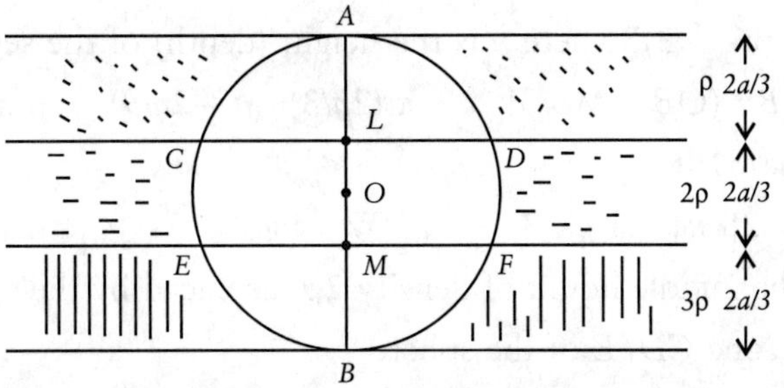

Let a be the radius of the sphere. Let σ be the density of the sphere. Let the densities of the three liquids from the top be ρ, 2ρ and 3ρ. Let CD and EF be the surfaces of separation of the three liquids meeting the vertical diameter AB at L and M respectively. Then,

$AL = LM = MB = 2a/3$

$OA = OB = a$

$OL = OA - AL = a - 2a/3 = a/3$

$OM = OB - BM = a - 2a/3 = a/3$

Volume of the sphere $(v) = 4/3\ \pi a^3$

$\therefore$ Weight of the sphere $(W) = 4/3\ \pi a^3 g\sigma$ | '$W = vg\rho$' ... (1)

Volume (v_1) of the upper most liquid of density ρ displaced by the sphere

= volume of the segment CAD of the sphere

= '$1/3\ \pi(d_1 - d_2)\ \{3a^2 - (d_1^2 + d_1 d_2 + d_2^2)\}$'

$= 1/3\ \pi(a - a/3)\ \{3a^2 - (a^2 + a\ .\ a/3 + a^2/9)\}$ | $d_1 = OA = a$, $d_2 = OL = a/3$

$= 1/3\ \pi\ .\ 2a/3\ \{3a^2 - (a^2 + a^2/3 + a^2/9)\}$ $\quad = 1/3\ .\ \pi\ .\ 2a/3\ (3a^2 - 13/9\ a^2)$

$= 1/3\ .\ \pi\ .\ 2a/3\ .\ 14a^2/9$ $\quad = 28\ \pi a^3/81$

Aliter: v_1 = '$\pi h^2\ (r - h/3)$' where h is the height (depth) of the segment of the sphere

$= \pi\ .\ (2a/3)^2\ (a - 2a/9)$ | $h = AL = 2a/3$

$= \pi\ .\ 4a^2/9.\ 7a/9 = 28\ \pi a^3/81$

Its weight $(W_1) = 28\ \pi a^3/81\ g\rho$ | '$W = vg\rho$' ... (2)

Volume (v_2) of the lower most liquid of density 3ρ displaced by the sphere

= volume of the segment *EBF* of the sphere

= '$1/3\ \pi(d_1 - d_2)\ \{3r^2 - (d_1^2 + d_1 d_2 + d_2^2)\}$'

= $1/3\ \pi\ (a - a/3)\ \{3a^2 - (a^2 + a\ .\ a/3 + a^2/9)\}$ | $d_1 = OB = a$, $d_2 = OM = a/3$ | as above

= $28\ \pi a^3/81$

Aliter: v_1 = '$\pi\ h^2\ (r - h/3)$' where h is the height (depth) of the segment of the sphere

$= \pi\ (MB)^2\ (OB - MB/3) \quad = \pi\ (2a/3)^2\ (a - 2a/9) \quad = \pi\ .\ 4a^2/9\ .\ 7a/9$

$= 28\ \pi a^3/81$

Its weight (W_2) = $28\ \pi a^3/81\ g\ .\ 3\rho$ | '$W = vg\rho$' ... (3)

Volume (v_3) of the middle liquid of density 2ρ displaced by the sphere

= volume of the zone *CDFE* of the sphere

$= v - (v_1 + v_2) \quad = 4/3\ \pi a^3 - (28\ \pi a^3/81 + 28\ \pi a^3/81) \quad = 4/3\ \pi a^3 - 56\ \pi a^3/81$

$= 52\ \pi a^3/81$

Its weight (W_3) = $52\ \pi a^3/81\ g2\rho$ | '$W = vg\rho$' ... (4)

For equilibrium,

Weight of the sphere = Sum of the weights of the three liquids displaced by it

⇒ $W = W_1 + W_2 + W_3$

⇒ $4/3\ \pi a^3\ g\sigma = 28\ \pi a^3/81\ g\rho + 52\ \pi a^3/81\ g\ .\ 2\rho + 28\ \pi a^3/81\ g\ .\ 3\rho$

⇒ $108\ \sigma = 28\ \rho + 104\ \rho + 84\ \rho \quad \Rightarrow 108\ \sigma = 216\ \rho \quad \Rightarrow \sigma = 2\rho$

⇒ Density of the liquid in the middle is equal to the density of the sphere.

EXERCISE 2

1. An alloy of gold (sp.gr. 19) and silver (sp.gr. 10·5) floats with 47/48th of its volume in mercury (sp.gr. 13·6) and its remainder in a liquid of sp.gr. 0·8. Find the proportion of gold and silver in the alloy.

   Ans. 1 : 2

   [**Hint:** Refer Ex. 1]

2. A piece of wood floats with 9/10th of its volume in water and the remaining in air. Find the additional volume which will be immersed when the vessel is placed in vacuum, the sp. of air being 0·0013.

   Ans. 0·0013

3. If a body be floating partly immersed is a liquid and the air in contact with it be suddenly removed, will the body rise or sink? Give reasons.

   Ans. The body will sink.

4. A vertical cylinder of density $7\rho/4$ floats completely immersed in two liquids. The density of the upper liquid is ρ, and that of the lower is 2ρ. Find the position of the equilibrium.

   Ans. 1/4 of the cylinder is in upper liquid.

4. Tension in the String Supporting a Body

A body is entirely immersed in a liquid, being supported by a string. To find the tension of the string.

Let the weight of the body which is completely immersed in a liquid be w. Let the body be supported by a string at A. Let T be the tension in the string in equilibrium position. Let w' be the weight of the liquid displaced.

The body is in equilibrium under the action of the following forces:

1. w, the weight of the body acting vertically downwards at G, the C.G. of the body.
2. w', the force of buoyancy (equal to the weight of the liquid displaced) acting vertically upwards at H, the centre of buoyancy.

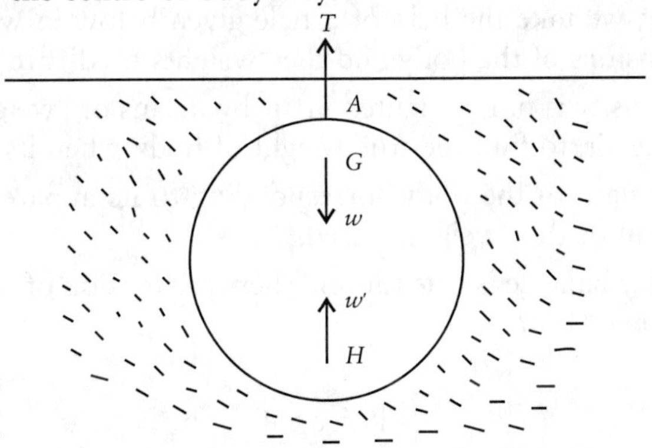

Hence, for equilibrium, we have

$$T + w' = w$$

$$\Rightarrow T = w - w'$$

$\Rightarrow$ Tension in the string = weight of the body – weight of the liquid displaced

Note 1: When the body is completely immersed in the liquid, the tension T in the string is called the apparent weight of the body. w is called the true weight of the body.

Note 2: If the density of the body < density of the liquid, then the tension T in the string acts vertically downwards and therefore for equilibrium in this case, we have,

$$T + w = w'$$

$$\Rightarrow T = w' - w$$

$\Rightarrow$ Tension in the string = weight of the liquid displaced – weight of the body

Note 3: Let ρ be the density of the body and σ be the density of the liquid. Then,

w' = weight of the liquid displaced

= volume of the liquid displaced × Density of the liquid × g | '$W = vg\rho$'

= (volume of the body) $\sigma\, g$ | By Principle of Archimedes

= $(W/g\rho)\, \sigma g$ | $\because$ weight of the body (W) = volume of the body (v) $\times \rho \times g$ $\therefore$ volume = $W/\rho g$

$\Rightarrow w' = W\, \sigma/\rho$

5. Weighting is a liquid: If a body, completely immersed in a liquid, is weighted by means of a balance, then the tension T in the string supporting the body will be equal to the true weight w of the body minus the weight w' of the liquid displaced, i.e., $T = w - w'$

$\Rightarrow w = T + w'$

$\Rightarrow$ True weight of the body = Apparent weight of the body + Weight of the liquid displaced

6. Weighing in air: In order to obtain a perfectly accurate weight of the body, it should be weighted in vacuum. However, it is not practically possible. The reason for doing so is that the quantities of the air displaced by the body and by the "weights used" are not the same. To remove this discrepancy, we take the help of article given below in which the true weight can be determined if the densities of the body and the "weights used" are given.

7. A body, whose density is ρ, is weighted in air by means of "weights whose" density is ρ'. If σ be the density of the air, to find the true weight of body when its apparent weight is w_0.

Let w be the true weight of the body corresponding to its apparent weight w_0. [Note that w_0 is nothing but the sum of the "weights" used.]

Let us assume that the balance is true (pure). Then, the tension of the two supporting strings of the scale-pans are equal, say T.

For one scale-pan,

$T = w - w'$ | Refer § 4

= weight of the body – weight of the air it displaces

$= w - w\, \sigma/\rho$ | Refer Note § 4 ... (1)

For the other scale-pan,

T = weight of the "weights" used – weight of the air the "weights" used displace

$= w_0 - w_0\, \sigma/\rho$ ... (2)

Equating the two values of T from (1) and (2), we get

$$w - w\,\sigma/\rho = w_0 - w_0\,\sigma/\rho' \quad \Rightarrow w\,(1 - \sigma/\rho) - w_0\,(1 - \sigma/\rho')$$

$$\Rightarrow w = \frac{w_0\left(1-\frac{\sigma}{\rho'}\right)}{1-\frac{\sigma}{\rho}} = w_0\,(1 - \sigma/\rho')\,(1 - \sigma/\rho)^{-1} = w_0\left(1-\frac{\sigma}{\rho'}\right)\left\{1+\frac{\sigma}{\rho}+\left(\frac{\sigma}{\rho}\right)^2+\ldots\right\}$$

| expanding by binomial theorem

Since, in general, σ is they small compared to ρ and ρ', therefore, neglecting the square and higher powers of σ/ρ, we have

$\Rightarrow w = w_0\,(1 + \sigma/\rho - \sigma/\rho')$ nearly

ILLUSTRATIVE EXAMPLES

Example 1: A body, whose weight is 100 kg and whose sp.gr. is 10, is suspended by a string in a liquid of sp.gr. 7. What is the tension of the string?

Sol. Let w be the weight per unit volume of water and v be the volume of the body. Then, weight of the body $(W) = 100\ g = v \cdot 10 \cdot w$ | '$W = vsw$' ... (1)

Weight of the liquid displaced by the body $(W') = v\ .\ 7\ .\ w$ | '$W = vsw$' ... (2)

Let T be the tension in the string. Then,

$$T = W - W'$$

$$= v\ .\ 10\ .\ w - v\ .\ 7\ .\ w \qquad \text{| From (1) and (2)}$$

$$= 3\ vw \qquad = 3\ .\ 10\ g \qquad = 30\ g \qquad = 30\ \text{kg. wt.}$$

Example 2: A piece of silver and a piece of copper fastened to the ends of a string passing over a pulley, hang in equilibrium when suspended in a liquid of density 1·15. Determine the relative volumes of the masses, the densities of silver and copper being 10·47 and 8·89 respectively.

Sol. Let v_1 and v_2 be the volumes of the pieces of silver and copper respectively. Since the pieces of silver and copper are suspended in a liquid of density 1·15, therefore, tension in the string supporting the piece of silver = tension in the string supporting the piece of copper = weight of the piece of silver – weight of the liquid of density 1·15 displaced by the piece of silver = weight of the piece of copper – weight of the liquid of density 8.89 displaced by the piece of copper

$$\Rightarrow v_1\ g\ (10{\cdot}47) - v_1\ g\ (1{\cdot}15) = v_2\ g\ (8{\cdot}89) - v_2\ g\ (1{\cdot}15)$$

$$\Rightarrow v_1\ (10{\cdot}47 - 1{\cdot}15) = v_2\ (8{\cdot}89 - 1{\cdot}15)$$

$$\Rightarrow v_1\ (9{\cdot}32) = v_2\ (7{\cdot}74)$$

$$\Rightarrow v_1/v_2 = 7.74/9.32 = 387/466$$

Example 3: A body immersed in a liquid is balanced by a weight P to which it is attached by a thread passing over a fixed pulley, and when half immersed, is balanced in the same manner by a weight 2ρ. Prove that the densities of the body and the liquid are as 3 : 2.

Sol. Let v be the volume of the body. Let σ, ρ be the densities of the body and the liquid respectively.

We know that when a body is wholly immersed in a liquid, then,

tension in the string = weight of the body – weight of the liquid displaced by the body

$$\therefore\ P = vg\sigma - vg\rho$$

$$\Rightarrow P = vg\ (\sigma - \rho) \qquad \text{... (1)}$$

In case when the body is half immersed, we similarly obtain,

$$2P = vg\sigma - v/2\ g\rho$$

$$\Rightarrow 2P = vg\ (\sigma - \rho/2) \qquad \text{... (2)}$$

From (1) and (2), we get

$$2\ vg\ (\sigma - \rho) = vg\ (\sigma - \rho/2)$$

$$\Rightarrow 2\ (\sigma - \rho) = \sigma - \rho/2 \qquad \Rightarrow \sigma = 3\rho/2 \qquad \Rightarrow \sigma/\rho = 3/2$$

Example 4: A solid whose specific gravity is 1·85 is weighted into a mixture of alcohol of sp.gr. 0·82 and water. It weighs 28·8 grammes in vacuum and 14·1 grammes in the mixture. Find the proportion of the alcohol present.

Sol. Let v c.c. be the volume of the solid. Let s be the sp.gr. of the mixture of alcohol and water. We know that

true weight of the solid = weight of the solid in vacuum

$\Rightarrow vg\,(1{\cdot}85) = 28{\cdot}8$ g

$\Rightarrow v = 28{\cdot}8/1{\cdot}85 = 2880/185 = 576/37$ c.c.

Again,

weight of the solid in mixture = true weight of the solid – weight of the mixture displaced by it

$\Rightarrow 14{\cdot}1\, g = 28{\cdot}8\, g - vgs$

$\Rightarrow vs = 28{\cdot}8 - 14{\cdot}1 = 14{\cdot}7$

$\Rightarrow s = 14{\cdot}7/v = 14{\cdot}7 \times 37/576$

$\Rightarrow 147 \times 37/10 \times 576$ ... (1)

Let v_1 and v_2 be the volumes of the alcohol and water in the mixture. Then, the sp.gr. of the mixture

$= `v_1 s_1 + v_2 s_2/ v_1 + v_2'$ $\quad \Rightarrow s = v_1\,(0{\cdot}82) + v_2\,(1)/\,v_1 + v_2$

Let $v_1 + v_2 = v_3$ (Whole volume)

Then, $v_2 = v_3 - v_1$

$\therefore\ s = v_1\,(0{\cdot}82) + v_3 - v_1/v_3 \quad \Rightarrow s = v_3 - v_1\,(0{\cdot}18)/v_3$

$\Rightarrow 147 \times 37/\,10 \times 576 = v_3 - v_1\,(0{\cdot}18)/v_3$ | From (1)

$\Rightarrow 147 \times 37\, v_3 = 5760\, v_3 - v_1 \times 576 \times 1{\cdot}8$

$\Rightarrow v_1\,(576 \times 1{\cdot}8) = (5760 - 5439)v_3$

$\Rightarrow v_1\,(576 \times 1{\cdot}8) = 321\, v_3$

$\Rightarrow v_1 = \dfrac{321\times10}{576\times1{\cdot}8}\, v_3 \quad \Rightarrow v_1 = 0{\cdot}31\, v_3$ (nearly) $\quad \Rightarrow v_1 = \dfrac{31}{100}\, v_3$

$\Rightarrow$ volume of alcohol = 31% of the whole volume

Example 5: The apparent weight of a piece of platinum when weighed in water is 20·6 gms; when weighed in mercury, it is 8 gms, the density of the mercury being 13·6 gm/c.c. Find the volume and true weight the piece of platinum.

Sol. Let v c.c. be the volume and w_0 be the true weight of the piece of platinum.

We know that

apparent weight = true weight – weight of liquid displaced

In water,

$20{\cdot}6\, g = w_0 - v\,.\,1\,.\,g$ ... (1)

In mercury,

$$8\,g = w_0 - v\,.\,13{\cdot}6\,g \qquad \text{... (2)}$$

From (1) and (2), we get

$$20{\cdot}6\,g + v\,g = 8\,g + 13{\cdot}6\,v\,g \qquad \Rightarrow 20{\cdot}6 + v = 8 + 13{\cdot}6\,v \qquad \Rightarrow 12{\cdot}6 = 126{\cdot}v$$

$$\Rightarrow v = 1 \text{ c.c.} \qquad \text{... (3)}$$

From (1) and (3), we get

$$w_0 = 20{\cdot}6\,g + 1\,g \qquad = 21{\cdot}6\text{ g} \qquad = 21{\cdot}6 \text{ gm. wt.}$$

Example 6: A body is weighted in two liquids, the first of sp.gr. 0·8 and the other of sp.gr. 1·2. In the two cases its apparent weights are 18 and 12 grammes respectively. Find its true weight and specific gravity.

Sol. Let v c.c. be the volume and s be the sp.gr. of the body. Let w be weight per c.c. of the standard substance.

We know that

apparent weight of a body = true weight – weight of liquid displaced

In first liquid,

$$18 = v\,s\,w - v\,(0{\cdot}8)\,w$$

$$\Rightarrow 18 = v\,w\,(s - 0{\cdot}8) \qquad \text{... (1)}$$

In second liquid,

$$12 = v\,s\,w - v\,(1{\cdot}2)\,w$$

$$\Rightarrow 12 = v\,w\,(s - 1{\cdot}2) \qquad \text{... (2)}$$

Dividing (1) by (2), we get

$$18/12 = s - 0.8/s - 1{\cdot}2 \qquad \Rightarrow 3/2 = s - 0{\cdot}8/s - 1{\cdot}2 \qquad \Rightarrow 3(s - 1{\cdot}2) = 2\,(s - 0{\cdot}8)$$

$$\Rightarrow s = 3{\cdot}6 - 1{\cdot}6 = 2$$

Hence, the sp.gr. of the body is 2.

From (1),

$$18 = v\,(2 - 0{\cdot}8)\,w \qquad \Rightarrow \qquad vw = 18/1{\cdot}2 = 15$$

$\therefore$ True weight of the body = $v\,s\,w$ = 15 × 2 = 30 grammes wt.

Example 7: A body is found to sink in each of liquids whose sp.gr. are 7 and 1·3 and its apparent weight in the two liquids is found to be 14 and 8 grammes respectively. Find the sp.gr. of the body and show that it will float half immersed when placed in a liquid of sp.gr. 4·2.

Sol. Let v be the volume and s be the sp.gr. of the body. Let ρ be the density of the standard substance. Then the densities of the two liquids are 0·7 ρ and 1·3 ρ respectively.

We know that

apparent weight of the body = true weight of the body – weight of liquid displaced

$$\Rightarrow 14 = v\,g\,s\,\rho - v\,(0{\cdot}7)\,g$$

and $8 = v\,g\,s\,\rho - v\,(1{\cdot}3)\,g$

$$\Rightarrow 14 = v\,g\,\rho\,(s - 0{\cdot}7) \qquad \text{...(1)}$$

and $8 = v\,g\,\rho\,(s - 1{\cdot}3)$ ...(2)

Dividing (1) by (2), we get

$14/8 = s - 0{\cdot}7/s - 1{\cdot}3 \quad \Rightarrow 7/4 = s - 0{\cdot}7/s - 1{\cdot}3 \quad \Rightarrow 7(s - 1{\cdot}3) = 4\,(s - 0{\cdot}7)$

$\Rightarrow\ 7s - 9{\cdot}1 = 4s - 2{\cdot}8 \quad \Rightarrow 3s = 6{\cdot}3$

$\Rightarrow\ s = 2{\cdot}1$ ... (3)

Hence, the sp.gr. of the body is 2·1.

Let v' be the volume of the body immersed in the liquid of density 4·2 ρ.

We know that

weight of the body = weight of liquid displaced

$\Rightarrow\ v\,g\,s\,\rho = v'\,g\,4{\cdot}2\,\rho \quad \Rightarrow v\,s = v'\,(4{\cdot}2) \quad \Rightarrow v\,(2{\cdot}1) = v'\,(4{\cdot}2)$ | From (3)

$\Rightarrow\ v' = v/2$

Hence, the body will float half immersed in a liquid of sp.gr. 4·2.

Example 8: A body is floating partly immersed in a liquid and the air in contact with the body is suddenly removed. State giving reasons whether the body will rise or sink.

Sol. Let v_1 and v_2 be the volumes of the body immersed in the liquid and air respectively. Let ρ_1 and ρ_2 be the densities of liquid and air respectively. Let w be the weight of the body. Then,

weight of the body = weight of liquid displaced by the body + weight of air displaced by the body

$\Rightarrow\ w = v_1\,\rho_1\,g + v_2\,\rho_2\,g$ ... (1)

Let v_3 be the volume of the body immersed in the liquid when air is removed. Then

weight of the body = weight of liquid displaced by the body

$\Rightarrow\ w = v_3\,\rho_1\,g$ ... (2)

From (1) and (2), we get

$v_1\,\rho_1\,g + v_2\,\rho_2\,g = v_3\,\rho_1\,g \quad \Rightarrow v_1\,\rho_1 + v_2\,\rho_2 = v_3\,\rho_1 \quad \Rightarrow (v_3 - v_1)\,\rho_1 = v_2\,\rho_2$

Now, $v_2\,\rho_2$ is positive

$\Rightarrow$ $(v_3 - v_1)\,\rho_1$ is also positive

$\Rightarrow$ $v_3 - v_1$ is positive

$\Rightarrow$ $v_3 > v_1$

$\Rightarrow$ The volume of the body immersed when air in removed > the volume of the body immersed when the air is not removed

$\Rightarrow$ The body will sink when the air is removed.

Example 9: The apparent weight of a sinker weighed in water is four times the weight in vacuum of a piece of material, whose sp.gr. is required, that of the sinker and piece together is three times that weight. Show that the sp.gr. of the material is 0·5.

Sol. Let v_1 and v_2 be the volumes of the piece of material and sinker respectively. Let their sp.gr. be s_1 and s_2 respectively. Let ρ be the density of water (the standard substance). Then,

true weight of the piece of material = weight of the piece of material in vacuum

$= v_1\,g\,s_1\,\rho$

We know that

apparent weight of a body = true weight – weight of liquid displaced by the body

$\therefore$ Apparent weight of the sinker

$= $ true weight of the sinker – weight of water displaced by it

$= v_2 g s_2 \rho - v_2 g \rho$

According to the question,

apparent weight of the sinker = 4 weight of the piece of material in vacuum

$\Rightarrow v_2 g s_2 \rho - v_2 g \rho = 4 v_1 g s_1 \rho$

$\Rightarrow v_2 s_2 - v_2 = 4 v_1 s_1$ ... (1)

Also apparent weight of the sinker and the piece of material together = true weight of sinker and piece of material together – weight water displaced by them

$= (v_2 g s_2 \rho + v_1 g s_1 \rho) - (v_2 g \rho + v_1 g \rho)$ ... (2)

According to the question,

apparent weight of the sinker and the piece of material together = 3 weight of the piece of material in vacuum

$= 3 v_1 g s_1 \rho$ ... (3)

From (2) and (3), we get

$(v_2 g s_2 \rho + v_1 g s_1 \rho) - (v_2 g \rho + v_1 g \rho) = 3 v_1 g s_1 \rho$

$\Rightarrow (v_2 s_2 + v_1 s_1) - (v_2 + v_1) = 3 v_1 s_1$ ... (4)

Subtracting (1) from (4), we get

$v_1 s_1 - v_1 = - v_1 s_1 \quad \Rightarrow s_1 - 1 = - s_1 \quad \Rightarrow 2 s_1 = 1 \quad \Rightarrow s_1 = 1/2 = 0{\cdot}5$

Example 10: If W and w be the weights of a body in vacuo and water respectively, prove that the weight in the air of sp.gr. σ will be $W - \sigma (W - w)$

Sol. True weight of the body = weight of the body in vacuo

$= W$ | Given ... (1)

Weight of body in water = True weight of the body – weight of water displaced by the body

$\Rightarrow W = w - v g \rho$ | From (1)

where v is the volume of the body and ρ is the density of water (standard substance)

$\Rightarrow v g \rho = W - w$ ... (2)

Now, weight of the body in air = True weight of the body – weight of air displaced by the body

$= W - v g \sigma \rho$

$= W - \sigma (W - w)$ | From (2)

Example 11: A solid of weight W is weighed in air by a spring balance and its weight is found to be W'. Prove that $W = W' (\rho/\rho - \sigma)$, where σ and ρ are the specific gravities of the air and the solid respectively.

Sol. We know that

apparent weight of a solid in air = its true weight – weight of air displaced by it

$\Rightarrow W' = W - v\, g\, \sigma\, \rho_1$ ... (1)

where v is the volume of the solid and ρ_1 is the density of the standard substance.

Weight of the body = $v\, g\, \rho\, \rho_1$

$\Rightarrow W = v\, g\, \rho\, \rho_1$

$\Rightarrow v\, g\, \rho_1 = W/\rho$ ... (2)

From (1) and (2), we get,

$W' = W - W\sigma/\rho$

$\Rightarrow W' = W(1 - \sigma/\rho) \qquad \Rightarrow W' = W\dfrac{\rho-\sigma}{\rho} \qquad \Rightarrow W = W'\rho/\rho - \sigma$

Example 12: A body of specific gravity σ when weighed against a "weight" of sp.gr. ρ in water (the whole balance being immersed), appears to have a weight W. Show that its true weight is

$$\frac{\sigma}{\sigma-1}\cdot\frac{\rho-1}{\rho}w$$

Sol. True weight of "weights" = W

∴ Volume of "weights" = $W/g\,\rho$

Apparent weight of "weights" in water

= True weight of "weights" – weights of water displaced by "weights"

= $W - W/g\,\rho\,.\,g\,.\,1$

= $W(1 - 1/\rho)$ ... (1)

Let W be the true weight of the body of sp.gr. σ.

Then, volume of the body = $w/g\,\sigma$

Apparent weight of the body in water

= True weight of the body – weight of water displaced by the body

= $w - w/g\,\sigma\,.\,g\,.\,1$

= $w(1 - 1/\sigma)$ ... (2)

When the whole balance is immersed in water, apparent weight of "weights" in water = apparent weight of the body in water

$\Rightarrow W(1 - 1/\rho) = w(1 - 1/\sigma)$

$$\Rightarrow W\frac{\rho-1}{\rho} = w\frac{\sigma-1}{\sigma} \qquad \Rightarrow w = W\left(\frac{\rho-1}{\rho}\right)\left(\frac{\upsilon}{\sigma-1}\right)$$

Example 13: A given mass weighed in air on a spring balance indicates a weight W. It is then compressed to $1/n^{th}$ of its former volume and appears to weigh W'. Prove that its weight in vacuo is

$nW' - W/n - 1$

Sol. Let v be the volume of the given mass. Let W'' be the weight of the given mass in vacuo. Let σ be the density of the air.

We know that

weight of the body in air = True weight of the body – Weight of the air displaced by the body

$\Rightarrow W = W'' - v\,g\,\sigma$

$\Rightarrow v\,g\,\sigma = W'' - W$ ... (1)

When the given mass is compressed to $1/n^{th}$ of its former volume, then new volume $= v/n$ and now apparent weight is W'.

$\therefore$ As above,

$W' = W'' - v/n\,g\,\sigma$

$\Rightarrow v\,g\,\sigma/n = W'' - W'$

$\Rightarrow v\,g\,\sigma = n\,(W'' - W')$ ... (2)

Equating the two values of $v\,g\,\sigma$ from (1) and (2), we get

$W'' - W = n(W'' - W')$

$\Rightarrow W''(n - 1) = nW' - W$

$\Rightarrow W'' = nW' - W/n - 1$

Example 14: A body floats in a fluid of sp.gr. ρ with as much of its volume out of the fluid as would be immersed in a second fluid of sp.gr. σ, if it floated in that fluid. Prove that the sp.gr. of the body is

$\rho\sigma/\ \rho + \sigma$

Sol. Let s be the sp.gr. of the body. Let v_1 and v_2 be the volumes of the body immersed in the fluids of sp.gr. ρ and σ respectively.

Then, volume of the whole body $= v_1 + v_2$ | $\because$ volume out of the first fluid = volume immersed in the second fluid

We know that

weight of the body = weight of the fluid displaced by it

$\therefore\ (v_1 + v_2)\,s\,g = v_1\,\rho\,g$

and, $(v_1 + v_2)\,s\,g = v_2\,\sigma\,g$

$\Rightarrow (v_1 + v_2)\,s = v_1\,\rho$ ... (1)

and, $(v_1 + v_2)\,s = v_2\,\sigma$ ... (2)

$\Rightarrow v_1\,\rho = v_2\,\sigma$

$\Rightarrow v_1/v_2 = \sigma/\rho$ ... (3)

Now, from (1), $s = v_1 + \rho/v_1 + v_2$

$$= \frac{\rho}{1+\frac{v_2}{v_1}} = \frac{\rho}{1+\frac{\rho}{\sigma}} \qquad \text{| From (3)}$$

$$= \frac{\rho\sigma}{\rho+\sigma}$$

Example 15: If a body of mass m_1 and density ρ_1 is balanced by a mass m_2 and density ρ_2 when placed on the pans of a common balance, show that

$$m_1 = m_2 \frac{\rho_1}{\rho_2}\left(\frac{\rho_2 - \sigma}{\rho_1 - \sigma}\right)$$

where σ is the density of the air.

Sol. Volume (v_1) of the body of mass m_1 and density $\rho_1 = m_1/\rho_1 g$

Volume (v_2) of the body of mass m_2 and density $\rho_2 = m_2/\rho_2 g$

We know that

apparent weight of the body = true weight of the body – weight of liquid displaced by it

$\Rightarrow$ apparent weight of the body of mass m_1

$m_1 g - (m_1/\rho_1 g)\, g\sigma$

and, apparent weight of the body of mass m_2

$= m_2 g - (m_2/\rho_2 g)\, g\sigma$

Since the two bodies balance each other when placed on the pans of a common balance, therefore,

apparent weight of the body of mass m_1 = apparent weight of the body of mass m_2

$$\Rightarrow m_1 g - \left(\frac{m_1}{\rho_1 g}\right) g\sigma = m_2 g - \left(\frac{m_2}{\rho_2 g}\right) g\sigma \quad \Rightarrow m_1 - \frac{m_1}{\rho_1}\sigma = m_2 - \frac{m_2}{\rho_2}\sigma$$

$$\Rightarrow m_1\left(1 - \frac{\sigma}{\rho_1}\right) = m_2\left(1 - \frac{\sigma}{\rho_2}\right) \quad \Rightarrow m_1\left(\frac{\rho_1 - \sigma}{\rho_1}\right) = m_2\left(\frac{\rho_2 - \sigma}{\rho_2}\right) \quad \Rightarrow m_1 = m_2 \frac{\rho_1}{\rho_2}\left(\frac{\rho_2 - \sigma}{\rho_1 - \sigma}\right)$$

Example 16: If the sp.gr. of air be s and w, w' be the weights of a body in air and water respectively, prove that its weight in vacuo is

$$\frac{w - sw'}{1 - s} \text{ or } w + \frac{s}{1 - s}(w - w')$$

Sol. Let σ be the sp.gr. of the body. Let w'' be the weight of the body in vacuo. Then,

Volume of the body = $w''/\sigma g$

Apparent weight of the body in air

= true weight of the body – weight of air displaced by it

$$\Rightarrow w = w'' - \frac{w''}{\sigma g} s g$$

$$\Rightarrow w = w''\,(1 - s/\sigma) \qquad \text{... (1)}$$

Also, apparent weight of the body in water

= True weight of the body – weight of water displaced by it

$$\Rightarrow w' = w'' - \frac{w''}{\sigma g}.1.g$$

$\Rightarrow w' = w''(1 - 1/\sigma)$ ... (2)

From (1),

$$w''\frac{s}{\sigma} = w'' - w \quad \text{... (3)}$$

From (2),

$$w''/\sigma = w'' - w' \quad \text{... (4)}$$

Dividing (3) by (4), we get

$s = w'' - w/w'' - w'$ $\quad \Rightarrow s(w'' - w') = w'' - w \quad \Rightarrow w''(s - 1) = sw' - w$

$\Rightarrow w''(1 - s) = w - sw'$ $\quad \Rightarrow w'' = \dfrac{w - sw'}{1 - s} \quad = \dfrac{w - ws + ws - sw'}{1 - s}$

$= \dfrac{w(1-s) + s(w - w')}{1 - s} \quad = w + \dfrac{s}{1-s}(w - w')$

Example 17: If w_1, w_2, w_3 be the apparent weights of a given body in fluids whose specific gravities are s_1, s_2, and s_3 respectively, then show that

$$w_1(s_2 - s_3) + w_2(s_3 - s_1) + w_3(s_1 - s_2) = 0$$

Sol. Let v be the volume of the body. Let w be the true weight of the body. Let ρ be the density of the standard substance (say water).

We know that

Apparent weight of a body in a fluid

= true weight of the body – weight of the fluid displaced by it

$\therefore\ w_1 = w - v\, s_1\, g\, \rho$ ... (1)

$w_2 = w - v\, s_2\, g\, \rho$ ... (2)

$w_3 = w - v\, s_3\, g\, \rho$ ... (3)

$\Rightarrow w_1 + v\, s_1\, g\, \rho - w = 0$

$w_2 + v\, s_2\, g\, \rho - w = 0$

$w_3 + v\, s_3\, g\, \rho - w = 0$

Eliminating 1, $v\, g\, \rho$ and $(-w)$ determinantically, we get

$$\begin{vmatrix} w_1 & s_1 & 1 \\ w_2 & s_2 & 1 \\ w_3 & s_3 & 1 \end{vmatrix} = 0$$

$\Rightarrow w_1(s_2 - s_3) + w_2(s_3 - s_1) + w_3(s_1 - s_2) = 0$

Example 18: A body floating in water has volumes v_1, v_2, v_3 above the surface when the densities of the surrounding air are respectively ρ_1, ρ_2, ρ_3. Prove that

$$\rho_2 - \rho_3/v_1 + \rho_3 - \rho_1/v_2 + \rho_1 - \rho_2/v_3 = 0$$

Sol. Let v be volume of the body. Let w be the weight of the body. Let ρ be the density of water. Then, weight of the body = weight of air displaced by the body + weight of water displaced by the body

$$\therefore \quad w = v_1 \rho_1 g + (v - v_1) \rho g \qquad \text{... (1)}$$

$$w = v_2 \rho g + (v - v_2) \rho g \qquad \text{... (2)}$$

$$w = v_3 \rho_3 g + (v - v_3) \rho g \qquad \text{... (3)}$$

$$\Rightarrow w - v \rho g - v_1 \rho_1 g + v_1 \rho g = 0$$

$$w - v \rho g - v_2 \rho_2 g + v_2 \rho g = 0$$

$$w - v \rho g - v_3 \rho_3 g + v_3 \rho g = 0$$

Eliminating $w - v \rho g$, $- g$ and ρg determinantically, we get

$$\begin{vmatrix} 1 & v_1\rho_1 & v_1 \\ 1 & v_2\rho_2 & v_2 \\ 1 & v_3\rho_3 & v_3 \end{vmatrix} = 0 \quad \Rightarrow \begin{vmatrix} 1 & 1 & 1 \\ v_1\rho_1 & v_2\rho_2 & v_3\rho_3 \\ v_1 & v_2 & v_3 \end{vmatrix} = 0 \quad \Rightarrow \begin{vmatrix} v_1\rho_1 & v_2\rho_2 & v_3\rho_3 \\ 1 & 1 & 1 \\ v_1 & v_2 & v_3 \end{vmatrix} = 0$$

$$\Rightarrow v_1 \rho_1 (v_3 - v_2) + v_2 \rho_2 (v_1 - v_3) + v_3 \rho_3 (v_2 - v_1) = 0$$

$$\Rightarrow v_2 v_3 (\rho_3 - \rho_2) + v_3 v_1 (\rho_1 - \rho_3) + v_1 v_2 (\rho_2 - \rho_1) = 0$$

$$\Rightarrow v_2 v_3 (\rho_2 - \rho_3) + v_3 v_1 (\rho_3 - \rho_1) + v_1 v_2 (\rho_1 - \rho_2) = 0$$

$$\Rightarrow \rho_2 - \rho_3/v_1 + \rho_3 - \rho_1/v_2 + \rho_1 - \rho_2/v_3 = 0 \qquad | \text{ dividing by } v_1 v_2 v_3$$

Example 19: If a body floats in a liquid with volumes v_1, v_2 and v_3 above the surface when barometric heights are h_1, h_2 and h_3, prove that

$$h_1v_1 (v_2 - v_3) + h_2v_2 (v_3 - v_1) + h_3v_3 (v_1 - v_2) = 0$$

Sol. Let ρ_1, ρ_2 and ρ_3 be the densities of the atmospheric (surrounding) air when the barometric heights are h_1, h_2 and h_3 respectively. Then,

$$\rho_1 = \lambda h_1, \quad \rho_2 = \lambda h_2, \quad \rho_3 = \lambda h_3$$

| since the densities of the atmospheric air are proportional to barometric heights

where λ is the constant of proportionality.

Let w be the true weight of the body.

Let v be the volume of the body. Let the density of the liquid be ρ. We know that for equilibrium of a body floating in a liquid,

weight of the body = weight of the liquid displaced + weight of the air displaced

$$\Rightarrow w = (v - v_1) \rho g + v_1\rho_1 g \qquad | \text{ case 1}$$

$$\Rightarrow w = (v - v_1) \rho g + v_1\lambda h_1 g$$

$$\Rightarrow v_1\lambda h_1 g + (v\rho g - w) - v_1\rho g = 0 \qquad \text{... (1)}$$

Similarly,

$$v_2\lambda h_2 g + (v\rho g - w) - v_2\rho g = 0 \qquad | \text{ case 2} \qquad \text{... (2)}$$

$$\text{and } v_3\lambda h_3 g + (vg\rho - w) - v_3\rho g = 0 \qquad | \text{ case 3} \qquad \text{... (3)}$$

Eliminating λg, $(vg\rho - w)$ and $-\rho g$ from (1), (2) and (3), we get

$$\begin{vmatrix} v_1h_1 & 1 & v_1 \\ v_2h_2 & 1 & v_2 \\ v_3h_3 & 1 & v_3 \end{vmatrix} = 0 \quad \Rightarrow \begin{vmatrix} v_1h_1 & v_2h_2 & v_1h_3 \\ 1 & 1 & 1 \\ v_1 & v_2 & v_3 \end{vmatrix} = 0$$

$$\Rightarrow v_1h_1 (v_3 - v_2) + v_2h_2 (v_1 - v_3) + v_3h_3 (v_2 - v_1) = 0$$

$$\Rightarrow h_1v_1 (v_2 - v_3) + h_2v_2 (v_3 - v_1) + h_3v_3 (v_1 - v_2) = 0$$

Example 20: Two solids are each weighed in succession in three homogeneous liquids of different densities. If the weight of one are w_1, w_2 and w_3 and those of the other are W_1, W_2 and W_3, prove that

$$w_1 (W_2 - W_3) + w_2 (W_3 - W_1) + w_3 (W_1 - W_2) = 0$$

Sol. Let the volumes of the two solids be v and V respectively. Let their true weights be w and W respectively. Let ρ_1, ρ_2 and ρ_3 be the densities of the three liquids.

We know that

apparent weight of a body in a liquid

= true weight of the body – weight of the liquid displaced by it

$\therefore$ For the first solid, we have

$$w_1 = w - vg\rho_1 \quad \text{... (1)}$$

$$w_2 = w - vg\rho_2 \quad \text{... (2)}$$

and, $$w_3 = w - vg\rho_3 \quad \text{... (3)}$$

From (1), (2) and (3), we get

$$w_1 - w_2 = vg (\rho_2 - \rho_1) \quad \text{... (4)}$$

$$w_2 - w_3 = vg (\rho_3 - \rho_2) \quad \text{... (5)}$$

and, $$w_3 - w_1 = vg (\rho_1 - \rho_3) \quad \text{... (6)}$$

Similarly, for the second solid, we have

$$W_1 - W_2 = Vg (\rho_2 - \rho_1) \quad \text{... (7)}$$

$$W_2 - W_3 = Vg (\rho_3 - \rho_2) \quad \text{... (8)}$$

and, $$W_3 - W_1 = Vg (\rho_1 - \rho_3) \quad \text{... (9)}$$

Dividing (7) by (4), (8) by (5) and (9) by (6), we get

$$W_1 - W_2/w_1 - w_2 = V/v \quad \text{...(10)}$$

$$W_2 - W_3/w_2 - w_3 = V/v \quad \text{... (11)}$$

$$W_3 - W_1/w_3 - w_1 = V/v \quad \text{... (12)}$$

From (10), (11) and (12), we have

$$\frac{W_1 - W_2}{w_1 - w_2} = \frac{W_2 - W_3}{w_2 - w_3} = \frac{W_3 - W_1}{w_3 - w_1} = \frac{V}{v} = \lambda \text{ (say)}$$

$$\Rightarrow W_1 - W_2 = \lambda (w_1 - w_2) \quad \text{... (13)}$$

$$W_2 - W_3 = \lambda (w_2 - w_3) \quad ... (14)$$
$$W_3 - W_1 = \lambda (w_3 - w_4) \quad ... (15)$$

Multiplying (13), (14) and (15) by w_3, w_1 and w_2 respectively and adding, we get

$$w_3 (W_1 - W_2) + w_1 (W_2 - W_3) + w_2 (W_3 - W_1) = 0$$
$$\Rightarrow w_1 (W_2 - W_3) + w_2 (W_3 - W_1) + w_3 (W_1 - W_2) = 0$$

Example 21: A body consists of an alloy of two metals of sp.gr. s_1 and s_2 respectively; its weight is vacuo is w and in water is W. Show that the proportion of two metals by volume is

$$s_2 W - (s_2 - 1) w : (s_1 + 1) w - s_1 W$$

Sol. Let v_1 and v_2 be the volumes of the two metals in the alloy (body). Let ρ be the density of water.

Since the sp.gr. of the two metals are s_1 and s_2 respectively, therefore their densities are $s_1 \rho$ and $s_2 \rho$ respectively.

Now, w = Weight of the body in vacuo = True weight of the body = weight of one metal + weight of other metal

$$= v_1 s_1 \rho g + v_2 s_2 \rho g$$
$$\Rightarrow w = v_1 s_1 \rho g + v_2 s_2 \rho g \quad ... (1)$$

Again,

Weight of the body in water

= True weight of the body – weight of the water displaced by the body

$$\Rightarrow W = w - (v_1 + v_2) \rho g$$
$$\Rightarrow W = (v_1 s_1 \rho g + v_2 s_2 \rho g) - (v_1 + v_2) \rho g \quad | \text{ From (1)}$$
$$\Rightarrow W = (v_1 s_1 + v_2 s_3 - v_1 - v_2) \rho g \quad ... (2)$$

Dividing (1) by (2), we get

$$\frac{w}{W} = \frac{v_1 s_1 + v_2 s_2}{v_1 s_1 + v_2 s_2 - v_1 - v_2}$$

$$\Rightarrow w v_1 s_1 + w v_2 s_2 - w v_1 - w v_2 = W v_1 s_1 + W v_2 s_2$$
$$\Rightarrow (w s_1 - w - W s_1) v_1 = (w s_2 - W s_2 + W) v_2$$

$$\Rightarrow \frac{v_1}{v_2} = \frac{w s_2 - w s_2 + w}{w s_1 - w - w s_1} \qquad \Rightarrow \frac{v_1}{v_2} = \frac{W s_2 - w(s_2 - 1)}{w(s_1 - 1) - W s_1}$$

Example 22: Prove that, if volumes v and V of two different substances balance in vacuum and volumes v', V′ balance when weighed in a liquid, the densities of the substances and the liquid are as

$$\frac{v' - V'}{v} : \frac{v' - V'}{V} : \left(\frac{v'}{v} - \frac{V'}{V}\right)$$

Sol. Let ρ_1 and ρ_2 be the densities of the two substances respectively and σ be the density of the liquid.

Since the volumes v and V of the two substances balance in vacuum, therefore, the true weights of the two substances are equal

$\therefore\ vg\rho_1 = Vg\rho_2$

$\Rightarrow\ v\rho_1 = V\rho_2$ ... (1)

Since the volumes v' and V' of the two substances balance when weighed in a liquid of density σ, therefore, the apparent weights of the two substances in the given liquid are equal. Therefore,

true weight of the first substance – weight of the liquid displaced by it

= true weight of the second substance – weight of the liquid displaced by it

$\Rightarrow\ v'\rho_1 g - v'\sigma g = V'\rho_2 g - V'\sigma g$

$\Rightarrow\ v'\rho_1 - v'\sigma = V'\rho_2 - V'\sigma$

$\Rightarrow\ v'\rho_1 - V'\rho_2 = (v' - V')\,\sigma$ ... (2)

$\Rightarrow\ v'\rho_1 - \dfrac{v\rho_1}{V}v' = (v'-V')\sigma$ | From (1)

$\Rightarrow\ \left(v' - \dfrac{vv'}{V}\right)\rho_1 = (v'-V')\sigma \quad \Rightarrow\ \dfrac{\rho_1}{\sigma} = \dfrac{v'-V'}{v\left(\dfrac{v'}{v} - \dfrac{v'}{V}\right)}$

$\Rightarrow\ \rho_1 : \sigma = \dfrac{v'-V'}{v} : \left(\dfrac{v'}{v} - \dfrac{v'}{V}\right)$ ... (3)

Again, from (2),

$v'\rho_1 - V'\rho_2 = (v' - V')\,\sigma$

$\Rightarrow\ v'\dfrac{V\rho_2}{v} - V'\rho_2 = (v'-V')\sigma$ | From (1)

$\Rightarrow\ \left(\dfrac{v'V}{v} - V'\right)\rho_2 = (v'-V')\sigma \quad \Rightarrow\ \dfrac{\rho_2}{\sigma} = \dfrac{v'-V'}{\dfrac{v'V}{v} - V'} = \dfrac{v'-V'}{V\left(\dfrac{v'}{v} - \dfrac{v'}{V}\right)}$

$\Rightarrow\ \dfrac{\rho_2}{\sigma} = \dfrac{\dfrac{v'-V'}{V}}{\dfrac{v'}{v} - \dfrac{V'}{V}}$

$\Rightarrow\ \rho_2 : \sigma = v' - V'/V : v'/v - V'/V$ ... (4)

From (3) and (4), we get

$$\frac{\rho}{\dfrac{v'-V'}{v}} = \frac{\rho_2}{\dfrac{v'-V'}{v}} = \frac{\sigma}{\left(\dfrac{v'}{v} - \dfrac{V'}{V}\right)} \Rightarrow \rho_1 : \rho_2 : \sigma = \frac{v'-V'}{v} : \frac{v'-V'}{v} : \left(\frac{v'}{v} - \frac{V'}{V}\right)$$

Aliter: $v\rho_1 = V\rho_2$ ... (1)

$v'\rho_1 - V'\rho_2 = (v' - V')\,\sigma$ ... (2)

From (1),

$$\frac{\rho_1}{V} = \frac{\rho_2}{v} = \frac{v'\rho_1 - V'\rho_2}{v'V - V'v} = \frac{(v'-V')\,\sigma}{v'V - V'v} \qquad \text{| From (2)}$$

$$\Rightarrow \frac{\rho_1}{V(v'-V')} = \frac{\rho_2}{v(v'-V')} = \frac{\sigma}{v'V - V'v} \Rightarrow \frac{\rho_1}{\dfrac{v'-V'}{v}} = \frac{\rho_2}{\dfrac{v'-V'}{V}} = \frac{\sigma}{\dfrac{v'V-V'v}{vV}}$$

$$\Rightarrow \frac{\rho_1}{\dfrac{v'-V'}{v}} = \frac{\rho_2}{\dfrac{v'-V'}{V}} = \frac{\sigma}{\dfrac{v'}{v} - \dfrac{V'}{V}}$$

Example 23: A substance, whose density is ρ, is weighed by means of weights, the density of which is ρ'. If σ be the density of air, find what is the true weight corresponding to any apparent weight. If the density of air increases from σ to σ', prove that the apparent weight of the body in less than its former weight by a fraction.

$$\frac{(\rho'-\rho)\,(\sigma'-\sigma)}{(\rho-\sigma)\,(\rho'-\sigma')} \text{ of the latter, } \rho' \text{ being greater than } \rho.$$

Sol. Let W be the true weight of the body corresponding to any apparent weight W_0 (say). Then,

$$W\left(1-\frac{\sigma}{\rho}\right) = W_0\left(1-\frac{\sigma}{\rho'}\right) \qquad \text{| Refer § 7}$$

$$\Rightarrow W\left(\frac{\rho-\sigma}{\rho}\right) = W_0\left(\frac{\rho'-\sigma}{\rho'}\right)$$

$$\Rightarrow W = \frac{W_0\,\rho(\rho'-\sigma)}{\rho'(\rho-\sigma)} \qquad \text{... (1)}$$

If the density of air increases from σ to σ', then let the new apparent weight corresponding to true weight W be W_1. Then, from (1), we get

$$W = W_1\,\frac{\rho}{\rho'}\,\frac{(\rho'-\sigma')}{(\rho-\sigma')} \qquad \text{... (2)}$$

From (1) and (2), equating the two values of W, we get

$$\frac{W_0\,\rho(\rho'-\sigma)}{\rho'(\rho-\sigma)} = \frac{W_1\,\rho(\rho'-\sigma')}{\rho'(\rho-\sigma')} \Rightarrow W_0\,\frac{\rho'-\sigma}{\rho-\sigma} = W_1\,\frac{\rho'-\sigma'}{\rho-\sigma'}$$

$$\Rightarrow \frac{W_1}{W_0} = \frac{(\rho'-\sigma)(\rho-\sigma')}{(\rho-\sigma)\,(\rho'-\sigma')} \qquad \text{... (3)}$$

$\therefore$ The new apparent weight (W_1) is less than the former apparent weight (W_0) by

$$W_0 - W_1 = W_0\left(1-\frac{W_1}{W_0}\right) = \left(1-\frac{W_1}{W_0}\right)W_0$$

$$= \left\{1 - \frac{(\rho' - \sigma)(\rho - \sigma')}{(\rho - \sigma)(\rho' - \sigma')}\right\} W_0 \qquad | \text{ From (3)}$$

$$= \left\{\frac{\rho\rho' - \rho\sigma' - \sigma\rho' + \sigma\sigma' - \rho'\rho + \rho'\sigma' + \rho\sigma - \sigma\sigma'}{(\rho - \sigma)(\rho' - \sigma')}\right\} W_0 \quad = \left\{\frac{-\rho\sigma' - \sigma\rho' + \rho'\sigma' + \rho\sigma}{(\rho - \sigma)(\rho' - \sigma')}\right\} W_0$$

$$= \frac{\rho(\sigma - \sigma') - \rho'(\sigma - \sigma')}{(\rho - \sigma)(\rho' - \sigma')} W_0 \quad = \frac{(\rho' - \rho)(\sigma' - \sigma)}{(\rho - \sigma)(\rho' - \sigma')} W_0$$

EXERCISE 3

1. A solid (sp.gr. 21) is held suspended by a string so that 19/24th of its volume is immersed in a liquid of sp.gr. 13 and the remainder in water. Prove that the tension of the string is half the weight of the solid.
2. A mixture of gold (sp.gr. 19·25) and silver (sp.gr. 10·25) lost 1/14th of its weight when weighted in water. Prove that the ratio of the volumes of the two metals is 2 : 3.
3. An iron (sp.gr. 7·2) spheroid shell is found to lose half its weight when weighed in water. If the external radius is 12 cm, find out the thickness of the metal of the shell.

   Ans. 1·23
4. A piece of wood floats in a vessel of water with 4/5th of its volume immersed. Find the additional volume which will be immersed when the vessel is placed in vacuum, the sp.gr. of the air being 0·0013.

   Ans. 0·00026 of its volume
5. A piece of lead and a piece of wood balance one another when weighed in air. Which will really weigh the more and why?

   Ans. wood

8. (1) A body is floating partly immersed is a liquid and with a string attached to a point of it. To find the conditions of equilibrium when the specific gravity of the body > the specific gravity of the liquid.

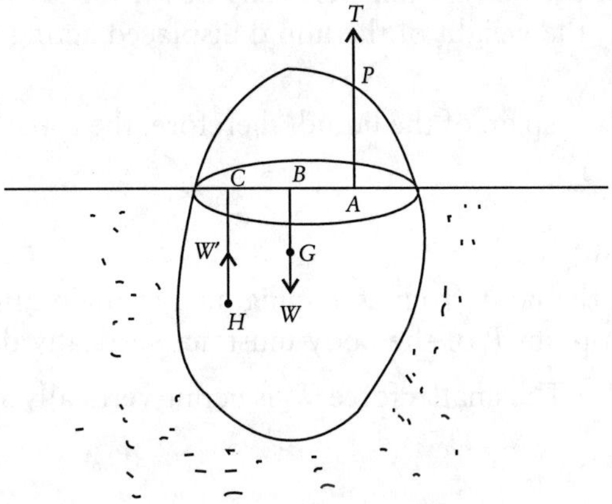

Let P be a point at which the strong is attached to the body. Let W be the weight of the body acting vertically downwards through its C.G.G. Let W' be the force of buoyancy, i.e., the weight of the liquid displaced acting vertically upwards through H, its centre of buoyancy.

Let the vertical lines through P, G and H meet the surface of the liquid in the points A, B, and C respectively.

Since sp.gr. of the body > sp.gr. of the liquid and volume of the liquid displaced is less than the volume of the body because the body is partly immersed in the liquid, therefore,

$W > W'$ | $\because$ weight = volume × sp.gr. × g × density of water

Hence, the body will have a tendency to descend. Consequently, for equilibrium, the tension T of the string must be acting vertically upwards.

| $\because$ The smaller force W' is acting vertically upwards

$\therefore\ W = T + W'$

$\Rightarrow\ T = W - W'$ ... (1)

Clearly B must lie between A and C because the resultant of two like parallel forces W' at C and T at A must act at a point between them.

Since the vertical forces acting on the body are T, W and W' and the body is in equilibrium, therefore, the algebraic sum of the moments of the forces about any point (Here A) is zero. [The point A is chosen in order to eliminate T.] | By statics

$\therefore\ W.\,AB - W'AC = 0$

$\Rightarrow\ W.\,AB = W'\,.\,AC$ ... (2)

Hence, the conditions of equilibrium in this case are as follows:

(1) $T = W + W'$

(2) $W\,.\,AB = W'AC$

8. (2) A body is completely immersed is a liquid whose sp.gr. is > sp.gr. of the body. To find the conditions of equilibrium.

Let W be the weight of the body acting vertically downwards through its C.G. G. Let W' be the force of buoyancy, i.e., the weight of the liquid displaced acting vertically upwards through H, its centre of buoyancy.

Since sp.gr. of the body < sp.gr. of the liquid, therefore, the force of buoyancy will be greater than the weight of the body.

Hence, $W < W'$

$\therefore$ The body will ascend.

$\therefore$ In order to prevent the body from ascending, i.e., to maintain equilibrium, the tension T of the string attached to a point P of the body must act vertically downwards.

| $\because$ The smaller force W is acting vertically downwards

$\therefore\ W' = T + W$

$\Rightarrow\ T = W' - W$ ... (3)

Let the vertical lines through P, G and H meet the surface of the liquid in the points A, B and C respectively.

In this case, C must lie between A and B, because the resultant of two like parallel forces W at B and T at A must act at a point between them.

Since the vertical forces acting on the body are T, W and W' and the body is in equilibrium, therefore, the algebraic sum of the moments of these forces about any point (here A) is zero. [The point A is chosen in order to eliminate T.] | By Statics

$\therefore\ W \,.\, AB - W' \,.\, AC = 0$

$\Rightarrow\ W.\, AB = W' \,.\, AC$ ... (4)

Hence, the conditions of equilibrium in this case are as follows:

(1) $W' = T + W$ (2) $W \,.\, AB = W' \,.\, AC$

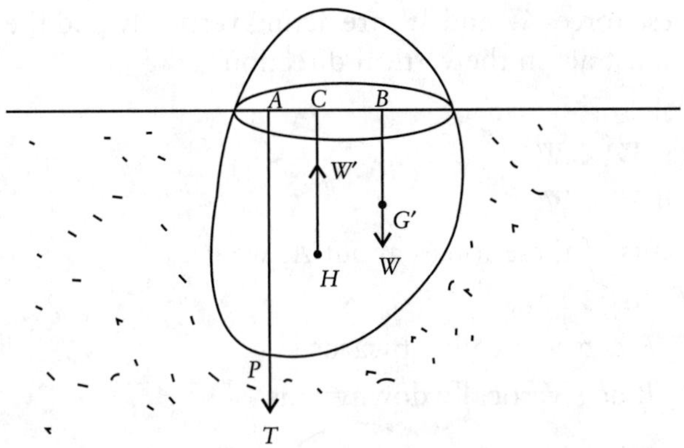

Note: The ascent of balloon depends upon § 8(2). A balloon consists of a very large envelope made of silk or some other light but strong substance. It is filled with a gas or hydrogen. Since the weight of the displaced air is greater than the weight of the balloon, i.e., $W' > W$, therefore, the balloon ascends.

The force with which the balloon ascends is called the ascensional force. It is equal to $W' - W$, i.e., equal to weight of the air displaced weight of the ballon.

9. A body floating under the constraint that one point of the body is fixed

Let the point P of the body be fixed, i.e., the body is capable of moving about this point. The forces acting on the body acting are as follows:

1. W, the weight of the body acting vertically downwards through its C.G. G.
2. W', the force of buoyancy which is equal to the weight of the liquid displaced acting vertically upwards through H, its centre of buoyancy.
3. R, the reaction at the point P.

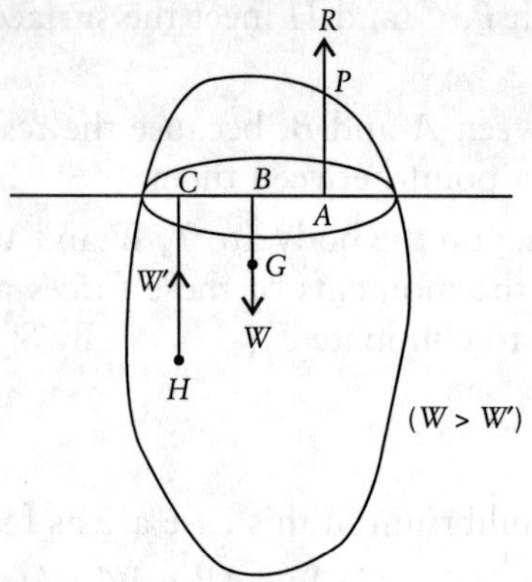

Since the two of these forces W and W' are acting vertically and the body is in equilibrium, therefore, the reaction must act in the vertical direction.

For equilibrium,

$$W = W' + R \text{ if } W' < W$$

and $$W' = W + R \text{ if } W < W'$$

Again taking moments of these forces about A, we get

$$W \,.\, AB = W' \,.\, AC$$

Note 1: If $W > W'$, R acts vertically upwards

Note 2: If $W < W'$, R acts vertically downwards

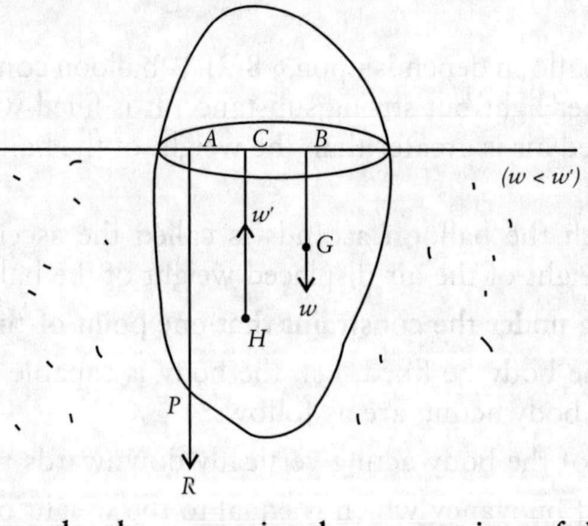

10. A body is floating under the constraint that two points of the body are fixed

Let two points A and B of the body be fixed. It means that a line AB is fixed in the body. In this case the necessary and sufficient condition of equilibrium is that the sum of the moments of the weight of the body and the force of buoyancy about this line AB must be zero. In other words, we can say that for equilibrium, the moments of the weight of the body and the force of buoyancy about the line must be equal and opposite. This condition can be easily obtained by taking moments about line AB to avoid reactions at A and B as the moments of the reactions at the points A and B about the line AB are zero each.

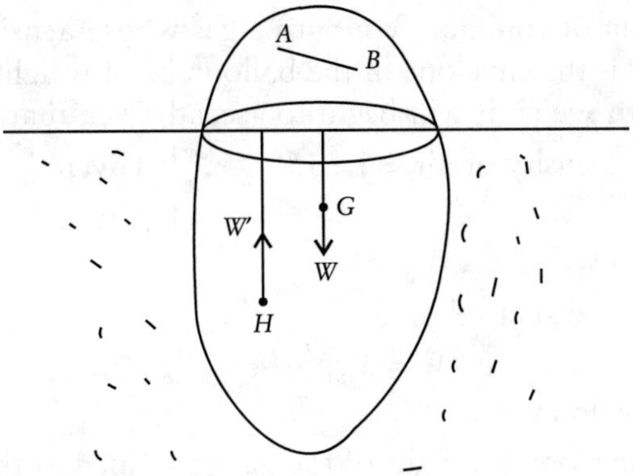

ILLUSTRATIVE EXAMPLES

Example 1: The mass of a balloon and the gas which it contains is 3500 gms. If the balloon displaces 48000 cm³ of air and the mass of a cu. cm. air be 1·25 gm, find the acceleration with which the balloon commences to ascend.

Sol. W' = weight of the air displaced by the balloon

$= 48000 \times 1{\cdot}25 \quad = 3750$ gms. wt.

W = weight of the balloon

= 3500 gms. wt.

$\therefore$ The ascensional force (i.e., the force with which the balloon ascends)

$= W' - W$ | Refer § 8. (2)

= weight of air displaced – weight of balloon

= 3750 – 3500

= 250 gms. wt.

$\therefore$ Initial acceleration of the balloon (i.e., the acceleration with which the balloon ascends)

$= \dfrac{\text{Force}}{\text{Mass}}$ | '$P = mf$'

$= \dfrac{250g}{3500} = \dfrac{g}{14}$

Example 2: Two cubic feet of cork of sp.gr. 0·24 is kept below water by a rope fastened to the bottom. Prove that the tension of the rope is 95 pounds.

(weight of 1 cube ft. of water = 62·5 lbs. wt.)

Sol. Weight of cork $= 2 \times 0{\cdot}24\, w \quad = 0{\cdot}48\, w$

where w is the weight per cubic ft. of water.

For equilibrium,

$T = W' - W$ | Refer § 8. (2)

$= 2w - 0{\cdot}48\, w = 1{\cdot}52\, w \quad = \dfrac{152}{100} \times 62{\cdot}5 \; = 95$ lbs.wt.

Example 3: A balloon of volume V contains a gas whose density is to that of the air on the earth's surface as 1 : 15. If the envelope of the balloon be of weight w but of negligible volume, find the acceleration with which it will begin to ascend given that σ is the density of the gas.

Sol. Density of gas : Density of air = 1 : 15 | Given

Density of gas = σ | Given

$\therefore$ Density of air = 15σ

W' = weight of the air displaced

$= V.g.\ (15\sigma)$ | '$W = Vg\rho$' ... (1)

W = weight of the balloon

= weight of the envelope + weight of the gas contained in the envelope

$= w + V\sigma g$ ... (2)

$\therefore$ The force with which the balloon ascends

$= W' - W$ | Refer § 8. (2)

$= 15\,Vg\sigma - (w + V\sigma g)$ | From (1) and (2)

$= 14\,Vg\sigma - w$

Mass of the balloon $= \dfrac{W}{g} = \dfrac{w + V\sigma g}{g} = \dfrac{w}{g} + V\sigma$

Let the balloon begin to ascend with the acceleration f. Then,

$$14\,Vg\sigma - w = \left(\frac{w}{g} + V\sigma\right).f \qquad | \text{ “}P = mf\text{”}$$

$$\Rightarrow \quad f = \frac{14V\sigma g - w}{\dfrac{w}{g} + V\sigma}$$

Example 4: A uniform rod, capable of turning about one end which is out of the water, rests inclined to the vertical with one-third of its length in some water. Prove that the sp.gr. of the rod is 5/9.

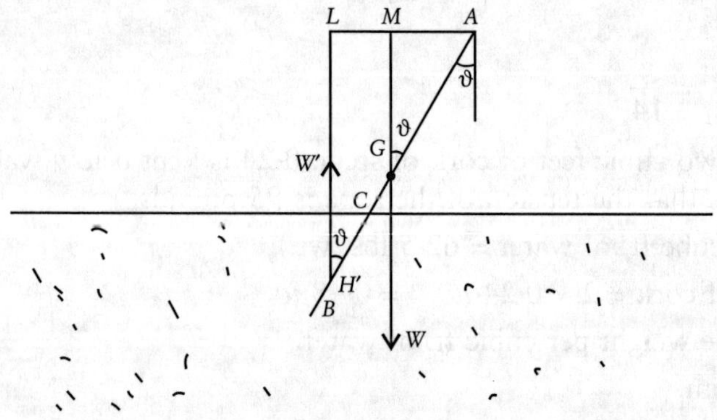

Let AB be a uniform rod of length $2a$ (say), which is capable of turning about one end A, rest with length BC immersed in water. Then,

$BC = 1/3\ AB = 2a/3$

Let S be the area of cross-section of the rod and σ be the sp.gr. of the rod. Let ρ be the density of water.

The rod AB is in equilibrium under the action of the following forces.

(1) Weight $W = 2a\, S\, \sigma\, g\, \rho$ acting vertically downwards through G, the C.G. of the rod.

(2) Force of buoyancy (or the weight of water displaced by the rod) $W' = BC.\ S.\ g\, \rho$ $= 2a/3\ S\ g\ \rho$, acting vertically upwards through H, the centre of buoyancy (C.G. of the water displaced)

(3) Reaction R at the fixed end A

Let ϑ be the inclination of the rod to the vertical. Then, taking moments of these forces about A, we have

$$2a\, S\, \sigma\, g\, \rho\, AM = 2a/3\ S\ g\ \rho\ AL$$

$\Rightarrow \sigma\, AM = AL/3 \qquad \Rightarrow \sigma\, AG \sin\vartheta = AH \sin\vartheta/3$

$\Rightarrow \sigma\, a \sin\vartheta = (AC + CH) \sin\vartheta/3 \qquad \Rightarrow \sigma\, a = (2/3\ AB + 1/2\ BC)\ 1/3$

$\Rightarrow \sigma\, a = (2/3\ 2a + 1/2\ 2a/3)1/3 \qquad \Rightarrow \sigma\, a = (4a/3 + a/3)\ 1/3$

$\Rightarrow \sigma\, a = 5a/9 \qquad \Rightarrow \sigma = 5/9$

Example 5: A thin rod of sp.gr. 3/4 and length 4 metres floats partly immersed in water, being supported by a strong fastened to one end of the rod. Find how much of the rod is immersed? If the upper end of the rod be 1 m above the surface of water, prove that the rod is inclined at an angle 30° to the horizontal.

Sol.

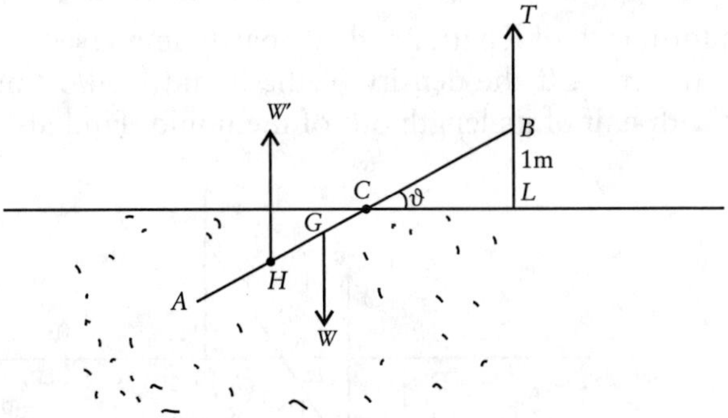

Let AB be a thin rod of sp.gr. 3/4 and length 4 metres, floating partly immersed in water, being supported by a strong fastened to one end B of the rod. Then,

$$AB = 4 \text{ m}$$

Let $AC = 2x$

Let ρ be the density of water. Then,

density of the rod $= 3\rho/4$

Let S be the area of cross section of the rod.

The rod AB is in equilibrium under the action of the following forces:

(1) Weight of the rod $W = 4\,S\,3\rho/4\,g$, acting vertically downwards through G, the middle point of the rod.

(2) W' = Force of buoyancy

= Weight of water displaced

= $2\,x\,S\,\rho\,g$, acting vertically upwards through H, the centre of buoyancy (C.G. of the water displaced)

(3) Tension T at B.

Taking moments of all these forces about B, we have

$$4\,S\,3\rho/4\,g\,.\,2\cos\vartheta = 2x\,S\,\rho\,g\,(4 - x)\cos\vartheta \qquad \left|\begin{array}{l} AG = GB = 2 \\ AH = HC = x \\ BH = AB - AH = (4-x) \end{array}\right.$$

$\Rightarrow 3 = x\,(4 - x) \Rightarrow x^2 - 4x + 3 = 0$

$\Rightarrow (x - 1)\,(x - 3) = 0 \qquad \Rightarrow x = 1$ or 3

But $x = 3$ is in admissible as then $2x = 6$, i.e., $AC = 6 \Rightarrow AC > AB$ which is not possible as $AC < AB$.

$\therefore\ x = 1 \quad \therefore\ AC = 2$ m

Hence, the rod is half immersed, i.e., Length of rod immersed = 2 m

Also, in $\Delta\,BLC$, $\sin\vartheta = BL/BC = 1/2 = \sin 30^\circ \Rightarrow \vartheta = 30^\circ$

Hence, the rod is inclined at an angle 30° to the horizontal.

Example 6: A uniform rod, of length $2a$, floats partly immersed in a liquid, being supported by a string to one of the ends. If the density of the liquid be 4/3 times that of the rod, prove that the rod will rest with half of its length out of the liquid. Find also the tension of the string.

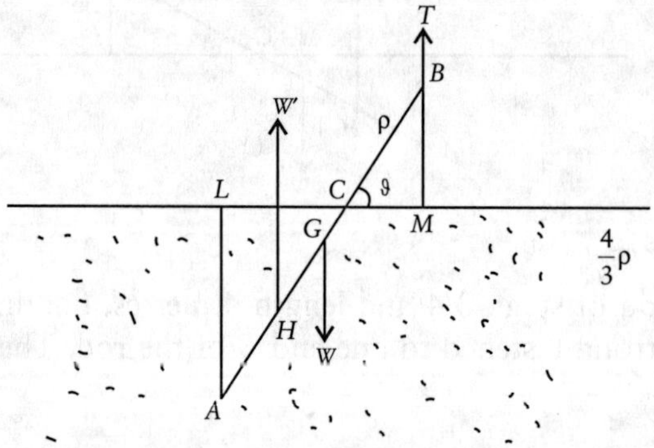

Let AB be a uniform rod of length $2a$ floating partly immersed in a liquid, being supported by a string to one of the ends B.

$AB = 2a \qquad |$ Given

Let S be the area of cross-section of the rod. Let ρ be the density of the rod. Then, density of the liquid = $4\rho/3$

Let $AC = 2x$

Let G be the C.G. of the rod. Then,

$$AG = GB = a$$

Let G be the C.G. of the liquid displaced.

Then, $AH = HC = x$

$\therefore\ BH = BC + CH$

$= (2a - 2x) + x \qquad = 2a - x$

The rod AB is in equilibrium under the action of the following forces:

(1) Weight of the rod $W = 2aS\rho g$ acting vertically downwards through G, the C.G. of the rod.

(2) Force of buoyancy (i.e., weight of the liquid displaced) $W' = 2x\,S\,4/3\,\rho g$ acting vertically upwards through H, the centre of buoyancy (i.e., the C.G. of the liquid displaced).

(3) Tension T at B acting vertically upwards Let the rod AB be inclined at an angle ϑ to the horizon. Then, taking moments of all the forces about B, we get

$W'\ BH \cos\vartheta = W\ .\ BG \cos\vartheta$

$\Rightarrow\ 2xS\ 4/3\ \rho\ g\ .\ (2a - x) = 2aS\rho ga \qquad \Rightarrow 4x\ (2a - x) = 3a^2 \qquad \Rightarrow 8ax - 4x^2 = 3a^2$

$\Rightarrow\ 4x^2 - 8ax + 3a^2 = 0 \quad \Rightarrow 4x^2 - 2ax - 6ax + 3a^2 = 0 \quad \Rightarrow 2x\ (2x - a) - 3a\ (2x - a) = 0$

$\Rightarrow\ (2x - a)\ (2x - 3a) = 0 \quad \Rightarrow 2x - a = 0 \text{ or } 2x - 3a = 0 \quad \Rightarrow x = a/2 \text{ or } x = 3a/2$

$\Rightarrow\ x = a/2,\ 3a/2$

$x = 3a/2$ is in admissible as then

$2x = 3a \qquad \Rightarrow AC = 3a \qquad \Rightarrow AC > AB$ which is impossible

$\therefore\ x = a/2 \qquad \Rightarrow 2x = a \qquad \Rightarrow AC = a \qquad \Rightarrow AC = BC = a$

Hence, the rod will be at rest with half of its length out of the liquid.

Again, for equilibrium,

$T = W - W' \quad = 2aS\rho g - 2xS\ 4/3\ \rho g \quad = 2aS\rho g - aS4/3\rho g \qquad | \because x = a/2$

$= 2a/3S\rho g \quad = 1/3\ (2aS\rho g) \quad = 1/3\ W \quad = 1/3$ weight of the rod

Example 7: A rod of small section and of density ρ, has a small particle of weight $1/n^{th}$ that of the rod attached to one extremity. Prove that the rod will float at any inclination in a liquid of density σ, if $(n + 1)^2\ \rho = n^2\ \sigma$.

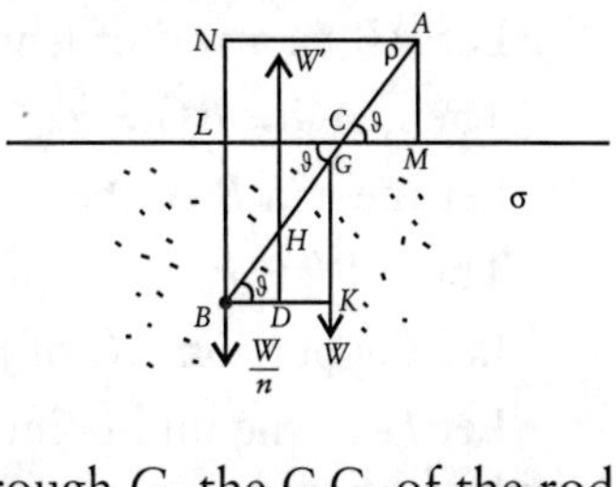

Sol. Let AB be a rod of small section S and of density ρ. Let $AB = 2a$. Let G be the C.G. of the rod. Then, $AG = BG = a$.

Let $BC = 2x$ be the length of the rod immersed in a liquid of density σ. Let H be the centre of buoyancy. Then,

$BH = CH = x$

Let the rod be inclined at any angle ϑ to the horizontal.

W = weight of the rod = $2aS\rho g$ acting vertically downwards through G, the C.G. of the rod

W' = Force of buoyancy = $2xS\sigma g$ acting vertically upwards through H, the centre of buoyancy

Let a small particle of weight of W/n be attached to one extremity B of the rod.

Resolving the forces vertically, we get

$$W + W/n = W' \quad \Rightarrow W(1 + 1/n) = W' \quad \Rightarrow 2aS\rho g(1 + 1/n) = 2xS\sigma g$$

$$\Rightarrow a\rho(1 + 1/n) = x\sigma$$

$$\Rightarrow a^2\rho^2(1 + 1/n)^2 = x^2\sigma^2 \quad \text{... (1)}$$

Again, taking the moments of all these forces about B, we get

$$W.\ BK = W'.\ BD \quad \Rightarrow W.\ BG\cos\vartheta = W'.\ BH\cos\vartheta \quad \Rightarrow W\ BG = W'\ BH$$

$$\Rightarrow Wa = W'.\ x \quad \Rightarrow 2aS\rho ga = 2xS\sigma gx$$

$$\Rightarrow a^2\rho = x^2\sigma \quad \text{... (2)}$$

Dividing (1) by (2), we get

$$\rho(1 + 1/n)^2 = \sigma \quad \Rightarrow (n + 1)^2\rho = n^2\sigma$$

Example 8: A uniform rod of length $2a$, can turn freely about one end which in fixed at a height $h(< 2a)$ above the surface of a liquid. If the densities of the rod and the liquid be ρ and σ, show that the rod can rest either in a vertical position or inclined at an angle ϑ to the vertical such that

$$\cos\vartheta = \frac{h}{2a}\sqrt{\frac{\sigma}{\sigma-\rho}}$$

Sol.

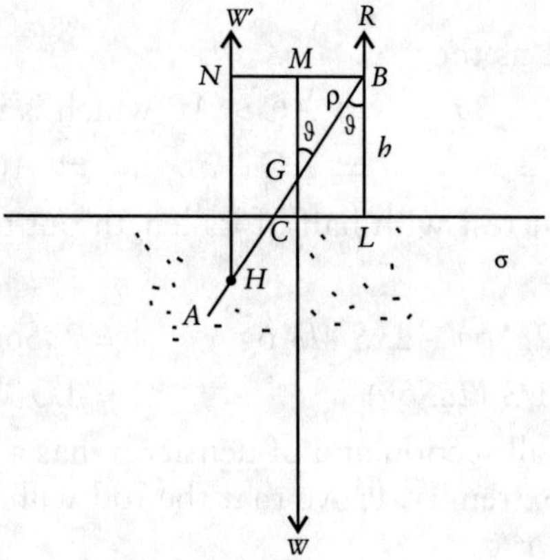

Let AB be a rod of length $2a$. Let G be its C.G.

Then, $AG = BG = a$.

Let the end B of the rod be fixed at a height h above the surface of the liquid.

Then, $BL = h$

Let the portion AC of the rod AB be immersed in a liquid of density σ.

Let H be the mid-point of AC. Then, H is the centre of buoyancy, i.e., H is the C.G. of the displaced liquid

From $\Delta\ BLC$,

$\cos \vartheta = BL/BC = h/BC \quad \Rightarrow BC = h \sec \vartheta$

$\therefore \; AC = AB - BC \quad \Rightarrow 2a - h \sec \vartheta \quad (= 2x)$

Also, $AH = HC = 1/2\ AC = 1/2\ (2a - h \sec \vartheta)$

$BH = BC + CH = h \sec \vartheta + 1/2\ (2a - h \sec \vartheta) \quad = 1/2\ (2a + h \sec \vartheta)$

Let S be the area of cross-section of the rod.

The rod AB is in equilibrium under the action of the following three vertical forces:

(1) W = weight of the rod AB

$= AB.\ S\ .\ \rho\ .\ g \quad = 2a\ .\ S\ .\ \rho\ .\ g$

acting vertically downwards through G.

(2) W' = Force of buoyancy

= weight of liquid displaced $\quad = AC.\ S\ .\ \sigma\ .\ g$

$= 2x\ S\ \sigma\ g$ acting vertically upwards through H.

(3) The reaction R at B acting vertically upwards Taking moments about B (to avoid reaction at B), we get

$WBG \sin \vartheta = W'BH \sin \vartheta \quad \Rightarrow \sin \vartheta\ (WBG - W'BH) = 0$

$\Rightarrow \sin \vartheta = 0$ or $WBG - W'BH = 0$

$\Rightarrow \vartheta = 0$ or $2aS\rho ga - 2xS\sigma g\ 1/2\ (2a + h \sec \vartheta) = 0$

$\Rightarrow \vartheta = 0$ or $2a^2\ \rho - (2a - h \sec \vartheta)\ \sigma\ 1/2\ (2a + h \sec \vartheta) = 0$

$\Rightarrow \vartheta = 0$ or $2a^2\rho - \dfrac{(4a^2 - h^2 \sec^2 \vartheta)\sigma}{2} = 0$

$\Rightarrow \vartheta = 0$ or $4a^2\ \rho - (4a^2 - h^2 \sec^2 \vartheta)\sigma = 0$

$\Rightarrow \vartheta = 0$ or $4a^2\ (\sigma - \rho) = h^2 \sec^2 \vartheta\ \sigma$

$\Rightarrow \vartheta = 0$ or $\sec^2 \vartheta = \dfrac{4a^2(\sigma - \rho)}{\sigma h^2}$

$\Rightarrow \vartheta = 0$ or $\cos^2 \vartheta = \dfrac{\sigma h^2}{4a^2(\sigma - \rho)}$

$\Rightarrow \vartheta = 0$ or $\cos^2 \vartheta = \dfrac{h}{2a}\sqrt{\dfrac{\sigma}{\sigma - \rho}}$

$\Rightarrow$ Either $\vartheta = 0$, then the rests $h/2a$ vertical position

or, $\cos \vartheta = \dfrac{h}{2a}\sqrt{\dfrac{\sigma}{\sigma - \rho}}$

Example 9: A uniform rod of length $2a$ and density ρ is movable in a vertical plane about one end which is fixed in a liquid of density ρ' at a depth $2h$ below the surface. A liquid of smaller density σ is added on the top of the first liquid. If in the oblique position of equilibrium, the rod is just covered by the liquid, prove that its inclination to the vertical is

$$\cos^{-1}\left\{\frac{h}{a}\left(\frac{\rho'-\sigma}{\rho-\sigma}\right)^{\frac{1}{2}}\right\}.$$

Sol. Let AB be the given rod. Then,

$AB = 2a$

Let G be the C.G. of the rod AB. Then,

$AG = GB = a$

A is the fixed end at a depth $2h$ below the surface of liquid of density ρ′.

$\therefore\ AL = 2h$

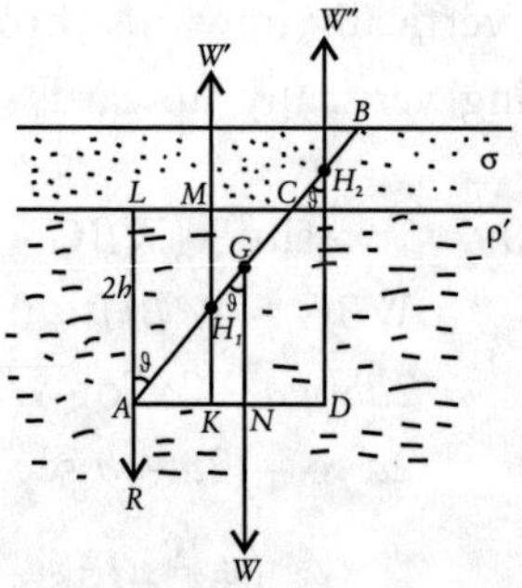

Let S be the area of cross-section of the rod. Let ϑ be the inclination of the rod to the vertical.

Let H_1 be the middle point of AC. Then,

$AC = 2h \sec \vartheta$

$\therefore\ AH_1 = CH_1 = 2h \sec \vartheta/2 = h \sec \vartheta$

$\therefore\ BC = AB - AC = 2h - 2h \sec \vartheta$

The rod AB is in equilibrium under the action of the following forces.

(1) W = weight of the AB

$= 2a\, S\, \rho\, g$ acting vertically downwards through G.

(2) W'= Force of buoyancy (i.e., weight of liquid of density ρ′ displaced)

$= 2h \sec \vartheta\, S\, \rho'\, g$

acting vertically upwards through H_1, the centre of buoyancy (C.G. of the liquid of density ρ′ displaced)

(3) W''= Force of buoyancy (i.e., weight of liquid of density σ displaced)

$= BC.\ S\, \sigma\, g = (2a - 2h \sec \vartheta)\, S\sigma g$

acting vertically upwards through H_2, the centre of buoyancy (C.G. of the liquid of density σ displaced)

(4) Reaction R at the fixed end A.

Taking moments of all these forces about A, we get

$W. AN = W' AK + W' AD$

$\Rightarrow 2a\ S\ \rho\ g\ .\ a \sin \vartheta = 2h \sec \vartheta\ S\ \rho'\ g\ h \sec \vartheta \sin \vartheta +$

$(2a - 2h \sec \vartheta)\ S\ \sigma\ g\ \{2h \sec \vartheta + (a - h \sec \vartheta) \sin \vartheta$

$$\begin{aligned} & AD \\ &= (AC+CH_2)\sin\vartheta \\ &= \left\{2h\sec\vartheta+\frac{1}{2}BC\right\}\sin\vartheta \\ &= \left\{2h\sec\vartheta+\frac{2a-2h\sec\vartheta}{2}\right\}\sin\vartheta \\ &= \{2h\sec\vartheta+(a-h\sec\vartheta)\}\sin\vartheta = \{a+h\sec\vartheta\}\sin\vartheta \end{aligned}$$

$\Rightarrow a^2\ \rho = h^2 \sec^2 \vartheta\ \rho' + (a - h \sec \vartheta)\ \sigma\ (a + h \sec \vartheta)$

$= h^2 \sec^2 \vartheta\ \rho' + (a^2 - h^2 \sec^2 \vartheta)\ \sigma \qquad = \sigma\ a^2 + h^2 (\rho' - \sigma) \sec^2 \vartheta$

$\Rightarrow a^2 (\rho - \sigma) = h^2 (\rho' - \sigma) \sec^2 \vartheta \qquad \Rightarrow \cos^2\vartheta = \dfrac{h^2(\rho'-\sigma)}{a^2(\rho-\sigma)}$

$$\Rightarrow \cos\vartheta = \frac{h}{a}\left(\frac{\rho'-\sigma}{\rho-\sigma}\right)^{\frac{1}{2}}$$

$$\Rightarrow \vartheta = \cos^{-1}\left\{\frac{h}{a}\left(\frac{\rho'-\sigma}{\rho-\sigma}\right)^{\frac{1}{2}}\right\}$$

Example 10: A uniform rod, loaded at one end so as to float upright, floats in a liquid of density ρ_1 with a length a unimmersed, in a liquid of density ρ_2 with a length b unimmersed. Show that the length that is unimmersed when the rod floats in a liquid of density ρ is

$$\frac{a\rho_1(\rho-\rho_2)-b\rho_2(\rho-\rho_1)}{\rho(\rho_1-\rho_2)}$$

Sol.

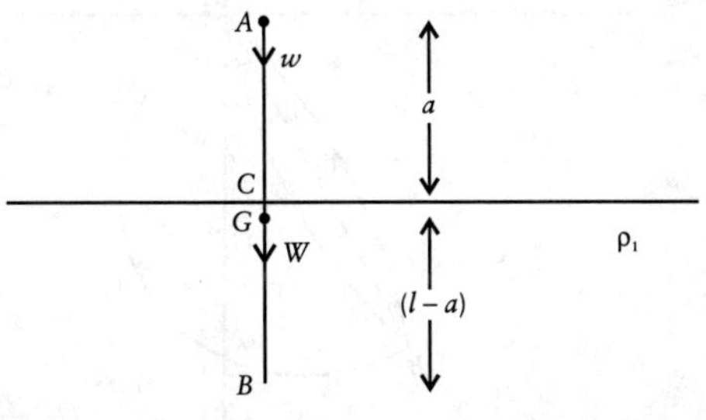

Let AB be a rod of length l. Let W be the weight of the rod AB. Let S be the area of cross-section of the rod AB. Let w be the load which is loaded at one end say A.

When it floats in a liquid of density ρ_1, then, for equilibrium,

$$w + W = (l - a)\, Sg\rho_1 \qquad \ldots (1)$$

| Weight of body = weight of liquid displaced by it

When it floats in a liquid of density ρ_2, then, for equilibrium,

$$w + W = (l - b)\, S\, g\, \rho_2 \qquad \ldots (2)$$

From (1) and (2), we get,

$$(l - a)\, S\, g\, \rho_1 = (l - b)\, S\, g\, \rho_2 \quad \Rightarrow (l - a)\, \rho_1 = (l - b)\, \rho_2 \quad \Rightarrow l(\rho_1 - \rho_2) = a\, \rho_1 - b\, \rho_2$$

$$\Rightarrow \quad l = \frac{a\rho_1 - b\rho_2}{\rho_1 - \rho_2} \qquad \ldots (3)$$

Let c be the length of the rod unimmersed when it floats in a liquid of density ρ. Then,

$$w + W = (l - c)\, S\, g\, \rho \qquad \ldots (4)$$

From (1) and (4), we get

$$(l - a)\, S\, g\, \rho_1 = (l - c)\, S\, g\, \rho \quad \Rightarrow (l - a)\, \rho_1 = (l - c)\rho \quad \Rightarrow c\rho = l\, (\rho - \rho_1) + a\, \rho_1$$

$$= \frac{a\rho_1 - b\rho_2}{\rho_1 - \rho_2}(\rho - \rho_1) + a\rho_1 \qquad = \frac{a\rho_1(\rho - \rho_1) - b\rho_2(\rho - \rho_1) + a\rho_1(\rho_1 - \rho_2)}{\rho_1 - \rho_2}$$

$$= \frac{a\rho_1(\rho - \rho_2) - b\rho_2(\rho - \rho_1)}{\rho_1 - \rho_2} \qquad \Rightarrow c = \frac{a\rho_1(\rho - \rho_2) - b\rho_2(\rho - \rho_1)}{\rho(\rho_1 - \rho_2)}$$

Example 11: A solid hemisphere floats in a liquid, completely immersed with a point of the rim joined to a fixed point by means of a string. Prove that the inclination of the base to the vertical is $\tan^{-1} 3/8$. Also prove that the tension of the string is $2/3\, \pi\, (\rho - \sigma)\, a^3\, g$, where ρ and σ are the densities of the solid and the liquid respectively.

Sol.

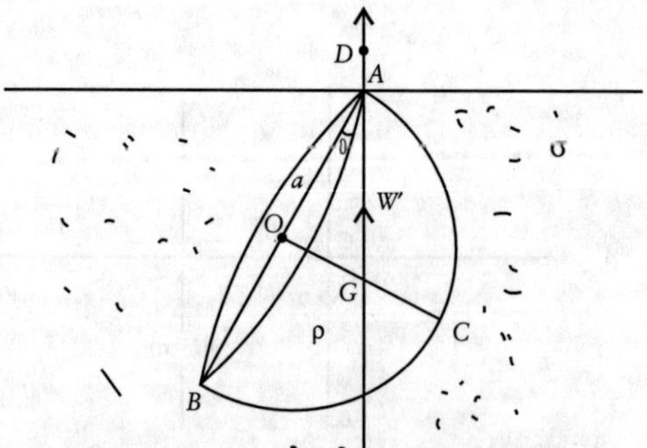

Let the base of the hemisphere be inclined at an angle ϑ to the vertical. Let the point A of the rim be joined to a fixed point by, say D, means of a string. Let a be the radius of the hemisphere. Let O be the centre of the base of the hemisphere. Let OC be the radius of the hemisphere perpendicular to the diameter AB.

Let G be the C.G. of the hemisphere. Then,

$$OG = 3a/8$$

The hemisphere is in equilibrium under the action of the following forces:

(1) W = weight of the hemisphere

$= 2/3\ \pi a^3 \rho g$

acting vertically downwards through G.

(2) W'= The force of buoyancy

= weight of the liquid displaced

$= 2/3\ \pi a^3 \sigma g$

acting vertically upwards through H (here G), the centre of buoyancy i.e., the C.G. of the liquid displaced

| ∵ The hemisphere is completely immersed in the liquid

(3) The tension T of the string attached at B.

Since the hemisphere is in equilibrium and the two of the forces W and W' are vertical and acting at G, therefore, the third force T must also the vertical and its line of action will also pass through G.

| By Statics

Resolving the forces vertically, we get

$$T + W' = W$$

$$\Rightarrow T = W - W' = 2/3\ \pi a^3 \rho g - 2/3\ \pi a^3 \sigma g = 2/3\ \pi\ (\rho - \sigma)\ a^3 g$$

Example 12: An equilateral triangular lamina of sp.gr. n^2 (<1) is movable about a fixed hinge at A and is in the equilibrium when the corner C is immersed in water and AB horizontal and above water. If the triangle be turned about A in its vertical plane and kept in the position with the side BC immersed and horizontal, prove that the action of the hinge in this position is $2/n\ (1 - n)W$, where W is the weight of the lamina.

Sol. Let ABC be the equilateral triangular lamina in which $AB = BC = CA = 2a$

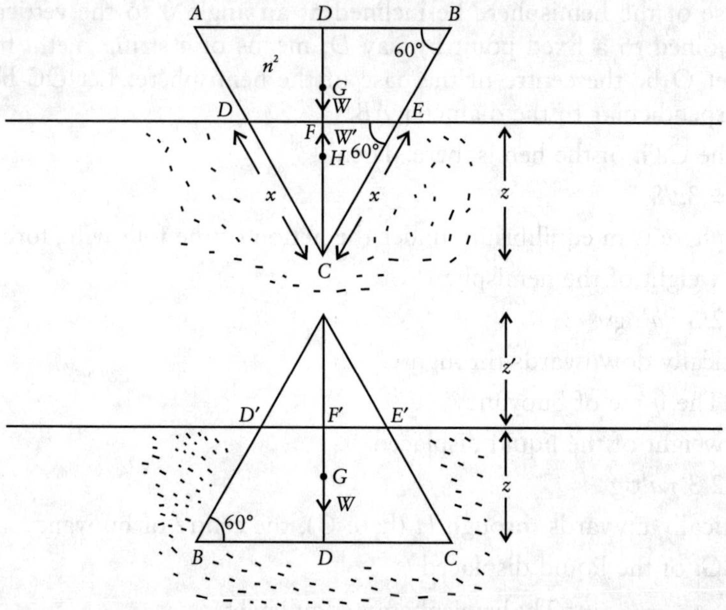

Let D be the middle point of AB. Join CD, then G, the C.G. of $\Delta\, ABC$ and H, the C.G. of the immersed portion $\Delta\, CDE$ of $\Delta\, ABC$ will both lie on CD.

Let x be the length of a side of $\Delta\, ABC$ immersed in water. Then, by geometry, $CE = CD = DE = x$

$\Rightarrow$ $\Delta\, CDE$ is also equilateral

$$\sin 60° = CD/CB \quad \Rightarrow \frac{\sqrt{3}}{2} = \frac{CD}{2a} \quad \Rightarrow CD = a\sqrt{3}$$

$$\therefore \text{ Area of } \Delta\, ABC = \frac{AB.CD}{2} = \frac{2a.a\sqrt{3}}{2} = a^2\sqrt{3}$$

Let $CF = z$

In $\Delta\, CFE$,

$$\cot 60° = FE/CF \quad \Rightarrow \frac{1}{\sqrt{3}} = \frac{DE}{2CF} = \frac{DE}{2z} \quad \Rightarrow DE = \frac{2z}{\sqrt{3}}$$

$$\therefore \text{ Area of } \Delta\, CDE = \frac{DE.CF}{2} = \frac{2z}{\sqrt{3}} \cdot \frac{z}{2} = \frac{z^2}{\sqrt{3}}$$

Let ρ be the density of water. Then, the forces acting on the triangle ABC are

(i) Its weight W = (Area of $\Delta\, ABC$) $n^2 g\rho = a^2\sqrt{3}\, n^2 g\rho$ acting vertically downwards through G, of $\Delta\, ABC$

(ii) W' = Force of buoyancy = weight of liquid displaced

$$= (\text{Area of } \Delta\, CDE)\ 1.\ g\rho = \frac{z^2 g\rho}{\sqrt{3}}$$

(iii) Reaction R at the hight A

Taking moments of all these forces about A, we get

$$a^2\sqrt{3}\,n^2 g\rho\, AD = \frac{z^2}{\sqrt{3}}\, g\rho\, AD \qquad \Rightarrow 3a^2n^2 = z^2$$

$$\Rightarrow \; z = \sqrt{3}\,an \qquad \text{... (1)}$$

In the second position, when BC is horizontal and BC is wholly immersed in water, depth of BC below the free surface of water = z

$\therefore$ Height of A above the free surface of water

$$= 2a \sin 60° - z \qquad = a\sqrt{3} - z \qquad = a\sqrt{3} - \sqrt{3}\,an \qquad \text{| From (1)}$$

$$= a\sqrt{3}\,(1-n) \qquad = z' \text{ (say)}$$

Also, action of the hight in the position

= weight of $\Delta\, ABC$ – weight of displaced water

= W – (Area BCE' D′) $g\rho$

= W – {Area of $\Delta\, ABC$ – Area of $\Delta\, AD'E'$} $g\rho$

$$= W - \left\{2a\cdot\frac{a\sqrt{3}}{2} - \frac{D'E'.z'}{2}\right\} g\rho \text{ where } z' = a\sqrt{3}(1-n)$$

$$= W - \left(a^2\sqrt{3} - \frac{1}{2}\cdot 2z'\cot 60°.\, z'\right) g\rho \quad = W - \left[a^2\sqrt{3} - \left\{a\sqrt{3}\,(1-n)\right\}^2 \frac{1}{\sqrt{3}}\right] g\rho$$

$$= W - \left[a^2\sqrt{3} - a^2\sqrt{3}\,(1-n)^2\right] g\rho \quad = W - \left[a^2\sqrt{3}\,g\rho - a^2\sqrt{3}(1-n)^2 g\rho\right]$$

$$= W - \left[\frac{W}{n^2} - \frac{W}{n^2}(1-n)^2\right] \qquad \Big|\because W = a^2\sqrt{3}\,n^2 g\rho$$

$$= W\left[1 - \frac{1}{n^2} + \frac{1}{n^2}\left(1 + n^2 - 2n\right)\right] = W\left(2 - \frac{2}{n}\right) = 2W\left(1 - \frac{1}{n}\right) = -2W\left(\frac{1}{n} - 1\right) = -2W\left(\frac{1-n}{n}\right)$$

The negative sign shows that the direction of the reaction at the hight A is vertically downwards.

Aliter: Let a be the side of the triangle and x be the length of a side immersed in water. Let h be the height of the lamina. Then,

$$\text{area of } \Delta\, ABC = \frac{\sqrt{3}}{4}\, a^2$$

$$\text{area of } \Delta\, CDE = \frac{\sqrt{3}}{4}\, x^2$$

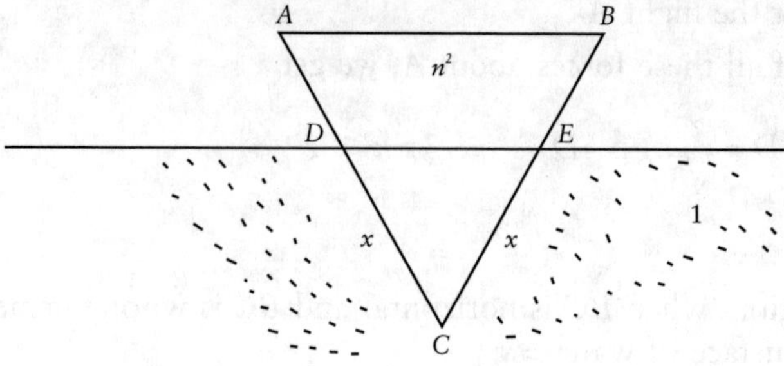

For equilibrium,

W = weight of the lamina = weight of water displaced

$$\Rightarrow \frac{\sqrt{3}}{4}a^2 hgn^2 = \frac{\sqrt{3}}{4}x^2 hg.1 \qquad \Rightarrow a^2 n^2 = x$$

$$\Rightarrow x = an \qquad \text{... (1)}$$

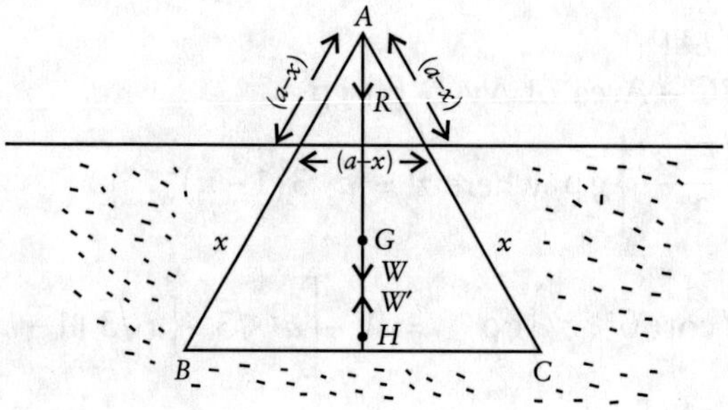

Now, when the lamina is rotated about A, let R be the reaction at A and W' be the weight of the water displaced. Then

$$R + W = W'$$

$$= \frac{\sqrt{3}}{4}\left\{a^2 - (a-x)^2 hg\right\} \quad = \frac{\sqrt{3}}{4}\left(2ax - x^2\right)hg \quad = \frac{\sqrt{3}}{4}\left(2a^2 n - a^2 n^2\right)hg \quad | \text{ From (1)}$$

$$= \frac{\sqrt{3}}{4}a^2 n^2\left(\frac{2}{n} - 1\right)hg \quad = W\left(\frac{2}{n} - 1\right)$$

$$\therefore \text{ R} = W\left(\frac{2}{n} - 1\right) - W \quad = W\left(\frac{2}{n} - 2\right) \quad = \frac{2}{n}(1-n)W$$

EXERCISE 4

1. The mass of a balloon and the gas which it contains is 3500 pounds. If the balloon displaces 48000 cubic ft. of air and the mass of a cubic ft. of air is 1.25 ozs., find the acceleration with which the balloon commences to ascend.

   (16 ozs. = 1 *lb.*)

   Ans. $g/14$ ft./sec^2 or 16/7 ft./sec^2

   (**Hint:** Refer Ex. 1)

2. The mass of a balloon and the gas it contains is 6·5 gm. If the balloon displaces 49000 c.c. of air and the mass of a c.c. of air be 0·001 gm, find the acceleration of the balloon with which it begins to ascend.

   Ans. 5 cm/sec^2

3. Two cubic cm of cork of sp.gr. 0·24 is kept below the water by a rope fastened to the bottom. Find the tension of the rope.

   Ans. 1·52 gm. wt.

   (**Hint:** Refer Ex. 2)

4. A uniform rod, eight metres long, can move about a fulcrum which is above the surface of water. In the position of equilibrium, six metres of the rod are immersed. Find its specific gravity.

   Ans. 15/16

   (**Hint:** Refer Ex. 4)

5. A thin rod of sp.gr. 3/4 and of length 40 cm floats partly immersed in water, being supported by a string fastened to one end of the rod. How much of the rod is immersed?

   If the upper end of the rod be 10 cm above the surface of water, prove that the rod is inclined at an angle of 30° to the horizontal.

   Ans. 20 cm

   (**Hint:** Refer Ex. 5)

6. A uniform rod 6 metres long can move about a fulcrum one metre above the surface of water. In the position of equilibrium, four metres of the rod is immersed. Prove that the sp.gr. of the rod is 8/9.

   (**Hint:** Refer Ex. 4)

7. A uniform rod 3 metres long is free to turn about one of its ends which is out of water and is in equilibrium being inclined to the vertical. If the specific gravity is 5/9, find the length immersed.

   Ans. 1 metre

   (**Hint:** Refer Ex. 4)

8. A uniform rod rests in position inclined to the vertical, with half its length immersed in water, and can turn about a point in it at a distance equal to 1/6th of the length of the rod from the extremity below the water. Prove that the sp.gr. of the rod is 0·125.

   (**Hint:** Refer Ex. 4)

9. A piece of wood of weight $1\frac{1}{2}$ gms. and sp.gr. 3/4 is tied by a string to the bottom of a vessel of water so as to be totally immersed. Prove that the tension of the rod is 1/2 gm. wt.

10. A cylinder of weight 15 gms. and length 3 cm floats in water with its axis vertical and half immersed in water. Prove that a force of 5 gms. wt. will be required to depress it 1/2 cm more.

11. A uniform rod is suspended by two vertical strings attached to its extremities and half of it is immersed is water. If its sp.gr. be 2·5, prove that the tension of the string will be as 9 : 7.

12. A uniform rod, of length $2a$, can turn freely about one end which is fixed at a height a above the surface of the liquid. If the densities of the rod and the liquid be 1/3 and 1/2, show that the rod can rest in a vertical position or inclined at an angle 30° to the vertical.

13. A heavy uniform rod of small cross-section and of length l is movable in a vertical plane about a hinge at one of its extremities situated at a depth h ($< l$) below the surface of a liquid of density ρ. If the density of the rod be σ ($< \rho$), prove that the inclination of the rod to the vertical in equilibrium is

$$\cos^{-1}\left(\frac{h}{l}\sqrt{\frac{\rho}{\sigma}}\right)$$

[Hint :

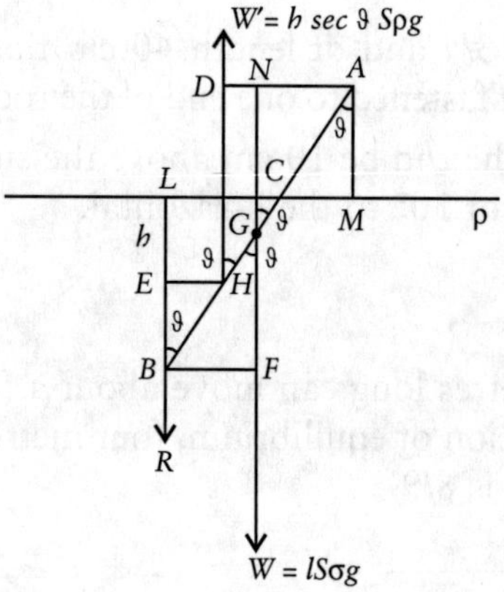

Taking moments about B, we get,

$W. BF = W'EH$

$\Rightarrow W. BG \sin\vartheta = W'BH \sin\vartheta$

$\Rightarrow WBG = W'BH$ $\quad | \because \sin\vartheta \neq 0$

$\Rightarrow l S \sigma g\, l/2 = h \sec\vartheta\, S \rho g \dfrac{h\sec\vartheta}{2} \quad \Rightarrow l^2 \sigma = h^2 \rho \sec^2\vartheta \quad \Rightarrow \cos^2\vartheta = \dfrac{h^2\rho}{l^2\sigma}$

$\Rightarrow \cos\vartheta = \dfrac{h}{l}\sqrt{\dfrac{\rho}{\sigma}} \quad \Rightarrow \vartheta = \cos^{-1}\left(\dfrac{h}{l}\sqrt{\dfrac{\rho}{\sigma}}\right)$]

14. A rod of length $2l$ and sp.gr. ρ is hinged at one end at a height h above the surface of liquid of sp.gr. $n\,\rho (n > 1)$. The rod remains in equilibrium in an inclined position partly immersed in the liquid. Prove that the inclination of the rod to the vertical is

$$\cos^{-1}\left(\frac{h}{2l}\sqrt{\frac{n}{n-1}}\right)$$

and the length immersed is

$$2l\left(1-\sqrt{1-\frac{1}{n}}\right)$$

[**Hint:** Refer Ex. 8

Put $a = l$ and $\sigma = n\,\rho$. Then,

$$\cos\vartheta=\frac{h}{2a}\sqrt{\frac{\sigma}{\sigma-\rho}} \quad =\frac{h}{2l}\sqrt{\frac{n\rho}{n\rho-\rho}} \quad =\frac{h}{2l}\sqrt{\frac{n}{n-1}} \quad \Rightarrow \vartheta=\cos^{-1}\left\{\frac{h}{2l}\sqrt{\frac{n}{n-1}}\right\}$$

Also, length immersed

= 'AC'

$$= \text{'}2a - h\sec\vartheta\text{'} \quad = 2l - h\sec\vartheta \quad = 2l-h\cdot\frac{2l}{h}\sqrt{\frac{n-1}{n}} \quad = 2l\left[1-\sqrt{1-\frac{1}{n}}\right]]$$

15. A uniform rod is bent in the form of the three sides *AB*, *BC*, *CD* of a square. *A* is attached to a hinge fixed on the surface of water and the frame rests in a vertical plane with *AB*, *BC* and half of *CD* immersed in water. Prove that the sp.gr. of the rod is 31/40.

[**Hint:**

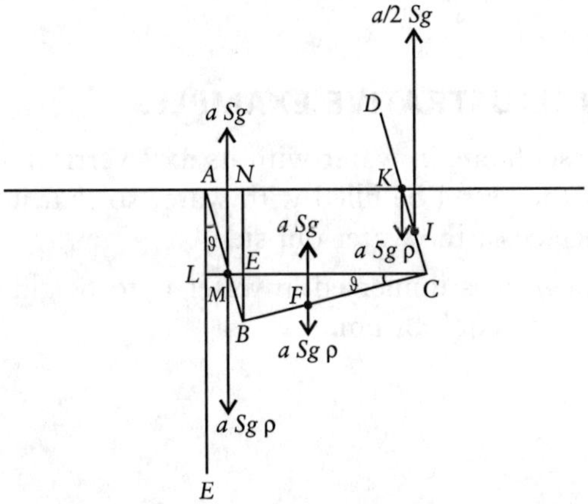

Let the side of the square be *a*. Let ρ be the density of the rod. Let *S* be the area of cross-section of the rods.

$\angle BAE = \vartheta$

$\therefore\ \angle BCM = \vartheta$

$BM = a/2$ | by geometry

$\therefore\ \tan\vartheta = BM/BC = 1/2$ ... (1)

Taking moments about *A*, we get

$$aSg\rho\frac{a}{2}\sin\vartheta+agS\rho\left(a\sin\vartheta+\frac{a}{2}\cos\vartheta\right)+ag\rho S\left(a\sin\vartheta+a\cos\vartheta-\frac{a}{2}\sin\vartheta\right)$$

$$= agS.1\frac{a}{2}\sin\vartheta + agS.1\left(a\sin\vartheta + \frac{a}{2}\cos\vartheta\right) + \frac{ag}{2}S.1\left(a\sin\vartheta + a\cos\vartheta - \frac{a}{4}\sin\vartheta\right)$$

$$\Rightarrow \rho\left(2\sin\vartheta + \frac{3}{2}\cos\vartheta\right) = \frac{15}{8}\sin\vartheta + \cos\vartheta \quad \Rightarrow \rho = \frac{15\tan\vartheta + 8}{16\tan\vartheta + 12}$$

$$\Rightarrow \frac{15\cdot\frac{1}{2}+8}{16\cdot\frac{1}{2}+12} \qquad | \text{ From (1)}$$

= 31/40]

16. An equilateral triangle ABC of weight W and sp.gr. σ is movable about a hinge at A and is in equilibrium when the angle C is immersed in water and the side AB is horizontal and above water. It is then turned about A in its own vertical plane until the whole of the side BC is in the water and horizontal. Prove that the pressure on the hinge in this position is

$$2\left(\frac{1-\sqrt{\sigma}}{\sqrt{\sigma}}\right)w$$

[**Hint:** Refer Ex. 12 Put $n^2 = \sigma$]

HARDER ILLUSTRATIVE EXAMPLES

Example 1: A hollow conical vessel floats in water with its axis vertical and $1/n$ of its axis immersed. Find to what depth must the vessel be filled with water so that it may just sink till its mouth is on the level with the surface of the water out side.

Sol. Let VAB be a cone of height h. It is immersed in water upto height h/n. Let α be the semi-vertical angle of the cone. Then, for equilibrium,

weight of the cone

= weight of water displaced

$\Rightarrow$ weight of cone

= $1/3\ \pi\ (h/n)^3\ \tan^2\alpha g\rho$ ... (1)

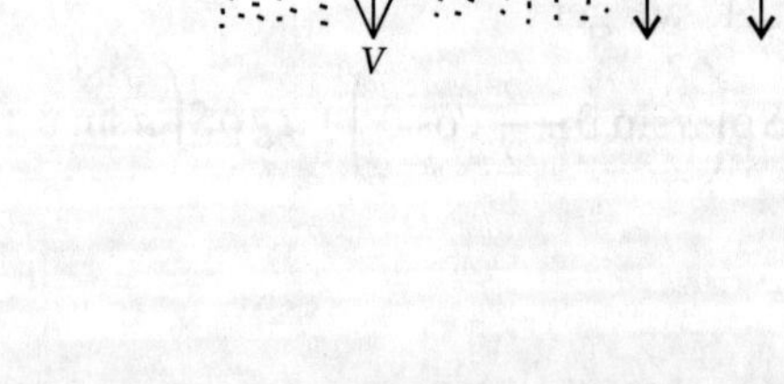

Let the vessel be filled with water upto depth x so that it may just sink till its mouth is on the level with the surface of the water out side. Then, for equilibrium, weight of the cone + weight of water poured into the cone = weight of water displaced

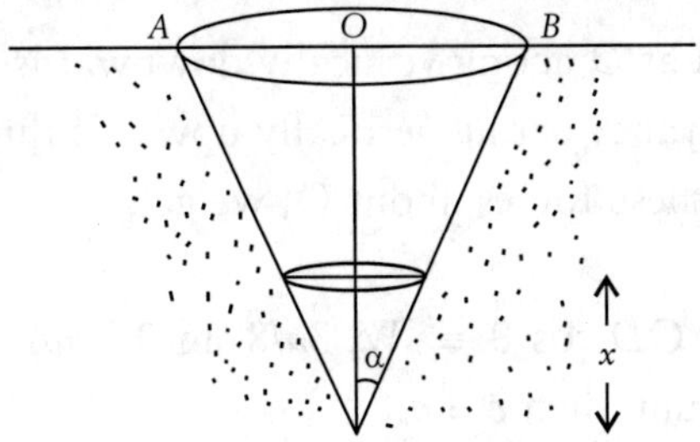

$$\Rightarrow \frac{1}{3}\pi\left(\frac{h}{n}\right)^3 \tan^2\alpha g\rho + \frac{1}{3}\pi x^3 \tan^2\alpha g\rho = \frac{1}{3}\pi h^3 \tan^2\alpha g$$

$$\Rightarrow \frac{h^3}{n^3} + x^3 = h^3 \qquad \Rightarrow x^3 = h^3\left(1 - \frac{1}{n^3}\right) = h^3\left(\frac{n^3-1}{n^3}\right) \qquad \Rightarrow x = \frac{h(n^3-1)^{\frac{1}{3}}}{n}$$

Example 2: A solid hemisphere of radius a and weight W is floating in a liquid, and at a point on the base at a distance c from the centre rests a weight w. Show that the tangent of the inclination of the axis of the hemisphere to the vertical for the corresponding position of equilibrium, assuming the base of the hemisphere entirely out of the liquid, is

$$\frac{8c}{3a}\frac{w}{W}$$

If w is attached to the rim, then prove that $\tan\vartheta = \frac{8}{3}\frac{w}{W}$

Sol.

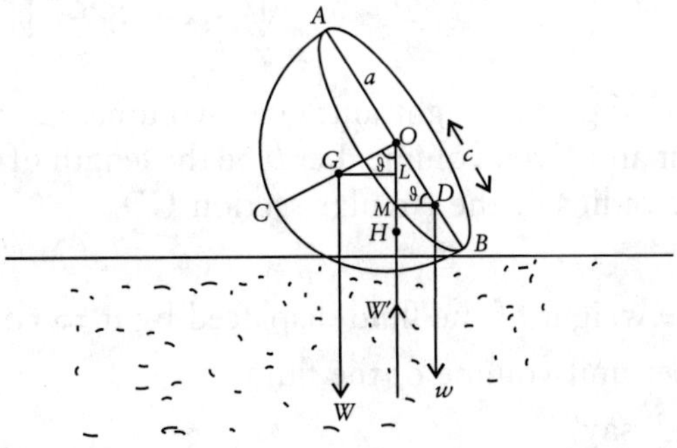

Let a weight w rest at a point D on the base of the hemisphere such that its distance from the centre O is c. Then, $OD = c$. Let OC be the axis of symmetry of the hemisphere. Then, G, the C.G. of the hemisphere will be on OC such that $OG = 3a/8$. Let OC make an angle ϑ with the vertical line through O.

Since the thrust at each point of the curved surface of the hemisphere passes through the centre of the base O of the hemisphere, therefore, the force of buoyancy W' also passes through the centre O. Moreover, W' is vertical as W and w are both vertical so that equilibrium is established. [By Statics]. Note that centre of buoyancy lies on the vertical line through O.

The hemisphere is in equilibrium under the action of the following forces:

(1) W = weight of the hemisphere acting vertically downwards through G, the C.G. of the hemisphere

(2) w = weight placed at D acting vertically downwards

(3) W' = force of buoyancy acting vertically upwards through O.

Taking moments of all these forces about O, we get,

$W.\ GL = w.\ DM$

$\Rightarrow W.\ OG \sin \vartheta = w\ OD \cos \vartheta \Rightarrow W.\ 3a/8 \sin \vartheta = w.\ c.\ \cos \vartheta \Rightarrow \tan \vartheta = 8w/3W\ c/a$

If w is attached to the ruin, then $c = a$.

$\therefore \quad \tan \vartheta = 8w/3W$

Example 3: A cone of given weight and volume, floats in a given fluid with vertex downwards. Show that the surface of the cone in contact with the fluid is least when the vertical angle of the cone is

$$2 \tan^{-1}\left(\frac{1}{\sqrt{2}}\right)$$

Sol.

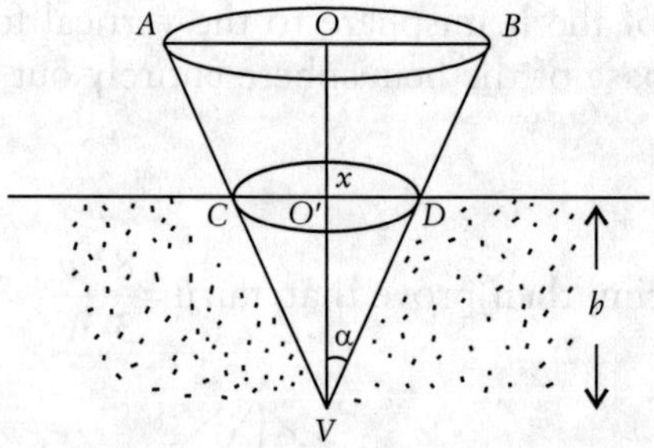

Let VAB be the cone of given weight and given volume. Clearly the density of the cone is constant for given weight and given volume. Let h be the length of the axis of the cone immersed in the fluid. Let x be the radius of the circular section CD.

For floatation,

Weight of the cone = weight of the fluid displaced by it $\Rightarrow$ constant = $1/3\ \pi x^2 hw$

where w is the weight per unit volume of the fluid

$= x^2 h$ = constant = c^3 (say) ... (1)

Let S be the surface area of the cone in contact with the fluid. Then,

$S = \text{`}\pi r l\text{'}$

$$= \pi x\ OD \quad = \pi x\sqrt{h^2 + x^2} \quad = \pi x\sqrt{\frac{c^6}{x^4} + x^2} \qquad | \text{ From (1)}$$

$$= \pi\sqrt{\frac{c^6}{x^2} + x^4}$$

Clearly S is minimum when $z = \frac{c^6}{x^2} + x^4$ is minimum

Now,

$$\frac{dz}{dx} = \frac{-2c^6}{x^3} + 4x^3 \qquad \text{... (2)}$$

and $$\frac{d^2z}{dx^2} = \frac{6c^6}{x^4} + 12x^2 \qquad \text{... (3)}$$

For maximum or minimum,

$$dz/dx = 0$$

$$\Rightarrow \frac{-2c^6}{x^3} + 4x^3 = 0 \quad \Rightarrow \frac{2c^6}{x^3} = 4x^3 \quad \Rightarrow \frac{c^6}{x^3} = 2x^3 \quad \Rightarrow 2x^6 = c^6$$

$$\Rightarrow \sqrt{2}\,x^3 = c^3 \qquad \text{... (4)}$$

For this value of x, d^2z/dx^2 is a positive quantity (> 0)

Hence, S is minimum when $\sqrt{2}\,x^3 = c^3$.

Again, from (1),

$$x^2h = \sqrt{2}\,x^3 \quad \Rightarrow \frac{x}{h} = \frac{1}{\sqrt{2}} \quad \Rightarrow \tan\alpha = \frac{1}{\sqrt{2}} \quad \Rightarrow \alpha = \tan^{-1}\left(\frac{1}{\sqrt{2}}\right)$$

$$\Rightarrow 2\alpha = 2\tan^{-1}\left(\frac{1}{\sqrt{2}}\right)$$

Hence, the required vertical angle of the cone is $2\tan^{-1}\left(\frac{1}{\sqrt{2}}\right)$

Example 4: A rod of density σ and length l is freely movable about one end fixed at a depth h below the surface of a liquid of density ρ. Show that the rod may rest in a position inclined to the vertical provided that

$$1 < \rho/\sigma < l^2/h^2$$

Sol.

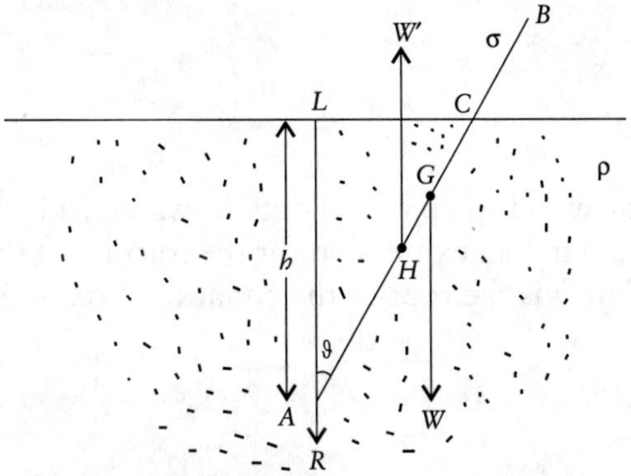

Let AB be a rod of density σ and length l, movable about one end A fixed at a depth h below the surface of a liquid of density ρ. Let S be the area of cross-section of the rod. Let the rod be inclined at an angle ϑ to the vertical. Let G and H be the mid-points of AB and AC respectively. Then,

$AG = GB = l/2$

$AH = 1/2\ AC = 1/2\ h \sec \vartheta$

The rod AB is in equilibrium under the action of the following forces:

(1) W = weight of the rod AB acting vertically downwards through G, the C.G. of the rod $AB = l\ S\ \sigma\ g$

(2) W' = force of buoyancy
= weight of the liquid displaced
$= AC.\ S.\ \rho\ g$
$= h \sec \vartheta\ S\ \rho\ g$

acting vertically upwards through H (the centre of buoyancy), i.e., the C.G. of the liquid displaced

(3) Reaction R at A

Taking moments of all these forces about A, we have

$w.\ AG \sin \vartheta = w'\ AH \sin \vartheta$

$\Rightarrow l\ S\ \sigma\ g\ l/2 = h \sec \vartheta\ S\ \rho\ g\ 1/2\ h \sec \vartheta$

$\Rightarrow l^2\ \sigma = h^2\ \rho \sec^2 \vartheta$

$\Rightarrow \cos^2 \vartheta = h^2\ \rho/l^2\ \sigma$

Since the rod rests partly immersed in the liquid, therefore,

$\sigma < \rho$

$\Rightarrow \sigma/\rho < 1$

$\Rightarrow 1 < \rho/\sigma$ ... (1)

Since the rod is at rest in a position inclined with the vertical, therefore,

$\vartheta \neq 0$

$\therefore \cos^2 \vartheta < 1$

$\Rightarrow h^2\ \rho/l^2\ \sigma < 1$

$\Rightarrow \rho/\sigma < l^2/h^2$ ... (2)

Combining (1) and (2), wc havc,

$1 < \rho/\sigma < l^2/h^2$

Example 5: A sphere of radius r made of wood of sp.gr. σ floats in water with the lowest point at a depth r/h below the surface of water. Find an expression for the fraction of the volume of the sphere which is immersed and show that h is the root of the equation $4\ \sigma x^3 - 3x + 1 = 0$

Sol. Volume of the sphere $= 4/3\ \pi r^3$

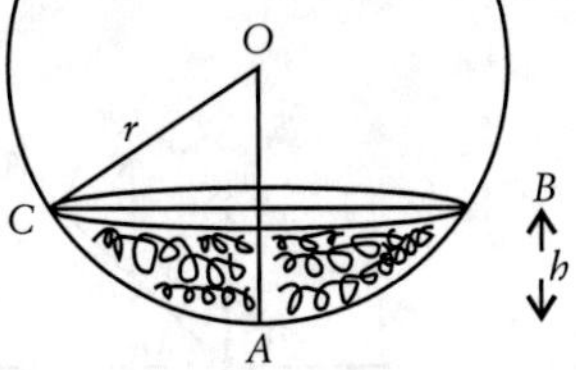

Volume of the segment BAC = '$\pi h^2 (r - h/3)$'

where h is the height (or depth) of the segment from the surface of the sphere

$$= \pi\left(\frac{r}{h}\right)^2\left\{r - \frac{\left(\frac{r}{h}\right)}{3}\right\}$$

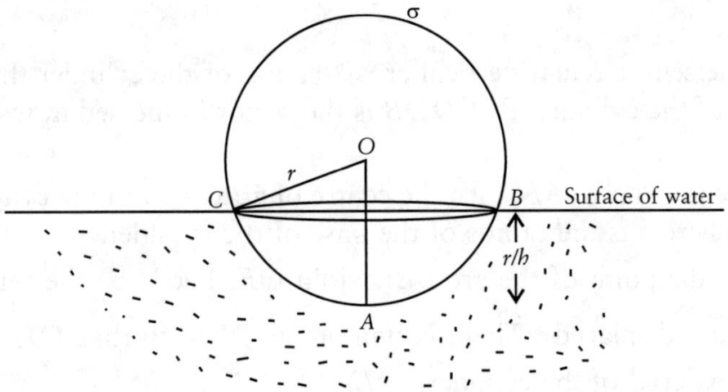

$$= \frac{\pi r^3}{h^2}\left(1 - \frac{1}{3h}\right) \qquad \dots (1)$$

$$= \frac{3}{4h^2}\left(1 - \frac{1}{3h}\right)\left(\frac{4}{3}\pi r^3\right)$$

$$= \frac{3\left(1 - \frac{1}{3h}\right)}{4h^2} \text{ Volume of the sphere} \qquad \dots (2)$$

For floatation in equilibrium, weight of the sphere = weight of water displaced by it

$$\Rightarrow \frac{4}{3}\pi r^3 \sigma g = \frac{\pi r^3}{h^2}\left(1 - \frac{1}{3h}\right) \cdot g \cdot 1 \qquad | \because \text{ density of water} = 1$$

$$\Rightarrow \frac{4}{3}\sigma = \frac{1}{h^2}\left(1 - \frac{1}{3h}\right) \quad \Rightarrow \frac{4\sigma}{3} = \frac{3h - 1}{3h^3} \quad \Rightarrow 4\sigma = \frac{3h - 1}{h^3}$$

$$\Rightarrow 4\sigma h^3 - 3h + 1 = 0 \qquad \dots (3)$$

Equation (3) shows that h is a root of the equation $4\sigma x^3 - 3x + 1 = 0$

Example 6: A semi-circular cylinder floats in water with its axis fixed in the surface of water. If this cylinder be movable about the fixed axis and if its density be half of that of water, show that it will be in equilibrium in any position.

Sol.

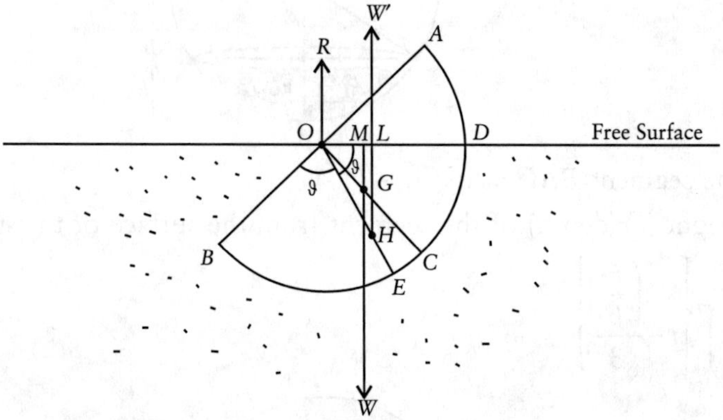

Let *ABCA* be the semi-circular vertical cross-section of the cylinder through O, the middle point O of the axis of the cylinder. *BECDOB* is the sector immersed in water of density ρ such that ∠BOD = 2ϑ.

Let *OC* be perpendicular to *AB*, then the centre of gravity *G* of the cylinder lies on *OC* such that *OG* = 4a/3π, where *a* is the radius of the base of the cylinder.

Let *E* be the middle point of the arc *BED*. Join *OE*. Let *H* be the centre of the buoyancy (i.e., C.G. of the water displaced). Then, *H* will lie on *OE* such that $OH = \frac{2}{3}\frac{a\sin\vartheta}{\vartheta}$

Density of the material of the cylinder = ρ/2.

Let *h* be the length of the cylinder. The forces acting on the cylinder are:

(1) W = weight of the cylinder

$= \frac{\pi a^2 h}{2} g \frac{\rho}{2} = \frac{\pi a^2 h g\rho}{4}$ acting vertically downwards through G.

(2) W′ = force of buoyancy

= weight of the water displaced

$= \left\{\frac{1}{2}a^2(2\vartheta)\right\} h g\rho$

$= a^2 \vartheta h g\rho$

acting vertically upwards through *H*.

(3) Reaction *R* of the axis of rotation acting at O.

Let the lines of action of *W* and *W′* meet the free surface in *L* and *M* respectively.

Taking moments of all these forces about the axis of rotation through O, the algebraic sum of the moments of these forces.

$= W'\,OL - W.\,OM = W'\,OH\cos\vartheta - W.\,OG\cos(2\vartheta - 90°)$

$$= W'\frac{2}{3}\frac{a\sin\vartheta}{\vartheta}\cos\vartheta - W\cdot\frac{4a}{3\pi}\sin 2\vartheta \;=\; a^2\vartheta h g\rho\cdot\frac{a\sin 2\vartheta}{3\vartheta} - \frac{\pi a^2 h g\rho}{4}\cdot\frac{4a}{3\pi}\sin 2\vartheta$$

$$= \frac{a^3 h g\rho\sin 2\vartheta}{3} - \frac{a^3 h g\rho}{3}\sin 2\vartheta = 0$$

Hence, the cylinder is in equilibrium for all values of ϑ.

Example 7: An equilateral lamina suspended freely from *A*, rests with the side *AB* vertical and the side bisected by the surface of a heavy fluid. Prove that the density of the lamina is to that of the fluid as 15 : 16.

Sol.

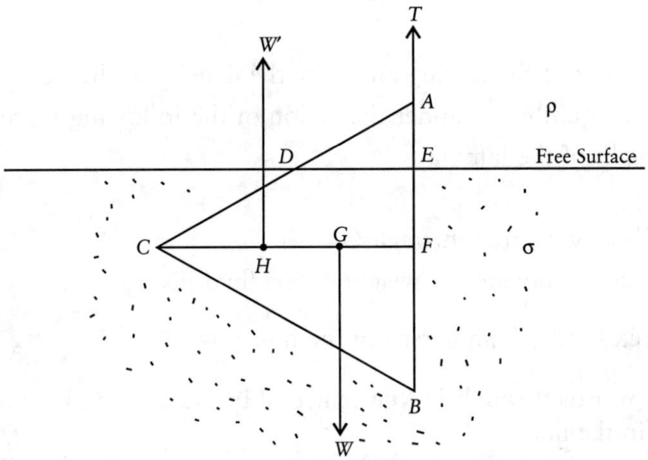

Let *ABC* be the equilateral triangular lamina, suspended freely from *A* with the side *AB* vertical. Let *D* be the mid-point of *AC*. Let the horizontal line through *D* meet *AB* in the point *E*. Then, *DE* lies in the free surface of the fluid. The area *DEBCD* is immersed in the fluid.

Draw $CF \perp AB$. Let $CF = h$ be the height of the $\Delta\, ABC$

$\because$ $\Delta\, ABC$ is equilateral

$\therefore$ *CF* is also a median of $\Delta\, ABC$.

In $\Delta\, ABC$,

$\because$ *D* is the mid-point of *AC* and *DE* is parallel to *CF* ($\because$ *DE* and *CF* both are perpendicular to *AB*)

$\therefore$ $DE = 1/2\ CF = 1/2\ h$ | by geometry ... (1)

Also, $AE = 1/2\ AF$ | by geometry

$= 1/2\ 1/2\ AB = 1/4\ AB$ ... (2)

Let *S* be the area of $\Delta\, ABC$. Then,

area of $\Delta\, ADE$ $= AE\,.\,DE/2 = \dfrac{\frac{1}{4}AB\cdot\frac{1}{2}h}{2}$ | From (1) and (2)

$= AB\ h/16 \quad = \dfrac{1}{8}\left(\dfrac{AB.h}{2}\right) \quad$ = 1/8 area of $\Delta\ ABC$ = S/8

∴ Area of the portion *DEBCD* of the lamina immersed in the fluid

= Area of quadrilateral *DEBC*

= Area of $\Delta\ ABC$ – Area of $\Delta\ ADE$

$= S - S/8 = 7S/8$

Distance of the C.G. g of $\Delta\ ABC$ from side $AB = h/3$

Distance of the C.G. of $\Delta\ ADE$ from side AB = 1/3 DE

= 1/3 1/2 h | From (1)

= 1/6 h

Let ρ be the density of the lamina and σ be the density of the fluid.

The lamina is in equilibrium under the action of the following forces:

(1) W = weight of the lamina

$= S\rho g$

acting vertically downwards through G

(2) W' = Force of buoyancy = weight of the fluid displaced

= (Area *DEBCD* immersed in the fluid) $\sigma g \quad = \dfrac{7S}{8}\sigma g$

acting vertically upwards through H (the centre of buoyancy, i.e., the C.G. of the quadrilateral *DEBC* immersed in the fluid)

∴ Distance of H from AB

= '$W_1 x_1 - W_2 x_2 / W_1 - W_2$' | Formula of remainder

$$= \frac{(S\rho g)\dfrac{h}{3} - \dfrac{S\rho g}{8}\dfrac{h}{6}}{S\rho g - \dfrac{S\rho g}{8}} = \frac{\dfrac{S\rho g h}{48}(16-1)}{\dfrac{7S\rho g}{8}} = 5h/14$$

(3) Tension T in the string at A

Taking moments of all these forces about A (in order to eliminate T), we get

W . distance of g from AB

= W' distance of H from $AB \Rightarrow W \,.\, GF = W'\ HF$

$\Rightarrow S\ \rho\ g\ .\ h/3 = 7\ S\ \sigma\ g/8\ 5h/14 \Rightarrow \rho/3\sigma = 5/16$

$= \rho/\sigma = 15/16$

Example 8: A body is composed of a hemisphere of radius r with a cone of height $2r$, on the same plane base. In a liquid of density ρ, it floats with the whole body first immersed, and in a liquid of density ρ_2 with the hemisphere just immersed, prove that $\rho_2 = 2\rho_1$.

Sol.

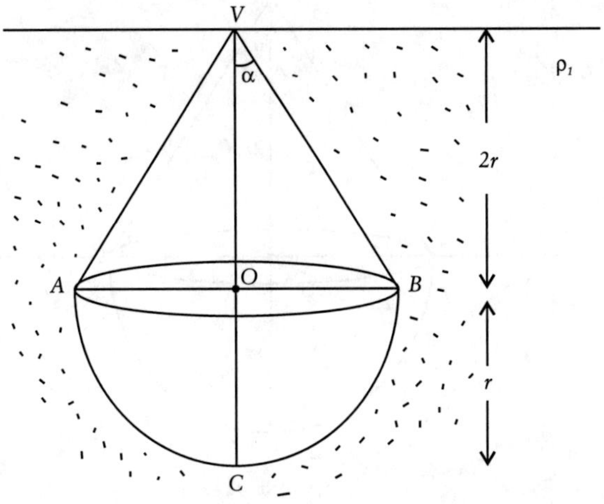

(Case I)

Let α be the semi-vertical angle of the cone.

Then, $\tan \alpha = r/2r = 1/2$ ... (1)

Volume of the cone

$= 1/3\ \pi r^2\ (r \tan \alpha)$ | '$V = 1/3\ \pi r^2 h$'

$\Rightarrow \pi\ r^3 \tan \alpha/3$

$\Rightarrow 2\ \pi r^3/3$ | From (1) ... (2)

Volume of the hemisphere $= 2/3\ \pi r^3$ ... (3)

$\therefore$ Total volume of the body

= Volume of the cone + volume of the hemisphere

$$= \frac{2\pi r^3}{3} + \frac{2\pi r^3}{3} = \frac{4\pi r^3}{3} \quad \text{... (4)}$$

Let W be the weight of the body.

In the liquid of density ρ, it is completely immersed

$\therefore\ W = 4/3\ \pi r^3 g\rho_1$ | For floatation in equilibrium; "$W = vg\rho$" using (4) ... (5)

In the liquid of density ρ_2, the hemisphere is just immersed

$\therefore$ weight of the body = weight of the liquid of density ρ_2 displaced by hemisphere

| For floatation in equilibrium

$\Rightarrow W = 2/3\ \pi\ r^3\ g\ \rho_2$ ... (6)

Equating the two values of W from (5) and (6), we get

$$\frac{4}{3}\pi r^3 g\rho_1 = \frac{2}{3}\pi r^3 g\rho_2 \qquad \Rightarrow 2\rho_1 = \rho_2$$

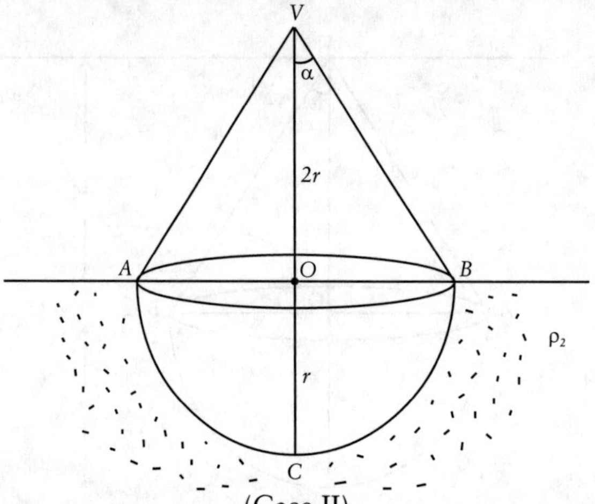

(Case II)

Example 9: A triangular lamina ABC, of which the sides AB, AC are equal, floats in water with BC vertical and three quarters of its length immersed, being kept in equilibrium in this position by means of a string fastened to A in the bottom of the vessel. Find the sp.gr. of the lamina, and show that the tension of the string is $1/27^{th}$ of the weight of the lamina.

Sol. Let T be the tension in the string fastened to A in the bottom of the vessel. Let W be the weight of the triangular lamina ABC acting vertically downwards through G, the C.G. of $\Delta\ ABC$. Let W' be the force of buoyancy acting vertically upwards through H, the centre of buoyancy, i.e., the C.G. of water displaced.

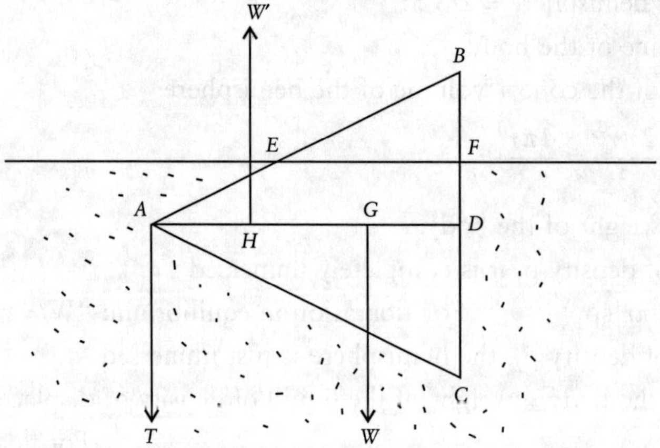

Let D be the mid-point of BC. Join AD.

Then, AD is horizontal as BC is vertical.

Since three quarters of BC is immersed in liquid, therefore,

$$BF = 1/4\ BC \qquad = 1/4\ (2\ BD) \qquad = 1/2\ BD$$

$\Rightarrow$ F is the middle point of BD.

$\therefore$ E is the middle point of AB as FE is parallel to DA and $EF = 1/2\ AD$

| by geometry

Let $AD = h$

Then, $EF = 1/2\ h$

Area of $\Delta\ ABC = \dfrac{BC.AD}{2} = \dfrac{2.BD.AD}{2} = BD.AD = BD.h$

Area of $\Delta\ BEF = \dfrac{BF.EF}{2} = \dfrac{\frac{1}{2}BD.\frac{1}{2}AD}{2} = \dfrac{BD.AD}{8} = \dfrac{BD.h}{8}$

$\therefore$ Area of the figure $EFDCAE$ = Area of $\Delta\ ABC$ – Area of $\Delta\ BEF = BD.h - BD\ .h/8 = 7BDh/8$

Horizontal distance of the C.G. g of $\Delta\ ABC$ from

$BC = GD$

$= h/3$

Horizontal distance of C.G. of $\Delta\ BEF$ from BC

$= 1/3\ EF = 1/3\ 1/2\ AD = AD/6 = h/6$

$\therefore$ Horizontal distance of H, the C.G. of area $FDCAEF$ from $BC = \dfrac{(BD.h)\frac{h}{3} - \left(\frac{BD.h}{8}\right)\frac{h}{6}}{BD.h - \frac{BD.h}{8}}$

$\Rightarrow 5/14\ h$

$\therefore$ Distance from $A = h - \dfrac{5h}{14} = \dfrac{9h}{14}$

Let the sp.gr. of the lamina be s. Let ρ be the density of water.

The forces acting on the lamina are:

(1) W = weight of the lamina = $BD\ .\ h\ S\ g\ \rho$, acting vertically downwards through G

(2) W'= Force of buoyancy

$= \text{weight of the water displaced} = \dfrac{7BD.h}{8} g\rho$

(3) Tension T in the string at A acting in the vertical direction as W and W' are acting vertically

Taking moments of all these forces about A, we have,

$W\ .\ 2h/3 = W'\ .\ 9h/14$

$\Rightarrow BD.hSg\rho\dfrac{2h}{3} = \dfrac{7BDh}{8} g\rho\dfrac{9h}{14} \quad \Rightarrow S = 27/32$

which gives the sp.gr. of the lamina

Also, from figure,

$T + W = W'$

$\Rightarrow T = W' - W = \dfrac{7BD.h}{8} g\rho - BD.hSg\rho = BD.h.g\rho\left(\dfrac{7}{8} - \dfrac{27}{32}\right)$

$= \dfrac{BD.h.g.\rho}{32}$ ... (1)

Again, $W = BD\ hSg\rho$

$$= \frac{27}{32} BD.hg\rho \qquad \text{... (2)}$$

From (1) and (2), we get

$$T = 1/27\ W$$

Hence, the tension of the string is 1/27th of the weight of the lamina.

Example 10: *ABC* an isosceles triangle right angled at A, composed of two heavy rods *AB*, *AC* hinged together at *A* and *BC* is a light string. It floats with the angle *A* immersed in water. Show that the tension of the string is $\frac{a-b}{2a}W$, where $2a$ is the length of each rod, $2b$ is the length immersed and W is the weight of each rod.

Sol. $AB = AC = 2a$

$AD = AE = 2b$

Let S be the area of cross section of the rods and ρ be the density of the rods.

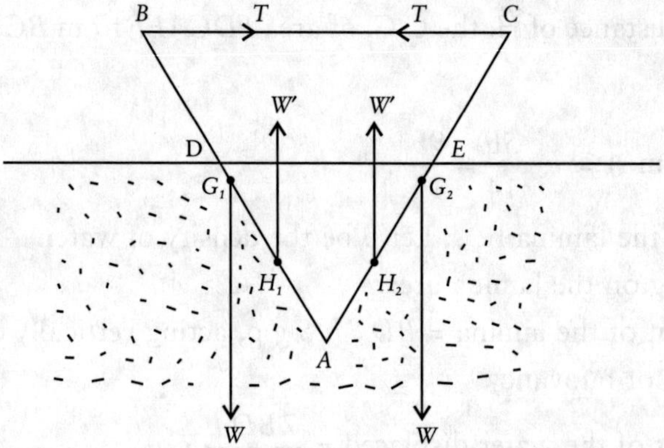

Weight of a rod = $W = 2aSg\rho$ ... (1)

Weight of water displaced = $W' = 2bSg$ ... (2)

For floatation in equilibrium,

$$W = W'$$

$$\Rightarrow 2aSg\rho = 2bSg$$

$$\Rightarrow a\rho = b \qquad \text{... (3)}$$

Taking moments about A,

$$W.AG_1 \cos 45° = W'\ AH_1 \cos 45° + T.AB \sin 45°$$

$$\Rightarrow W\ AG_1 = W'\ AH_1 + TAB \quad \Rightarrow Wa = W'b + T.\ 2a \quad \Rightarrow Wa = 2bSgb + T\ .\ 2a$$

$$= 2a\rho Sgb + T\ .\ 2a \quad | \text{ From (3)}$$

$$= Wb + T\ .\ 2a \quad \Rightarrow W\ (a - b) = T\ .\ 2a \quad \Rightarrow T = \left(\frac{a-b}{2a}\right) W$$

Example 11: A thin uniform rod AB, of length $2a$ and weight W has particle of weight W' attached to a point D near the end A, AD being of length b. The rod floats, in an inclined position, freely in water with a length AC (= $2x$) immersed. Prove that

$$x = \frac{aW + bW'}{W + W'}$$

and that the sp.gr. σ of the rod is given by $a\sigma(W + W')^2 = W(aW + bW')$.

Sol.

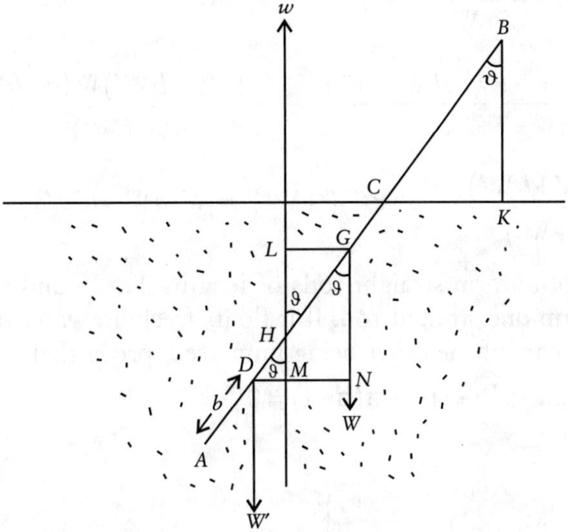

Let G be the C.G. of the rod AB. Then,

$$AG = GB = a$$

Let the rod be inclined at an angle ϑ to the vertical.

$$AC = 2x$$

Let H be the mid-point of A.C. Then,

$$AH = HC = x \qquad \Rightarrow HG = AG - AH = a - x$$

$$HD = AH - AD = x - b$$

Taking moments about H, we get

$W.\ HG \sin \vartheta = W'\ HD \sin \vartheta$

$\Rightarrow W.\ HG = W'\ HD \quad \Rightarrow \quad W(a - x) = W'(x - b)$

$\Rightarrow x(W + W') = aW + bW'$

$$\Rightarrow x = \frac{aW + bW'}{W + W'} \qquad \text{... (1)}$$

Let S be the area of cross-section of the rod. Then,

$$W = 2aSg\sigma$$

$$W = 2xSg\ 1$$

Taking moments about D, we get

$W \cdot DN = w \cdot DM$

$\Rightarrow W \cdot DG \sin \vartheta = w\, DH \sin \vartheta \quad \Rightarrow \quad W\, DG = w\, DH$

$\Rightarrow W\,(AG - AD) = w\,(AH - AD) \quad \Rightarrow \quad W\,(a - b) = w\,(x - b)$

$\Rightarrow 2aSg\sigma\,(a - b) = 2xSg(x - b) \quad \Rightarrow \quad a\sigma\,(a - b) = x\,(x - b)$

$$= \frac{aW + bW'}{W + W'}\left(\frac{aW + bW'}{W + W'} - b\right) \qquad \text{| From (1)}$$

$$= \frac{(aW + bW')(aW + bW' - bW - bW')}{(W + W')^2} = \frac{(aW + bW')W(a - b)}{(W + W')^2}$$

$$\Rightarrow a\sigma = \frac{W(aW + bW')}{(W + W')^2} \Rightarrow a\sigma(W + W')^2 = W(aW + bW')$$

Example 12: Two uniform straight rods of lengths $2a$, $2b$ and sp.gr. ρ, σ respectively, are joined together to form one straight rod. If it floats freely in water in an inclined position, the rod of length $2a$ and part of the other being immersed, prove that

$$a^2\rho(\rho - 1) + 2ab\sigma(\rho - 1) + b^2\sigma(\sigma - 1) = 0$$

Sol.

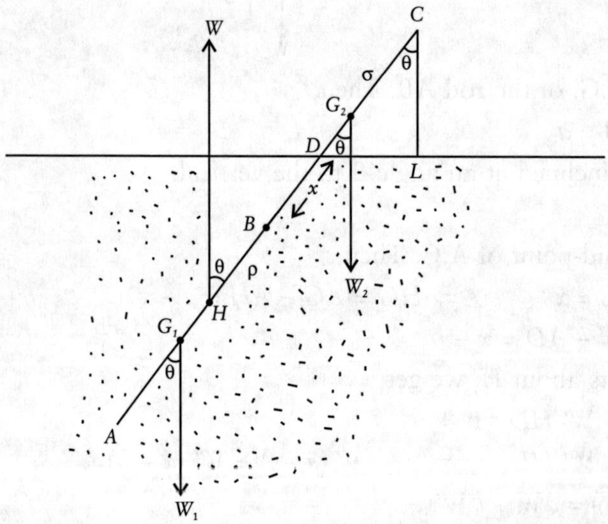

Let AB be a uniform rod of length $2a$ and BC be a uniform rod of length $2b$. Their sp.gr. are ρ and σ respectively. Let the length x of the rod BC be immersed in water. Let S be the area of cross-section of each rod.

W_1 = weight of rod AB $= 2aSg\rho$

W_2 = weight of rod BC $= 2bSg\sigma$

For equilibrium,

$W = W_1 + W_2 \quad = 2aSg\rho + 2bSg\sigma$

$= 2Sg\,(a\rho + b\sigma)$ ... (1)

Also, W = weight of the water displaced

$= 2a\,S\,g\,1 + S\,x\,g\,1$

$= (2a + x)\,Sg$ ... (2)

From (1) and (2), we get

$2Sg\,(a\rho + b\sigma) = (2a + x)\,Sg \qquad \Rightarrow 2\,(a\rho + b\sigma) = 2a + x$

$\Rightarrow \; a\rho + b\sigma = 2a + x/2$ ... (3)

Taking moments about B,

$W_1G_1B \sin\vartheta = W.HB\sin\vartheta + W_2G_2B\sin\vartheta$

$\Rightarrow \; W_1G_1B = WHB + W_2G_2B$

$$\Rightarrow \; W_1\,a = W\left(2a - \frac{2a+x}{2}\right) + W_2\,b$$

$$\begin{aligned} HB &= AB - AH \\ &= AB - \frac{AD}{2} \\ &= AB - \frac{AB+BD}{2} \\ &= 2a - \frac{2a+x}{2} \end{aligned}$$

$\Rightarrow \; 2aSg\rho a = 2Sg(a\rho + b\sigma)\{2a - (a\rho + b\sigma)\} + 2bSg\sigma b$

$\Rightarrow \; a^2\rho = (a\rho+b\sigma)(2a - a\rho - b\sigma) + b^2\sigma \quad = 2a^2\rho - a^2\rho^2 - ab\rho\sigma + 2ab\sigma - ab\rho\sigma - b^2\sigma^2 + b^2\sigma$

$\Rightarrow \; a^2\rho(\rho-1) + 2ab\sigma(\rho-1) + b^2\sigma(\sigma-1) = 0$

Example 13: A semi-circular lamina has one of the ends of its diameter smoothly hinged to a fixed point above the surface of a liquid and floats with its plane vertical and its diameter half immersed. If the inclination of the diameter to the horizontal in $\pi/4$, prove that the ratio of the density of the liquid to that of the lamina is $4(3\pi - 4) : (9\pi - 8)$.

Sol.

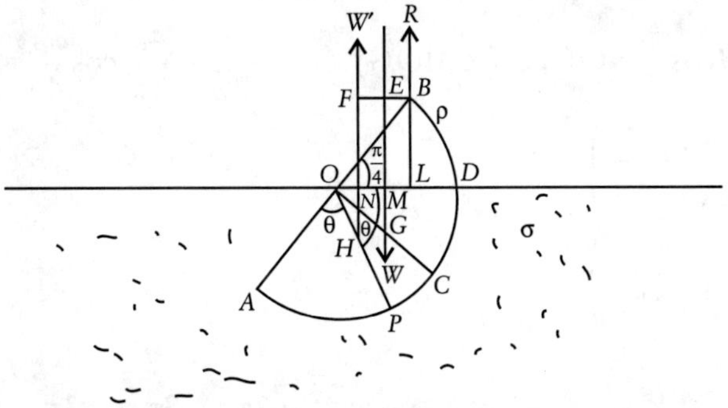

Let $ABC\,A$ be the given semi-circular lamina of radius a (say), centre O and the end B of the diameter AB fixed above the free surface of the liquid. The sector $APC\,DOA$ is immersed in the liquid. Let OC be the radius of the semi-circle perpendicular to the diameter AB. Then, the centre of gravity G of the semi-circle lies on OC such that $OG = 4a/3\pi$. Let ρ be the density of the material of the lamina and σ be the density of the liquid. Let P be the mid-point of the are ACD. Join OP. Then, H the centre of buoyancy i.e., the C.G. of the liquid displaced lies on OP such that $OH = \frac{2}{3}\frac{a\sin\theta}{\theta}$, where 2θ is the angle AOD

Here,

$$\angle AOD = \pi - \pi/4 = 3\pi/4 \quad \Rightarrow 2\theta = 3\pi/4 \quad \Rightarrow \theta = 3\pi/8$$

$$\therefore \quad OH = \frac{2}{3}\frac{a\sin\left(\frac{3\pi}{8}\right)}{\left(\frac{3\pi}{8}\right)} = \frac{16a}{9\pi}\sin\left(\frac{3\pi}{8}\right)$$

Area of the sector $APCDOA$ = '$1/2\ r^2\ \theta$' where θ is the angle of the sector

$$= 1/2\ a^2\ (2\theta) \qquad = a^2\,\theta$$

The lamina is in equilibrium under the action of the following forces:

(1) W = weight of the semi-circle = $\frac{\pi a^2}{2} g\rho$

acting vertically downwards at G

(2) W' = Forces of buoyancy

= weight of the liquid displaced

$$= (a^2\,\theta)\,\sigma g \qquad = \frac{3\pi a^2}{8}\sigma g$$

acting vertically upwards at H

(3) Reaction R at B

Taking moments about B (to avoid R), we get

$$W\,.\,BE - W'\,.\,BF = 0 \quad \Rightarrow \quad W\,.\,BE = W'\,.\,BF$$

$$\Rightarrow\ W\,.\,LM = W'\,LN \qquad \text{...(1)}$$

Now,

$$LM = OL - OM$$

$$= OB\cos\frac{\pi}{4} - OG\cos(\angle MOG) \quad = \frac{a}{\sqrt{2}} - \frac{4a}{3\pi}\cos\frac{\pi}{4} \quad = \frac{a}{\sqrt{2}} - \frac{4a}{3\pi}\cdot\frac{1}{\sqrt{2}}$$

$$= \frac{a}{\sqrt{2}}\left(1 - \frac{4}{3\pi}\right)$$

$$= \frac{a}{\sqrt{2}}\left(\frac{3\pi - 4}{3\pi}\right) \qquad \text{... (2)}$$

$$LN = OL - ON$$

$$= OB\cos\frac{\pi}{4} - OH\cos\angle NOH \quad = \frac{a}{\sqrt{2}} - \frac{16a}{9\pi}\sin\frac{3\pi}{8}\cos\frac{3\pi}{8} \quad \left|\because \angle NOH = \theta = \frac{3\pi}{8}\right.$$

$$= \frac{a}{\sqrt{2}} - \frac{8a}{9\pi}\sin\frac{3\pi}{4} = \frac{a}{\sqrt{2}} - \frac{8a}{9\pi}\frac{1}{\sqrt{2}} = \frac{a}{\sqrt{2}}\left(1 - \frac{8}{9\pi}\right)$$

$$= \frac{a}{\sqrt{2}}\left(\frac{9\pi - 8}{9\pi}\right) \qquad \text{... (3)}$$

Using (2) and (3), (1) gives

$$\frac{\pi a^2}{2} g\rho \frac{a}{\sqrt{2}}\left(\frac{3\pi - 4}{3\pi}\right) = \frac{3\pi a^2}{8}\sigma g \frac{a}{\sqrt{2}}\left(\frac{9\pi - 8}{9\pi}\right)$$

$$\Rightarrow (3\pi - 4)\rho = \frac{(9\pi - 8)\sigma}{4} \qquad \Rightarrow \frac{\sigma}{\rho} = \frac{4(3\pi - 4)}{9\pi - 8}$$

Note: In case of a body completely immersed in a liquid, C.G. of the liquid displaced is the same as the C.G. of the body.

Example 14: A rod floats upright partially immersed in a homogeneous liquid. Prove that a small increase of atmospheric density will produce a small rise of the rod proportional to the square of the length of the unimmersed portion.

Sol.

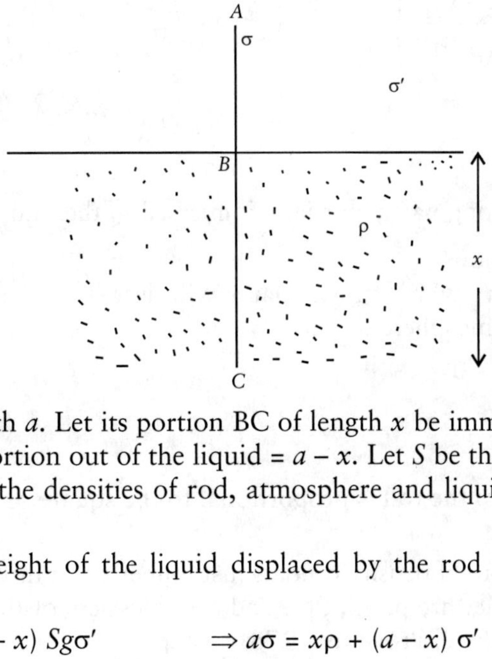

Let AC be a rod of length a. Let its portion BC of length x be immersed in a homogeneous liquid. Then, length of its portion out of the liquid = $a - x$. Let S be the area of cross-section of the rod. Let σ, σ' and ρ be the densities of rod, atmosphere and liquid respectively. Then, for floatation in equilibrium,

Weight of the rod = weight of the liquid displaced by the rod + weight of atmosphere displaced by the rod

$$\Rightarrow aSg\sigma = xSg\rho + (a - x)\,Sg\sigma' \qquad \Rightarrow a\sigma = x\rho + (a - x)\,\sigma'$$

$$\Rightarrow a(\sigma - \sigma') = x(\rho - \sigma')$$

$$\Rightarrow x = \frac{a(\sigma - \sigma')}{\rho - \sigma'} \qquad \text{... (1)}$$

$\therefore$ Length of the immersed portion AB of the rod

$$= a - x \quad = a - \frac{a(\sigma - \sigma')}{\rho - \sigma'} \quad = a\left\{1 - \frac{\sigma - \sigma'}{\rho - \sigma'}\right\}$$

$$= a\left(\frac{\rho - \sigma}{\rho - \sigma'}\right) \qquad \text{... (2)}$$

We know that by changing the atmospheric density σ', the length x of the unimmersed portion changes. Hence,

$$x = f(\sigma')$$

We have,

$$a - x = a\left(\frac{\rho - \sigma}{\rho - \sigma'}\right) \qquad | \text{ From (2)}$$

Differentiating this equation w.r.t. σ', we get

$$-\frac{dx}{d\sigma'} = a(\rho - \sigma)\left\{-\frac{1}{(\rho - \sigma')^2}\right\}(-1) \Rightarrow \frac{dx}{d\sigma'} = \frac{-a(\rho - \sigma)}{(\rho - \sigma')^2}$$

$$= -a(\rho - \sigma)\left\{\frac{(a - x)^2}{a^2(\rho - \sigma)^2}\right\} \qquad | \text{ From (2)}$$

$$= -\frac{(a - x)^2}{a(\rho - \sigma)} \qquad \text{... (3)}$$

Since the rod is floating partly partially immersed in the liquid, therefore,

$$\sigma < \rho$$

$\therefore$ $dx/d\sigma'$ is negative which shows that x will decrease by increasing σ', i.e., the rod will rise by increasing the atmosphere density.

Also, from (3),

$$\frac{dx}{d\sigma'} \propto (a - x)^2$$

Hence, a small rise of the rod is proportional to the square of the length of the unimmersed portion.

Example 15: A sphere of density σ floats just immersed in three liquids. The densities of the liquid in descending order are ρ, 4ρ, 9ρ , and the thickness of the two upper liquid layers are each one-third of the sphere. Prove that $27\sigma = 122\rho$.

Sol. Let the radius of the sphere be a. Then, volume of the sphere = $4/3\ \pi a^3$

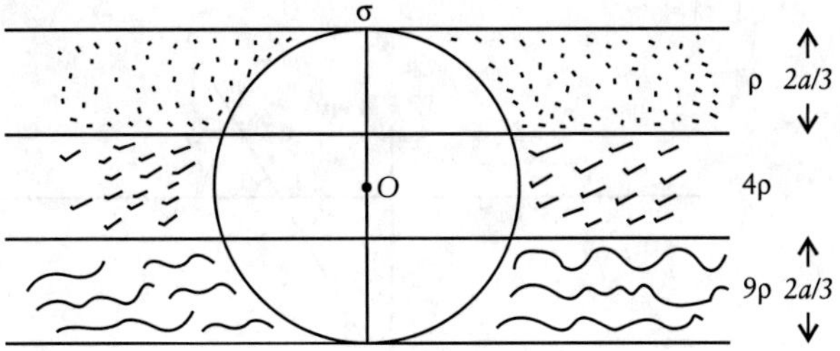

Volume of the upper segment

= '$\pi h^2 (r - h/3)$' Where h is depth (or height) of the segment from the surface of the sphere

$$= \pi\left(\frac{2a}{3}\right)^2\left\{a-\frac{\left(\frac{2a}{3}\right)}{3}\right\} = \pi\,\frac{4a^2}{9}\left(a-\frac{2a}{9}\right) = \pi\,\frac{4a^2}{9}\cdot\frac{7a}{9} = \frac{28\pi a^3}{81}$$

Similarly, volume of the lower segment

$$\frac{28\pi a^3}{81}$$

$\therefore$ volume of the middle portion (zone) of the sphere

$$= \frac{4}{3}\pi a^3 - \left(\frac{28\pi a^3}{81}+\frac{28\pi a^3}{81}\right) = \frac{52}{81}\pi a^3$$

For floatation in equilibrium,

weight of the sphere = sum of the weights of the three liquids displaced

$$\Rightarrow \frac{4}{3}\pi a^3 g\sigma = \frac{28\pi a^3}{81}g\rho+\frac{52}{81}\pi a^3 g4\rho+\frac{28\pi a^3}{81}g9\rho$$

$$\Rightarrow \frac{4}{3}\sigma = \frac{28}{81}\rho+\frac{52}{81}4\rho+\frac{28}{81}9\rho \quad \Rightarrow \sigma = \frac{7}{27}\rho+\frac{52}{27}\rho+\frac{63}{27}\rho \quad \Rightarrow \sigma = \frac{(7+52+63)\rho}{27}$$

$$\Rightarrow 27\,\sigma = 122\,\rho$$

Example 16: A thin uniform rod of weight W has a particle of weight w attached to one end. It is floating, in an inclined position, in water with this end immersed. Prove that the length of the rod above water is $\frac{w}{w+W}$ times its whole length and that the sp.gr. of the rod is $\frac{W^2}{(w+W)^2}$.

Sol. Let AB be a rod of length a. Let x^{th}

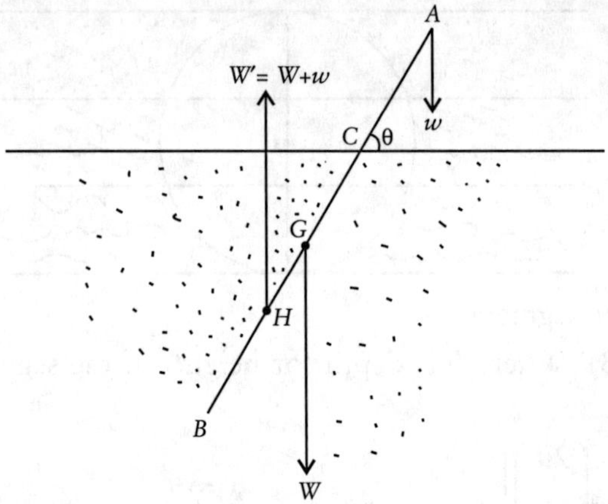

part of the rod be above water. Then,

$AC = ax$;

$BC = a - ax$

Let the rod make an angle θ with the horizontal. Then, for equilibrium,

$W' = W + w$ where W' is the force of buoyancy,

Taking moments about A,

$$W\frac{a}{2}\cos\theta = (W+w)\left(\frac{a-xa}{2}\right)\cos\theta \quad \Rightarrow W = (W+w)(1-x) \quad \Rightarrow x = 1-\frac{W}{W+w}$$

$$\Rightarrow \; x = \frac{w}{W+w}$$

∴ Length of the rod above water

$$= x\,a \quad = \frac{w}{W+w}a$$

$$= \frac{W}{W+w} \text{ times whole length}$$

Let σ be the sp.gr. of the rod. Then,

$$W = aS\sigma g \qquad \text{... (1)}$$

where S is the area of cross section of the rod.

For equilibrium,

$$w + W = (a - xa)\,Sg \quad = a\,(1 - x)\,Sg \quad = a\frac{W}{w+W}Sg$$

$$= \frac{W}{S\sigma g}\frac{W}{w+W}Sg \qquad | \text{ From (1)}$$

$$= \frac{W^2}{\sigma(w+W)} \quad \Rightarrow \sigma = \frac{W^2}{(w+W)^2}$$

Example 17: A thin uniform wooden rod AB is in equilibrium in an inclined position with one end A immersed in a bowl of water, one point D supported on the edge of the bowl. Show that the sp.gr. σ of the rod is

$$\frac{AC}{AB} = \frac{2AD - AC}{2AD - AB}$$

Also prove that the greatest fraction of the length of the rod which can remain immersed is $1 - \sqrt{(1-\sigma)}$.

Sol. First Part

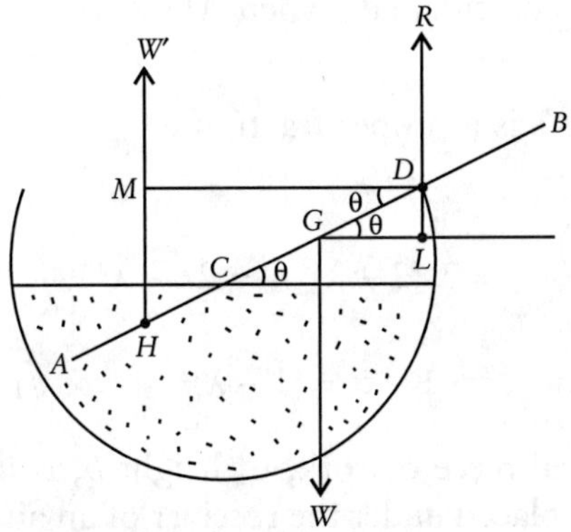

AB is the rod with one point D of it supported on the edge of the bowl and portion AC immersed in water. Let S be the area of cross section of the rod and σ be the sp.gr. of the rod. Let θ be the inclination of the rod to the horizontal. Let G be the C.G. of the rod AB. Let H be the centre of buoyancy. Let ρ be the density of water. Then,

$$AG = GB = 1/2\ AB$$

$$AH = HC = 1/2\ AC$$

The rod AB is in equilibrium under the action of the following forces:

(1) W = weight of the rod AB

$= AB\ S\ \sigma\ g\rho$

acting vertically downwards through G.

(2) W' = Force of buoyancy

= weight of the displaced water

$= AC\ S\ g\ \rho$

acting vertically upwards through H, the centre of buoyancy, i.e., the C.G. of the displaced water.

(3) Reaction R at D

Taking moments of all these forces about D, we get,

$$W.\ GL = W'.\ DM$$

$$\Rightarrow W.\ DG \cos\theta = W'\ DH \cos\theta \quad \Rightarrow W\ .\ DG = W'.\ DH$$

$$\Rightarrow W(AD - AG) = W'(AD - AH) \quad \Rightarrow W\left(AD - \frac{1}{2}AB\right) = W'\left(AD - \frac{1}{2}AC\right)$$

$$\Rightarrow W(2AD-AB) = W'(2AD-AC) \quad \Rightarrow ABS\sigma g\rho(2AD-AB) = ACSg\rho(2AD-AC)$$

$$\Rightarrow AB\sigma(2AD-AB) = AC(2AD-AC)$$

$$\Rightarrow \sigma = \frac{AC(2AD-AC)}{AB(2AD-AB)} \quad \dots (1)$$

Second Part

Since the weight of the portion BD of the rod AB has a tendency to lift the rod out of water, therefore, the length immersed of the rod will be the greatest when BD is the least, i.e., when $BD = 0$, i.e., when B and D coincide, i.e., when $AD = AB$.

Let $AC/AB = \lambda$

Then, $AC = \lambda\, AB$ where λ is a proper fraction.

Then, from (1), we get

$$\sigma = \frac{\lambda(2AB-\lambda AB)}{(2AB-AB)} = \lambda(2-\lambda) = 2\lambda - \lambda^2$$

$$\therefore\ 1-\sigma = 1-2\lambda+\lambda^2 \quad \Rightarrow 1-\sigma = (1-\lambda)^2 \quad \Rightarrow \sqrt{1-\sigma} = 1-\lambda \quad \Rightarrow \lambda = 1-\sqrt{1-\sigma}$$

Example 18: A cylindrical piece of cork, of height h, is floating with its axis vertical in a basin of water. If the basin is placed under the receiver of an air-pump and the air is pumped out, prove that the cork will sink through a distance $\frac{\sigma}{1-\sigma}(1-\rho)h$, where σ and ρ are respectively the sp. gravities of air and cork.

Sol.

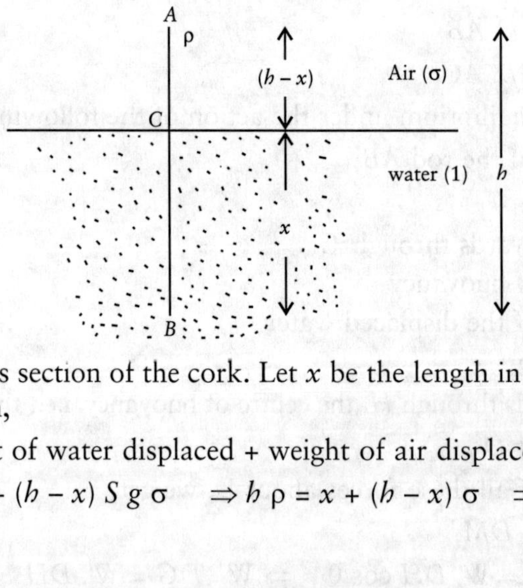

Let S be the area of cross section of the cork. Let x be the length in the water. Then, $(h-x)$ is the length in the air. Then,

Weight of cork = weight of water displaced + weight of air displaced

$$\Rightarrow h\,S\,g\,\rho = x\,S\,g\,1 + (h-x)\,S\,g\,\sigma \quad \Rightarrow h\rho = x + (h-x)\,\sigma \quad \Rightarrow h(\rho-\sigma) = x(1-\sigma)$$

$$\Rightarrow x = \frac{h(\rho-\sigma)}{1-\sigma} \quad \dots (1)$$

Let y be the length to which the cork sinks, when the air is pumped out. Then, weight of the cork = weight of the water displaced

$\Rightarrow hSg\rho = (x + y)\ Sg1 \qquad \Rightarrow h\rho = x + y$

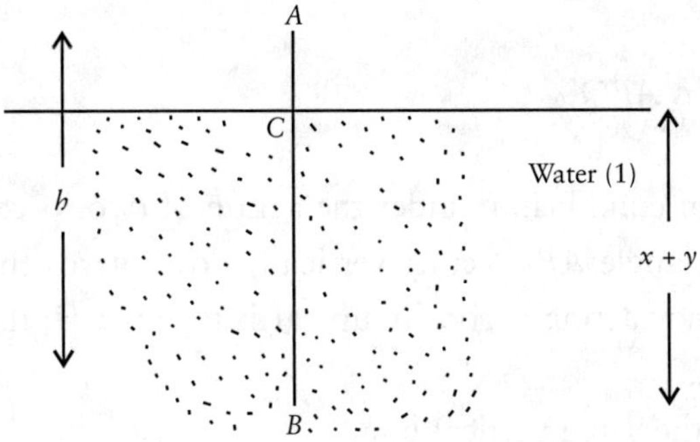

$\Rightarrow h\rho - x = y$

$\Rightarrow h\rho - \dfrac{h(\rho-\sigma)}{1-\sigma} = y$ | From (1)

$\Rightarrow h\left\{\dfrac{\rho(1-\sigma)-(\rho-\sigma)}{1-\sigma}\right\} = y \qquad \Rightarrow y = h\sigma\dfrac{(1-\rho)}{(1-\sigma)}$

Example 19: A uniform isosceles triangular lamina (sp.gr. σ) floats in water with its plane vertical, its vertical angle (which is equal to 2α) immersed and the base wholly above the water. Prove that in the position of equilibrium in which the base is not horizontal, the sum of the lengths of the immersed portions of the two sides is $2a\cos^2\alpha$, where a is one of the equal sides.

Also prove that $\sigma < \cos^4\alpha$ as well as $\cos 2\alpha$.

Sol. First Part

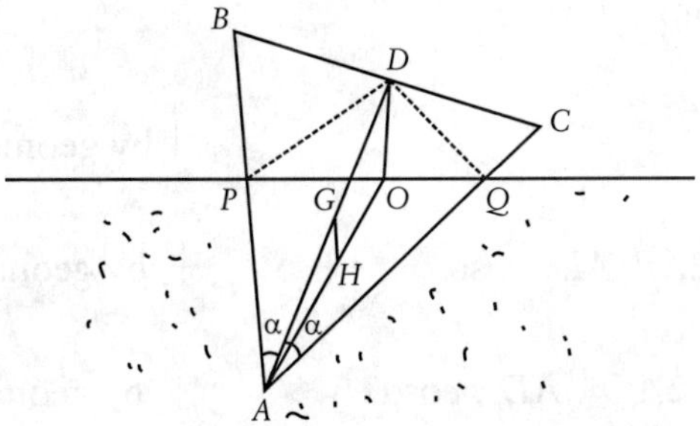

ABC is the triangular lamina whose portion APQ is immersed in water.

$AB = AC = (= a)$

$\angle BAC = 2\alpha$

The base BC is not horizontal.

Let D be the mid-point of BC.

Join AD.

Then, $\angle BAD = \angle CAD = \alpha$ | by geometry

Let G be the C.G. of $\Delta\, ABC$. Then, G will lie on AD such that $AG = 2/3\ AD$

Let O be the mid-point of PQ.

Join A to O.

Let H be the C.G. of $\Delta\, APQ$.

Then, $AH = 2/3\ AO$

The triangle ABC is in equilibrium under the action of two forces:

(1) Weight of the triangle ABC acting vertically downwards through G.

(2) Force of buoyancy acting vertically upwards through H, the centre of buoyancy.

$\therefore$ For equilibrium,

G and H must be in the same vertical line

Also, in $\Delta\, ADO$,

$$\frac{AG}{AH} = \frac{\frac{2}{3}AD}{\frac{2}{3}AO} = \frac{AD}{AO}$$

$\therefore$ DO is parallel to GH | by geometry

$\because$ GH is vertical

$\therefore$ DO is vertical

Join DP and DQ.

$\therefore$ PQ is horizontal

and DO is vertical

$\therefore$ $\angle DOP = \angle DOQ = 90°$

Also, $OP = OQ$

$\therefore$ $DP = DQ$ | by geometry ... (1)

In $\Delta\, APD$,

$DP^2 = AP^2 + AD^2 - 2AP.\ AD.\ \cos\alpha$ | by geometry ... (2)

In $\Delta\, AQD$,

$DQ^2 = AQ^2 + AD^2 - 2AQ\ .\ AD\ .\ \cos\alpha$ | by geometry ... (3)

But $DP = DQ$ | From (1)

$\therefore$ $AP^2 + AD^2 - 2AP.\ AD.\ \cos\alpha = AQ^2 + AD^2 - 2AQ\ .\ AD\ .\ \cos\alpha$

$\Rightarrow$ $\left(AP^2 - AQ^2\right)\ 2AD\cos\alpha(AP - AQ) = 0$ $\Rightarrow (AP - AQ)(AP + AQ - 2AD\cos\alpha) = 0$

$\Rightarrow$ $AP - AQ = 0$

or

$AP + AQ - 2AD\ \cos\alpha = 0$

But $AP \neq AQ$, since then BC becomes horizontal which is against hypothesis.

$\therefore$ $AP + AQ - 2AD\ \cos\alpha = 0$

$\Rightarrow$ $AP + AQ = 2AD\ \cos\alpha$

$= 2\ (AB\cos\alpha)\cos\alpha$ $\quad | \because AD = AB\cos\alpha$

$= 2AB\cos^2\alpha$

$= 2a\cos^2\alpha$ ... (4)

Second Part

For floatation in equilibrium,

weight of the lamina = weight of the water displaced by it

$\Rightarrow$ (Area of $\Delta\ ABC$) $\sigma g\rho$ = (Area of $\Delta\ APQ$) $1g\rho$

where ρ is the density of water

$\Rightarrow\ 1/2\ AB\ .\ AC\sin 2\alpha\sigma g = 1/2\ AP.\ AQ\sin 2\alpha g$ $\quad | \because \rho = 1$

$\Rightarrow\ AB.\ AC\ \sigma = AP.\ AQ \quad \Rightarrow a^2\sigma = AP.\ AQ$ $\quad | \because AB = AC = a$

$$\Rightarrow\ \sigma = \frac{AP.AQ}{a^2} \qquad \text{... (5)}$$

Now we know that Arithmetic mean > geometric mean

$$\therefore\ \frac{AP+AQ}{2} > \sqrt{AP.AQ} \quad \Rightarrow \sqrt{AP.AQ} < \frac{AP+AQ}{2} \quad \Rightarrow AP.AQ < \frac{(AP+AQ)^2}{4}$$

$$\Rightarrow\ \sigma a^2 < \frac{\left(2a\cos^2\alpha\right)^2}{4} \qquad | \text{ From (5) and (4)}$$

$\Rightarrow\ \sigma < \cos^4\alpha$

Third Part

We Know that as σ increases, the area of the immersed portion increases, i.e., the area of $\Delta\,APQ$ increases. But, by hypothesis, the base BC must always be wholly above water. Therefore, σ will be the greatest when PQ passes through the lower point C, i.e., when C is on the surface of water, i.e., when $AQ = AC = a$.

From (4),

$AP = 2a\cos^2\alpha - AQ$

$\quad = 2a\cos^2\alpha - a$

$\therefore$ From (5),

$$\text{the greatest value of } \sigma = \frac{\left(2a\cos^2\alpha - a\right)a}{a^2} = 2\cos^2\alpha - 1 = \cos 2\alpha$$

Hence, $\sigma \le \cos 2\alpha$

Example 20: A triangular lamina ABC of sp.gr. σ floats with its plane vertical in water, A being outside the surface and BC not horizontal. The angle A is a right angle and $AC = AB$. If θ be the inclination of AB to the vertical, prove that

$$\sin 2\theta = \frac{2-2\sigma}{2\sigma-1} \text{ given } \sigma > 3/4.$$

Sol.

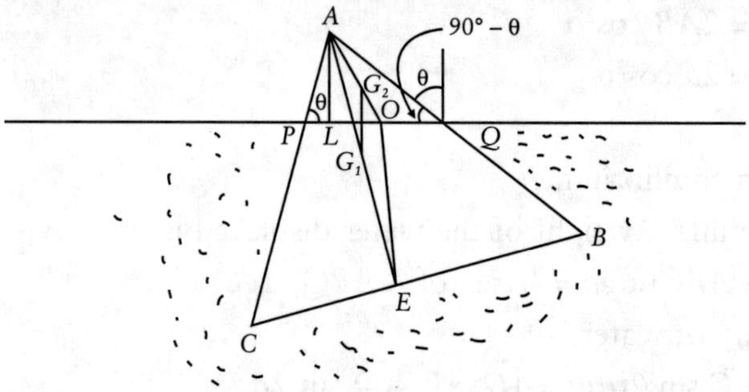

ABC is the given triangular lamina with its portion $PQBCP$ immersed in water. Let ρ be the density of water. Let the side AB be inclined at an angle θ to the vertical. Then, the side AC will be inclined at angle θ to the horizontal since $\angle A = 90°$ (given).

Let E and O be the mid-points of CB and PQ respectively. Let G_1 and G_2 be the C.G.'s of $\Delta\, ABC$ and $\Delta\, APQ$ respectively. Then, G_1 and G_2 will lie on AE and AO respectively such that

$AG_1 = 2/3\ AE$

and $AG_2 = 2/3\ AO$

In Δ 's $AG_1\ G_2$ and $\Delta\, AEO$, we have

$$\frac{AG_1}{AG_2} = \frac{\frac{2}{3}AE}{\frac{2}{3}AO} = \frac{AE}{AO}$$

$\therefore$ $G_1\, G_2$ is parallel to OE. | by geometry

Since $\Delta\, ABC$ floats freely, therefore, G_1 and G_2 must be in the same vertical line. Hence, OE is also vertical.

For floatation in equilibrium,

weight of the lamina = weight of the water displaced by it

$\Rightarrow$ $(\Delta\, ABC)\ S\sigma g\rho$ = (Area $PQBCP$) $1\ \rho g \Rightarrow$ $(\Delta\, ABC)\ \sigma = (\Delta\, ABC - \Delta\, APQ)$

$\Rightarrow$ $(1 - \sigma)\ \Delta\, ABC = \Delta\, APQ$ ... (1)

Draw perpendicular from A to the free surface. Let $AL = h$

Then, $AP = h \operatorname{cosec} \theta$

and, $AQ = h \sec \theta$

$$\therefore \text{ Area of } \Delta\, APQ = \frac{AP.AQ}{2} \qquad | \because AP \perp AQ$$

$$= \frac{h \operatorname{cosec}\theta\ h \sec\theta}{2}$$

$$= \frac{h^2}{\sin 2\theta} \qquad \text{... (2)}$$

In $\Delta\, APQ$,

$\therefore\ \angle PAQ = 90°$

and O is the mid-point of PQ

$\therefore\ OA = OP = OQ$ | by geometry

$\therefore\ OA = 1/2\, PQ = 1/2\, AP \sec\theta = 1/2\, .\, h \operatorname{cosec}\theta \sec\theta = h/\sin 2\theta$

By geometry,

AE is the bisector of $\angle CAB$

$\therefore\ \angle CAE = \angle BAE = 45°$ | $\because\ \angle CAB = 90°$ (Given)

$\therefore\ AE = AB \cos 45° = \dfrac{AB}{\sqrt{2}}$

From (1),

$$1-\sigma = \frac{\Delta\, APQ}{\Delta\, ABC} = \frac{\dfrac{h^2}{\sin 2\theta}}{\dfrac{1}{2}\, AB.AC} \qquad |\ \because\ AB \perp AC$$

$$= \frac{2h^2}{AB.AC\sin 2\theta}$$

$$= \frac{2h^2}{AB^2 \sin 2\theta} \qquad |\ \because\ AB = AC \text{ (Given)} \qquad \ldots (3)$$

Now,

$\angle LAE = \angle BAL - \angle BAE = \theta - \pi/4$

$\therefore\ \angle OEA = \angle EAL = \theta - \pi/4$

In $\Delta\, AOQ$,

$\angle AQO = 90° - \theta$

and, $OA = OQ$

$\therefore\ \angle OAQ = \angle OQA = 90° - \theta$

$\therefore$ Exterior $\angle AOL$

$= \angle OAQ + \angle OQA = (90° - \theta) + (90° - \theta) = 180° - 2\theta$

$\therefore\ \angle AOE = \angle AOL + \angle LOE = (180° - 2\theta) + 90° = 270° - 2\theta$

In $\Delta\, AOE$,

$$\frac{AE}{\sin\angle AOE} = \frac{AO}{\sin\angle OEA} \qquad |\ \text{by sine formula}$$

$$\Rightarrow\ \frac{AE}{AO} = \frac{\sin\angle AOE}{\sin\angle OEA} = \frac{\sin(270° - 2\theta)}{\sin\left(\theta - \dfrac{\pi}{4}\right)} = \frac{-\cos 2\theta}{\sin\left(\theta - \dfrac{\pi}{4}\right)}$$

$$\Rightarrow \frac{\frac{AB}{\sqrt{2}}}{\frac{h}{\sin 2\theta}} = \frac{-\cos 2\theta}{\sin\theta\cos\frac{\pi}{4} - \cos\theta\sin\frac{\pi}{4}} \Rightarrow \frac{AB\sin 2\theta}{h\sqrt{2}} = \frac{-\sqrt{2}\cos 2\theta}{\sin\theta - \cos\theta} \Rightarrow \frac{AB\sin 2\theta}{h\sqrt{2}} = \frac{\sqrt{2}\cos 2\theta}{\cos\theta - \sin\theta}$$

$$\Rightarrow AB = \frac{2h\cos 2\theta}{\sin 2\theta(\cos\theta - \sin\theta)} = \frac{2h\left(\cos^2\theta - \sin^2\theta\right)}{\sin 2\theta(\cos\theta - \sin\theta)} = \frac{2h(\cos\theta + \sin\theta)}{\sin 2\theta}$$

From (3),

$$1 - \sigma = \frac{2h^2}{AB^2 \sin 2\theta} = \frac{2h^2 \sin^2 2\theta}{4h^2(\cos\theta + \sin\theta)^2 \sin 2\theta} = \frac{\sin 2\theta}{2(1 + \sin 2\theta)}$$

$$\Rightarrow 2(1-\sigma) + 2(1-\sigma)\sin 2\theta = \sin 2\theta \quad \Rightarrow 2(1-\sigma) = (2\sigma - 1)\sin 2\theta$$

$$\Rightarrow \sin 2\theta = \frac{2(1-\sigma)}{2\sigma - 1} = \frac{2 - 2\sigma}{2\sigma - 1}$$

Now, $\sin 2\theta \le 1 \Rightarrow \frac{2-2\sigma}{2\sigma-1} \le 1 \quad \Rightarrow 2 - 2\alpha \le 2\sigma - 1 \quad \Rightarrow 3 \le 4\sigma \Rightarrow \frac{3}{4} \le \sigma \Rightarrow \sigma \ge \frac{3}{4}$

Example 21: A hollow hemispherical shell has a heavy particle fixed to its rim and floats in water with the particle just above the surface and with the plane of the rim inclined at an angle 45° to the surface. Show that the weight of the hemisphere: the weight of the water which it would contain as $\left(4\sqrt{2} - 5\right) : 6\sqrt{2}$

Sol.

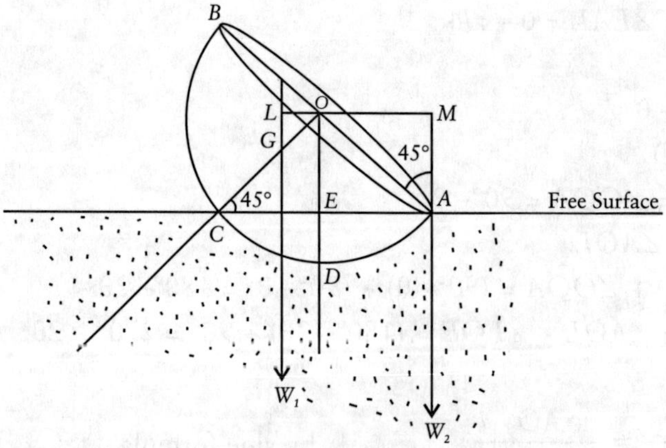

AC is the free surface

$\angle OAC = 45°$ | Given

$\because \; OA = OC$

$\therefore \; \angle OAC = \angle OCA = 45°$

$\therefore \; \angle AOC = 90° \quad \Rightarrow$ C is the vertex of the shell

Let W_1 be the weight of the hemisphere and W_2 be the weight of the particle.

For equilibrium,

$W_1 + W_2$ = weight of the water displaced by the volume ADC ... (1)

Taking moments about O, we get

$$W_1 OL = W_2 AM$$

$$W_1 OG \sin 45° = W_2 OA \cos 45°$$

$$\Rightarrow W_1 OG = W_2 OA \quad \text{... (2)}$$

Now, volume AMC

$$= \int_{\frac{a}{\sqrt{2}}}^{a} \pi y^2 dx = \int_{\frac{a}{\sqrt{2}}}^{a} \pi\left(a^2 - x^2\right) dx = \pi\left(a^2 x - \frac{x^3}{3}\right)_{\frac{a}{\sqrt{2}}}^{a} = \pi\left[a^2\left(a - \frac{a}{\sqrt{2}}\right) - \frac{1}{3}\left(a^3 - \frac{a^3}{2\sqrt{2}}\right)\right]$$

$$= \pi a^3\left[\left(1 - \frac{1}{\sqrt{2}}\right) - \frac{1}{3}\left(1 - \frac{1}{2\sqrt{2}}\right)\right] = \pi a^3\left[\frac{\sqrt{2}-1}{\sqrt{2}} - \frac{2\sqrt{2}-1}{6\sqrt{2}}\right] = \pi a^3 \frac{6\left(\sqrt{2}-1\right) - \left(2\sqrt{2}-1\right)}{6\sqrt{2}}$$

$$= \pi a^3\left(\frac{4\sqrt{2}-5}{6\sqrt{2}}\right) \quad \text{... (3)}$$

From (1) and (3),

$$W_1 + W_2 = \frac{\pi a^3}{6\sqrt{2}}\left(4\sqrt{2}-5\right) g\rho \quad \text{... (4)}$$

From (2),

$$W_1 \frac{a}{2} = W_2 a$$

$$\Rightarrow W_1 = 2W_2 \quad \text{... (5)}$$

From (4) and (5), we get

$$W_1 + \frac{W_1}{2} = \frac{\pi a^3}{6\sqrt{2}}\left(4\sqrt{2}-5\right) g\rho \Rightarrow \frac{3W_1}{2} = \frac{\pi a^3}{6\sqrt{2}}\left(4\sqrt{2}-5\right) g\rho \Rightarrow W_1 = \frac{\pi a^3\left(4\sqrt{2}-5\right) g\rho}{9\sqrt{2}}$$

$$\therefore \frac{W_1}{\frac{2}{3}\pi a^3 g\rho} = \frac{\frac{\pi a^3\left(4\sqrt{2}-5\right) g\rho}{9\sqrt{2}}}{\frac{2}{3}\pi a^3 g\rho} = \frac{4\sqrt{2}-5}{6\sqrt{2}}$$

Example 22: Show that a right circular cone of density ρ and semi-vertical angle α can float vertex downwards in a liquid of density ρ with one generator vertical and base just out (clear) of the liquid if

$$\rho = \sigma(\cos 2\alpha)^{3/2}$$

Sol.

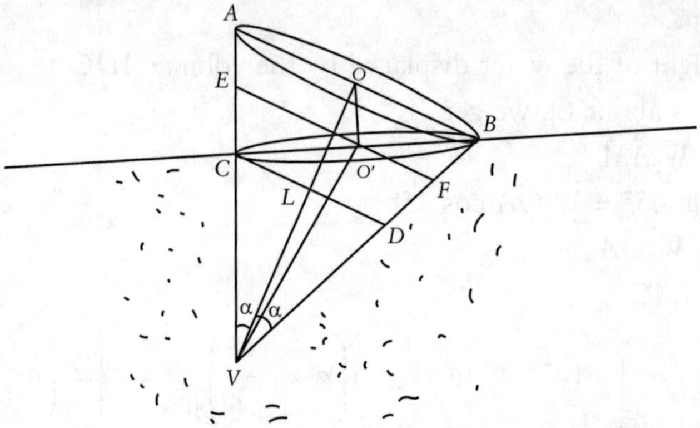

Let O be the centre of the base and h be the height of the cone. Let BC denote the surface of floatation.

Draw perpendicular OO' from O to the surface of floatation BC. Then, O' is the mid-point of BC as BO' is parallel to AC, both being vertical. | by geometry

weight of the cone VAB

$$= \frac{1}{3}\pi h^3 \tan^2 \alpha\, g\rho \quad \text{... (1)}$$

where ρ is the density of material of the cone. Now, the section of the cone VAB by the surface of floatation BC is an ellipse whose centre is O', major axis is BC. Let the lengths of its axes be $2a$ and $2b$ respectively. Then,

$$2a = BC = VB \sin 2\alpha = h \sec \alpha \sin 2\alpha$$

$$= 2h \sin \alpha \quad \text{... (2)}$$

Through O' and C draw EF and CD respectively parallel to AB. Then,

$$b^2 = EO' \times O'F$$

$$= \frac{1}{2} AB \times \frac{1}{2} CD$$

| In $\Delta ABC, O'$ is the mid = point of BC and EO' is parallel to AB; Similarly, in ΔBCD, O' is the mid-point of BC and $O'F$ is parallel to CD

$$= \frac{1}{2}(2h\tan\alpha)\left(\frac{1}{2}CD\right) \quad \text{... (3)}$$

$$\because \Delta VCD \sim \Delta VAB$$

$$\therefore \frac{CD}{AB} = \frac{VC}{VA} \quad \text{| by geometry}$$

$$\Rightarrow CD = AB\frac{VC}{VA} = 2h\tan\alpha \frac{VC}{h\sec\alpha} = 2VC \sin \alpha$$

$= 2VB\cos 2\alpha \sin\alpha$ | In ΔVBC, $VC = VB\cos 2\alpha$

$= 2h\sec\alpha\cos 2\alpha\sin\alpha$

$= 2h\tan\alpha\cos 2\alpha$ ... (4)

$\therefore\ b^2 = \frac{1}{2}(2h\tan\alpha)\left(\frac{1}{2}\cdot 2h\tan\alpha\cos 2\alpha\right)$ | From (3) and (4)

$= h^2\tan^2\alpha\cos 2\alpha$

$= b = h\tan\alpha\sqrt{\cos 2\alpha}$ ... (5)

Area of the ellipse with centre O' and BC as major axis

$=$ "$\pi a b$" $= \pi(h\sin\alpha)\left(h\tan\alpha\sqrt{\cos 2\alpha}\right)$ | From (2) and (5)

$= \pi h^2\sin\alpha\tan\alpha\sqrt{\cos 2\alpha}$ ... (6)

$\therefore$ Volume of the cone immersed in the liquid

= 1/3 Area of the base × Height of the vertex V above the base

$= \frac{1}{3}\pi h^2\sin\alpha\tan\alpha\sqrt{\cos 2\alpha}\ VC$

$= \frac{1}{3}\pi h^2\sin\alpha\tan\alpha\sqrt{\cos 2\alpha}\,(h\sec\alpha\cos 2\alpha)$ | $VC = VB\cos 2\alpha = h\sec\alpha\cos 2\alpha$

$= \frac{1}{3}\pi h^3\tan^2\alpha\,(\cos 2\alpha)^{3/2}$ ... (7)

For floatation in equilibrium,

weight of the cone = weight of the liquid displaced

$\Rightarrow \frac{1}{3}\pi h^3\tan^2\alpha\, g\rho = \frac{1}{3}\pi h^3\tan^2\alpha(\cos 2\alpha)^{3/2} g\sigma \quad \Rightarrow \rho = \sigma(\cos 2\alpha)^{3/2}$

EXERCISE 5

1. Two equal and similar rods AB, BC, fixed at an angle α at B rest in a fluid of twice the sp.gr. with the angle B out of the fluid, and the bisector of the angle ABC makes an angle θ with the horizon. Prove that

   $\cos 2\theta + \sec\alpha = 2$

2. A graduated glass vessel is in the form of a cylinder with vertical walls containing water to a height of 20 cm from the bottom. A body weighing 20 gms. is placed in it and floats, and the water level rises to 25 cm. The body in then completely submerged with some force, the water level rises to 25 cm. Find the mean sp.gr. of the body, the force required to submerge it and volume of water in the vessel.

   Ans. 0·8; weight of 5 gms.; 100 c.c.

3. A rectangle movable about an angular point floats with its area immersed in a liquid. If the angular point lies outside the liquid, and if the rectangle floats with its sides equally inclined to the vertical, show that the ratio of the density of the rectangle to that of the liquid is $3a + b : 4a$, where a and b are the sides of the rectangle.
4. A rectangle, movable about an angular point which is fixed below the surface of a liquid, floats with its sides equally inclined to the vertical and with half its area immersed in the liquid. If the lengths of the sides be a and b and one of the sides of length b be entirely immersed in the liquid, show that the ratio of the density of the body to that of the liquid is $(a - b) : 4a$.
5. A life-belt, in the form of an anchor-ring generated by a circle of radius a, floats in water with its equatorial plane horizontal. Show that z, the depth immersed, is given by $z = a(1-\cos\beta);\ 2\pi s = 2\beta - \sin 2\beta$, where s is the sp.gr. of the material of belt.
6. A vessel containing water, the whole weighing 3 kgs, is suspended from a spring balance. A man of iron weighing 1·3 kgs is suspended by a thin string from a second spring balance, which is moved so that the iron is immersed in the water, without touching the sides or bottom of the vessel. State what variations and their amounts, you would expect to observe in the readings on the spring balances and explain why they occur. The sp.gr. of the iron may be taken as 7·7.

   Ans. $3\frac{13}{17}$ Kgs. ; $\left(1\cdot 3-\frac{13}{77}\right)$Kgs.
7. ABC is a right-angled triangular lamina and floats with its plane vertical, and the right angle immersed in water. Prove that if its sp.gr. be to that of the water as 2 : 5 and $CB : CA = 5 : 4$, CB is cut by the surface of the water at a distance from C equal to CA.
8. If the height of a right circular cone be equal to the diameter of the base, prove that it will float with its slant side horizontal in any liquid of greater density.
9. The vertical angle of a solid right circular cone is 60°. Prove that it can just float in a liquid with its vertex above the surface and one point of its base in the surface, if the densities of the cone and the liquid are in the ratio of $2\sqrt{2}-1 : 2\sqrt{2}$.
10. A glass, partly filled with a liquid, is placed on one pan of a common balance and is counter poised by suitable weight pieces placed on the other. A solid which is suspended by means of a string is now brought and immersed in the liquid without touching the sides of the glass being still in suspension from the string. Neglecting the effect of the atmosphere, explain why some more weights are now needed on the second pan to balance the glass of water.
11. A prism of square section floats in water with its long edges horizontal and the centre line of one of its faces hinged to an axis fixed in the surface of the water. Show that, if the sp.gr. of the prism is 39/64, the opposite face of the prism will be 3/4 immersed.
12. A uniform log whose cross-section is a square floats horizontally with one edge in the surface and one edge above the surface of a homogeneous liquid. Show that the ratio of their sp.gr. is as 3 is to 4.

Index

E

F

G

H

I

L

U

V

W